Quantum Flavordynamics, Quantum Chromodynamics, and Unified Theories

NATO ADVANCED STUDY INSTITUTES SERIES

A series of edited volumes comprising multifaceted studies of contemporary scientific issues by some of the best scientific minds in the world, assembled in cooperation with NATO Scientific Affairs Division.

Series B: Physics

RECENT VOLUMES IN THIS SERIES

Volume 46 – Nondestructive Evaluation of Semiconductor Materials and Devices
edited by Jay N. Zemel

Volume 47 – Site Characterization and Aggregation of Implanted Atoms in Materials
edited by A. Perez and R. Coussement

Volume 48 – Electron and Magnetization Densities in Molecules and Crystals
edited by P. Becker

Volume 49 – New Phenomena in Lepton-Hadron Physics
edited by Dietrich E. C. Fries and Julius Wess

Volume 50 – Ordering in Strongly Fluctuating Condensed Matter Systems
edited by Tormod Riste

Volume 51 – Phase Transitions in Surface Films
edited by J. G. Dash and J. Ruvalds

Volume 52 – Physics of Nonlinear Transport in Semiconductors
edited by David K. Ferry, J. R. Barker, and C. Jacoboni

Volume 53 – Atomic and Molecular Processes in Controlled Thermonuclear Fusion
edited by M. R. C. McDowell and A. M. Ferendeci

Volume 54 – Quantum Flavordynamics, Quantum Chromodynamics, and Unified Theories
edited by K. T. Mahanthappa and James Randa

Volume 55 – Field Theoretical Methods in Particle Physics
edited by Werner Rühl

Volume 56 – Vibrational Spectroscopy of Molecular Liquids and Solids
edited by S. Bratos and R. M. Pick

This series is published by an international board of publishers in conjunction with NATO Scientific Affairs Division

A Life Sciences B Physics	Plenum Publishing Corporation London and New York
C Mathematical and Physical Sciences	D. Reidel Publishing Company Dordrecht, Boston and London
D Behavioral and Social Sciences E Applied Sciences	Sijthoff & Noordhoff International Publishers Alphen aan den Rijn and Germantown U.S.A.

Quantum Flavordynamics, Quantum Chromodynamics, and Unified Theories

Edited by
K. T. Mahanthappa
and
James Randa
University of Colorado
Boulder, Colorado

PLENUM PRESS • NEW YORK AND LONDON
Published in cooperation with NATO Scientific Affairs Division

Library of Congress Cataloging in Publication Data

Nato Advanced Study Institute on Quantum Flavordynamics, Quantum Chromodynamics, and the Unified Theories, University of Colorado, 1979.
Quantum flavordynamics, quantum chromodynamics, and unified theories.

(NATO advanced study institutes series: Series B, Physics; v. 54)
Includes index.
1. Quantum chromodynamics–Congresses. 2. Quantum flavordynamics–Congresses. 3. Unified field theories–Congresses. I. Mahanthappa, K. T. II. Randa, James. III. North Atlantic Treaty Organization. Division of Scientific Affairs. IV. Title. V. Series.
QC793.3.Q35N37 1979 539.7'2 80-12289
ISBN 0-306-40436-2

Lectures presented at the NATO Advanced Study Institute on Quantum Flavordynamics, Quantum Chromodynamics, and Unified Theories, held at the University of Colorado, Boulder, Colorado, July 9–27, 1979.

A Division of Plenum Publishing Corporation
227 West 17th Street, New York, N.Y. 10011

Printed in the United States of America

PREFACE

The Advanced Study Institute on Quantum Flavordynamics, Quantum Chromodynamics and Unified Theories was held on the campus of the University of Colorado at Boulder from July 9th through July 27th of 1979.

There has been a rapid progress in the understanding of weak, electromagnetic and strong interactions and their unification during the past few years. The purpose of the Institute was to have a group of lecturers active in these areas of research give a series of lectures on various aspects of these topics beginning at the elementary level and ending with the up-to-date developments.

There were three lecturers, Professors S. Ellis, R. Field and C.H. Llewellyn Smith who covered the different but related aspects of Quantum Chromodynamics. Their lectures were well coordinated, but some overlap was inevitable. Dr. Buras gave two lectures on QCD corrections beyond the leading order. Professor D. Gross covered the nonperturbative aspects and a possible mechanism of quark confinement. At a more phenomenological level, Professor C. De Tar covered the bag models. The subject matter of electroweak interactions was covered by Professor G. Altarelli. Professor J. Wess gave six lectures on supersymmetry and supergravity. All these lectures with the exception of those of Professor D. Gross are incorporated in this volume. The contents of Professor Gross' lectures are available elsewhere and therefore only references and problems are included here.

In addition to the above lectures, there were workshop-like discussion sessions. There were also seminars by participants, Drs. G. Bhanot, V. Chang, T. Clark, A. Davidson, H. Haber, S. Gottlieb and P.J. Mulders, on their recent research.

The Institute was sponsored by the North Atlantic Treaty Organization, which provided generous financial support which enabled many young physicists from the United States of America and abroad to participate in the Institute.

Additional co-sponsors were the U.S. Department of Energy and the University of Colorado. The former offered further financial assistance and the latter furnished clerical and technical services and its campus facilities for the purpose of the organization and running of the Institute.

The International Organizing Committee consisted of Professors J.D. Bjorken, N. Cabibbo, R. Dalitz, K.T. Mahanthappa (Director), A. Salam and S. Treiman. The Local Committee consisted of Professors A.O. Barut, J. Dreitlein, K.T. Mahanthappa, U. Nauenberg and J. Randa.

David Unger and Marc Sher assisted in the day-to-day functioning of the Institute. Jeannette Royer organized the social programs. Joe Dreitlein and Jim Smith led the mountain hikes.

We thank the lecturers for their cooperation in the preparation of this volume and Harriet Ortiz for her meticulous typing of the manuscript.

K.T. Mahanthappa
J. Randa

Boulder
November, 1979

LECTURERS

G. Altarelli
Istituto di Fisica dell' Universita-Roma, Italy

A.J. Buras
Fermi National Accelerator Laboratory, Batavia, Illinois

C. De Tar
University of Utah, Salt Lake City, Utah

S. Ellis
University of Washington, Seattle, Washington

R. Field
California Institute of Technology, Pasadena, California

D. Gross
Princeton University, Princeton, New Jersey

C.H. Llewellyn Smith
University of Oxford, Oxford, England

J. Wess
University of Karlsruhe, Karlsruhe, West Germany

CONTENTS

PHENOMENOLOGY OF FLAVORDYNAMICS

G. Altarelli

Istituto di Fisica dell'Università - Roma

Istituto Nazionale di Fisica Nucleare - Sezione di Roma

1. INTRODUCTION

The present lectures are devoted to a summary of our present understanding of the weak and electromagnetic interactions. The standard $SU(3)_{color} \times SU(2) \times U(1)$ gauge theory of strong and electroweak interactions is the general framework for the present discussion. We assume in particular that the t quark will eventually be discovered, thus filling the remaining gap in the three generations of fermions known so far. We shall mainly discuss the phenomenological aspects of flavordynamics in the standard model with particular emphasis on the more recent elements of experimental support for this theory and the empirical determination of its general parameters. We do not aim at a complete systematic review of the numerous interesting issues in the subject. We only intend to present a concise outline of the structure of charged and neutral currents without going into the detailed applications to various processes. The last two lectures are devoted to an elementary introduction to unified theories and their implications for experiment.

We assume a previous knowledge of the formalism of spontaneously broken gauge theories and of the more elementary aspects of the phenomenology of weakly interacting particles.

2. THE STANDARD MODEL OF ELECTROWEAK INTERACTIONS

In this section the structure of the electroweak Lagrangian is recalled[1,2] and our notations are specified.

For this discussion we split the Lagrangian into two parts by separating the Higgs boson couplings:

$$L = L_{\text{SYMM}} + L_{\text{HIGGS}} \; . \tag{1}$$

We start by specifying L_{SYMM} which only involves gauge bosons and fermions:

$$L_{\text{SYMM}} = -\frac{1}{4}\sum_{A=1}^{3} F^{A}_{\mu\nu}F^{A\mu\nu} - \frac{1}{4}B_{\mu\nu}B^{\mu\nu} + \bar{\psi}_L i\gamma^{\mu}D_{\mu}\psi_L + \\ + \bar{\psi}_R i\gamma^{\mu}D_{\mu}\psi_R \; . \tag{2}$$

This is the Yang-Mills Lagrangian for the gauge group SU(2) × U(1) with fermion matter fields. Here

$$B_{\mu\nu} = \partial_{\mu}B_{\nu} - \partial_{\nu}B_{\mu} \\ F^{A}_{\mu\nu} = \partial_{\mu}W^{A}_{\nu} - \partial_{\nu}W^{A}_{\mu} - g\varepsilon^{ABC}W^{B}_{\mu}W^{C}_{\nu} \tag{3}$$

are the gauge antisymmetric tensors constructed out of the gauge field B_{μ}, associated to the Abelian group U(1), and W^{A}_{μ} (A = 1,2,3) corresponding to the three SU(2) generators. ε^{ABC} are the group structure constants that for SU(2) coincide with the totally antisymmetric Levi-Civita tensor. The normalization of the SU(2) gauge coupling g is therefore specified by Eq. (3).

The fermion fields are described through their left and right handed components:

$$\psi_{L,R} = \frac{1 \mp \gamma_5}{2}\psi, \quad \bar{\psi}_{L,R} = \bar{\psi}\,\frac{1 \pm \gamma_5}{2} \; . \tag{4}$$

Because of the fact that in the symmetric limit no mass terms appear in the fermion sector, the left and right fermions can be given different transformation properties under the gauge group and a chiral structure results. By $\psi_{L,R}$ we mean a column vector including all fermions in the theory which span a reducible representation of SU(2) × U(1). The gauge transformations are

$$\psi'_L = \exp i[g\alpha_L(x) + g'\beta_L(x)]\psi_L \; , \\ \psi'_R = \exp i[g\alpha_R(x) + g'\beta_R(x)]\psi_R \; , \tag{5}$$

with

$$\alpha_{L,R}(x) = \sum_{A=1}^{3} t^A_{L,R}\alpha^A(x) ,$$
$$\beta_{L,R}(x) = \frac{1}{2} Y_{L,R}\beta(x) , \tag{6}$$

where $\alpha^A(x)$ and $\beta(x)$ are the space-time dependent parameters of the transformation and $t^A_{L,R}(Y_{L,R})$ are a reducible representation of the SU(2) (U(1)) generators on $\psi_{L,R}$. The commutation relations of the SU(2) generators are given by

$$[t^A_L, t^B_L] = i\varepsilon^{ABC} t^C_L ,$$
$$[t^A_R, t^B_R] = i\varepsilon^{ABC} t^C_R , \tag{7}$$

with the normalization choice

$$\mathrm{Tr}\, t^A_L t^B_L = \mathrm{Tr}\, t^A_R t^B_R = \frac{1}{2}\,\delta^{AB} . \tag{8}$$

The electric charge generator is given by

$$Q = t^3_L + \frac{1}{2}Y_L = t^3_R + \frac{1}{2}Y_R . \tag{9}$$

Finally the covariant derivatives D_μ are explicitly determined by

$$D_\mu\psi_{L,R} = \left[\partial_\mu + ig \sum_A t^A_{L,R}W^A_\mu + ig'\frac{1}{2}Y_{L,R}B_\mu\right]\psi_{L,R} . \tag{10}$$

All fermion couplings to the gauge bosons are now fixed. The charged current couplings are simplest. From

$$g(t^1W^1_\mu + t^2W^2_\mu) = g\left(\frac{t^1+it^2}{\sqrt{2}}\,\frac{W^1_\mu-iW^2_\mu}{\sqrt{2}} + \text{h.c.}\right) =$$
$$= g\left(\frac{t^+}{\sqrt{2}}\,W^-_\mu + \text{h.c.}\right) , \tag{11}$$

where $t^\pm = t^1 \pm it^2$, $W^\mp = (W^1 \mp iW^2)/\sqrt{2}$, we obtain the vertex

$$g\bar{\psi}\gamma^\mu\left[\frac{t^+_L}{\sqrt{2}}\,\frac{1-\gamma_5}{2} + \frac{t^+_R}{\sqrt{2}}\,\frac{1+\gamma_5}{2}\right]\psi W^-_\mu + \text{h.c.} \quad . \tag{12}$$

In the neutral sector the photon A_μ and the mediator of the neutral weak current Z_μ are orthogonal and normalized linear combinations of B_μ and W^3_μ

$$
\begin{aligned}
A_\mu &= \cos\theta_W B_\mu + \sin\theta_W W^3_\mu \, , \\
Z_\mu &= -\sin\theta_W B_\mu + \cos\theta_W W^3_\mu \, .
\end{aligned}
\tag{13}
$$

The photon is identified by being coupled equally to left and right fields with a coupling strength equal to the electric charge. This leads to the relations

$$
\begin{aligned}
\left(\sin\theta_W g t^3_L + \cos\theta_W g' \tfrac{1}{2} Y_L\right)\gamma_\mu \frac{1-\gamma_5}{2} &= eQ\gamma_\mu \frac{1-\gamma_5}{2} \, , \\
\left(\sin\theta_W g t^3_R + \cos\theta_W g' \tfrac{1}{2} Y_R\right)\gamma_\mu \frac{1+\gamma_5}{2} &= eQ\gamma_\mu \frac{1+\gamma_5}{2} \, ,
\end{aligned}
\tag{14}
$$

where Q is the charge matrix in units of the proton charge e. Recalling Eq. (9) for Q we immediately obtain

$$
\sin\theta_W g = \cos\theta_W g' = e \, , \tag{15}
$$

or equivalently

$$
\begin{aligned}
\tan\,\theta_W &= g'/g \\
e &= gg'/(g^2 + g'^2)^{1/2} \, .
\end{aligned}
\tag{16}
$$

With the weak mixing angle θ_W fixed as above, it is a simple matter of algebra to write down the neutral current couplings, with the result

$$
\frac{g}{2\cos\theta_W}\, \bar\psi\gamma^\mu\left[t^3_L(1-\gamma_5) + t^3_R(1+\gamma_5) - 2Q\sin^2\theta_W\right]\psi Z_\mu \, . \tag{17}
$$

In order to derive effective four fermion interactions that are equivalent at low energies to the charged and neutral current couplings as given in Eqs. (12) and (17), we anticipate that spontaneous symmetry breaking will lead to large masses M_W and M_Z for $W^\pm$ and Z. By neglecting the momentum transfer squared with respect to M_W^2 in the propagator of Born diagrams with single W exchange in charged current interactions, from Eq. (12) we are led for left-left couplings to

$$\frac{g^2}{8M_W^2}\bar{\psi}\gamma_\mu(1-\gamma_5)t_L^+\psi\bar{\psi}\gamma^\mu(1-\gamma_5)t_L^-\psi \quad . \tag{18}$$

By further specializing to the case of doublet fields as $\nu_e - e^-$ or $\nu_\mu - \mu^-$, we obtain the relation of the gauge coupling g to the Fermi coupling constant G,

$$\frac{G}{\sqrt{2}} = \frac{g^2}{8M_W^2} \quad . \tag{19}$$

By recalling that $g\sin\theta_W = e$ we can also cast this relation in the form

$$M_W = \left(\frac{\pi\alpha}{\sqrt{2}G}\right)^{1/2}\frac{1}{\sin\theta_W} \simeq \frac{38\ \text{GeV}}{\sin\theta_W} \quad . \tag{20}$$

The present experimental evidence (that we shall discuss shortly) leads to $\sin^2\theta_W \simeq 0.23$, so that a value of $M_W \simeq 80$ GeV is derived from the previous Eq. (20).

In the same way, for neutral currents, one obtains from Eq. (17) the effective four fermion interaction, given by

$$\sqrt{2}\ G\rho\bar{\psi}\gamma_\mu[\ldots]\psi\bar{\psi}\gamma^\mu[\ldots]\psi \quad , \tag{21}$$

where

$$[\ldots] \equiv t_L^3(1-\gamma_5) + t_R^3(1+\gamma_5) - 2Q\sin^2\theta_W \quad , \tag{22}$$

and

$$\rho = \frac{M_W^2}{M_Z^2\cos^2\theta_W} \quad . \tag{23}$$

We see that a study of charged current couplings tells us about the transformation properties of the different species of fermions and the structure of multiplets. From neutral current couplings we learn in addition the magnitude of θ_W and the ratio M_W/M_Z. It goes without saying that the possibility of describing all fermion neutral current couplings in terms of θ_W and ρ provides us with a very stringent test of the $SU(2) \times U(1)$ theory.

We now turn to the Higgs sector in the electroweak Lagrangian. The general form of L_{HIGGS} is specified by the gauge principle and the requirement of renormalizability to be

$$L_{HIGGS} = (D_\mu \vec{\phi}^\dagger)(D^\mu \vec{\phi}) - V(\vec{\phi}^\dagger \vec{\phi}) - \bar{\psi}_L \vec{\Gamma} \psi_R \vec{\phi} - \bar{\psi}_R \vec{\Gamma}^\dagger \psi_L \vec{\phi}^\dagger , \qquad (24)$$

where $\vec{\phi}$ is a column vector including all Higgs bosons. $\vec{\phi}$ transforms as a reducible representation of the gauge group. The quantities $\vec{\Gamma}$ include all coupling constants and the matrices that make the Yukawa couplings invariant under the gauge group. The symmetric potential contains at most quartic couplings. Spontaneous symmetry breaking is induced if the minimum of V, the vacuum, is obtained for nonvanishing $\vec{\phi}$ values. Precisely, we denote the vacuum expectation value of $\vec{\phi}(x)$ by $\vec{\eta}$,

$$\langle 0 | \vec{\phi}(x) | 0 \rangle = \vec{\eta} \neq 0 . \qquad (25)$$

The fermion mass matrix is obtained from the Yukawa couplings by replacing $\vec{\phi}(x)$ by $\vec{\eta}$:

$$M = \bar{\psi}_L M \psi_R + \bar{\psi}_R M^\dagger \psi_L , \qquad (26)$$

where

$$M = \vec{\Gamma} \cdot \vec{\eta} . \qquad (27)$$

In the standard model all ψ_L are doublets of SU(2) and all ψ_R are singlets. In this case the only Higgs that can possibly contribute to fermion masses are doublets. On the other hand there are enough free coupling constants in $\vec{\Gamma}$ so that one single complex Higgs doublet is indeed sufficient to generate the most general fermion mass matrix.

It is important to observe that by a suitable change of basis one can always make the matrix M hermitian, γ_5-free and diagonal. In fact we can make separate unitary transformations on ψ_L and ψ_R according to

$$\psi_L' = U\psi_L \ ; \quad \psi_R' = V\psi_R \ ; \qquad (28)$$

and consequently

$$M \to M' = U^\dagger M V . \qquad (29)$$

Such a transformation does not alter the general structure of the fermion couplings in L_{SYMM}.

If only one Higgs doublet is present it is simple to realize that the change of basis that makes M diagonal at the same time also diagonalizes the fermion-Higgs Yukawa couplings. Thus in this case no flavor changing neutral Higgs exchanges are present. This is not true in general for more Higgs doublets. For several Higgs doublets it is also possible to have complex phases in the fermion-Higgs couplings, thereby generating CP violations in the Higgs sector[3]. In the presence of six quark flavors this mechanism for CP violation is not necessary. At present the simplest model with only one Higgs doublet seems adequate to describe the observed phenomena.

Turning now to the gauge boson masses, these are generated by the $(D_\mu\vec{\phi}^\dagger)(D^\mu\vec{\phi})$ terms in L_{HIGGS}, Eq. (24), where

$$D_\mu\vec{\phi} = (\partial_\mu + ig \sum_A t^A W^A_\mu + ig'\frac{1}{2}YB_\mu)\vec{\phi}, \tag{30}$$

with t^A and Y/2 being the generators of the gauge group in the reducible representation spanned by $\vec{\phi}$. All possible Higgs representations, not only doublets, can contribute in this case. The condition that the photon remains massless is equivalent to the condition that the electric charge annihilates $\vec{\eta}$,

$$Q\vec{\eta} = (t^3 + Y/2)\vec{\eta} = 0 . \tag{31}$$

The charged W mass is given by the quadratic terms in the W field arising from L_{HIGGS} when $\vec{\phi}(x)$ is replaced by $\vec{\eta}$. One immediately obtains

$$M_W^2 W_\mu W^{\dagger\mu} = g^2 \left|\frac{t^+}{\sqrt{2}}\vec{\eta}\right|^2 W_\mu W^{\dagger\mu} . \tag{32}$$

As for the Z mass, one obtains (by recalling Eq. (13)):

$$\frac{1}{2}M_Z^2 Z_\mu Z^\mu = \left|(g\cos\theta_W t^3 - g'\sin\theta_W \frac{Y}{2})\vec{\eta}\right|^2 Z_\mu Z^\mu , \tag{33}$$

where the factor of 1/2 on the left hand side is the correct normalization for the definition of mass of a neutral field. By using Eq. (31) relating the action of t^3 and $\frac{1}{2}Y$ on $\vec{\eta}$, and Eqs. (15) and (16) for θ_W, we obtain

$$\begin{aligned}\frac{1}{2}M_Z^2 Z_\mu Z^\mu &= (g\cos\theta_W + g'\sin\theta_W)^2 |t^3\vec{\eta}|^2 Z_\mu Z^\mu \\ &= \frac{g^2}{\cos^2\theta_W}|t^3\vec{\eta}|^2 Z_\mu Z^\mu .\end{aligned} \tag{34}$$

For Higgs doublets,

$$\vec{\phi} = \begin{pmatrix} \phi^+ \\ \\ \phi^0 \end{pmatrix} , \qquad \vec{\eta} = \begin{pmatrix} 0 \\ \\ \eta \end{pmatrix} , \tag{35}$$

one has

$$|\vec{t}\vec{\eta}|^2 = \eta^2 , \qquad |t^3\vec{\eta}|^2 = \frac{1}{4}\eta^2 \quad . \tag{36}$$

In this case one obtains from Eqs. (32), (34), (36) the important result

$$\rho = \frac{M_W^2}{M_Z^2 \cos^2\theta_W} = 1 \quad . \tag{37}$$

This relation is typical of one or more Higgs doublets and would be spoiled by the existence of Higgs triplets etc., as is clear from the general expressions in Eqs. (32) and (34) for the W and Z masses. Recall that the parameter ρ controls the relative strength of charged versus neutral current interactions (see Eqs. (21), (22) and (23)). Present evidence strikingly confirms Eq. (37) for ρ, so that the existence of only Higgs doublets is indicated.

Note that if only one doublet of Higgs is present then the fermion-Higgs couplings are in proportion to the fermion masses. In fact we have from the above derivation of mass formulae that

$$\begin{aligned} & g^2\eta^2 \sim M_W^2 , \\ & g_{\phi\bar{f}f}\cdot\eta \sim m_f , \end{aligned} \tag{38}$$

and consequently

$$g_{\phi\bar{f}f} \sim \frac{m_f}{\eta} \sim \frac{m_f}{M_W} g \sim \sqrt{G}\, m_f \quad . \tag{39}$$

These order of magnitude relations may be invalidated if more Higgs doublets are present.

With only one complex Higgs doublet, three out of the four hermitian fields are eaten up by the Higgs mechanism and turn into the longitudinal modes of W^+, W^- and Z. The fourth neutral Higgs is physical and should be found. If more doublets are present two more charged and two more neutral Higgs should be around for each additional doublet.

Little is known about the mass of the physical Higgs boson (or bosons). The effective potential has the form

$$V(\phi) = \mu^2\phi^\dagger\phi + \lambda(\phi^\dagger\phi)^2 + \text{LOOPS} \quad , \tag{40}$$

where the contributions beyond the tree diagram semi-classical approximation have also been indicated. At the tree level, for one doublet, one obtains from the minimum condition

$$\begin{aligned} \lambda\eta^2 &= -\mu^2 \quad , \\ m_H^2 &= -2\mu^2 \qquad (\mu^2 < 0) \quad . \end{aligned} \tag{41}$$

Since the magnitude of η is fixed (see Eq. (38)) increasing $|\mu|$ makes λ larger. In order for a meaningful perturbation theory of weak processes not to be invalidated, it has been observed[4] that m_H must be smaller than a few hundred GeV. On the other hand, for the one loop diagram contribution to $V(\phi)$ not to wash out the non-trivial minimum at $\eta \neq 0$ one needs[5,6]

$$m_H^2 \gtrsim \frac{3\alpha^2}{8\sqrt{2}G}\left\{\frac{2 + \sec^4\theta_W}{\sin^4\theta_W} - 0\left(\frac{m_f^4}{M_W^4}\right)\right\} \quad . \tag{42}$$

If $m_f \ll M_W$ that would imply $m_H \gtrsim 10$ GeV. The lower bound is in particular obtained for $\mu = 0$, that is, for a spontaneous symmetry breaking only induced through radiative corrections.[5] This latter possibility was recently discussed in detail.[6]

In conclusion it appears that the mass of the physical Higgs is in practice completely unpredicted.[7] It is clear that the discovery of a Higgs boson, identified through its pattern of couplings to fermions in proportion to their masses, would be a major experimental breakthrough, which would definitely establish perhaps the newest and most striking theoretical aspect of the theory.

Before concluding this section we discuss the issue of mixing angles and the constraints imposed by the requirement of natural flavor conservation for the neutral current. In the standard version of the theory all ψ_L are doublets and all ψ_R are singlets. In particular for quarks one has the multiplets:

$$\begin{pmatrix} u \\ d' \end{pmatrix}_L \begin{pmatrix} c \\ s' \end{pmatrix}_L \begin{pmatrix} L \\ b' \end{pmatrix}_L \quad \cdots ; \qquad q_R: \text{ all singlets.} \tag{43}$$

The reason for the primes is that the states d', s', b' ... with definite weak interaction properties do not in general coincide with the eigenstates of the mass matrix (which the unprimed quark labels refer to). In general a change of basis, i.e. a unitary transformation, connects the primed states with definite weak couplings and the unprimed states which diagonalize the mass matrix,

$$\begin{pmatrix} d' \\ s' \\ b' \\ \vdots \end{pmatrix} = U \begin{pmatrix} d \\ s \\ b \\ \vdots \end{pmatrix} \tag{44}$$

Since U is unitary and commutes with T^2, T_3 (the weak isospin generators) and Q, the neutral current couplings are diagonal both in the primed and unprimed basis. This is the GIM mechanism[8] that automatically ensures flavor conservation in the neutral current couplings. Conversely it is simple to realize that the general criterion for obtaining natural flavor conservation in the neutral current is[9] that all states with the same Q also have the same transformation properties under the weak group, i.e. the same T^2 and T_3.

For N generations of quarks (i.e. 2N flavors) U is an $N \times N$ unitary matrix that can in general be parametrized in terms of N^2 real numbers. However, the 2N-1 relative phases of quarks are not observable and can be fixed arbitrarily (the overall phase has no effect on U). Therefore $N^2 - 2N + 1 = (N-1)^2$ real numbers are left. On the other hand the most general $N \times N$ orthogonal matrix involves N(N-1)/2 real numbers. Thus the $(N-1)^2$ parameters can be split into N(N-1)/2 mixing angles and (N-1)(N-2)/2 phases. For N=2, as in the old four quark model, one had one mixing angle (the Cabibbo angle) and no phases. For N=3, as in the present orthodoxy, one obtains three mixing angles and one phase. This phase is important and welcome, because it allows for an elegant accounting of CP violation in the theory.[10,11] For N=4, if one more generation would be found, one would have six mixings and three phases, and so on. We shall discuss the phenomenology of mixings and phases in the case of N=3 in Sec. 3. Note that the known weak properties of light quarks and leptons plus the condition of natural flavor conservation of the neutral current by itself imposes the standard classification of all ψ_L in doublets and all ψ_R in singlets. In turn the model with three quark doublets and three lepton doublets is automatically free of γ_5

anomalies[12] through the cancellation of quark anomalies with lepton anomalies.

Having thus completed the general discussion of the Weinberg-Salam model and the description of the standard version of it, we now consider in the following sections the comparison of the model with the data.

3. NEUTRAL CURRENT COUPLINGS

In this section we summarize the experimental results on neutral current interactions which provide quite convincing evidence in favor of the standard model. We divide the discussion of neutral current couplings by considering experiments in the neutrino-quark, neutrino-electron and electron-quark sectors. It has become, in fact, widely customary[13-16] to analyze the data in a model independent way by first writing down general four fermion effective Lagrangians for each sector, then extracting from the data the values of the various coupling constants and finally comparing the results with the predictions of the standard model. We shall follow this line in our discussion.

3.1 Neutrino-Quark Sector

The most general effective Lagrangian for neutrino quark interactions from a flavor conserving neutral interaction can be written in the form

$$\begin{aligned} L_{EFF} = \frac{G}{\sqrt{2}} \bar{\nu}_\mu \gamma^\alpha (1-\gamma_5) \nu_\mu \{ & u_L \bar{u} \gamma_\alpha (1-\gamma_5) u + d_L \bar{d} \gamma_\alpha (1-\gamma_5) d + \\ & + u_R \bar{u} \gamma_\alpha (1+\gamma_5) u + d_R \bar{d} \gamma_\alpha (1+\gamma_5) d + \ldots \} \end{aligned} \qquad (45)$$

where the dots stand for sea quark terms which, for simplicity, will not be explicitly considered here, but are of course taken into account in the actual analysis of the data. The four coupling constants $u_{L,R}$ and $d_{L,R}$ are to be determined from experiment. The overall sign of the four couplings is beyond reach and therefore we take u_L to be positive by convention.

Present data allow a complete determination of $(u,d)_{L,R}$. This is obtained by a number of steps that we can organize as follows.

The first important piece of information is obtained from data on ν_μ or $\bar{\nu}_\mu$ inclusive scattering on isoscalar targets at high energy. The measurement of the total cross sections from ν and $\bar{\nu}$ beams can be directly translated into a separate measurement of $u_L^2 + d_L^2$ and $u_R^2 + d_R^2$. In fact the familiar parton model formulae lead to

$$\sigma^{\nu} \sim \int_0^1 dx\ x\ [u(x) + d(x)]\{(u_L^2+d_L^2) + \tfrac{1}{3}(u_R^2+u_R^2)\} + \ldots ,$$
$$\sigma^{\bar{\nu}} \sim \int_0^1 dx\ x\ [u(x) + d(x)]\{\tfrac{1}{3}(u_L^2+d_L^2) + (u_R^2+u_R^2)\} + \ldots , \qquad (46)$$

where u(x) and d(x) are the valence parton densities and the dots stand here too for terms proportional to sea densities. Contributions from sea densities and corrections for QCD scaling violations are included in the analysis. The results from different high energy experiments are in perfect agreement with one another and read

$$\begin{array}{llll} u_L^2+d_L^2 = 0.29 \pm 0.02 & u_R^2+u_R^2 = 0.03 \pm 0.01 & (\text{CDHS}^{17}) & \\ 0.29 \pm 0.06 & 0.02 \pm 0.06 & (\text{HPWF}^{18}) & (47) \\ 0.32 \pm 0.03 & 0.04 \pm 0.03 & (\text{BEBC}^{19}) & \end{array}$$

leading to the average

$$u_L^2+d_L^2 = 0.299 \pm 0.016 , \qquad u_R^2+d_R^2 = 0.031 \pm 0.009 . \qquad (48)$$

This list includes results up to some months ago. Recently the CDHS analysis was somewhat improved.[20] The new figures for R_{ν} and $R_{\bar{\nu}}$ (ratios of neutral current to charged current cross sections) are reported to be (the data refer to $E_H > 10$ GeV)

$$R_{\nu} = 0.307 \pm 0.008 \qquad (0.29 \pm 0.01) ,$$
$$R_{\bar{\nu}} = 0.373 \pm 0.025 \qquad (0.35 \pm 0.025) , \qquad (49)$$

where the previous results are also shown in brackets. Also new data from the CHARM collaboration at CERN-SPS have become available[21] (for $E_H > 8$ GeV),

$$R_{\nu} = 0.30 \pm 0.02 ,$$
$$R_{\bar{\nu}} = 0.39 \pm 0.02 . \qquad (50)$$

These new results are in agreement with the previous ones and do not substantially alter the couplings as given above in Eq. (48).

A separate determination of all the four squared couplings is made possible by measurements of pion inclusive production from ν_μ and $\bar{\nu}_\mu$ beams on isoscalar targets $(\nu(\bar{\nu})+N \to \nu(\bar{\nu})+\pi^\pm + \text{all})$ and of totally inclusive scattering from ν_μ and $\bar{\nu}_\mu$ on protons and neutrons. The former process is analyzed in the parton model in terms of fragmentation functions for u and d quarks into pions

$$\sigma^\nu \sim \int_0^1 dx\ x\ [u(x)+d(x)]\{(u_L^2+\tfrac{1}{3}u_R^2)D_u^{\pi\pm} + (d_L^2+\tfrac{1}{3}d_R^2)D_d^{\pi\pm}\} + \ldots ,$$

$$\sigma^{\bar{\nu}} \sim \int_0^1 dx\ x\ [u(x)+d(x)]\{(\tfrac{1}{3}u_L^2+u_R^2)D_u^{\pi\pm} + (\tfrac{1}{3}d_L^2+d_R^2)D_d^{\pi\pm}\} + \ldots . \tag{51}$$

Integrated fragmentation functions are extracted from charged current and electroproduction experiments. Until recently the only available data were from low energy Gargamelle data,[22] which made the use of parton formulae suspect. But recently the low energy data were confirmed by new data at high energy, so that this step of the analysis also appears on a firm ground now. The available data are[23]

$$\left(\frac{\pi^+}{\pi^-}\right)_\nu = \begin{cases} 0.77 \pm 0.14 & \text{GGM-PS} \quad \text{low energy} \\ 0.81 \pm 0.20 & \text{BEBC} \quad \text{high energy} \end{cases}$$

$$\left(\frac{\pi^+}{\pi^-}\right)_{\bar{\nu}} = \begin{cases} 1.64 \pm 0.36 & \text{GGM-PS} \quad \text{low energy} \\ 1.21 \pm 0.29 & \text{FNAL 15'} \quad \text{high energy} \end{cases} \tag{52}$$

The analysis of totally inclusive scattering on proton and neutron targets only requires the input of the quantity

$$r = \int_0^1 dx\ x\ d(x) \Big/ \int_0^1 dx\ x\ u(x) \simeq 0.509 , \tag{53}$$

where the quoted number is taken from an estimate of Feynman and Field. Then for the ratios of neutral current to charged current cross sections on protons one has

$$R_p = d_L^2 + \tfrac{1}{3}d_R^2 + \tfrac{1}{r}u_L^2 + \tfrac{1}{3r}u_R^2 + \ldots ,$$

$$\bar{R}_p = rd_L^2 + 3rd_R^2 + u_L^2 + 3u_R^2 + \ldots , \tag{54}$$

and similarly for the ratios on neutrons. The available data are

$$
\begin{aligned}
R_p &= \begin{cases} 0.48 \pm 0.17 & \text{FNAL-BHM} \\ 0.52 \pm 0.04 & \text{BEBC} \end{cases} \\
\bar{R}_p &= 0.42 \pm 0.13 \quad \text{BEBC} \\
R_{n/p} &= \frac{\sigma(\nu n)_{NC}}{\sigma(\nu p)_{NC}} = \begin{cases} 0.76 \pm 0.15 & \text{GGM} \\ 1.22 \pm 0.35 & \text{FNAL} \end{cases} \\
\bar{R}_{n/p} &= \frac{\sigma(\bar{\nu} n)_{NC}}{\sigma(\bar{\nu} p)_{NC}} = 0.53 \pm 0.39 \quad \text{FNAL} \quad .
\end{aligned}
\tag{55}
$$

From the previous results separate values of the squared couplings are extracted

$$
\begin{aligned}
u_L^2 &= 0.101 \pm 0.032 \ , \qquad u_R^2 = 0.03 \pm 0.015 \ , \\
d_L^2 &= 0.189 \pm 0.036 \ , \qquad d_R^2 = 0 \pm 0.015 \ .
\end{aligned}
\tag{56}
$$

Because u_L is positive by convention and d_R is zero by experiment, a four-fold ambiguity remains depending on the signs of u_R and d_L. This ambiguity can be resolved by the data on elastic scattering of ν and $\bar{\nu}$ on protons ($\nu(\bar{\nu}) + p \to \nu(\bar{\nu}) + p$), and on exclusive pion production on isoscalar targets ($\nu(\bar{\nu}) + N \to \nu(\bar{\nu}) + \pi^{\pm} + N$). In this last process what is important is the observation of a strong Δ signal[25] that excludes a predominantly isoscalar solution for the neutral current couplings. This is relevant because a detailed analysis of π exclusive production must include nuclear physics corrections, for example as treated by Adler,[26] which are to some extent model dependent. We also quote a recent measurement[27] at reactor energies of $\bar{\nu}_e + D \to \bar{\nu}_e + n + p$ (D: deuterium). At these low energies a nonrelativistic treatment is perfectly adequate. This is a pure Gamow-Teller transition that measures the isovector axial coupling of the neutral current (which is independent of $\sin^2\theta_W$). The result can be quoted as a ratio of the experimental to the predicted value in the standard model,

$$
\frac{\sigma_{exp}}{\sigma_{TH}} = 0.8 \pm 0.2 \quad . \tag{57}
$$

The experiments listed above allow the resolution of the remaining four-fold ambiguity, and the determination of the valence quark couplings to the neutral current is thus completed with the results

$$\begin{aligned} u_L &= \ \ 0.35 \pm 0.07 \quad (\ 0.35) \ , \\ u_R &= -0.19 \pm 0.06 \quad (-0.15) \ , \\ d_L &= -0.40 \pm 0.07 \quad (-0.42) \ , \\ d_R &= \ \ 0.0 \ \ \pm 0.11 \quad (\ 0.08) \ . \end{aligned} \tag{58}$$

The numbers in brackets are the predictions from the standard model with $\sin^2\theta_W = 0.23$. In fact taking for granted that ν_μ has $t^3 = 1/2$, a general SU(2) x U(1) model predicts, according to Eqs. (17) and (45) that

$$(u,d)_{L,R} = \rho[t^3_{L,R} - Q\sin^2\theta_W] \ . \tag{59}$$

In particular, in the standard model $\rho=1$, $t^3_L = \pm 1/2$ and $t^3_R = 0$. Note that changing a singlet into a doublet makes a difference of 0.5 in the couplings, which is much larger than the errors. It is thus established that the right handed u and d quarks are singlets. This is independent of the identity and the mass of the possible partners in a multiplet, because the neutral current couplings are diagonal and there are no threshold problems, contrary to the case of charged currents.

3.2 Neutrino-Electron Sector

Assuming $\nu_e - \nu_\mu$ universality the general effective Lagrangian for ν-e interactions can be written down in terms of two coupling constants,

$$L_{EFF} = \frac{G}{\sqrt{2}} \left[\bar{\nu}_\mu \gamma^\alpha (1-\gamma_5)\nu_\mu + \nu_\mu \leftrightarrow \nu_e\right]\left[g_V \bar{e}\gamma_\alpha e - g_A \bar{e}\gamma_\alpha\gamma_5 e\right] \ . \tag{60}$$

The experimental data are from $\nu_\mu(\bar{\nu}_\mu)$ elastic scattering on electrons and from $\bar{\nu}_e e$ scattering at reactor energies. The cross sections for $\nu_\mu(\bar{\nu}_\mu)e$ scattering are related to the electron couplings by

$$\begin{aligned} \sigma_{\nu_\mu e} &\simeq g_L^2 + \tfrac{1}{3} g_R^2 = \tfrac{4}{3}(g_V^2 + g_A^2 + g_V g_A) \ , \\ \sigma_{\bar{\nu}_\mu e} &\simeq \tfrac{1}{3} g_L^2 + g_R^2 = \tfrac{4}{3}(g_V^2 + g_A^2 - g_V g_A) \ . \end{aligned} \tag{61}$$

The experimental data are listed[28] in Table I. The different experiments are now in reasonable agreement with each other. Note that the data on $\bar{\nu}_\mu e$ are still very poor.

TABLE I

EXP	$(\sigma_{\nu_\mu e}/E_\nu)10^{42}\mathrm{cm}^2/\mathrm{GeV}$	$(\sigma_{\bar{\nu}_\mu e}/E_\nu)10^{42}\mathrm{c},^2/\mathrm{GeV}$
GGM-PS	3.9	$1 \pm {}^{0.9}_{0.6}$
A-P	1.1 ± 0.6	2.2 ± 1.0
15' FNAL	1.8 ± 0.8	< 2.1
GGM-SPS	$2.4 \pm {}^{1.2}_{0.9}$	< 2.7
CHARM-SPS	2.6 ± 1.4	< 4.2

The cross section for $\bar{\nu}_e e$ scattering, when the electron mass is neglected, is given by

$$\sigma_{\bar{\nu}_e e} \simeq (1+g_V)^2 + (1+g_A)^2 - (1+g_V)(1+g_A) \quad , \tag{62}$$

where the terms not proportional to g_V or g_A are from the charged current contribution. The mass of the electron is not negligible at reactor energies and mass corrections must in reality be kept. The only data available for this reaction are from the old Savannah reactor experiment[29] and no news is to be reported. Note that all the cross sections for $\nu_\mu e$, $\bar{\nu}_\mu e$ and $\bar{\nu}_e e$ are symmetric under the exchange $g_V \leftrightarrow g_A$ as is apparent from Eqs. (61) and (62). Thus two solutions are obtained for g_V and g_A that differ by an exchange. One solution is

$$\begin{aligned} g_V &= 0 \pm 0.18 \qquad (-0.04) \\ g_A &= -0.56 \pm 0.14 \qquad (-0.50) \quad , \end{aligned} \tag{63}$$

where the values in brackets are the standard model predictions for $\sin^2\theta_W \simeq 0.23$. We shall come back later to the removal of the $g_V \leftrightarrow g_A$ ambiguity.

3.3 Electron-Quark Sector

In this sector only parity violating effects from weak neutral currents can be experimentally observed. We therefore write down a general parity violating effective Lagrangian. We adopt the notations of Hung-Sakurai[15] in terms of vector and axial isoscalar and isovector couplings:

$$L_{EFF}^{PV} = -\frac{G}{\sqrt{2}}\left\{\bar{e}\gamma^{\alpha}\gamma_5 e\left[\frac{1}{2}\tilde{\alpha}(\bar{u}\gamma_{\alpha}u - \bar{d}\gamma_{\alpha}d) + \frac{1}{2}\tilde{\gamma}(\bar{u}\gamma_{\alpha}u + \bar{d}\gamma_{\alpha}d)\right] + \right.$$

$$\left. + \bar{e}\gamma^{\alpha}e\left[\frac{1}{2}\tilde{\beta}(\bar{u}\gamma_{\alpha}\gamma_5 u - \bar{d}\gamma_{\alpha}\gamma_5 d) + \frac{1}{2}\tilde{\delta}(\bar{u}\gamma_{\alpha}\gamma_5 u + \bar{d}\gamma_{\alpha}\gamma_5 d)\right]\right\} . \quad (64)$$

The available data are from parity violation in atoms and from the measurement of left-right asymmetry in electron-deuterium (or proton) scattering. The latter experiment is the most valuable in that the results are clear cut and the interpretation is direct. We shall discuss the two classes of experiments in turn in the following.

Parity violation in atomic physics is induced in lowest order by one Z exchange between an electron and the quarks in the nucleus. In principle there are two contributions: one from the axial coupling to the electron times the vector coupling to the quarks and one from the vector coupling to the electron and the axial coupling to the quarks. Available data only refer to heavy atoms. In this case the first alternative is dominant and produces a detectable effect. In fact the vector charges of all quarks in the nucleus add up coherently. On the other hand the axial couplings to quarks, which correspond to charge times spin, give a negligible effect because the nucleus spin is small and different quarks with the same charge add up with different signs according to their spins. Thus in heavy atoms the measurement of parity violation determines a quantity given by the vector coupling of u quarks times the number of u quarks in the nucleus plus the same quantity for d quarks. In terms of the parameters in L_{EFF}^{PV} (see Eq. (64)) the factor that can be extracted from experiment is

$$\begin{aligned} Q_W &= -[(\tilde{\gamma}+\tilde{\alpha})(2Z+N) \quad + (\tilde{\gamma}-\tilde{\alpha})(Z+2N)] \\ &= -\ [\tilde{\alpha}(Z-N) + 3\tilde{\gamma}(Z+N)] \quad , \end{aligned} \quad (65)$$

where Z and N are the number of protons and neutrons in the nucleus.

The experiments on Bismuth (Z=83, N=126) measure the rotation of the polarization of light when it goes through Bi vapour, due to the difference in refractive index for left and right polarizations. Three different groups have reported results on two different wavelengths. The results and the predictions of the standard model on R (the ratio of the parity conserving to the parity violating amplitude) are reported in Table II. Taken at face value these results indicate a marked experimental contradiction. Over the last year the Novosibirsk experiment has been continued and improved. With respect to the standard model with $\sin^2\theta_W = 1/4$

TABLE II

	$R \times 10^8$–8757 Å	$R \times 10^8$–6477 Å
Oxford[30]		2.7 ± 4.7
Seattle[31]	−0.7 ± 3.2 −0.5 ± 1.7	
Novosibirsk[32]		−18 ± 5
Theory (Standard model: $\sin^2\theta_W = 1/4$)		
Rel. Central Field[33]	−18	−24
Same plus shielding[33]	− 8	−11
Semiempirical[34]	−14	−18

in the semiempirical approach (that is so called because of the use of as much experimental information as possible) they now quote the improved result[32]

$$\frac{(Q_W)_{EXP}}{(Q_W)_{TH}} = 1.07 \pm 0.14 \quad . \tag{66}$$

Also a new experiment[35] on Thallium has been completed recently. Thallium (Z=81, N=123) is simpler to analyze theoretically than Bismuth. What is measured is the ratio $(\sigma_L - \sigma_R)/(\sigma_L + \sigma_R)$ for absorption of left and right handed polarized light for a specific, suitable line of the atom. The quoted result for the ratio of parity violating to parity conserving amplitudes is given by

$$\begin{aligned} \delta_{TH} &= (2.3 \pm 0.3)10^{-3} \quad , \\ \delta_{EXP} &= (5.2 \pm 2.4)10^{-3} \quad , \end{aligned} \tag{67}$$

where the theoretical prediction refers to the standard model with $\sin^2\theta_W = 1/4$.

In conclusion it is clear that more work is needed in order to completely clarify the issue of parity violation in atomic physics. Also it is apparent that theoretical uncertainties for complex atoms make a precise determination of neutral current parameters from these experiments difficult. This makes even more important the results from the SLAC experiment on polarized electron scattering on deuterium (or protons) which we now consider.

The SLAC experiment[36] measures the parity violating asymmetry for the cross sections of left-handed and right-handed electrons off deuterium, defined by

$$A(x,y,Q^2) = \frac{\sigma_R - \sigma_L}{\sigma_R + \sigma_L} \quad . \tag{68}$$

Here $-Q^2$ is the virtual photon mass squared, x is the Bjorken variable and $y = (E-E')/E$, where E and E' are the initial and final electron energies in the lab system. The dependence of A on x,y and Q^2 is of the form

$$A(x,y,Q^2) = Q^2\left[A_1(x) + A_2(x)\frac{1-(1-y)^2}{1+(1-y)^2}\right] \quad . \tag{69}$$

$A_{1,2}$ are proportional to the interference of the weak and electromagnetic amplitudes. In the valence approximation for deuterium the x dependence of $A_{1,2}$ drops away. A_1 is proportional to the axial coupling of the electron times the vector coupling of the quarks, while A_2 is proportional to the vector coupling of the electron times the axial coupling of the quarks. A simple calculation in the valence approximation leads to

$$A_1 \sim A^e V^q = \frac{G}{\sqrt{2}e^2}\,\frac{9}{5}(\tilde{\alpha} + \tfrac{1}{3}\tilde{\gamma}) \quad ,$$
$$A_2 \sim V^e A^q = \frac{G}{\sqrt{2}e^2}\,\frac{9}{5}(\tilde{\beta} + \tfrac{1}{3}\tilde{\delta}) \quad . \tag{70}$$

Note that A_1 involves the same couplings $\tilde{\alpha}$ and $\tilde{\gamma}$ as for parity violation in heavy atoms (although of course in a different combination). At $y = 0$ only A_1 survives, so that the asymmetry at $y = 0$ can be directly related to the atomic physics measurements, while A_2 contains new independent information. The above valence approximation can be easily corrected for sea effects, violations of the Callan-Gross relation, etc.

The first experiment was at $y = 0.21$ with $\langle Q^2\rangle = 1.4$ GeV^2 and $\langle x\rangle = 0.15$. More recently the experiment was improved by including results in a range of y between 0.15 and 0.35. This allows the separate determination of A_1 and A_2 with the result

$$A_1 = (-9.70 \pm 2.57)10^{-5}\ \text{GeV}^{-2} ,$$
$$A_2 = (4.9 \pm 8.1)10^{-5}\ \text{GeV}^{-2} . \qquad (71)$$

A measure of the asymmetry on protons has also been performed, though with less precision:

$$A(x \simeq 0.2,\ y \simeq 0.21)/Q^2 = (-9.7 \pm 2.7)10^{-5}\ \text{GeV}^{-2} . \qquad (72)$$

The deuterium asymmetry implies the following values for the relevant combinations of couplings,

$$\tilde{\alpha} + \frac{1}{3}\tilde{\gamma} = -0.60 \pm 0.16 \qquad (-0.49) ,$$
$$\tilde{\beta} + \frac{1}{3}\tilde{\delta} = 0.31 \pm 0.51 \qquad (0.15) , \qquad (73)$$

where the values in brackets are the standard model predictions with $\sin^2\theta_W = 0.23$. If we add the atomic physics results from Novosibirsk and from the Thallium experiment a separate determination of $\tilde{\alpha}$ and $\tilde{\gamma}$ is possible

$$\tilde{\alpha} = -0.72 \pm 0.25 \qquad (-0.54) ,$$
$$\tilde{\gamma} = 0.38 \pm 0.28 \qquad (0.15) . \qquad (74)$$

This concludes the discussion of the three sectors taken separately. We see that we are left with some ambiguities. These are: a) the overall sign of the four couplings in the ν-q sector; b) the $g_V \leftrightarrow g_A$ ambiguity in the ν-e sector; c) the separate determination of $\tilde{\beta}$ and $\tilde{\delta}$ in the e-q sector. In order to go further we must now discuss the relations among the three sectors that would follow by an assumption of factorization in the neutral current couplings, as discussed in Ref. 15.

3.4 Relations Among the Three Sectors

If a single Z boson is exchanged the couplings in each sector are products of two vertices. For example in the ν-q sector the couplings can be split into $Z_{\nu\bar{\nu}}$ times $Z_{q\bar{q}}$. Then, given $Z_{\nu\bar{\nu}}$, the couplings in the e-q sector can be derived from those in the ν-q and ν-e sectors. Without factorization there are ten independent couplings. With factorization the number of independent couplings is reduced to seven (4 for ν-q, 2 for ν-e and 1 for ν-ν), which implies three relations. In Fig. 1 we display one of these factorization tests, which appears to be fairly well fulfilled. A second

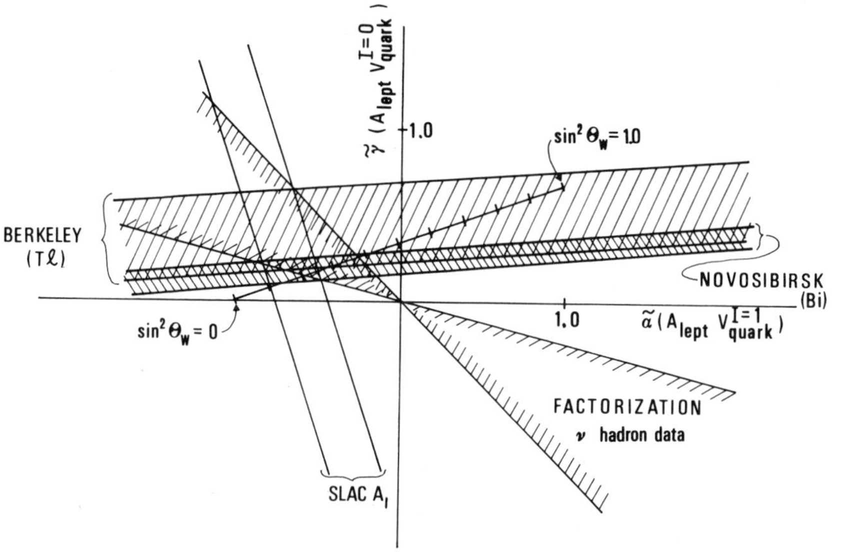

Fig. 1. Test of factorization (from Ref. 15).

test of factorization is displayed in Fig. 2. Here the interesting result is obtained that factorization is valid provided the $g_V \leftrightarrow g_A$ ambiguity is resolved in favor of an axial dominant solution as in Eq. (63). Finally by using the factorization assumption (which, as we have seen, is tested in the data to some extent) we can determine the separate values of $\tilde{\beta}$ and $\tilde{\delta}$. According to Ref. 15 one finds

$$\tilde{\beta} = 0.06 \pm 0.21 \qquad (0.15) \quad ,$$
$$\tilde{\delta} = 0 \pm 0.02 \qquad (0) \quad . \qquad (75)$$

We also remark that the signs of the couplings in the ν-e sector are fixed (by interference with the charged current contribution in $\bar{\nu}_e$e scattering). The same is true for the couplings in the e-q sector (because of interference with electromagnetism).

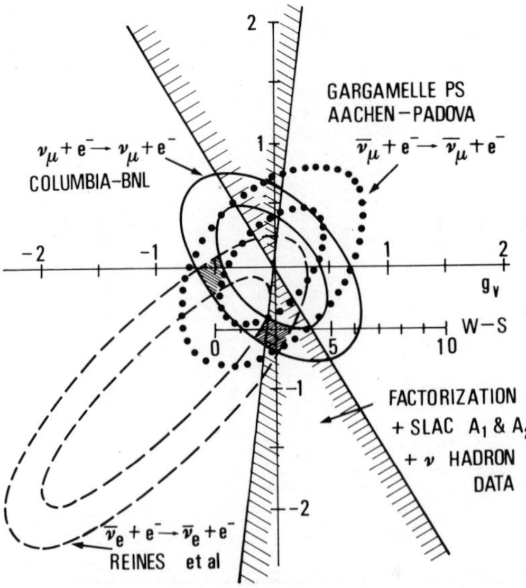

Fig. 2. Test of factorization (from Ref. 15).

Thus the factorization hypothesis allows us to also specify the signs of the couplings in the ν-q sector, which turn out to correspond to $u_L > 0$ (because the $\nu\bar{\nu} \to \nu\bar{\nu}$ amplitude is positive).

Finally an overall fit of all available data can be done in the standard model in terms of the single parameter $\sin^2\theta_W$. We already noticed that the various couplings as given by experiment turn out to be all in fair agreement with $\sin^2\theta_W \simeq 0.23$. In fact the best fit value of $\sin^2\theta_W$, according to Ref. 16 is given by

$$\sin^2\theta_W = 0.229 \pm 0.014 \quad , \tag{76}$$

with $\chi^2/\langle\chi^2\rangle = 10.5/19$. Alternatively all data can be fitted in terms of two parameters, $\sin^2\theta_W$ and ρ (see Eq. (23)). In this case the same authors find

$$\sin^2\theta_W = 0.245 \pm 0.027 \; , \qquad \rho = 1.04 \pm 0.04 \quad , \tag{77}$$

with $\chi^2/\langle\chi^2\rangle = 8.8/18$. It is remarkable that experiment confirms that the ratio of M_W and M_Z is in agreement with the simplest assignment of Higgs bosons to doublets (see Eq. (37)). Also it is difficult to imagine a different origin for the relation $\rho = 1$ other than the Higgs mechanism. Thus the experimental value of ρ provides us with a test of both the Higgs machinery and the Higgs classification.

Actually ρ is so close to one experimentally that one can use this fact to put bounds on the mass of fermions. In fact the tree level relation that corresponds to $\rho = 1$ is corrected by radiative corrections with a fermion loop in the W and Z propagators. For example a doublet of fermions with masses m_1 and m_2 leads to a correction to ρ given by[37]

$$\rho = 1 + \begin{pmatrix}1\\3\end{pmatrix} \frac{G}{8\sqrt{2}\pi^2} \left[\frac{2m_1^2 m_2^2}{m_1^2 - m_2^2} \ln \frac{m_2^2}{m_1^2} + m_1^2 + m_2^2 \right] \quad , \tag{78}$$

where the factors 1 and 3 (color) apply to leptons and quarks respectively. In particular, for a lepton doublet $\binom{\nu}{L^-}$ with $m_\nu = 0$ we obtain

$$\rho = 1 + \frac{Gm^2}{8\sqrt{2}\pi^2} \quad . \tag{79}$$

The present limits on ρ imply that $m \lesssim 10^3$ GeV.

In conclusion the experimental study of neutral couplings provides a very solid evidence in favour of the standard model. The success of the standard model is quite significant because the selective power of the data is indeed high by now. This is proved by the failure of a large number of alternative models that have been proposed in recent years to explain one or the other of some alleged deviations of the experiments from the standard model. These models have all been proven wrong in a short time, often by experiments on processes other than those that at a given time were supposed to provide evidence against the standard model. No equally simple alternative to the standard model appears to survive at present, even if, of course, more complicated models involving more parameters are still certainly possible.

4. QUARK MIXINGS

We have seen in Sec. 2 that for three doublets of quarks the unitary matrix providing the change of basis from the eigenstates of mass to the states with definite weak couplings involves three mixing angles and one phase, as first observed in Ref. 10. In this section we summarize our knowledge on these parameters. There has been some recent theoretical work[38-40] on this subject, so that a discussion of this problem is particularly appropriate.

We start by a suitable definition of the mixing angles and the CP violating phase. In the current literature the definitions of Ref. 10 are usually adopted. We prefer a different set of definitions, due to Maiani,[41] which we find more convenient for reasons to be explained below. As in Eq. (43) we define as d', s', b' the three down partners of u, c, t respectively in the left doublets. All right quarks are assumed to be singlets. We write down d' in the following form

$$d' = \cos\varepsilon \; d_c + \sin\varepsilon \; b \quad , \tag{80}$$

where d_c is the Cabibbo rotated down quark,

$$d_c = \cos\theta_c \; d + \sin\theta_c \; s \quad . \tag{81}$$

Note that in a four quark model the Cabibbo angle fixes both the ratio of the u to d coupling with respect to the ν_μ to μ coupling and the ratio of the u to s and u to d couplings. In a six quark model one has to choose whether to keep the first or the second definition. Here the second is taken and in fact the u to d coupling is given by $\cos\varepsilon \cos\theta_c$, i.e. it is no longer completely specified by θ_c. Also note that we can certainly fix the phases of u, d, s and b so that only real coefficients appear in d'.

We now construct two orthonormal vectors, both orthogonal to d'. They can be chosen as

$$e^{i\psi}[-\sin\varepsilon\ d_c + \cos\varepsilon\ b] \ , \qquad (82)$$

$$e^{i(\delta+\psi)} s_c \ ,$$

where s_c is the Cabibbo rotated strange quark,

$$s_c = -\sin\theta_c\ d + \cos\theta_c\ s \ . \qquad (83)$$

The general form of s' and b' will be given by two orthonormal superpositions of the latter two vectors. Note that the common phase $e^{i\psi}$ can be reabsorbed in the c and t definitions. Thus we have

$$s' = \cos\alpha\ e^{i\delta} s_c + \sin\alpha(-\sin\varepsilon\ d_c + \cos\varepsilon\ b) \ , \qquad (84)$$

$$b' = -\sin\alpha\ e^{i\delta} s_c + \cos\alpha(-\sin\varepsilon\ d_c + \cos\varepsilon\ b) \ .$$

We can always arrange ε and α to vary between 0 and $\pi/2$, provided δ is allowed to vary between $-\pi$ and π.

The success of the Cabibbo theory poses a severe bound on ε. This has been recently reanalyzed in Ref. 38. One has

$$\mathcal{U}_{ud} = \cos\varepsilon\ \cos\theta_c = 0.974 \pm 0.0025 \ , \qquad (85)$$

$$\mathcal{U}_{us} = \cos\varepsilon\ \sin\theta_c = 0.219 \pm 0.002 \pm 0.011 \ ,$$

where the last error is from SU(3) breaking effects, radiative corrections (as discussed in Ref. 42) etc. It then follows for the u to b coupling that

$$\mathcal{U}_{ub} = \sin\varepsilon = 0.06 \pm 0.06 \ . \qquad (86)$$

We see that ε must be smaller than $\theta_c/2$ or so. A lower bound on ε is obtained if the observed CP violation is to be completely explained by the phase δ. In fact the CP violating parameter η_{+-} is proportional to

$$\eta_{+-} \sim \frac{\sin\varepsilon}{\sin\theta_c} \sin\alpha\ \cos\alpha\ \sin\delta \sim 10^{-3} \ . \qquad (87)$$

We therefore obtain for ε the allowed range

$$0.5 \times 10^{-3} \lesssim \sin\varepsilon \lesssim 0.12 \ . \qquad (88)$$

Since ε and θ_c are rather small we can simplify the following discussion by adopting a linear approximation in ε and θ_c (i.e. by neglecting terms of order ε^2, θ_c^2 and $\varepsilon\theta_c$). The possibility of doing so is a great advantage of the present definition in that the small coupling $\mathcal{U}_{ub}$ is denoted by $\sin\varepsilon$ and not by a product $\sin\beta_1 \sin\beta_2$ as in Ref. 10. In the linear approximation we obtain for the three doublets

$$\begin{pmatrix} u \\ d + \theta_c s + \varepsilon b \end{pmatrix}_L ,$$

$$\begin{pmatrix} c \\ -(\cos\alpha\, e^{i\delta}\theta_c + \sin\alpha\,\varepsilon)d + \cos\alpha\, e^{i\delta} s + \sin\alpha\, b \end{pmatrix}_L , \tag{89}$$

$$\begin{pmatrix} t \\ (\sin\alpha\, e^{i\delta}\theta_c - \cos\alpha\,\varepsilon)d - \sin\alpha\, e^{i\delta} s + \cos\alpha\, b \end{pmatrix}_L .$$

Constraints on α and δ can be derived from the observed values of the $K_S - K_L$ mass difference,

$$\delta m \simeq 3.5 \times 10^{-15}\ \text{GeV} , \tag{90}$$

and of the CP violating parameters η_{+-}, η_{00} etc.

The $K_L - K_S$ mass difference arises from second order weak interactions through the diagrams in Fig. 3. An explicit calculation for massless "old" quarks,[43,44] leads to the expression

$$\delta m = ABRe\left[\frac{\lambda_c^2}{\theta_c^2} + \frac{\lambda_t^2}{\theta_c^2}\frac{m_t^2}{m_c^2} + \frac{2\lambda_c\lambda_t}{\theta_c^2}\frac{m_t^2}{m_t^2 - m_c^2}\ln\frac{m_t^2}{m_c^2}\right] , \tag{91}$$

valid for $m_{c,t}^2 \ll M_W^2$. (A more general formula can be found in Ref. 40.) Here $\lambda_i = \mathcal{U}_{is}\mathcal{U}_{id}^*$, and

$$A = \frac{f_k^2 m_K (G/\sqrt{2})(\alpha/2\pi)\theta_c^2 m_c^2}{3M_W^2 \sin^2\theta_W} , \tag{92}$$

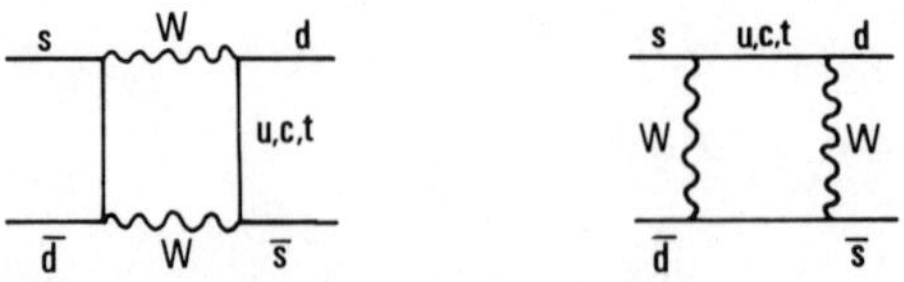

Fig. 3. Diagrams contributing to $\Delta S = 2$ transitions.

$$B = \frac{\langle K^o | (\bar{s}\gamma_\mu \frac{1}{2}(1-\gamma_5)d)^2 | \bar{K}^o \rangle}{B_o} , \tag{93}$$

with B_o being the same expression computed in free field theory with vacuum insertion. The value of A equals the experimental value of δm for m_c within its range of possible values around 1.5 GeV. The value of B is difficult to precisely and reliably estimate. In previous discussions of this argument B was taken equal to one.[44] Recently a bag model evaluation[45] of the matrix element led to the value $B = 0.4$. I feel that ambiguities of a factor of 2-3 on the size of B cannot be soundly removed at present. The effect of QCD corrections induced by gluon exchange between the quark lines in the diagrams of Fig. 3 are a priori large and are not included in the above estimates of B. However, an analysis of these corrections in the four quark model was carried through in Ref. 81. The result leads to a correction factor close to one. The inclusion of the t quark is not likely to substantially modify the previous estimate. In the four quark model the real part of the bracket in Eq. (91) is equal to one. The present argument was in fact used to correctly predict the charm quark mass in the few GeV range. Since the t mass is certainly much larger than the c mass, the bracket would become much larger than one (by orders of magnitude) if λ_t is not sufficiently small. Conversely, the same bracket never becomes much smaller than one for whatever allowed value of $\lambda_{c,t}$. Therefore I feel that by this argument one can find a reliable upper limit on $\sin\alpha$ (as a function of ε), but it would be unsafe to push the argument as far as to extract a more precise range of values for $\sin\alpha$, because this would require assuming a specific value for B.

The explicit form of λ_c and λ_t in terms of α, ε and δ is directly obtained from Eqs. (89):

$$\frac{\lambda_c}{\theta_c} = -\cos\alpha \left[\cos\alpha + \frac{\varepsilon}{\theta_c}\sin\alpha\, e^{i\delta}\right] ,$$
$$\frac{\lambda_t}{\theta_c} = -\sin\alpha \left[\sin\alpha - \frac{\varepsilon}{\theta_c}\cos\alpha\, e^{i\delta}\right] . \tag{94}$$

In some previous discussions[40] of this problem ε/θ_c was somewhat arbitrarily neglected, leading to a bound on $\sin\alpha$ independent of ε and δ. Actually, for ε close to its upper bound in Eq. (88) the contributions of the previously neglected terms are important. On the other hand $\sin\delta$ can be neglected for simplicity, because, for α near its upper bound, only if ε is very small can $\sin\delta$ be much larger than 10^{-3}; and in this case the ε terms can be dropped.

Thus for the sake of this argument $e^{i\delta}$ can be taken to be either plus or minus one. Then imposing on the bracket in Eq. (91) an upper bound for a few units, we obtain the bounds on $\sin^2\alpha$ as a function of ε shown in Fig. 4 which refers to $m_t \sim 15$ GeV. Note that the result that in any case $\sin^2\alpha \lesssim 0.4$ is in agreement with the experimental indications (to be discussed later) of a normal charm lifetime and of a substantial rate of dimuon production in antineutrino scattering on matter (through $\bar{\nu} + \bar{s} \to \mu^+ + \bar{c}_{\to\mu^-}$). Both these effects indicate that $\cos^2\alpha$ is rather large.

We see that the above argument on the K_S-K_L mass difference does not restrict $\sin\delta$. In principle δ can be constrained by a separate study of the CP violating parameters (as a function of ε and α). Here we shall only make a number of remarks on CP violation.

CP violation has only been observed so far in the K-$\bar{K}$ system. Of particular importance are the amplitudes for $K_{S,L} \to \pi\pi$ decays. In terms of

$$\eta_{+-} = \frac{A(K_L \to \pi^+\pi^-)}{A(K_S \to \pi^+\pi^-)}, \qquad \eta_{00} = \frac{A(K_L \to \pi^0\pi^0)}{A(K_S \to \pi^0\pi^0)}, \tag{95}$$

the experimental results are[46]

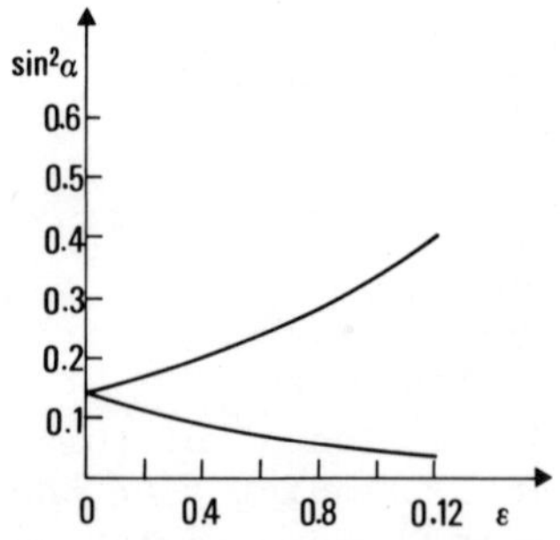

Fig. 4. Upper bounds for $\sin^2\alpha$ as a function of ε. The upper curve is for Re $e^{i\delta} = +1$, the lower curve for Re $e^{i\delta} = -1$. The t mass was taken as $m_t \simeq 15$ GeV.

$$|\eta_{+-}| = (2.30 \pm 0.035)10^{-3} \quad ,$$

$$|\eta_{+-}/\eta_{00}| = 1.05 \pm 0.046 \quad , \qquad (96)$$

$$\phi_{+-} = (46 \pm 1.7)^{o} \quad ; \qquad \phi_{00} = (43 \pm 19)^{o} \quad .$$

η_{+-} and η_{00} can be expressed in the form[47]

$$\eta_{+-} = \varepsilon + \frac{\varepsilon'}{1 + \omega/\sqrt{2}}$$

$$\eta_{00} = \varepsilon - \frac{2\varepsilon'}{1 - \sqrt{2}\omega} \quad , \qquad (97)$$

where

$$|\omega| = \left|\frac{\mathrm{ReA}_2}{\mathrm{ReA}_0}\right| \cong \frac{1}{20} \quad ,$$

$$|\varepsilon| = \frac{1}{\sqrt{2}}\left[\frac{\mathrm{ImM}}{8m} + \frac{\mathrm{ImA}_0}{\mathrm{ReA}_0}\right] \quad , \qquad (98)$$

$$|\varepsilon'| = \frac{|\omega|}{\sqrt{2}}\,\frac{\mathrm{ImA}_0}{\mathrm{ReA}_0} \quad .$$

In the previous formulae $A_{0,2}$ are the amplitudes for $K \to 2\pi$ with final state isospin zero or two. The value of $|\omega|$ is determined by the ratio of $\Delta T = 3/2$ and $1/2$ amplitudes. In the Kobayashi-Maskawa model it is natural to conventionally take $\mathrm{ImA}_2 = 0$, because the CP violating phase is associated with transitions involving c and t quarks that can only contribute to $\Delta T = 1/2$ transitions. ImM is the imaginary part of the bracket in Eq. (91) which takes into account the CP violation induced by the mass matrix.

In the superweak theory[48] all of the CP violation arises from the mass matrix. Thus $\mathrm{ImA}_0 = 0$, $\varepsilon' = 0$ and $\eta_{+-} = \eta_{00}$. We recall that the superweak theory predicts a bound for the neutron electric dipole moment given by $d_N^e \lesssim 10^{-29}$e cm. The present experimental bound[49] is

$$(d_N^e)_{EXP} \leq (0.40 \pm 0.75)10^{-24} \text{ em} \quad . \qquad (99)$$

In gauge theories CP violation can arise through: a) quark mixings as in the six quark model; b) Higgs couplings[3] (for more than one doublet); c) more gauge bosons[50] (for example $SU(2)_L \times SU(2)_R \times U(1)$; d) QCD instanton vacuum effects;[51] or a combination of the above mechanisms. The Higgs mechanism tends to produce a rather large $d_N^e \sim 10^{-25}$e cm (a size within experimental reach) and $|\eta_{+-}/\eta_{00}|$ compatible with experiment. The QCD vacuum effect poses a problem, because the natural size of the CP violation would be large,[51] and consequently an unnatural fixing of the relevant parameter to a very small value is required.

In the six quark model, d_N^e is predicted to be very small. In fact the single quark contribution vanishes up to two loops included.[52] It can arise from higher orders or from two quark contributions[53] and is expected to be of order 10^{-30}e cm or less. As for η_{+-} and η_{00} it was argued that $\mathrm{Im}A_0$ could be negligible because the phase appears in connection with heavy quarks and the corresponding amplitude would be Zweig suppressed. This argument was recently criticised by Gilman and Wise,[54] who observed that a quite sizable contribution to $\mathrm{Im}A_0$ can arise from "penguin diagrams" (see Sec. 7 below) known to be important in $K \to 2\pi$ decays. As a conclusion, deviations from the superweak relation $\eta_{+-} = \eta_{00}$ of the order of the present experimental limit may well appear.

We shall not pursue this subject further in the present lectures. In particular we shall not discuss the constraints imposed on δ by the assumption that the whole CP violation arises from quark mixings.

5. LEPTONIC MIXINGS

If all neutrinos are massless (or degenerate) there are no mixings in the leptonic sector. In fact in this case any linear combination of neutrinos is indifferent as far as the mass matrix is concerned. One can then arbitrarily call ν_L the state coupled to L^- $(L = e, \mu, \tau)$. But neutrinos may well be nondegenerate and in this case mixings would arise. One possibility considered in the past was that ν_τ could be heavier than τ.[55] This can now be excluded by experiment. For simplicity we take ν_e and ν_μ as massless. This degeneracy reduces the mixing angles to only two (and no CP violating phase), because ν_μ can be defined as the light component coupled to μ^-. The mixing angles can be defined as follows (a second order expansion was used):

$$\begin{pmatrix} (1 - \frac{1}{2}\beta^2)\nu_e - \beta\gamma\nu_\mu + \beta\nu_\tau \\ e^- \end{pmatrix}_L ,$$

$$\begin{pmatrix} (1 - \frac{1}{2}\gamma^2)\nu_\mu + \gamma\nu_\tau \\ \mu^- \end{pmatrix}_L , \qquad (100)$$

$$\begin{pmatrix} (1 - \frac{1}{2}\beta^2 - \frac{1}{2}\gamma^2)\nu_\tau - \beta\nu_e - \nu_\mu \\ \tau^- \end{pmatrix}_L .$$

For ν_τ heavier than τ the latter can only decay through the mixings β and γ with the light neutrinos ν_e and ν_μ. The lifetime would then be larger by a factor $(\beta^2+\gamma^2)^{-1}$ with respect to the expected lifetime for unit coupling, which can be readily estimated from μ decay,

$$\tau_0 = B_e \left(\frac{m_\tau}{m_\mu}\right)^5 \tau_\mu \simeq 2.7 \tag{101}$$

where B_e is the branching ratio for $\tau \to e\nu\bar{\nu}$. Thus the present limit[56] on the τ lifetime,

$$\tau_\tau \lesssim 2.3 \times 10^{-12} \text{ sec (DELCO)} \quad , \tag{102}$$

implies that

$$\frac{\tau_\tau}{\tau_0} \lesssim 8.5 \qquad \text{or} \qquad (\gamma^2 + \beta^2) \geq 0.12 \quad . \tag{103}$$

On the other hand there are upper bounds on $\nu_\mu N \to eX$. At low energies these are translated into limits on the direct $\nu_\mu - e$ coupling[57] (see Eq. (100)),

$$(\beta\gamma)^2 \lesssim 2 \times 10^{-3} \quad . \tag{104}$$

At high energy a bound is also implied on the $\nu_\mu - \tau$ coupling that could induce $\nu_\mu N \to \tau_{\hookrightarrow e} X$. On finds[58]

$$\begin{aligned} \gamma^2 &\lesssim 0.025 \qquad \text{(15' FNAL)} \quad , \\ &\lesssim 0.05 \qquad \text{(BEBC)} \quad . \end{aligned} \tag{105}$$

In order to satisfy Eqs. (103) β^2 should be substantially larger than γ^2. But this is excluded by the agreement of the standard prediction (based on $\mu \leftrightarrow e$ universality) with the observed ratio[59] of the $\pi \to \mu\nu$ and $\pi \to e\nu$ rates. Within two standard deviations this leads to

$$0.072 \geq \gamma^2 - \beta^2 \geq -0.006 \quad . \tag{106}$$

Thus β^2 cannot exceed γ^2 by the amount required by Eqs. (103-105) and one obtains the stated conclusion that ν_τ cannot be heavier than τ. Note that the above argument, being independent of the quark couplings, cannot be circumvented by a common mixing of quark and leptons preserving the universality of the u and μ couplings.

For light neutrinos (by light I mean for example $\nu_\tau \lesssim 250$ MeV as from the limit[56] on the mass of the second neutrino in $\tau \to e\nu\bar{\nu}$) there can be mixings inducing violations of the separate conservation of e, μ and τ lepton numbers.

There are recent limits[60] on $\mu \leftrightarrow e$ transitions,

$$B(\mu \to e\gamma) \lesssim 2 \times 10^{-10},$$

$$R_S = \frac{\sigma(\mu S \to eS)}{\sigma(\mu S \to \nu P^*)} \lesssim 1.5 \times 10^{-10} \tag{107}$$

(where R_S refers to the anomalous μ capture on sulphur), to be added to the known limits[61,62]

$$B(\mu \to 3e) \lesssim 1.9 \times 10^{-9} \quad ,$$

$$R_{Cu} \lesssim 1.6 \times 10^{-8} \quad . \tag{108}$$

In the standard model mixings among light neutrinos can only induce[55] $\mu \leftrightarrow e$ transitions many orders of magnitude weaker than the present limits in Eqs. (107) and (108). Actually in this case the only detectable effects of mixings can be observed through ν oscillations (including CP violating effects[63] if the three neutrinos are nondegenerate and massive). On the other hand (apart from the possible existence of heavy neutrinos still to be discovered) there is another mechanism that could produce relatively large $\mu \leftrightarrow e$ transitions. This may happen[64] through Higgs couplings if there are more than one Higgs doublets. (For only one doublet the coupling would automatically disappear when the mass matrix is diagonalized.) The natural order of magnitude for this effect was estimated to be as large as $B(\mu \to e\gamma) \sim (\alpha/\pi)^3 \sim 10^{-8}$ (and $R_S \sim R_{Cu} \sim 10^{-9}$). The present limits in Eqs. (107) now seem to make this possibility rather unlikely, perhaps suggesting the esitence of only one Higgs doublet.

6. REMARKS ON τ PHYSICS

The measured properties of τ[66] beautifully fit in the standard model. I shall mention some relatively recent data that have filled most of the remaining gaps.

The mass has been precisely measured,[56]

$$M_\tau = 1782\,^{+3}_{-4}\ \text{MeV} \quad . \tag{109}$$

Thus the famous awkward degeneracy was removed, and τ is definitely below charm threshold (M_{D^0} = 1863.3 ± 0.9 MeV).

The measurement of the Michel parameter ρ for $\tau \to e\nu_\tau\bar{\nu}_e$ decay confirms the V-A nature of the $\nu_\tau - \tau$ current. The predicted values of ρ for V-A, V+A or pure V or A are 3/4, 0 and 0.375 respectively. The measured value[56]

$$\rho_{EXP} = 0.72 \pm 0.15 \tag{110}$$

completely excludes V+A and strongly indicates V-A.

The measured leptonic branching ratio of τ is in very good agreement with the theoretical prediction[65] of about 17%. The free field theory value of 20% from the existence of 5 possible parton modes is in fact slightly reduced by QCD corrections, i.e. by the same effect that makes $R_{e^+e^-} = \sigma(\text{hadrons})/\sigma_{\mu^+\mu^-}$ larger, at finite energies, than the asymptotic free field result determined by the sum of squared charges. Conversely if one accepts that τ is a sequential lepton then the branching ratio confirms that quarks exist in three colors.

As for the hadronic modes, the decay $\tau \to \pi\nu$ was finally established with a branching ratio consistent with the firm theoretical prediction of about 10% (9.3 ± 3.9% MARK I; 9.0 ± 2.9 ± 2.5 PLUTO; $6.0 \pm 1.6^{+1.9}_{-1.2}$ DELCO; 8.0 ± 1.1 ± 1.5 and 8.2 ± 2.0 ± 1.5 MARK II).

Finally I mention that the bound on $\tau \to 3e$ of $\lesssim 0.6\%$[66] together with the limit on the τ lifetime in Eq. (102) completely rule out the possibility[67] that ν_τ does not exist and only 5 leptons are around.

In conclusion all known properties of τ fit beautifully in the standard model.

7. HEAVY QUARK DECAYS

In this section we discuss weak decays of heavy quarks, with special emphasis on charm meson decays, on which a wealth of experimental data already exist. The aim is to review the theoretical ideas involved, the different levels of model dependence in each step and the comparison with the available data. Some extrapolations to the physics of b and t quark decays are also included.

A quite clear theoretical prediction can be made on inclusive semileptonic decays of heavy flavored hadrons. The momentum to the lepton pair is sufficiently large with respect to the hadron binding energy as to make a parton approach justified. Thus the rate for semileptonic decay is reduced to the rate for the corresponding quark decay $Q \to q + \ell + \nu_\ell$. The latter can be estimated in first approximation by an extrapolation from muon decay,

$$\Gamma^{o}_{SL}(Q) = \Gamma_\mu \left(\frac{m_Q}{m_\mu}\right)^5 \sum_q |U_{Qq}|^2 \, I\left(\frac{m_q}{m_Q}, \frac{m_\ell}{m_Q}, 0\right) , \qquad (111)$$

where $\sum_q$ extends over all possible quark flavors in the final state, U_{Qq} is the entry corresponding to the $Q \to q$ transition in the mixing matrix (see Eq. (89)) and the function I, normalized to $I(0,0,0) = 1$, is a phase space factor (we neglect the electron mass in muon decay). This first approximation can be corrected for gluon effects arising from the diagrams in Fig. 5. The result is given by[68,69]

$$\Gamma_{SL}(Q) = \Gamma^{o}_{SL}(Q) \left[1 - \frac{2}{3} \frac{\alpha_s(m_Q^2)}{\pi} f\left(\frac{m_q}{m_Q}\right)\right] , \qquad (112)$$

where f(x) is a known function given in Ref. 68. For generally accepted values of $\alpha_s(m_Q^2)$, corresponding to $\Lambda = 0.5$ GeV, and for $m_c = 1.5$ GeV, $m_b = 4.5$ GeV, $m_t = 15$ GeV (the latter value being purely indicative) the gluon correction factor is quite substantial and is given in Table III for the various transitions.[70]

From the semileptonic rate, approximately the same for all charmed hadrons, the total rate (or the lifetime) can be directly obtained if the semileptonic branching ratio of the given hadron is known. For example the (average) semileptonic branching ratio of D mesons is known[71] to be

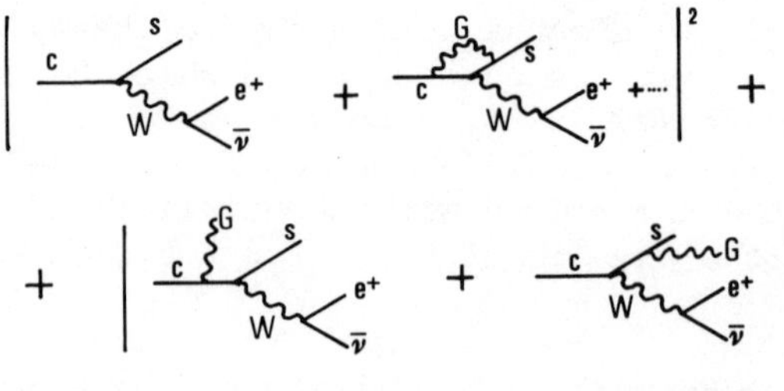

Fig. 5. Gluon corrections to charm semileptonic decays.

TABLE III

$Q \to q$	Γ/Γ^o
$c \to s$	0.65
$c \to d$	0.52
$b \to c$	0.81
$b \to u$	0.72
$L \to b$	0.86
$t \to s$	0.80
$t \to d$	0.79

$$B_e(D \to eX) \simeq (9.8 \pm 1.4)\% \quad . \tag{113}$$

This leads to the estimate

$$\tau(D) = B_e/\Gamma_{SL} \simeq (3.5 - 7)10^{-13}\text{sec} \quad , \tag{114}$$

where the bound $\cos^2\alpha \geq 0.6$ was taken into account. This prediction for $\tau(D)$, also expected to hold for other weakly decaying charmed hadrons, appears in fair agreement with a few charm decays recently observed in emulsions,[72] whose proper decay time was measured.

Turning now to nonleptonic decays, one starts by establishing an effective four quark Hamiltonian. To be specific we take the case of charm decays. In the free field limit the effective Lagrangian would be

$$L_{NL}^{free} = \frac{G}{\sqrt{2}} (\bar{c}s')(\bar{d}'u) + h.c. \quad , \tag{115}$$

where for brevity

$$(\bar{c}s')(\bar{d}'u) \equiv \bar{c}\gamma_\mu(1-\gamma_5)s' \; \bar{d}'\gamma^\mu(1-\gamma_5)u \tag{116}$$

and s',d' are the fields defined in Eq. (43) which are coupled to c and u by the charged current. This four quark Lagrangian is modified in an essential way by gluon corrections. These corrections, arising from diagrams where virtual gluons are exchanged among the quark lines, build up, in the limit of massless quarks, a series of terms of order $\alpha_s^n \ell n^n(M_W^2/\mu^2)$ (with μ an arbitrary reference mass) that cannot be neglected and must be summed up taking all orders in perturbation theory into account. This can be done by using renormalization group techniques applied to the coefficient functions in a short distance operator expansion[73] for the time ordered product of the two weak currents. Short distance here means distances of order $1/M_W$. The four quark operators appearing in the operator expansion can be written in the form

$$T(J_\mu(x)J^{\dagger\mu}(0)) = f_+(x^2)H_+(0) + f_-(x^2)H_-(0) + \ldots \quad , \qquad (117)$$

where the dots stand for other types of operators, and $H_\pm$ are given by

$$H_\pm = \frac{1}{2}\left[(\bar{c}s')(\bar{d}'u) \pm (\bar{c}u)(\bar{d}'s')\right] \quad . \qquad (118)$$

In free field theory $f_+ = f_- = 1$ and the terms in the dots vanish so that Eq. (117) reduces to the naive product of currents. $H_\pm$ are introduced because they are multiplicatively renormalizable under gluon corrections. This is clear because in the massless quark limit the QCD Lagrangian is SU(f) symmetric for f flavors. Thus the operators of definite (logarithmic) anomalous dimensions are those with definite SU(f) transformation properties. It is simple to realize that this is the case for H_+ and H_-. In fact the weak charged current is of the form $J = \bar{Q}\,U\,q$, with U transforming as the regular representation of SU(f). Then the time ordered product of two currents transforms as a sum of representations in the symmetric product of two regular representations:

$$T(JJ^\dagger) \sim \left[U^A_B \times U^{\dagger C}_D\right]_{SYMM} \approx 1^{AC}_{BD} + \{uu^\dagger\}^{AC}_{BD} + H^{[AC]}_{-[BD]} + H^{\{AC\}}_{+\{BD\}} \;, \qquad (119)$$

where A,B... are flavor indices. The singlet representation is irrelevant here because we are interested in a flavor changing component. The regular representation component from the anti-commutator is absent because U is unitary and $UU^\dagger = U^\dagger U = 1$. The two remaining representations are spanned by tensors either symmetric or antisymmetric in both the upper and lower indices, which are easily seen to correspond to the operators H_+ and H_-.

When gluon corrections are included the effective four quark Lagrangian is therefore changed into[74,75]

$$\begin{aligned} L_{NL} &= \frac{G}{\sqrt{2}}\left[f_+H_+ + f_-H_-\right] \\ &= \frac{G}{\sqrt{2}}\left[\frac{f_+ + f_-}{2}(\bar{c}s')(\bar{d}'u) + \frac{f_+ - f_-}{2}(\bar{c}u)(\bar{d}'s')\right] \quad . \end{aligned} \qquad (120)$$

A standard evaluation of the logarithmic exponents leads to

$$f_\pm = \left[\frac{\alpha_s(m_c^2)}{\alpha_s(M_W^2)}\right]^{\gamma_\pm} , \tag{121}$$

$$\gamma_- = -2\gamma_+ = \frac{12}{33-2f} ,$$

where f is the number of flavors excited up to the W mass. Note that $f_+ = (f_-)^{-1/2}$. The gluon correction, as given in Eq. (121), includes the contribution of virtual momenta in the range from m_c up to M_W. For $\Lambda = 0.5$ GeV, $m_c = 1.5$ GeV, $M_W \simeq 80$ GeV, one obtains from Eq. (121)

$$f_- \simeq 2.15 \quad , \qquad f_+ \simeq 0.68 \quad . \tag{122}$$

Totally inclusive nonleptonic rates can be computed from L_{NL} by again making a parton model approximation. The main decay mechanism is assumed to be $c \to s' + \bar{d}' + u$. In fact the annihilation channels $c + \bar{s}' \to \bar{d}' + u$ or $c + \bar{u} \to s' + d'$ that can in principle also contribute to the decay of pseudoscalar charmed mesons are thought to be negligible, because they are suppressed with respect to the first mechanism by an helicity factor of m_s^2/m_c^2, and by the (roughly) estimated[76] value of f_P, the axial current matrix element between the pseudoscalar meson and the vacuum. If these mechanisms were important the lifetime of D^0 and F^+ would be shorter than that of D^+. When the annihilation channel is tentatively neglected (or absent, as in baryon decays) the nonleptonic inclusive rate can be computed in analogy with the semileptonic case by

$$\Gamma_{NL}(Q) = \Gamma^o_{NL} \frac{2f_+^2 + f_-^2}{3} ,$$

$$\Gamma^o_{NL}(Q) = \Gamma_\mu \left(\frac{m_Q}{m_\mu}\right)^5 \sum_{q,q_\alpha,q_\beta} \left|U_{qQ}\right|^2 \left|U_{q_\alpha q_\beta}\right| I\left(\frac{m_q}{m_Q}, \frac{m_{q_\alpha}}{m_Q}, \frac{m_{q_\beta}}{m_Q}\right) . \tag{123}$$

Here Γ^o_{NL} is the rate for $Q \to q + q_\alpha + \bar{q}_\beta$ in the free field case. The factor $(2f_+^2 + f_-^2)/3$, which reduces to 1 in the free field limit, embodies the effect of gluon corrections. It arises as follows. The second term in L_{NL} can be Fierz rearranged in the form

$$(\bar{c}u)(\bar{d}'s') = \frac{1}{3}(\bar{c}s')(\bar{d}'u) + 2\sum_A (\bar{c}t^A s')(\bar{d}'t^A u) \quad . \tag{124}$$

Thus the total coefficient of $(\bar{c}s')(\bar{d}'u)$ becomes $\frac{2}{3}f_+ + \frac{1}{3}f_-$, while the coefficient of the color octet times octet part is $f_+ - f_-$. The incoherent sum of the two contributions then leads to

$$\left(\frac{2}{3}f_+ + \frac{1}{3}f_-\right)^2 + \frac{2}{9}(f_+ - f_-)^2 = \frac{2}{3}f_+^2 + \frac{1}{3}f_-^2 \quad , \tag{125}$$

where the factor 2/9 arises from the ratio of $\mathrm{Tr}\, I \times \mathrm{Tr}\, I = 9$ for the singlet-singlet contribution and $\sum_{AB}(\mathrm{Tr}\, t^A t^B)^2 = 2$ for the octet-octet term (recall Eq. (8)). The overall normalization was taken to reproduce the free field case for $f_+ = f_- = 1$. The color factor produces a ratio[70] $\Gamma_{NL}/\Gamma^o_{NL}$ of ~1.85 for charm, ~1.30 for bottom and ~1.10 for a top of 15 GeV mass.

It must be stressed that in the nonleptonic case the nonleading corrections vanishing with $\alpha_S(m_c^2)$ have not been computed so far. These corrections would be analogous to those included in the semileptonic case (which is simpler because of the presence of only two quark lines). This is relevant for the estimate of the semileptonic branching ratio of charm (which is independent of $\cos^2\alpha$),

$$B(C \to eX) = \frac{\Gamma_{SL}}{\Gamma_{SL} + \Gamma_{NL}} \quad . \tag{126}$$

If we include the factor 0.65 of nonleading gluon corrections in Γ_{SL} (see Eq. (112)), we obtain (neglecting phase space differences in the various channels)

$$B(C \to eX) \simeq \frac{0.65}{2(0.65) + 2f_+^2 + f_-^2} \simeq 10\% \quad . \tag{127}$$

However, this approximation is inconsistent because the nonleptonic nonleading corrections are not included. If we replace 0.65 by 1 we instead obtain ~13%. This measures the present level of uncertainty in the theoretical prediction. For the b mesons the prediction of 10% (with the same ingredients) becomes 12-14%. The branching ratio does not depend much on the ε and α mixings and remains in the mentioned range for both $b \to c$ and $b \to u$ dominant transitions.

Predictions for exclusive channels are far more model dependent. Matrix elements of the effective interaction between specific hadronic final states are in fact difficult to estimate. As an example let us consider the two body decays of charmed mesons. For Cabibbo allowed modes we have the following experimental data,[77]

$$B(D^0 \to K^-\pi^+) \simeq (2.2 \pm 0.6)\% \quad ,$$

$$\frac{B(D^+ \to \bar{K}^0\pi^+)}{B(D^0 \to K^-\pi^+)} = 0.68 \pm 0.33 \quad , \tag{128}$$

and the new result[78]

$$\frac{B(D^0 \to K^-\pi^+)}{B(D^0 \to \bar{K}^0\pi^0)} = 1.6 \pm 0.8 \quad . \tag{129}$$

For the Cabibbo suppressed modes we have the new measurements[78]

$$\frac{B(D^0 \to K^-K^+)}{B(D^0 \to K^-\pi^+)} = (11.3 \pm 3)\% \; , \quad \frac{B(D^0 \to \pi^-\pi^+)}{B(D^0 \to K^-\pi^+)} = (3.3 \pm 1.5)\% \quad . \tag{130}$$

A plausible theory[79] is based on the following steps: a) the effective Hamiltonian in Eq. (120), b) the dominance of the quark decay mode over the annihilation channels, c) the assumption that the amplitude for producing, say, a π^+ is proportional to the amplitude for producing a pair $u\bar{d}$ in a singlet color state and d) a pair in a color octet state has a negligible amplitude to rearrange color with the other pair. For example the amplitude for $D^0 \to K^-\pi^+$ is given by the diagram in Fig. 6. Then the term $(\bar{c}s)(\bar{d}u)$ in L_{NL} (see Eq. (120)) contributes with $(f_+ + f_-)/2$, while the term $(\bar{c}u)(\bar{d}s)$ after Fierz rearrangement (according to Eq. (124)) and dropping out the color octet-octet piece contributes with $(f_+ - f_-)/6$. Adding up the two terms we obtain

$$A(D^0 \to K^-\pi^+) = (\tfrac{2}{3}f_+ + \tfrac{1}{3}f_-)A \equiv X_+A \quad . \tag{131}$$

In the SU(3) limit, one finds by the same approach

$$A(D^0 \to \bar{K}^0\pi^0) = \frac{1}{\sqrt{2}}\left(\frac{2}{3}f_+ - \frac{1}{3}f_-\right)A \equiv \frac{1}{\sqrt{2}}X_-A \quad ,$$

$$A(D^+ \to \bar{K}^0\pi^+) = \frac{4}{3}f_+A = (X_+ + X_-)A \quad . \tag{132}$$

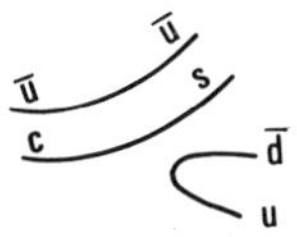

Fig. 6. Diagram for $D^0 \to K^-\pi^+$ decay.

The relation implied by Eqs. (131) and (132),

$$A(D^+ \to \bar{K}^0\pi^+) - A(D^0 \to K^-\pi^+) = \sqrt{2}\, A(D^0 \to \bar{K}^0\pi^0) \quad , \tag{133}$$

is more general because it follows from the effective Hamiltonian being an isospin vector. From the values in Eq. (122) for f_+ and f_- we find $X_+ \simeq 1.17$, $X_- \simeq -0.26$. Assuming equal lifetimes for D^0 and D^+, we see that the present approach is consistent with the experimental result in Eq. (128) but is completely at odds with the large rate for $D^0 \to \bar{K}^0\pi^0$ implied by the new result in Eq.(129). Thus at least one of the assumptions listed above must be blamed. Note that the general isospin relation in Eq. (133) is indeed consistent with the data.

In the Cabibbo suppressed sector a naive four quark model would predict

$$\begin{aligned} \frac{B(D^0 \to K^-K^+)}{B(D^0 \to K^-\pi^+)} &\simeq \theta_c^2\, 0.94 \simeq 4.7\% \quad , \\ \frac{B(D^0 \to \pi^-\pi^+)}{B(D^0 \to K^-\pi^+)} &\simeq \theta_c^2\, 1.10 \simeq 5.5\% \quad , \end{aligned} \tag{134}$$

where phase space corrections were also included. Two effects can be of help. One is simply SU(3) breaking corrections.[80] For example, in $D^0 \to K^-K^+$ analysed in the previously mentioned model, a kaon is produced from the vacuum, while in $D^0 \to K^-\pi^+$ and $D^0 \to \pi^-\pi^+$ a pion is produced from the vacuum. This leads to a factor f_k in one case and f_π in the two other cases. Then there are the mixing angles in the six quark model. A simple analysis leads to

$$\begin{aligned} \frac{B(D^0 \to \pi^-\pi^+)}{B(D^0 \to K^-\pi^+)} &\simeq \theta_c^2 \left| 1 + \frac{\varepsilon}{\theta_c} \tan\alpha\, e^{-i\delta} \right|^2 1.10 \quad , \\ \frac{B(D^0 \to K^-K^+)}{B(D^0 \to K^-\pi^+)} &\simeq \theta_c^2 \left(\frac{f_k}{f_\pi} \right)^2 0.94 \simeq 7.7\% \quad . \end{aligned} \tag{135}$$

Mixing angles are of no help to increase the K^-K^+ mode, but in this case the ratio $(f_k/f_\pi)^2$ may perhaps be enough. The mixing angles on the other hand can reduce the $\pi^+\pi^-$ mode, provided that $\mathrm{Re}\, e^{-i\delta} = -1$ and ε is not too small. This observation tends to favor the solution with $\mathrm{Re}\, e^{-i\delta} = -1$ for the bounds on mixing angles discussed in Section 4.

The above brief discussion of two body decays of $D^{+,0}$ illustrates how primitive is the state of the art of estimating hadron matrix elements.

I mention that the theory of ordinary strange particle non-leptonic decays is in a similar situation. We believe we know the effective Lagrangian[74,81] including resummation of virtual gluon exchanges. The effect of integration over virtual momenta below the charm mass is also roughly estimated, leading to a sum of operators with definite SU(3) properties.[81] The operators with $\Delta T = 1/2$ are enhanced by the gluon effects by the same exponent γ_- in Eq. (121), while $\Delta T = 3/2$ operators are suppressed by γ_+. The amount of the enhancement is not sufficient by itself to explain the observed strength of the $\Delta T = 1/2$ rule. The remaining enhancement is believed to arise from various low energy effects in the matrix elements. Also it was observed in Ref. 81 that the "penguin" diagrams (see Fig. 7) play an important role in explaining the observed decay rates. A reasonable quantitative description emerges, which however is not yet completely satisfactory.[82]

NOTE ADDED: After these lectures were delivered, it has become known that the D^0 lifetime is shorter by a factor greater than about five than the D^+ lifetime. This points toward the importance in charmed meson decays of the annihilation modes which were neglected in the approach commonly accepted up to now.

8. UNIFICATION OF STRONG AND ELECTROWEAK FORCES

In the previous lectures I presented a discussion of the structure and the phenomenological implications of the standard model based on the gauge group SU(3) × SU(2) × U(1). Certainly much more could be said on the subject and I only restricted my discussion to the general structure of the model, the main experimental confirmations and the available information on the basic parameters in the theory. In the remaining lectures I shall consider the exciting problem of unification of the strong and electroweak interactions to be achieved at a realm of energies far beyond the present domain. The standard model is remarkably successful and it certainly represents an enormous progress in our

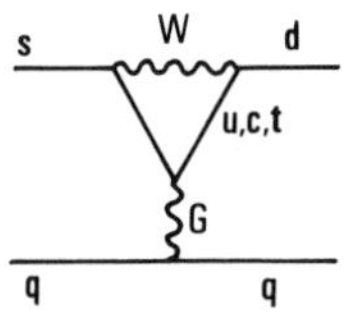

Fig. 7. Penguin diagram.

understanding of the fundamental interactions. However, it obviously contains many unsatisfactory aspects and leaves a lot of open questions that make quite natural the search for a more comprehensive theory. Examples are the proliferation of free parameters, the unexplained sequence of similar families of fermions, the total darkness on the origin of the observed mass ratios and their peculiarities and so on.

At least seventeen parameters must be specified from outside: three coupling constants, eight fermion mass ratios (even taking for granted the vanishing of neutrino masses — a puzzle in itself), four among mixing angles and the CP violating phase, and two parameters in the Higgs system for the vacuum expectation value and the physical mass if only one doublet is present.

Assuming that the t quark will be found, we do not even know for sure whether or not the iteration of families will continue. A cosmological argument[83] relates the observed Helium abundance to the number of massless neutrinos with standard interactions. The argument is valid within the big bang cosmology with the assumption of a $\nu_e - \bar{\nu}_e$ imbalance negligible[84] with respect to the number of photons (recall that the baryon number imbalance is $(n_B - n_{\bar{B}})/n_\gamma \sim 10^{-9}$). For an He abundance of 25%, as observed, there should be no more standard (almost massless) ν's. But for a He proportion in the universe of 29%, a figure not completely excluded, at most two more ν's could be allowed.

In an attempt to understand the mass matrix, one has tried to add discrete symmetries to generate constraints and thus explain the empirical relations between masses and couplings. For example the generalized Cabibbo angles couple the u quark with a strength which decreases with the masses of the d, s and b quarks. However no-go theorems[85] have been proven in $SU(2) \times U(1)$ that show it is impossible to obtain a nontrivial Cabibbo angle without at the same time introducing neutral flavor violating Higgs couplings, whose natural order of magnitude would require Higgs masses as high as 10^3 GeV to reproduce the known selection rules. Recently also the possibility of a trivial Cabibbo angle at the tree level with a nonzero Cabibbo angle arising from radiative corrections has been excluded.[86]

Other motivations for investigating theories with a unique interaction (apart from gravity, still left out for a further synthesis) are the possibility of understanding charge quantization, of deducing testable predictions for observed couplings at present energies (the value of θ_W, of α_s/α etc.), of guessing the properties of yet undiscovered new interactions (such as baryon and lepton number violating couplings) and of course the aesthetical appeal of a more comprehensive theory.

The problem is to find a gauge group G, either simple or a product of simple factors related by a discrete symmetry, thus leading to a single gauge coupling constant, that contains $SU(3) \times SU(2) \times U(1)$ as a subgroup. Fermions and Higgs should transform in as simple and elegant a way as possible. The observed pattern of symmetry breaking is to be realized by a hierarchy of intermediate vector boson masses.[87]

The best introduction to the problem of unification is the discussion of the SU(5) theory. This was first studied[88] long ago and still plays a central role in this subject and therefore is a good starting point for our considerations here.

As a preliminary observation we remark that the group G commutes with the Lorentz group, and consequently all states in a given multiplet of G must have the same helicity. It is therefore useful to describe fermions in terms of a pair of Weyl two component spinors, one for the lefthanded fermion and one for the lefthanded charge conjugated fermion (instead of the righthanded fermion). In this sense we shall replace, say, e_R^- with e_L^+ and so on, and the label L can simply be omitted.

SU(5) emerges as the minimal group that contains $SU(3) \times SU(2) \times U(1)$ and admits a spectrum of fermions with three (or more) identical generations (as in the standard model) and no other fermions. The content of one generation in terms of $SU(3) \times SU(2) \times U(1)$ is

$$\begin{matrix} \begin{pmatrix} u \\ d \end{pmatrix} & \begin{pmatrix} \nu \\ e^- \end{pmatrix} & \bar{u} & \bar{d} & e^+ \\ (3,2)_{2/3,-1/3} & (1,2)_{0,-1} & (\bar{3},1)_{-2/3} & (\bar{3},1)_{1/3} & (1,1)_1 \end{matrix} \tag{136}$$

where the charge Q has also been indicated. This makes fifteen Weyl spinors. Note that $\bar{\nu}$ has not been included (massless neutrinos). Then the rank of G must be not smaller than 4 and G must admit complex representations (that is non-self-conjugate representations, because the spectrum of fermions listed above is not self-conjugate). Now the only simple groups that admit complex representations are $SU(n)\,(n > 2)$, $SO(4n+2)$ and E_6. Of these groups SU(5) is of rank 4, SU(6) and SO(10) of rank 5, SU(7) and E_6 of rank 6 etc. The only other possibility of rank 4 is $SU(3) \times SU(3)$. But it is easy to show (and will be seen later) that it does not work. Thus SU(5) is the only candidate of rank 4. To make sure that it does indeed possess representations with the required content it is sufficient to work out the simplest representations and their content:

$$
\begin{aligned}
1 &= (1,1)_0 \quad , \\
5 &= (3,1)_{-1/3} + (1,2)_{1,0} \quad , \\
\bar{5} &= (\bar{3},1)_{1/3} + (1,2)_{0,-1} \quad , \\
(5\times 5)_{\text{ANTISYMM.}} &\equiv 10 = (\bar{3},1)_{-2/3} + (3,2)_{2/3,-1/3} + (1,1)_1 \quad , \\
(5\times 5)_{\text{SYMM.}} &\equiv 15 = (6,1) + (1,3) + (3,2) \quad , \\
5 \times \bar{5} &= 24 + 1 = (8,1)_0 + (1,1)_0 + (1,3)_{1,0,-1} + (3,2)_{-1/3,-4/3} \\
&\qquad + (\bar{3},2)_{1/3,4/3} + (1,1)_0 \quad .
\end{aligned}
\tag{137}
$$

We see that each family is to be assigned to a sum of representations $\bar{5}+10$, according to

$$
\bar{5} = \begin{pmatrix} \bar{d}_1 \\ \bar{d}_2 \\ \bar{d}_3 \\ \nu \\ e^- \end{pmatrix} , \qquad 10 = \begin{pmatrix} 0 & \bar{u}_3 & -\bar{u}_2 & u_1 & d_1 \\ & 0 & \bar{u}_1 & u_2 & d_2 \\ & & 0 & u_3 & d_3 \\ & & & 0 & e^+ \\ & & & & 0 \end{pmatrix} , \tag{138}
$$

where the indices 1,2,3 stand for color and the notation for the 10 is in terms of a 5×5 antisymmetric matrix (only one half of it is shown explicitly). Note that the quantization of charge is in fact implied. The d quark charge is 1/3 of the e^+ charge because there are three colors (and Q has zero trace). The gauge bosons are in the adjoint representation. Schematically,

$$
24 = \left(\begin{array}{ccc|cc} & & & X_1^{4/3} & Y_1^{1/3} \\ & \text{8 gluons} & & X_2^{4/3} & Y_2^{1/3} \\ & & & X_3^{4/3} & Y_3^{1/3} \\ \hline X_1^{-4/3} & X_2^{-4/3} & X_3^{-4/3} & W^0 & W^+ \\ Y_1^{-1/3} & Y_2^{-1/3} & Y_3^{-1/3} & W^- & B \end{array}\right) , \tag{139}
$$

where B is actually associated with the U(1) generator,

$$
U(1) \sim \frac{1}{\sqrt{15}} \begin{pmatrix} -2 & & & & \\ & -2 & & & \\ & & -2 & & \\ & & & 3 & \\ & & & & 3 \end{pmatrix} . \tag{140}
$$

The superstrong breaking from SU(5) to SU(3) × SU(2) × U(1) is most simply obtained by a 24 multiplet of Higgs bosons, with the vacuum expectation value transforming according to the U(1) generator in Eq. (140). Then $X^{\pm 4/3}$ and $Y^{\pm 1/3}$ become superheavy and eat up twelve Higgs. The remaining twelve are physical and superheavy. The standard breaking down to SU(3) × $U(1)_Q$ is induced by a different set of Higgs, which for economy color must also provide the fermion masses. The only representations that can contribute to fermion masses are those in $\bar{5} \times 10 = 5 + 45$. Both are needed for a general mass matrix. But an interesting possibility is to only include a 5 with vacuum expectation value given by

$$\begin{pmatrix} 0 \\ 0 \\ 0 \\ 0 \\ v \end{pmatrix} , \qquad (141)$$

where the nonvanishing entry corresponds to the only neutral component. This implies the same mass for d and e and

$$\frac{m_d}{m_e} = \frac{m_s}{m_\mu} = \frac{m_b}{m_\tau} \qquad (142)$$

in the symmetry limit.

We now consider the constraints imposed by the unifying group G on the couplings in the symmetric limit and at present energies. In fact those we measure are effective scale dependent couplings in the renormalization group sense. Otherwise a unique gauge coupling could not be reconciled with the observed strength of the different interactions. The symmetric limit applies at a scale of energy larger than the symmetry breaking superheavy masses of gauge bosons and Higgs. For the general case of a unifying group G, let us normalize the generators T_ℓ such that the charge generator is given by

$$Q = T_3 + b\, T_0 \quad , \qquad (143)$$

where T_3 corresponds to the weak isospin and T_0 is then proportional to the weak hypercharge. These generators are normalized to $\mathrm{Tr}(T_\ell T_m) = N\delta_{\ell m}$ where N does not depend on the representation. Consequently the color generators T_c are related to the Gell-Mann color matrices by

$$a\, T_c = \frac{\lambda_c}{2} \quad . \qquad (144)$$

In the symmetric limit only one coupling constant g_G exists, coupled to the generators T_ℓ, i.e. the product $g_G T_\ell$ specifies the couplings. When G is broken into SU(3) × SU(2) × U(1) three couplings can be defined through $g_3 T_c$, $g_2 T_{weak}$, $g_1 T_0$. When the scale of masses is increased up to the symmetric region $g_{1,2,3}$ all approach g_G. The previously defined physical couplings g_s, g_W and g'_W (the label W = weak is here added for sake of clarity with respect to the notation of Sect. 2) are related to $g_{3,2,1}$ by

$$g_s a = g_3 \ , \quad g_W = g_2 \ , \quad g'_W b = g_1 \ , \tag{145}$$

or equivalently in terms of squared couplings $\alpha = g^2/4\pi$

$$\alpha_s a^2 = \alpha_3 \ , \quad \alpha_W = \alpha_2 \ , \quad \alpha'_W b = \alpha_1 \ . \tag{146}$$

Recalling the relations between $\sin\theta_W, e^2$, g_W, g'_W we also have

$$\alpha_2 = \frac{\alpha}{\sin^2\theta_W} \ , \quad \alpha_1 = \frac{b^2\alpha}{\cos^2\theta_W} \ , \tag{147}$$

where α is the electromagnetic fine structure constant. The constants a and b only depend on the group G and on the embedding of SU(3) × SU(2) × U(1) in G and not on the representation. They are given by

$$\frac{\mathrm{Tr}\ T_3^2}{\mathrm{Tr}\ Q^2} = \frac{1}{1+b^2} \ , \quad \frac{\mathrm{Tr}\ T_3^2}{\mathrm{Tr}\ (\lambda_c/2)^2} = \frac{1}{a^2} \ . \tag{148}$$

For example in SU(5) from the content of the $\bar{5}$ in Eq. (138) we immediately obtain

$$a^2 = 1 \ , \quad b^2 = \frac{5}{3} \ . \tag{149}$$

In the symmetric limit α_1 and α_2 are equal. Thus we see from Eqs. (147) that

$$(\sin^2\theta_W)_{SYMM} = 1/(1+b^2) \tag{150}$$

and

$$\alpha_G = a^2(\alpha_s)_{SYMM} = (1+b^2)(\alpha)_{SYMM} \ . \tag{151}$$

In SU(5) at the symmetric point $\sin^2\theta_W = \frac{3}{8}$ and $\alpha_s = \frac{8}{3}\alpha$.

The behavior of the couplings when descending from the symmetric limit down to a scale of mass of order of the W and Z masses can be followed[89] by the renormalization group provided we assume that a) there is one step only of symmetry breaking from G down to SU(3) × SU(2) × U(1), b) the only multiplet of both light and superheavy particles is the gauge boson multiplet (if so then all other multiplets do not break the symmetry and do not contribute to differences of couplings), c) a smooth behavior at the threshold corresponding to the superheavy masses. If the sequence of fermions is limited to light masses the only danger with respect to point b) may arise from the Higgs sector. With the previous hypothesis the couplings extrapolation from $\mu \sim 2M_W$ up to M, the unification mass, is governed by the equations

$$\frac{1}{\alpha_3(M)} = \frac{1}{\alpha_3(\mu)} + \left(\frac{1}{a^2}\frac{33}{6\pi} + X\right) \ln\frac{M}{\mu} \,,$$
$$\frac{1}{\alpha_2(M)} = \frac{1}{\alpha_2(\mu)} + \left(\frac{22}{6\pi} + X\right) \ln\frac{M}{\mu} \,, \tag{152}$$
$$\frac{1}{\alpha_1(M)} = \frac{1}{\alpha_1(\mu)} + X \ln\frac{M}{\mu} \,,$$

where X is the contribution, equal for all couplings, of all multiplets but the gauge boson multiplet. At the symmetry point M $\alpha_3(M) = \alpha_2(M) = \alpha_1(M)$. By taking differences we can eliminate X and obtain two relations,

$$\ln\frac{M}{\mu} = \frac{6\pi}{11\alpha(\mu)} \frac{1 - \frac{1+b^2}{a^2}\frac{\alpha(\mu)}{\alpha_s(\mu)}}{3\,\frac{1+b^2}{a} - 2} \,, \tag{153}$$
$$\sin^2\theta_W = \frac{1}{1+b^2}\left[1 - b^2\,\frac{11}{3\pi}\,\alpha(\mu)\ln\frac{M}{\mu}\right] \,,$$

where Eqs. (147) have been used. Also the unified coupling $\alpha_G(M)$ can be evaluated from Eq. (146), by taking into account the familiar QCD formula

$$\alpha_s(M) = \frac{6\pi}{(33-2f)\ln M/\Lambda} \,. \tag{154}$$

In evaluating Eqs. (153) we take $\mu \simeq 2M_W \simeq 160$ GeV. Then $\alpha(\mu) \simeq 1/127.8$ since the value of fine structure constant is changed[90] from its definition point at $Q^2 \sim m_e^2$. For $f = 6 - 8$, $\Lambda \simeq 0.2 - 0.5$ GeV and the SU(5) values for a and b in Eqs. (149) one obtains

$$
\begin{aligned}
& M \simeq 2\times 10^{15} - 2\times 10^{16}\ \text{GeV} , \\
& \sin^2\theta_W \sim 0.19 - 0.20 , \\
& \alpha_G(M) \simeq 0.023 - 0.025 .
\end{aligned}
\tag{155}
$$

Various improvements have been recently added to this elementary treatment including two loop contributions to the β functions,[90] a careful evaluation of threshold effects,[91] and so on.[92] These refinements all lower the unifying mass and increase slightly $\sin^2\theta_W$, so that the accepted values now are

$$
\begin{aligned}
& M \simeq 2\times 10^{14} - 2\times 10^{15}\ \text{GeV} , \\
& \sin^2\theta_W \simeq 0.20 \pm 0.01 , \\
& \alpha_G(M) \simeq 0.024 \pm 0.001 .
\end{aligned}
\tag{156}
$$

The value for $\sin^2\theta_W$ is still somewhat smaller than the observed value but the discrepancy is too small to rule out the model. We see that from the observed values of the various coupling constants an enormous value for M follows, only a few orders of magnitude away from the Planck mass of $\sim 10^{19}$ GeV where the neglect of gravitation is no more conceivable. The necessity of such a large unification mass is also imposed by a different argument valid in SU(5) and in a large class of other models that has to do with proton stability. In a unified theory both quarks and leptons are in one multiplet of G. Then some of the gauge bosons turn quarks into leptons (in SU(5) this is the case for $X^{4/3}$ and $Y^{1/3}$). Higgs exchange also may induce baryon and lepton number nonconservation. Note that the unbroken symmetry $SU(3)_{color} \times U(1)_Q$ allows the transition $q+q \to \bar{q}+\ell$ because in color $3\times 3 \supset \bar{3}$. In SU(3) the following transitions are allowed:

$$
\begin{aligned}
& u+u \to \bar{d}+e^+ && (X^{1/3}) , \\
& u+d \to \bar{u}+e^+,\ \bar{d}+\bar{\nu} && (Y^{1/3},\ H^{1/3}) ,
\end{aligned}
\tag{157}
$$

where in brackets the gauge boson and the Higgs mediating that process are also indicated. In a model with no Higgs 45, $H^{1/3}$ is a superposition of 5 and 24, the orthogonal combination and $H^{4/3}$ being eaten up by $X^{4/3}$, $Y^{1/3}$. In practice the exchange of H should be suppressed by its small couplings to light fermions. The proton lifetime is clearly given by

$$
\tau(P) \simeq \frac{(\text{factors})}{\alpha_G^2(M)} \frac{M^4}{m_p^5} .
\tag{158}
$$

For $M \sim 5 \times 10^{14}$ GeV, $\alpha_G \sim 0.024$ and the factors estimated from X and Y exchanges, one finds[93]

$$\tau(P) \simeq 10^3 - 10^4 \frac{M^4}{m_p^5} \sim 10^{30} - 10^{32} \text{ years} \quad . \tag{159}$$

The present limits are $\tau(P) > 2 \times 10^{30}$ ys.[94] (if decays into final μ are dominant) or $\tau(P) > 1.3 \times 10^{29}$ ys.[95] (without this restriction). We see that M cannot be smaller than the order of magnitude obtained from the coupling constants. On the other hand the urge toward more sensitive searches for proton decay is enhanced by the closeness of the prediction in Eq. (159) to the present limits.

It has been observed that the existence of baryon number violating interactions, together with CP violation and the cessation of statistical equilibrium at a given stage of the big bang evolution, can generate a baryon number asymmetry[96] as present in the observed universe. In SU(5) with only Higgs 5 and 24 the estimated ratio of $(n_B - n_{\bar{B}})/n_\gamma$ is of order $\sim 10^{-16}$ (the observed value is $\sim 10^{-9}$) which seems to hint toward a more complex Higgs structure. In this respect we may add that the mass ratios in Eq. (142), predicted by light Higgs transforming as a 5 (and no 45), are renormalized according to[97]

$$\frac{m_b(\mu)}{m_\tau(\mu)} = \left[\frac{\alpha_s(\mu)}{\alpha_G(M)}\right]^{\frac{12}{33-2f}} \left[\frac{\alpha_1(\mu)}{\alpha_G(M)}\right]^{3/2f} \frac{m_b(M)}{m_\tau(M)} \quad . \tag{160}$$

With $m_b(M) = m_\tau(M)$ this leads to a value of about 3. This is good for b, disputable for s and certainly bad for d. The failure could be blamed on the too small value of the s and d masses, or taken as a further indication that a 5 is not sufficient.

What beyond SU(5)? First there are interesting generalizations. For example the fermion classification in $\bar{5} + 10$ may be considered ugly. Then we may add the $\bar{\nu}$ as a SU(5) singlet and go to SO(10).[98] In fact,

$$SO(10) \supset SU(5) \times U(1) \quad .$$

The lowest dimensional complex representation of SO(10) is the 16 with the SU(5) content

$$16 = 1 + \bar{5} + 10 \quad , \tag{161}$$

which beautifully accommodates one generation. The price is that the vanishing of ν masses is no more guaranteed. The fermion mass term can arise from $(16 \times 16)_{SYMM} = 126 + 10$ and the 10 is not sufficient. Note that SO(10) also admits the alternative breaking according to

$$SO(10) \supset SU(4) \times SU(2)_L \times SU(2)_R \quad , \tag{162}$$

which would implement the elegant idea of lepton number as a fourth color.[99]

Also some recent attempts should be mentioned of explaining the observed iteration of (at least) three families, by accommodating all of them in a set of multiplets (with no iteration of replicas) of a large group containing SU(5). Candidates could be SU(11)[100] (also involving superheavy fermions) with three generations of light fermions, or SO(16)[101] with four generations (and no superheavy fermions).

A completely different class of models that have been investigated are theories where preunification of the electroweak sector is a possible intermediate stage.[102-104] That is with a group G with the subgroup

$$G \supset SU(3)_{color} \times W \quad , \tag{163}$$

with W a preunifying group with one coupling constant containing SU(2) × U(1). Two general features of this class of models are that a) the sum of charges of all quarks, of all antiquarks and of all leptons must separately vanish; b) the charged current of W cannot all be pure V-A. The first conclusion is demonstrated as follows. W contains the Q generator. Generators of W must have zero trace in each representation of W. Quarks, antiquarks and leptons have different colors and therefore must be in separate multiplets of W because W commutes with color. The second statement follows because in order to not have V+A currents, all left handed antiquarks should be singlets of SU(2) × U(1). But then they should also be singlets of W and they cannot because of their charges.

The simplest possibility in this class is that W = SU(3).[103] Then quarks must be triplets (of charges 2/3, -1/3, -1/3),

$$\begin{aligned} q \sim 3 \sim \begin{pmatrix} u \\ d \\ b \end{pmatrix} , \begin{pmatrix} c \\ s \\ b' \end{pmatrix} \quad , \\ \bar{q} \sim \bar{3} \sim \begin{pmatrix} \bar{u} \\ \bar{b} \\ \bar{d} \end{pmatrix} , \begin{pmatrix} \bar{c} \\ \bar{b}' \\ s \end{pmatrix} \quad . \end{aligned} \tag{164}$$

Note that no t quark appears. The leptons must then be classified in triality zero representations, the simplest option being octets and singlets. However the model with W = SU(3) is directly ruled out because from Eq.(150) applied to a triplet of quarks it follows that $(\sin^2\theta_W)_{SYMM} = 3/4$. There is no chance of sufficiently modifying this too high value by renormalization. Note in particular that 3/4 applies at the mass where the W symmetry is implemented. At this stage a superheavy mass is not necessary to guarantee proton stability.

A possibility[104] not yet ruled out is that $W = SU(3)_L \times SU(3)_R$ (times a discrete symmetry). In this case quarks, antiquarks and leptons can be classified into two identical replicas of the scheme (under $SU(3)_{color} \times SU(3)_L \times SU(3)_R$),

$$\underset{\text{quarks}}{(3,\bar{3},1)} + \underset{\text{antiquarks}}{(\bar{3},1,3)} + \underset{\text{leptons}}{(1,3,\bar{3})} \quad . \tag{165}$$

In this case $(\sin^2\theta_W)_{SYMM} = 3/8$ because each multiplet is parity doubled and the right components contribute to Q^2 but not to T^2_{3L}. A natural choice for the unifying group G is in this case E_6, whose lowest dimensional representation is a 27 with precisely the required content as in Eq. (165),

$$E_6 \supset SU(3) \times SU(3) \times SU(3) \quad .$$

A common problem of the models of this class is the absence a priori of a natural mechanism for a flavor conserving neutral current. In fact quarks of the same charge have different $SU(2) \times U(1)$ classification. In principle a natural mechanism could be induced by a pattern of symmetry breaking and a suitable spectrum for the Higgs, but no convincing example has yet been discussed.

In the context of fractionally charged quarks all models of unified theories involve superheavy masses of order 10^{15} GeV. The only alternative seems to be integrally charged quarks with chiral color and partially confined quarks as in the models studied by Pati and Salam, as for example $[SU(4)]^4$.

In conclusion we started with a list of motivations for unified theories, but at the end we see that only a partial list of successes can be claimed: charge quantization, constraints among couplings at present energies, hints on new interactions to be discovered, perhaps aesthetic appeal. However the present unified models present us with even more free parameters than in the starting model and no sufficiently constraining limitations on the fermion spectrum, their masses, the mixing angles and the like. The problem has too many solutions. The enormous ratios of masses in unified theories pose the problem of the origin and the natural explanation for these hierarchies.[87] The relative closeness of the superheavy masses to the gravitational domain poses problems on the feasibility of a unification excluding gravity. The present situation suggests that some important element is still missing. However, it is a result in itself of the fact that our present theories are advanced enough to make it at least possible to conceive a synthesis of all forces among elementary particles.

REFERENCES

1. S. Weinberg, Phys. Rev. Letters 19, 1264 (1967); A. Salam in Elementary Particle Physics, ed. by N. Svartholm (Almquist and Wiksells, Stockholm, 1968), p. 367.

2. For reviews see E.S. Abers and B.W. Lee; Phys. Reports 9, 1 (1973); J.C. Taylor, "Gauge Theories of Weak Interactions" (Cambridge Univ. Press, 1976); S. Weinberg, Proceedings of the Hamburg Conference, 1977 and rapporteur talks at "Rochester" Conferences by B.W. Lee (1972), J. Iliopoulos (1974), A. Slavnov (1976) and S. Weinberg (1978).

3. T.D. Lee, Phys. Reports 9C, 143 (1974); S. Weinberg, Phys. Rev. Lett. 37, 657 (1976).

4. B.W. Lee, C. Quigg and H.B. Thacker, Phys. Rev. Letters 38, 883 (1977); M. Veltman, Acta Physica Polonica 88, 475 (1977).

5. S. Coleman and E. Weinberg, Phys. Rev. D7, 1888 (1973).

6. J. Ellis, M.K. Gaillard, D.V. Nanopoulos and C.T. Sachrajda, Nucl. Phys. (1979).

7. For bounds on Higgs masses in grand unified theories see, for example, N. Cabibbo, L. Maiani, G. Parisi and R. Petronzio, CERN-TH 2683 (1979), and references therein.

8. S.L. Glashow, J. Iliopoulos and L. Maiani, Phys. Rev D2, 1285 (1970).

9. S.L. Glashow and S. Weinberg, Phys. Rev. D15, 1968 (1977); E.A. Paschos, Phys. Rev. D15, 1966 (1977).

10. M. Kobayashi and K. Maskawa, Progress of Theor. Phys. 49, 652 (1973).

11. S. Pakvasa and H. Sugawara, Phys. Rev. D14, 305 (1976); L. Maiani, Phys. Letters 62B, 183 (1976).

12. S.L. Adler, Phys. Rev. 177, 2426 (1969); J.S. Bell and R. Jackiw, Nuovo Cimento 51, 47 (1969). See also C. Bouchiat, J. Iliopoulos and Ph. Meyer, Phys. Letters 38B, 519 (1972); D. Gross and R. Jackiw, Phys. Rev. D6, 477 (1972).

13. For previous summaries see for example: G Altarelli, Proceedings of the Tokyo Conference (1978); J.J. Sakurai, Proceedings of the Topical Conference on Neutrino Physics, Oxford (1978).

14. J.E. Kim, P. Langacker, N. Levine, H.H. Williams and D.P. Sidhu, Proceedings of the Neutrino '79 Conference, Bergen (1979).

15. P.Q. Hung and J.J. Sakurai, UCLA/79/TEP/9 (1979).

16. I. Liede and M. Roos, Phys. Letters 83B, 89 (1979) and Proceedings of the Neutrino '79 Conference, Bergen (1979).

17. M. Holder et al., Phys. Lett. 72B, 254 (1977).

18. P. Wanderer et al., HPWF-77/1 (to be published in Phys. Rev.).

19. P.C. Bosetti et al., Phys. Letters 76B, 505 (1978).

20. CDHS Collaboration, presented at the Neutrino '79 Conference, Bergen (1979).

21. U. Amaldi, Talk presented at the Neutrino '79 Conference, Bergen (1979).

22. H. Klutting et al., Phys. Lett. 71B, 446 (1977).

23. See for example F. Dydak, Proceedings of the EPS Conference, Geneva (1979), and Ref. 21.

24. See for example Refs. 21 and 23.

25. W. Krena et al., Nucl. Phys. B135, 45 (1978); O. Erriques et al., Phys. Letters 73B, 350 (1978).

26. S.L. Adler, Ann. of Phys. 50, 189 (1968) and Phys. Rev. D12, 2644 (1975); S.L. Adler et al., Phys. Rev. D13, 1216 (1976).

27. E. Pasierb et al., Phys. Rev. Lett. 43, 96 (1979).

28. J. Blietschau et al., Phys. Letters 73B, 232 (1978); H. Faissner et al., (Aachen-Padova) Phys. Rev. Lett. 41, 213 (1978); A.M. Cnops et al., Phys. Rev. Lett. 41, 357 (1978); P. Petiau in Proceedings of the Topical Conference on Neutrino

Physics, Oxford (1978); N. Armenise et al., CERN Preprint CERN/EP/79-38 (1979).

29. F. Reines, H.S. Gurr and H.W. Sobel; Phys. Rev. Lett. 37, 315 (1976).

30. P.E.G. Baird et al., Phys. Rev. Lett. 39, 798 (1977).

31. L.L. Lewis et al., Phys. Rev. Lett. 39, 795 (1977); N. Fortson in Proceedings of the "Neutrino '78" Conference, Purdue University (1978).

32. L.M. Barkov and M.S. Zolotorev, Pisma Zh.Eksp.Teor.Fiz (JETP Lett.) 26, 379 (1978); and L.M. Barkov, Proceedings of the EPS Conferences, Geneva (1979).

33. As quoted by P.G.H. Sandars in Proceedings of the Topical Conference on Neutrino Physics, Oxford (1978).

34. I.B. Khriplovich, V.N. Novikov and O.P. Suchkov, Sov. Phys. JETP 44, 872 (1976).

35. R. Conti et al., Phys. Rev. Lett. 42, 343 (1979).

36. C.Y. Prescott et al., Phys. Lett. 72B, 347 (1978) and 84B, 524 (1979).

37. M.S. Chanowitz, M.A. Furman and I. Hinchliffe LBL-8270 (1978) to be published in Nuclear Physics.

38. R.E. Shrock and L.L. Wang, Phys. Rev. Letters 41, 1692 (1978).

39. R.E. Shrock, S. Treiman and L.L. Wang, Phys. Rev. Lett. 42, 1589 (1979).

40. V. Barger, W.F. Long and S.P. Pakvasa, Phys. Rev. Lett. 42, 1585 (1979).

41. L. Maiani, Proceedings of the Hamburg Conference (1977).

42. A. Sirlin, Nucl. Phys. B100, 291 (1979) and to be published; S.S. Shei, A. Sirlin and H.S. Tsao NYU/TR9/78.

43. M.K. Gaillard and B.W. Lee, Phys. Rev. D10, 897 (1974).

44. J. Ellis, M.K. Gaillard, D. Nanopoulos and S. Rudaz, Nucl. Phys. B131, 285 (1977).

45. R. Shrock and S. Treiman, Phys. Rev. D19, 2148 (1979).

46. For a review see for example: E. Paul, Springer Tracts in Modern Physics 79, 53 (1976).

47. See for example T.D. Lee and C.S. Wu, Rev. Nucl. Science 16, 511 (1966).

48. L. Wolfenstein, Phys. Rev. Lett. 13, 562 (1964).

49. V. Lobashev et al., Leningrad Inst. Nucl. Phys. preprint (1978).

50. See for example R.N. Mohapatra and J.C. Pati, Phys. Rev. D8, 2317 (1973) and Phys. Rev. D11, 566 (1975).

51. A.A. Belavine et al., Phys. Letters 59B, 85 (1975); R. Jackiw and C. Rebbi, Phys. Rev. Lett. 37, 172 (1976); C. Callan, R. Dashen and D. Gross, Phys. Lett. 63B, 334 (1976); G. 't Hooft, Phys. Rev. Lett. 37, 8 (1976); W.A. Bardeen, Nucl. Phys. B75, 246 (1974).

52. E.P. Shabalin, ITEP-31 preprint (1978).

53. B.F. Morel, Nucl. Phys. B157, 23 (1979).

54. F.J. Gilman, Phys. Letters 83B, 83 (1979) and SLAC-PUB-2341 (1979).

55. B.W. Lee, S. Pakvasa, R.E. Schrock and H. Sugawara FNAL-Pub 77/20 THY; S.B. Treiman, F. Wilczek and A. Zee, Princeton Preprint (1977); H. Fritzsch, Phys. Lett. 67B, 451 (1977); G. Altarelli, L. Baulieu, N. Cabibbo, L. Maiani and R. Petronzio, Nucl. Phys. B125, 285 (1977).

56. W. Bacino et al., Phys. Rev. Lett. 42, 749 (1979).

57. E. Bellotti, D. Cavalli, E. Fiorini and M. Rollier, Nuovo Cim. Lett. 17, 553 (1976).

58. M. Murtagh, Proceedings of the Hamburg Conference (1977); K. Schultze, Proceedings of the Hamburg Conference (1977).

59. E. Di Capua et al., Phys. Rev. 133B, 1333 (1964); D.A. Bryman and C. Picciotto, Phys. Rev. D11, 1337 (1975).

60. For $B(\mu \to e\gamma)$: J.D. Bowman et al., Phys. Rev. Lett 42, 556 (1979). For R_S: A. Badertscher et al., Berne-SIN preprint (presented at the Tokyo Conference, 1978).

61. S.M. Korechenko et al., Journal Exp. and Theor. Phys. 70, 3 (1976).

62. D.A. Bryman et al., Phys. Rev. Lett 28, 1469 (1972).

63. N. Cabibbo, Phys. Lett. 72B, 333 (1978).

64. J.D. Bjorken and S. Weinberg, Phys. Rev. Lett. 38, 622 (1977); S. Weinberg, Proceedings of the Hamburg Conference (1977).

65. For a recent discussion see Y.S. Tsai, SLAC-PUB-2105 (1978).

66. For a review see for example G.J. Feldman, Proceedings of the Tokyo Conference (1978).

67. D. Horn and G.G. Ross, Phys. Letters 67B, 460 (1977); G. Altarelli, N. Cabibbo, L. Maiani and R. Petronzio, Phys. Lett. 67B, 463 (1977).

68. N. Cabibbo and L. Maiani, Phys. Lett. 79B, 109 (1978). See also N. Cabibbo, G. Corbò and L. Maiani, CERN-Preprint (1979).

69. M. Suzuki, Nucl. Phys. B145, 420 (1978).

70. V. Barger, W.F. Long and S. Pakvasa, Univ. of Wisconsin Preprint UW-COO-881-92 (1979).

71. Particle Data Group, Phys. Lett. 75B, 1 (1978).

72. For a review see for example M. Conversi, Proceedings of the EPS Conference, Geneva (1979).

73. K. Wilson, Phys. Rev. 179, 1499 (1969).

74. B.W. Lee and M.K. Gaillard, Phys. Rev. Lett. 33, 108 (1974); G. Altarelli, L. Maiani, Phys. Letters 52B, 351 (1974).

75. B.W. Lee, M.K. Gaillard and G. Rosner, Rev. Mod. Phys. 47, 277 (1975); G. Altarelli, N. Cabibbo and L. Maiani, Nucl. Phys. B88, 285 (1975), Phys. Rev. Lett. 35, 635 (1975), Phys. Lett. 57B, 27 (1975); S.R. Kingsley, S. Treiman, F. Wilczek and A. Zee, Phys. Rev. D11, 1919 (1975); M.B. Voloshin, V.I. Zakharov and L.B. Okun, JETP Lett. 21, 183 (1975); J. Ellis, M.K. Gaillard and D. Nanopoulos, Nucl. Phys. B100, 313 (1975).

76. R.N. Cahn and S.D. Ellis, Univ. of Michigan report UM-HE-76-45 (1976); S.S. Gerstein and M.Yu. Khlopov, Serpukov Preprint IHEP-76-73 (1976); V.A. Novikov et al., Phys. Rev. Lett. 38, 626 and 791 (E) (1977).

77. I. Peruzzi et al., Phys. Rev. Lett. 39, 1301 (1977).

78. Mark II Collaboration, Proceedings of the EPS Conference, Geneva (1979).

79. N. Cabibbo and L. Maiani, Phys. Lett. 73B, 418 (1978); D. Fakirov and B. Stech, Nucl. Phys. B133, 315 (1978).

80. V. Barger and S. Pakvasa, Phys. Rev. Lett. 43, 812 (1979).

81. M.A. Shifman, A.I. Vainshtein and V.I. Zacharov, Nucl. Phys. B120, 316 (1977) and ZHETF 72, 1275 (1977).

82. See for example J.F. Donoghue, E. Golowich, W.A. Ponce and B.R. Holstein, MIT Preprint CTP-799 (1979).

83. For a review see for example M. Rees, Proceedings of the EPS Conference, Geneva (1979).

84. A.D. Linde, Phys. Lett. 83B, 311 (1979).

85. R. Gatto, G. Morchio and F. Strocchi, Phys. Lett. 83B, 348 (1979); G. Segré, H.A. Weldon and J. Weyers, Phys. Lett. 83B, 351 (1979).

86. G. Segré, Proceedings of the EPS Conference, Geneva (1979) and references therein.

87. See for example: E. Gildener, Phys. Rev. D14, 1667 (1976); S. Weinberg, Phys. Lett. 82B, 387 (1979); K.T. Mahanthappa, M.A. Sher and D.G. Unger, Phys. Lett. 84B, 113 (1979). For a review and complete references see I. Bars, Orbis Scientiae 1979, Coral Gables, Florida.

88. H. Georgi and S.L. Glashow, Phys. Rev. Lett. 32, 438 (1974).

89. H. Georgi, H.R. Quinn and S. Weinberg, Phys. Rev. Lett. 33, 451 (1974).

90. T.J. Goldman and D.A. Ross, CALT 68-704 (1979).

91. D.A. Ross, Nucl. Phys. B140, 1 (1978).

92. K.T. Mahanthappa and M.A. Sher, COLO-HEP-13 (1979).

93. C. Jarlskog and F.J. Yndurain, CERN-TH-2526 (1978); C. Jarlskog, Proceedings of the Neutrino '79 Conference, Bergen (1979), and Ref. 90.

94. F. Reines and M.F. Crouch, Phys. Rev. Lett. 32, 493 (1974).

95. L. Bergamasco et al., Nuovo Cimento.

96. A. Yu. Ignatiev, N.V. Krosnikov, V.A. Kuzmin and A.N. Tavkhelidze, Phys. Lett. 76B, 436 (1978); M. Yoshimura, Phys. Rev. Lett. 41, 381 (1978); S. Dimopoulos and L. Susskind, Phys. Rev. D18, 4500 (1978) and Phys. Lett. 81B, 416 (1979); S. Weinberg, Phys. Rev. Lett. 42, 850 (1979); D. Toussaint and F. Wilczek, Phys. Lett. 81B, 238 (1979); J. Ellis, M.K. Gaillard and D.V. Nanopoulos, CERN-TH-2596 (1978).

97. A.J. Buras, J. Ellis, M.K. Gaillard and D.V. Nanopoulos, Nucl. Phys. B135, 66 (1978).

98. H. Fritzsch and P. Minkowski, Ann. of Phys. (N.Y.) 93, 193 (1975); Nucl. Phys. B103, 61 (1976).

99. J.C. Pati and A. Salam, Phys. Rev. D8, 1240 (1973), Phys. Rev. D10, 275 (1974), Phys. Rev. Lett. 31, 661 (1973), Phys. Lett. 58B, 333 (1975).

100. H. Georgi, Harvard Preprint (1979).

101. F. Wilczek and A. Zee, Princeton Preprint (1979).

102. F. Gursey and P. Sikivie, Phys. Rev. Lett. 36, 775 (1976); P. Ramond, Nucl. Phys. B110, 214 (1976).

103. See for example H. Fritzsch and P. Minkowski, Phys. Lett. 63B, 99 (1976).

104. Y. Achiman, Phys. Lett. 70B, 187 (1977); J.D. Bjorken and K. Lane, Proceedings of the Neutrino '77 Conference, Elbrus (1977); P. Minkowski, Nucl. Phys. B138, 527 (1978); Y. Achiman and B. Stech, Phys. Lett. 77B, 389 (1978).

TOPICS IN QUANTUM CHROMODYNAMICS

C.H. Llewellyn Smith

Department of Theoretical Physics
1 Keble Road
Oxford OX1 3NP England

These lectures were given in parallel with lectures by S.D. Ellis and R.D. Field, which they are intended to complement. Section 1 contains a general introduction to QCD. In Section 2 I present an overview of perturbative QCD assuming the essential property of factorization to which I return in Sections 6 and 7. The choice of topics in Sections 3-5 was dictated by the demands of the audience and of other lecturers. This is not a review article — I have not attempted to cover all points of view or to compile a complete set of references. (I apologise to authors I have ignored.) For more systematic expositions the reader is referred to the abundant pedagogical literature.[1]

1. INTRODUCTION TO QCD

If we assume that hadrons are made of coloured quarks whose interactions are described by a field theory, there is an almost unique route which leads to QCD. The assumption of quarks no longer needs discussion nor does the use of field theory, given its success in electromagnetic and weak interactions. There is evidence for the existence of field quanta (or gluons) from deep inelastic e, μ and ν experiments, which show that roughly 50% of a rapidly moving proton's momentum is carried by weak and electrically neutral particles. There are good reasons to assume that the gluons are vectors: 1) Fields with even spin couple to particles and antiparticles in the same way so they would make QQ states degenerate with $Q\bar{Q}$ states. This leaves $J = 1$ as the simplest possibility (the only tractable renormalizable one). 2) With $J = 1$, there is a limit ($M_{u,d,s} \to 0$) in which the theory is exactly chirally invariant in the light quark sector. This is needed since there is excellent evidence for approximate chiral symmetry. (See Section 3.) 3) With $J = 1$ we can make a gauge theory. This is highly desirable since the other interactions (gravity, weak and electromagnetic) are known to be gauge theories and it is necessary for any grand unification.

The evidence for colour is as follows. 1) It solves the spin statistics problem in the quark model for baryons e.g. it allows us to put three up quarks with $S_Z = 1/2$ into S wave states to construct a Δ^{++} with $J_Z = 3/2$ by making the state totally antisymmetric in the colour indices. 2) In the chiral limit, $\Gamma(\pi^0 \to \gamma\gamma)$ is exactly calculable. (See Section 3.) The amplitude is determined by the diagram

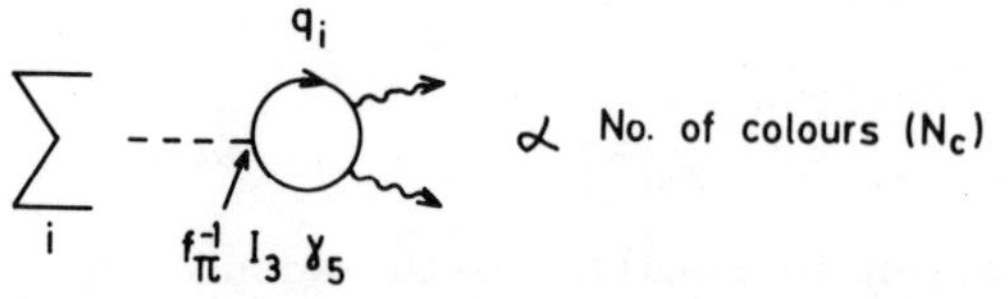

(The contribution of all other diagrams is forbidden by a low energy current-algebra theorem: this diagram — uniquely — escapes the assumptions of the theorem because it is so singular.) Assuming Gell-Mann Zweig non-integrally charged quarks, the result is

$$\Gamma(\pi^0 \to \gamma\gamma) = 7.87 \left(\frac{N_c}{3}\right)^2 \text{eV}$$

to be compared to the experimental value of $7.95 \pm .05$ eV. 3) If we assume that $\sigma(e^+e^- \to \text{hadrons})$ is given by

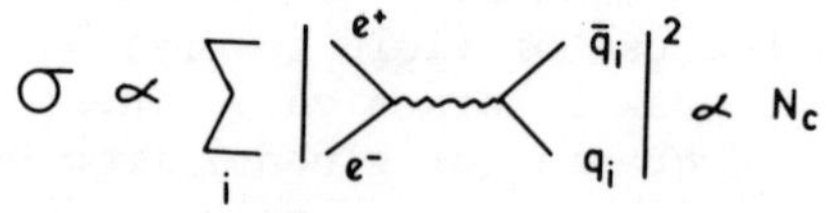

the data definitely require $N_c = 3$. (This assumption is actually only justified in QCD.) Likewise the Drell-Yan model for $pp \to \mu\bar{\mu}X$ (also only justified in QCD) gives a cross section inversely proportional to N_c and the data prefer $N_c = 3$ to $N_c = 1$.

Given coloured quarks, the forces must be colour dependent. (Otherwise there would be eight coloured pions degenerate with the "white" pion.) The simplest possibility is to let the gluons have eight colours and interact with the coloured quarks in an $SU(3)_{col.}$ invariant way

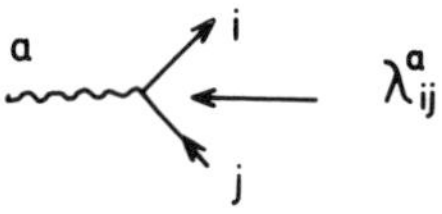

where a, i, j are colour indices (i,j = 1..3, a = 1..8) and λ^a are Gell-Mann matrices.*

Colour interactions of this type allow the forces to be attractive in the 3Q colour singlet state — yet repulsive for other 3Q states and for QQ etc.[2] This is possible because $[QQ] = 3^* + 6$ i.e. two quarks antisymmetrized in their colour indices behave (in colour) like an anti-quark and can attract a third quark. Clearly the non-Abelian nature of the forces is essential here. We can demonstrate that the chromoelectric forces could have the features needed to confine colour by assuming that the long range Q-Q forces are proportional to the colour charges i.e. that they have the same colour properties as one gluon exchange (1GE):

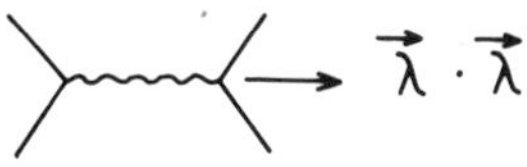

(Obviously other diagrams are important; it is only the colour properties which we abstract.) The eight gluons can be labelled by colour indices Red, Blue and Green thus

$$R\bar{B},\ R\bar{G},\ \ldots\ldots,\ (2R\bar{R} - B\bar{B} - G\bar{G})/\sqrt{6},\ (B\bar{B} - G\bar{G})/\sqrt{2}\ .$$

We can now calculate the colour factor to be abstracted from one gluon exchange (1GE) by reading off the colour colour content of the gluons which can be exchanged. Thus (in arbitrary units)

---

*Three colours of gluon with an SO(3) invariant interaction is also a possibility; I leave it as an exercise to show that this model does not share all the desirable features of the SU(3) model.

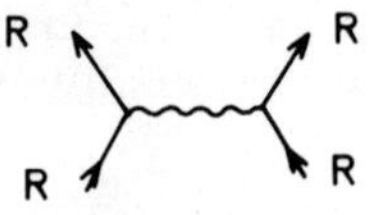

$\longrightarrow (2/\sqrt{6})^2$

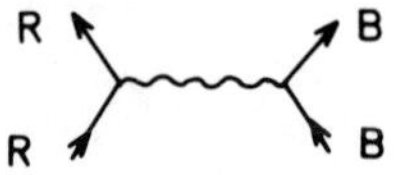

$\longrightarrow (2/\sqrt{6})(-1/\sqrt{6})$

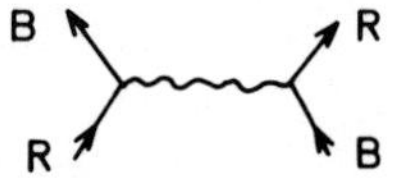

$\longrightarrow 1$

from which we deduce that

$$V_{QQ} = P - \frac{1}{3}$$

where P is the colour exchange operator:

P = +1 for states symmetric in colour exchange

P = −1 for states antisymmetric in colour exchange.

We see that the forces are most attractive in antisymmetric states. With the toy model

$$\langle E\rangle = C(\langle V_{QQ}\rangle + \frac{4}{3}\langle N_Q + N_{\bar{Q}}\rangle)$$

we find

$$\langle E\rangle_{[QQ]_{3^*}} = \frac{4c}{3}$$
$$\langle E\rangle_{[QQ]_6} = \frac{10c}{3}$$
$$\langle E\rangle_{[QQQ]_1} = 0$$
$$\langle E\rangle_{[QQQ]_{10}} = 6c$$

etc.

In fact $\langle E \rangle = 0$ for colour singlets but $\langle E \rangle > 0$ for all other states; if we let $C \to \infty$ we would banish all but colour singlets.

For $Q\bar{Q}$ states we obtain figures similar to the above but with the direction of the arrows reversed on the right hand side; the associated numerical factors are the same except for an overall minus sign which is due to the fact that vector exchange between particles and antiparticles is intrinsically attractive. Hence we easily find

$$V_{[Q\bar{Q}]_1} = -8/3$$

$$V_{[Q\bar{Q}]_8} = 1/3 \ .$$

The toy model again banishes all but colour singlets as $C \to \infty$.

This model shows clearly that it may be possible to explain the fact that only "white" hadrons are observed. The chromomagnetic forces abstracted from one gluon exchange also have features which fit the facts. Non-relativistic reduction of 1GE gives

$$g^2 \frac{\vec{\lambda}_1}{2} \cdot \frac{\vec{\lambda}_2}{2} \left[\frac{-8\pi}{3} \vec{\mu}_1 \cdot \vec{\mu}_2 \delta(\vec{r}) + \frac{1}{r^3} \left(\vec{\mu}_1 \cdot \vec{\mu}_2 - \frac{3\vec{r} \cdot \vec{\mu}_1 \vec{r} \cdot \vec{\mu}_2}{r^2} \right) \right] ,$$

where $\vec{\mu}_i$ are the chromomagnetic moments of the quarks. Apart from the colour factor, the same form occurs in QED where the $\vec{\mu}_1 \cdot \vec{\mu}_2 \delta(\vec{r})$ term makes $M(e^+e^-)_{S=0} < M(e^+e^-)_{S=1}$ and is responsible for the 21 cm. hyperfine line in hydrogen. (The second term vanishes in S wave states.) In an exactly analogous way, this term makes $M_\pi < M_\rho$ and $M_N < M_\Delta$ in QCD. Furthermore, introducing SU(3) breaking, it explains why $M_\Lambda < M_\Sigma$, which had long been a puzzle.[3]

Necessarily

$$\vec{\mu}_i = \Lambda_i \vec{S}_i$$

where $\vec{S}_i$ is the spin and Λ_i are constants with dimensions M^{-1}. We would obviously guess that $\Lambda_i \propto m_i^{-1}$ so that $\Lambda_s < \Lambda_u \simeq \Lambda_d$. Consider a u d s state for which the $\delta(\vec{r})$ term in the Hamiltonian gives

$$\delta M = c(\Lambda_u^2 \vec{S}_u \cdot \vec{S}_d + \Lambda_u \Lambda_s \vec{S}_s \cdot (\vec{S}_u + \vec{S}_d)) \ .$$

In a Σ, u and d couple to $I = 1$. Therefore they are symmetric in space and isospace and must also be symmetric in spin (being anti-symmetric in colour) i.e.

$$(\vec{S}_u + \vec{S}_d)^2_\Sigma = 2$$

$$(\vec{S}_u \cdot \vec{S}_d)_\Sigma = 1/4 \ .$$

Similarly in a Λ

$$(\vec{S}_u + \vec{S}_d)^2_\Lambda = 0$$

$$(\vec{S}_u \cdot \vec{S}_d)_\Lambda = -3/4 \ .$$

With the total spin $(\vec{S}_u + \vec{S}_d + \vec{S}_s)^2$ fixed we find

$$\begin{aligned} \vec{S}_s \cdot (\vec{S}_u + \vec{S}_d) &= 0 \quad \Lambda \\ &= -1 \quad \Sigma \\ &= 1/2 \quad \Sigma^* \ . \end{aligned}$$

Thus this mechanism makes $M_\Sigma < M_{\Sigma^*}$. (With $\lambda_s \to \lambda_u$ the formula for $M_\Sigma - M_{\Sigma^*}$ gives $M_N - M_\Delta$, which is also negative.) Furthermore with $\lambda_s < \lambda_u$ we find $M_\Lambda < M_\Sigma$ as anticipated.

Quantitatively

$$\begin{aligned} \frac{\lambda_u}{\lambda_s} &= 1 + \frac{3(\Sigma-\Lambda)}{2(\Sigma^*-\Sigma)} = 1.60 \\ &= \frac{M_\rho - M_\pi}{M_K^* - M_K} = 1.60 \ ! \end{aligned}$$

The simple minded model works better than we had any right to expect. Writing $\vec{\mu}_{em} = \lambda^{em} \vec{S}$ the data give

$$\frac{\lambda_u^{em}}{\lambda_s^{em}} \quad \frac{-\mu(p)}{3\mu(\Lambda)} = 1.52 \pm .01$$

which is also very satisfactory since a first guess would be that $\lambda_i^{em} = \lambda_i$.*

Isgur and Karl have used the Hamiltonian above to calculate chromomagnetic corrections to the masses of p-wave baryons in the harmonic oscillator quark model with remarkably successful results.[5] In the s-wave case, it is known that the bag model gives essentially the same results as the non-relativistic calculation,[6] despite the fact that bagged quarks are relativistic. Whether the bag model also describes p-wave baryons successfully is not known since no bag results are available.

The theory discussed above is not yet QCD. With only the quark-gluon coupling, it is non-unitary. If we consider Compton scattering in QED

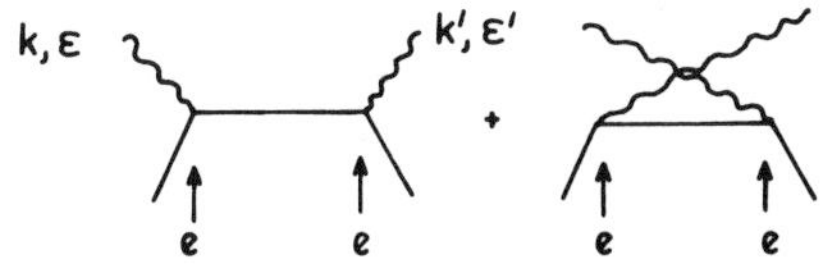

it is well known that when $\varepsilon \to k$ the two contributions cancel and the amplitude vanishes i.e. longitudinal photons decouple. In QCD

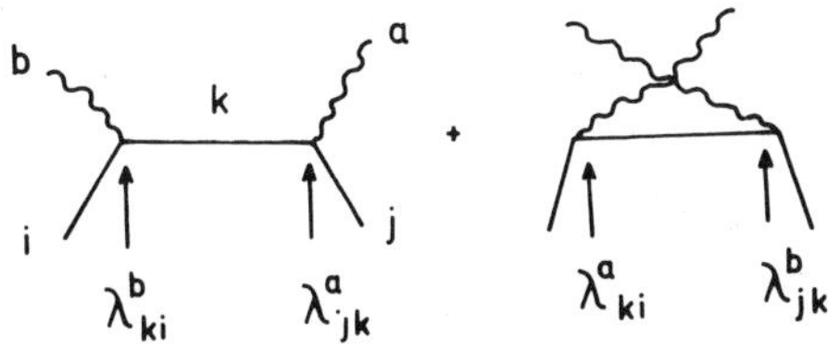

and when $\varepsilon^a \to k^a$ or $\varepsilon^b \to k^b$ the amplitude is proportional to

$$[\lambda^a, \lambda^b] \neq 0.$$

Another contribution from

---

*Lipkin has pointed out[4] that we can also calculate $\lambda_u \lambda_s^{-1}$ from strangeness mass splittings if we set $\mu_i = Q_i M_N m_i^{-1}$ nuclear magnetons and make the drastic assumption that $m_s - m_u = M_\Lambda - M_p$. This gives $\frac{\lambda_u}{\lambda_s} = \frac{m_s}{m_u} = 1 + \frac{(M_\Lambda - M_p)}{M_p} \mu_p = 1.53$. The extraordinary success of this naive calculation is unexpected and not understood.

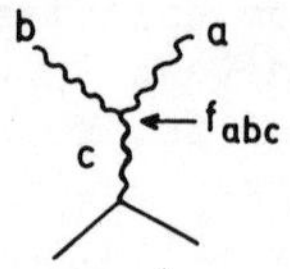

is needed to force the longitudinal gluon to decouple. With a suitable space-time structure for the three gluon vertex, a longitudinal gluon will decouple provided

$$[\lambda^a,\lambda^b] = 2f_{abc}\,\lambda^c .$$

Furthermore to make a longitudinal gluon decouple in gg → gg

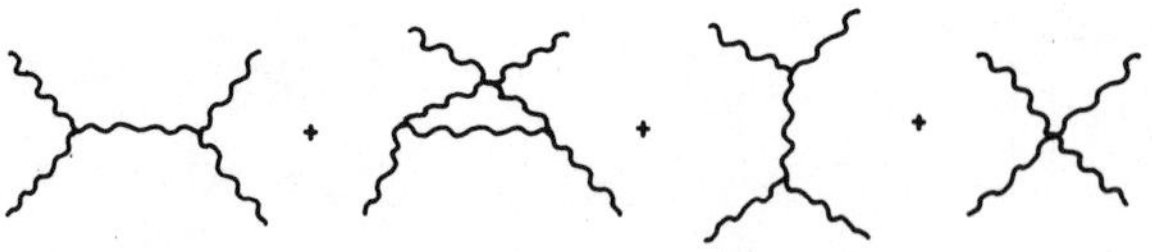

a four gluon coupling proportional to $f_{abe}f_{cde}$ is necessary and the f's must satisfy the Jacobi identity i.e. they are the structure constants of SU(3).

This leads us finally to the Lagrangian of QCD

$$L = \sum_i \bar{\psi}_i (i\partial\!\!\!/ - m_i + g\frac{\vec{\Lambda}\cdot\vec{A\!\!\!/}}{2})\psi_i - \frac{1}{4}[\partial_\mu \vec{A}_\nu - \partial_\nu \vec{A}_\mu + g\,\vec{A}_\mu \times \vec{A}_\nu]^2$$

where i is a flavour index, ψ_i is a spinor in colour space $[\psi_i^T = (q_R, q_B, q_G)]$ $\vec{A}_\mu$ is an eight component vector in colour space, $\vec{\Lambda}$ are the eight 3x3 Gell-Mann matrices and $\vec{A}_\mu \times \vec{A}_\nu \equiv f_{abc}\,A^b_\mu A^c_\nu$. This Lagrangian is, of course, invariant under <u>local</u> rotations in colour space

$$\psi \to \psi + ig\,\frac{\vec{\Lambda}\cdot\vec{\omega}(x)}{2}\,\psi \ldots .$$

$$\vec{A}_\mu \to \vec{A}_\mu + \partial_\mu \vec{\omega}(x) + g\,\vec{A}_\mu \times \vec{\omega}(x) \ldots$$

We now briefly consider the other symmetries of the Lagrangian. In flavour space, various vector currents are conserved in the limit of degeneracy between quark masses (e.g. with $m_u = m_d = m_s$ the eight $SU(3)_{flavour}$ vector currents are conserved). These currents and the associated axial vector currents satisfy Gell-Mann's equal time algebra even when the associated symmetry is broken by the quark masses.

When $m_{u,d} \to 0$ the $SU(2)_{flavour}$ axial currents

$$\bar{\psi}\, \gamma_\mu \gamma_5\, \vec{I}\, \psi$$

are exactly conserved. This chiral symmetry could be satisfied in two ways: either manifest symmetry, $M_p = 0$ (or there must be parity doublets), or hidden symmetry, $M_\pi = 0$ (and there are exact low energy theorems). The second alternative is an excellent approximation to the real world. (See Section 3.) In this limit the isoscalar axial current

$$\bar{u}\gamma_\mu\gamma_5 u + \bar{d}\gamma_\mu\gamma_5 d$$

is also conserved, which apparently implies $M_\eta = 0$. (With $m_{u,d} \neq 0$ it can be shown that $M_\eta < \sqrt{3}M_\pi$ to first order in symmetry breaking.[7]) This disaster is probably averted by instantons which break the isoscalar chiral symmetry.[8]

A most attractive feature of QCD is that it is asymptotically free.[9] To explain qualitatively what this means, we must first understand the notion of a scale dependent coupling constant. Consider placing a charge in a classical dielectric. To measure the charge we must surround it by a sphere of radius R and measure the flux $4\pi\int \vec{E} \cdot d\vec{S}$:

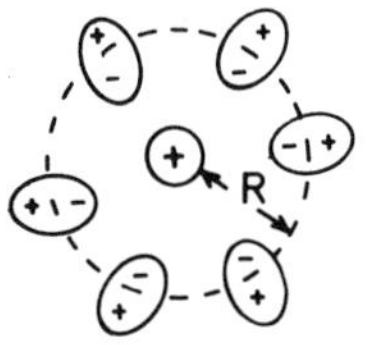

The result will depend on R

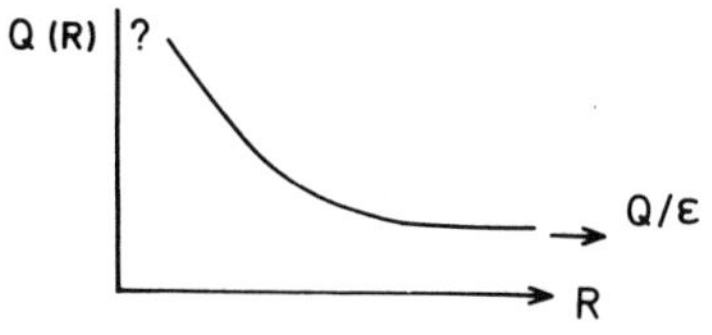

For R>> (the intermolecular radius), there is screening due to polarization and $Q \Rightarrow Q/\varepsilon$. For R<< (molecular radius) there can be no screening and Q has the same value as in free space. Likewise in field theories a distance is needed to define a charge. In QED there is also screening due to vacuum polarization

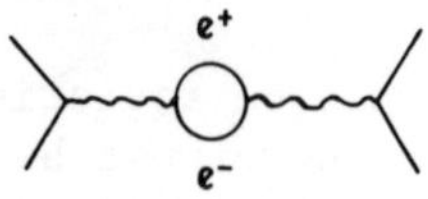

The physical charge e is usually defined as $Q(\infty)$. We have $Q(\infty) < Q(R)$ for $R < 1/m_e$ due to shielding. (As $R \to 0$ we do not really know what happens.)

In QCD there is antiscreening

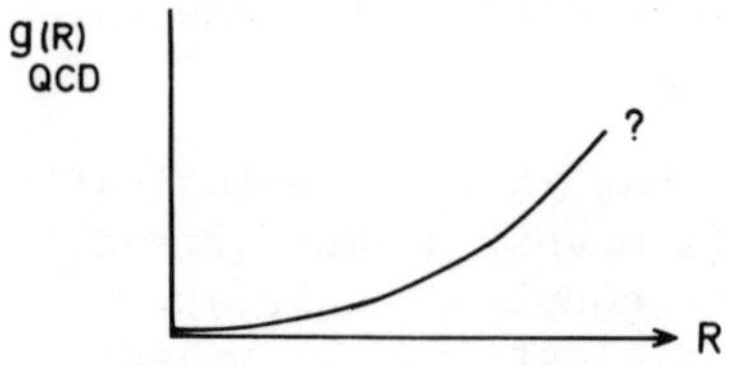

As $R \to 0$, $g^2(R) \sim 1/\ln R$. (As $R \to \infty$ we do not know what happens, but perhaps $g \to \infty$ leading to confinement.) The origin of this phenomenon is hard to trace since the diagrams responsible are different in different gauges. Very roughly speaking in diagrams such as

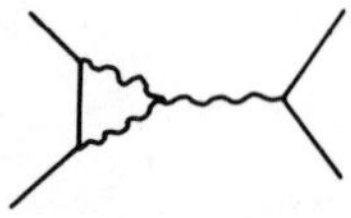

the bifurcated gluon sees the quark's colour charge spread out, leading to a decrease as $R \to 0$.

The fact that g depends on a distance or energy scale and that we can put $m_q = 0$ to first approximation in treating the light quarks (and neglect the heavy quarks) has the very remarkable consequence that there are no free parameters in QCD. With $m_q = 0$, there are no dimensional parameters in L_{QCD}. To get started we can introduce a mass scale μ at which we set (e.g.) $g(\mu) = 1$. If we were clever enough to solve the theory, presumably we would find that spontaneous chiral symmetry breaking occurs, in which case necessarily M_p = (calculable no.) × μ, and $g(M_p)$ would be calculable. All dimensionless quantities would be calculable in terms of no free parameters. This is a very attractive feature of QCD but it is also a source of difficulty in solving the theory.

In the absence of a solution, we need to take one parameter λ from experiment to set the scale of the coupling. Asymptotically

$$\alpha_s(Q^2) \equiv \frac{g^2(Q^2)}{4\pi} = \frac{12\pi}{(33-2N_f)\ln Q^2/\lambda^2}$$

where N_f is the number of flavours. λ is not yet well determined but probably 0.05 GeV $\lesssim \lambda \lesssim$ .5 GeV.[10] With λ = 500 MeV:

$Q^2 (\mathrm{GeV})^2$	1	5	10	10^2	10^3	10^5
$\alpha_s(Q^2)$	1.1	.50	.41	.25	.18	.12
$\frac{\alpha_s(Q^2)}{\pi}$	.35	.16	.13	.08	.06	.04 .

In the absence of detailed calculations, we do not know whether α_s or α_s/π is the "correct" expansion parameter (which need not be the same in each process[11]). In either case it is clear that we must be very cautious in using perturbation theory except at enormous energies. Furthermore, the variation of α_s with Q^2 is most evident in a region where α_s is large and perturbation theory cannot be trusted; to observe the variation of α_s in a region where it is small requires an enormous span of energy. Large energy is necessary but not sufficient to justify an expansion in α_s; the circumstances in which perturbation theory is allowed are discussed in the next section.

2. OVERVIEW OF PERTURBATIVE QCD

The fact that $g^2(E) \sim 1/\ln E$ is not sufficient to justify a perturbative treatment of high energy processes in general for two reasons: 1) Perturbation theory treats quarks and gluons, which are not in the physical final state. Therefore it certainly cannot be used in a straightforward way to calculate the distribution of individual particles or exclusive cross-sections. However, we might be able to calculate gross properties independent of the detailed nature of final states, e.g. $\sigma_{tot}^{e^+e^-}$, F_2^{eN}, $\sigma(e^+e^- \to \text{jets})$. We might think that total cross-sections for deep inelastic processes are determined by the initial response at short times/distances, which is controlled by QCD - what happens later may not change the total probability that something happens. (This is true of F_2 in potential models with confining potentials.[12])

Furthermore it turns out that in deep inelastic processes the bulk of the energy emerges from short distances in narrow jets according to perturbation theory; it is plausible that this is not upset by the confining forces. In addition we may be able to calculate the energy dependence of fragmentation functions and perhaps also of form factors and even of some exclusive cross-sections[13], which is determined by the dynamics at short distances - although the absolute value will depend on unknown "hadronic wavefunctions."

2) Given a large energy scale (E) we can always use g(E) but this is not sufficient to justify the use of perturbation theory. A measurable quantity W will generally depend on masses m as well as any other dynamical variable (E') involved, i.e. $W = W(g(E), m/E, E'/E)$. If we attempt to expand W in powers of g(E) we will generally encounter powers of $\ell n E/m$ which spoil the expansion. (It may also be spoiled by power of $\ell n E'/E$ if E'/E is not held fixed when E is made large.) Physically, sensitivity to m indicates that the large distance properties of the theory are involved which we cannot possibly hope to treat perturbatively. Therefore the use of perturbation theory requires

either that $\ell n E/m$ terms do not occur; this happens in $\sigma_{e^+e^-}$ and $\sigma(e^+e^- \to \text{jets})$, which are "mass finite" as $m \to 0$ and therefore cannot depend on $\ell n E/m$;

or that the logs can be summed up and the sensitivity to m can somehow be factored out; this "factorization of mass singularities" occurs (e.g.) in deep inelastic lepton scattering.

Consider a dimensionless observable $\psi(g(\mu), Q^2/\mu^2, m^2/Q^2)$ which factorizes:

$$\psi \underset{Q^2\to\infty}{\to} C(g(\mu), Q^2/\mu^2)\ D(g(\mu), m^2/\mu^2) + O(m^2/Q^2) \ .$$

The value of ψ cannot depend on the scale μ chosen to define g so that

$$\mu \frac{d\psi}{d\mu} = (\mu \frac{\partial}{\partial\mu} + \mu\frac{\partial g}{\partial\mu}\frac{\partial}{\partial g} + \mu\frac{\partial m}{\partial\mu}\frac{\partial}{\partial m})\psi = 0 \ ,$$

i.e. if we change μ, the same result will be obtained provided the parameters g and m are also changed appropriately. (I am here supposing that m is a parameter introduced to normalize propagators at $p^2 = -\mu^2$ - see Section 5.) This is the renormalization group equation (RGE) for an observable. If ψ factorizes we have (dropping terms of order m^2/Q^2):

$$\frac{\mu}{C}\frac{dC}{d\mu} = -\frac{\mu}{D}\frac{dD}{d\mu} = \gamma(g)$$

where the separation constant can only depend on g. This equation determines the behaviour of C, and hence ψ, as a function of Q^2, i.e. it sums up the logs above.

This result can be put in closer correspondence with the parton picture if we introduce a second large scale Q_0^2 which we use to eliminate D. Thus

$$\begin{aligned}\psi &= C(g(\mu),Q^2/\mu^2)\ C^{-1}(g(\mu),Q_0^2/\mu^2)\ \psi(Q_0^2)\\ &= C(g(Q^2))\ \{C^{-1}(g(Q^2),Q_0^2/Q^2)\ \psi(Q_0^2)\}\ ,\end{aligned}$$

where we have set $\mu^2 = Q^2$ in the second line. $C(g(Q^2))$ corresponds to the parton model "hard scattering cross-section", which can be calculated perturbatively, and $\{C^{-1}\psi\}$ is the "Q^2 dependent quark distribution", whose Q^2-dependence is controlled by the dynamics for $Q^2 > Q_0^2$ and can be calculated (using the RGE); the dependence on m (and long distance dynamics, confinement etc.) is factored into $\psi(Q_0^2)$ which must be taken from experiment.

We shall show (Section 7) that factorization occurs in all traditional parton model processes and that the same "universal" quark distributions (or fragmentation functions) occur in each case. This is illustrated in Fig. 1. In zeroth order we obtain the "naive parton" model results with scale dependent quark distributions and fragmentation functions and $g \to g(E)$. Neglecting terms of order m^2/Q^2, the corrections to the zeroth order picture can be calculated by expanding in $g^2(E)$.

Let us illustrate this in deep inelastic neutrino scattering

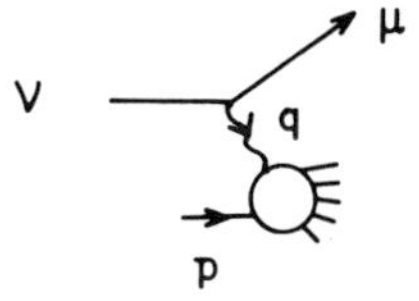

which is described in terms of the total cross-sections for left-handed, right-handed and helicity zero virtual W's or, equivalently, three structure functions $F_{1,2,3}$ which are functions of $Q^2 = -q^2 = \vec{q}^{\,2} - q_0^2$, and $x = Q^2/2\nu$ $(\nu = q \cdot p)$. It turns out (see Section 7) that the moments of F_3 factorize,

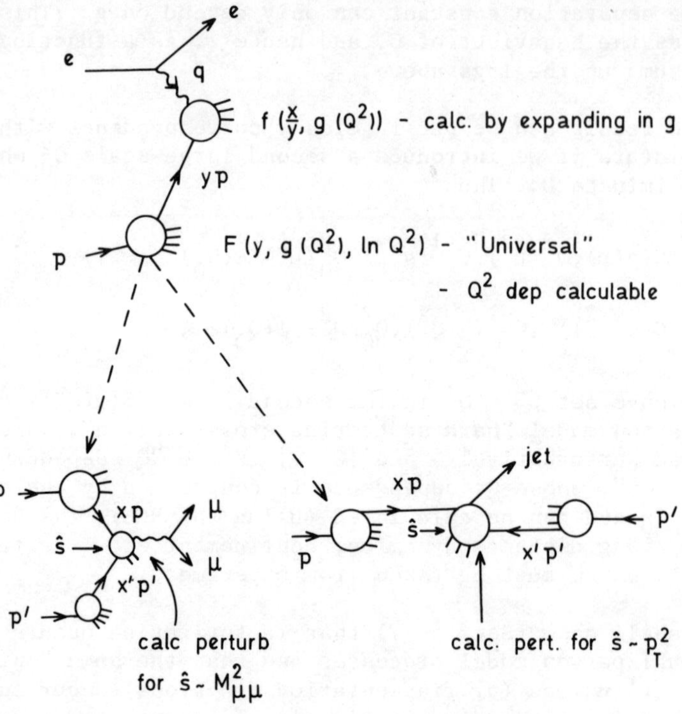

Fig. 1. Illustration of factorization (valid up to terms of order m^2/Q^2).

$$M_n(Q^2) \equiv \int x^n F_3(x,Q^2)dx$$

$$= C_n(g(Q^2),1)[C_n^{-1}(g(Q^2),Q^2/Q_0^2)M_n(Q_0^2)] + O(\frac{m^2}{Q^2}) .$$

Differentiating we obtain

$$\left.\frac{\partial M_n}{\partial \ell n Q^2}\right|_{Q^2 = Q_0^2} = A_n(Q_0^2)M_n(Q_0^2) ,$$

where

$$A_n(Q_0^2) = -C_n(g(Q_0^2),1)^{-1} \left. \frac{\partial C_n(g(Q_0^2),Q^2/Q_0^2)}{\partial \ln Q^2} \right|_{Q^2 = Q_0^2} .$$

A_n can only depend on $g(Q_0^2)$ and can therefore be calculated perturbatively if Q_0^2 is large. This is easily done by calculated $M_n(Q^2)$ in ordinary perturbation theory and factorizing the result by hand to obtain C_n and hence A_n:

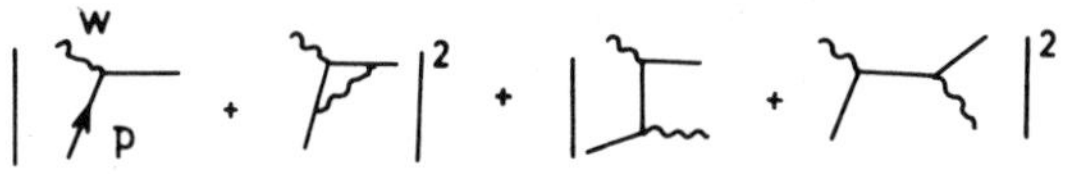

$$M_n(Q^2) = 1 + B_n g^2 \ln Q^2/m^2 + O(g^4)$$

$$= (1 + B_n g^2 \ln Q^2/\mu^2)(1 + B_n g^2 \ln \mu^2/m^2) + O(g^4) .$$

(Logs of Q^2/μ^2 only appear in $O(g^4)$ so factorization appears trivial to the order exhibited explicity; the B_n are given in Section 6.) Hence

$$A_n(Q^2) = -B_n g^2(Q^2) + O(g^4(Q^2)) .$$

Various remarks:

1) We see that assuming factorization (which will be proved in Section 7) we can, in principle, calculate A_n to arbitrary accuracy and hence find the Q^2 dependence of M_n up to terms of order m^2/Q^2.
2) The equation for M_n suggests a simple picture of deep inelastic scattering to leading order in $g^2(Q^2)$. First we recall that in the parton model $F_3(x,Q^2) \propto q_-(x,Q^2) = q(x,Q^2) - \bar{q}(x,Q^2)$, where $q(\bar{q})$ is the quark (antiquark) distribution.[14] The RGE for M_n can therefore be written

$$\frac{\partial q_-(x,Q^2)}{\partial \ln Q^2} = g^2(Q^2) \int_x^1 q_-(y,Q^2)\ f\left(\frac{x}{y}\right) \frac{dy}{y}$$

where f is a function such that

$$\int_0^1 z^n f(z)dz = -B_n .$$

The infinitesimal integral of the equation for q_- can be interpreted pictorially thus

$$q_-(x,Q^2 + \delta Q^2) = q_-(x,Q^2) + \frac{\delta Q^2}{Q^2}\, g^2(Q^2) \int_x^1 q_-(y,Q^2) f(\tfrac{x}{y}) \frac{dy}{y}$$

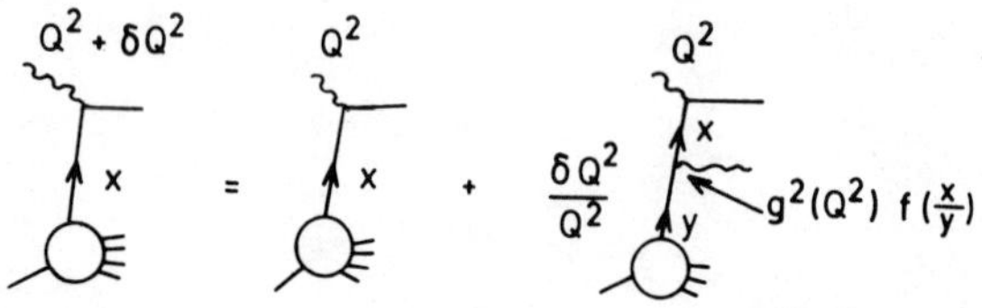

i.e. as Q^2 is increased quarks radiate gluons shedding a fraction z of their momentum with probability $(\delta Q^2/Q^2)\, g^2(Q^2) f(z)$ (leading to a softening of $F(x,Q^2)$ with increasing Q^2). This interpretation of the RGE was proposed by Altarelli and Parisi.[15] (The physical picture had been developed earlier by Kogut and Susskind[16]; we shall derive it in Section 6.) 3) Writing $g^2(Q^2) = b/\ell n(Q^2/\Lambda^2)$, the solution of the equation for M_n is

$$M_n(Q^2) = \left(\frac{\ell n Q^2/\lambda^2}{\ell n Q_0^2/\lambda^2}\right)^{-b\, B_n} M_n(Q_0^2)$$

to leading order (where b and B_n are of course known explicitly). It follows that

$$\ell n\, M_n(Q^2) = \frac{B_n}{B_{n'}}\, \ell n\, M_{n'}(Q^2) + \text{const. (indep of } Q^2) \ .$$

I.e. QCD predicts that if $\ell n\, M_n(Q^2)$ is plotted against $\ell n\, M_{n'}(Q^2)$ for large Q^2 the data should lie on straight lines with slopes equal to $B_n/B_{n'}$. This prediction has been tested by the BEBC experiment[17] using Nachtmann (N) moments of data with $1 < Q^2 < 100\ \text{GeV}^2$ and by the CDHS experiment[18] using Nachtmann (N) and Cornwall and Norton (C-N) moments of data with $5 < Q^2 < 100\ \text{GeV}^2$. Note that N and C-N moments are identical for $Q^2 \to \infty$ but differ by terms of order M_N^2/Q^2 for finite Q^2; N moments, which correspond to operators of definite spin in the operator product expansion, are simpler theoretically and it is possible that they incorporate all M_N^2/Q^2 terms, remaining m^2/Q^2 terms being characterised by a "dynamical" mass m, e.g. $m^{-1} = R_p$ (the radius of the proton) or $m = \langle p_T \rangle$

(a typical hadronic $\langle p_T \rangle$), or by m_q - since $m^2_{u,d} \ll \langle p_T \rangle^2 \sim 0.1$ GeV^2 $\ll M^2_N$, this would imply that it is better to use N moments.[19] The data can be fitted by straight lines in $\ln M_n - \ln M_{n'}$ plots with slopes given by:

n/n'	BEBC(N)	CDHS(N)	CDHS(C-N)	QCD	Scalar gluons
5/3	1.50 ± .08	1.34 ± .12	1.58 ± .12	1.46	1.12
6/4	1.29 ± .06	1.18 ± .09	1.34 ± .07	1.29	1.06
6/3	—	1.38 ± .15	1.76 ± .15	1.62	1.14
7/3	1.84 ± .20	—	—	1.76	1.16

The slopes are obviously compatible with QCD. To assess the significance of this result I have also listed the slopes obtained using perturbation theory in a model with scalar gluons. (The use of perturbation theory is not necessarily justified in scalar theories,* which are not asymptotically free, but the results provide a useful touchstone for the success of QCD.) The BEBC data obviously favour QCD over the scalar gluon model but caution in drawing conclusions is needed for several reasons: a) the CDHS data are more equivocal; b) some of the data used are at quite small Q^2 where we would not necessarily expect the leading order theory to work — a conservative attitude would be to take the difference of N and C-N moments as a measure of the uncertainty due to $1/Q^2$ effects (although N moments are preferable); higher order terms in $g^2(Q^2)$ should also be included.[20]

Nevertheless the data agree with QCD as well as could possibly be expected and this agreement is significant.** Apart from a

---

*A fixed point at $g = g^* \ll 1$ is needed to obtain approximate scaling; the existence of such a fixed point would mean that $\beta(g^*)$ cannot be calculated perturbatively and therefore we may doubt whether $\gamma(g^*)$ (and hence the slope in $\ln M_n - \ln M_{n'}$ plots) can be calculated perturbatively.

**Harari denies this,[21] alleging that "quite general" assumptions, which he implies are very likely satisfied in any field theory, restrict the slopes to be close to the QCD results, but this claim is invalidated by the fact that the predictions of the scalar gluon model do not satisfy his bounds. In fact Harari's assumptions severely restrict the x dependence of the scaling violations in a way which happens to agree with QCD. (That the parameterization used approximately satisfies the RGE of QCD was pointed out some time ago by Buras and Gaemers.[22])

brief discussion of the behaviour of $\int F_2(x,Q^2)dx$ in Section 5, I shall not discuss the complex and time-dependent subject of the comparison of QCD perturbation theory with experiment further in these lectures. The present situation is encouraging and more accurate data extending to higher Q^2 should provide decisive tests in the future.

3. CHIRAL SYMMETRY, CURRENT ALGEBRA AND $\pi^o \to \gamma\gamma$

In Section 1 we argued that the success of chiral symmetry requires the interquark force to be vector in nature and we used the result of calculating $\pi^o \to \gamma\gamma$ in the chiral limit as evidence for colour. We now review these "old" but important results.

We define the iso-triplet of axial currents by

$$A^i_\lambda(x) = \bar{\psi}(x)\gamma_\lambda\gamma_5 I^i\psi(x)$$

where $\psi = \begin{pmatrix} u \\ d \end{pmatrix}$. If $M_{u,d} = 0$ and there are only vector (or axial-vector) forces, it follows from the equations of motion that these axial currents are conserved

$$\partial^\lambda \, A^i_\lambda(x) = 0 \; .$$

The associated symmetry, called Chiral Symmetry, can be realized in two ways. 1) Manifest symmetry: in this case all hadrons would belong to degenerate "chiral multiplets" containing states of opposite parity (zero mass fermions being a special case). Evidently there is no sign of this in nature. 2) Hidden symmetry, realized in the Nambu-Goldstone mode. This requires the existence of a massless isotriplet of pseudoscalar mesons — presumably the pion.

Let us turn immediately to some classic tests of the hypothesis of Hidden (or Spontaneously Broken) Chiral Symmetry; we shall see the choice between realizations 1 and 2 in an explicit example. (For a general discussion of the Goldstone theorem see ref. 23; for a succinct review of classic current algebra results and references to the original literature see ref. 24.) We work in the chiral limit $\partial^\lambda A_\lambda = 0$ since exact results are easier to derive than approximate ones. 1) Consider

$$\langle p(p')|A^+_\lambda(0)|n(p)\rangle = \bar{u}(p')[-g_A\gamma_\lambda\gamma_5 + f_P q_\lambda\gamma_5]u(p)$$

where $q_\lambda = (p'-p)_\lambda$. With $\partial^\lambda A_\lambda$:

$$q^\lambda\langle p|A^+_\lambda|n\rangle = 0 = \bar{u}\gamma_5 u[-2M_N g_A + q^2 f_P] \qquad (1)$$

As $q^2 \to 0$, this can only be true if

<u>either</u> $M_N = 0 \to$ manifest symmetry,

<u>or</u> $f_p \sim 1/q^2 \to M_N \neq 0 \to$ hidden symmetry.

One possible diagram is

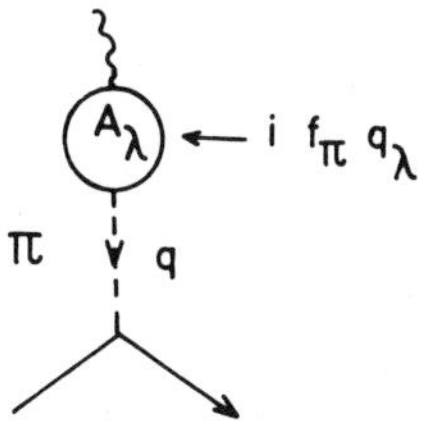

which contributes

$$\frac{q_\lambda f_\pi g_{np\pi}}{q^2 - M_\pi^2}$$

to f_p. Hence Eq. (1) can be satisfied provided $M_\pi = 0$ and

$$2M_N g_A + f_\pi g_{np\pi} = 0 \ .$$

This is known as the Goldberger-Treiman relation; experimentally it works well:

$$1 + \frac{2M_N g_A}{f_\pi g_{np\pi}} = 0.08 \pm 0.02 \ .$$

2) Next we consider low energy theorems for πN scattering. We start from the identity

$$0 = \int d^4x \frac{\partial}{\partial x_\mu} [e^{iq \cdot x} \langle p,f|T(A_\mu^\alpha(x)\ A_\nu^\beta(0))|p,i\rangle]$$

$$= q^\mu T_{\mu\nu} + \int d^4x e^{iq \cdot x} \langle p,f|\delta(x_o)\ [A_o^\alpha(x),A_\nu^\beta(0)]|p,i\rangle$$

where α, β, i and f are isospin indices, we have used $\partial^\mu A_\mu^\alpha = 0$ and

$$T_{\mu\nu} \equiv i\int d^4x \langle p,f|T(A^\alpha_\mu(x)A^\beta_\nu(0))|p,i\rangle \; .$$

The commutator, which is easily evaluated, is given by

$$\delta(x_o)[A^\alpha_o(x),A^\beta_\nu(0)] = i\varepsilon^{\alpha\beta\gamma}V^\gamma_\nu(0)\delta^4(x)$$

$$= -(I_\pi)^\gamma_{\alpha\beta}V^\gamma_\nu(0)\delta^4(x)$$

where I_π is the pion isospin matrix. Then using

$$\langle p,f|V^\gamma_\nu(0)|p,i\rangle = (I_N)^\gamma_{fi}2p_\nu \; ,$$

we obtain

$$q^\mu q^\nu T_{\mu\nu} = 2p\cdot q(\vec{I}_\pi)_{\alpha\beta}\cdot(\vec{I}_N)_{fi} \; .$$

Writing separately the terms with two, one and no pion poles respectively we have (suppressing isospin indices)*

$$T_{\mu\nu} = \frac{q_\mu q_\nu f^2_\pi}{(q^2)^2} M - i\frac{q_\mu f_\pi}{q^2}B_\nu + i\, C_\mu\frac{q_\nu f_\pi}{q^2} + R_{\mu\nu} \; ,$$

where M is the pion-nucleon scattering amplitude and

$$B_\nu = \langle\pi,N|A_\nu(0)|N\rangle - i\,\frac{q_\nu f_\pi}{q^2}M$$

$$C_\mu = \langle N|A_\mu(0)|\pi,N\rangle + i\,\frac{q_\mu f_\pi}{q^2}M \; .$$

Hence (using $\partial^\nu A_\nu = 0$ again)

---

*In ref. 25 (which is an excellent review of current algebra) on p. 228 the single pion pole terms are (incorrectly) ignored, the correct final answer being obtained nevertheless because of a sign mistake in the evaluation of the dipole term.

$$q^\mu q^\nu T_{\mu\nu} = -f_\pi^2 M + q^\mu q^\nu R_{\mu\nu}$$

$$= 2p \cdot q \ \vec{I}_\pi \cdot \vec{I}_N \ .$$

We interpret this as a theorem for M in the real ($M_\pi \neq 0$) world which should be valid up to terms of $0(M_\pi^2)$*. Evaluated at threshold with $q = (M_\pi, 0)$, $p = (M_N, 0)$ it leads to the following formula for the s-wave scattering lengths ($a \equiv M_{th}(8\pi[M_N + M_\pi])^{-1}$) due to Tomozawa and Weinberg

$$a = -\frac{M_N M_\pi}{8\pi(M_N + M_\pi) f_\pi^2} \vec{I}_\pi \cdot \vec{I}_N + 0(M_\pi^2)$$

which gives

$$a_{1/2} = 0.16/M_\pi \ (\text{expt} : 0.17 \pm .005/M_\pi)$$

and

$$a_{3/2} = -0.079/M_\pi \ (\text{expt} : -0.088 \pm .004/M_\pi) \ ,$$

or $a_{1/2} + 2a_{3/2} = 0$ (this is an "Adler zero") and a formula for $a_{1/2} - a_{3/2}$ which, when combined with a dispersion relation, is equivalent to the Adler-Weissberger relation. 3) As a last example consider the decay $K \to \pi e \nu$ which is described by

$$\langle \pi(q) | V_\mu(0) | K(k) \rangle = f_+(k+q)_\mu + f_-(k-q)_\mu \ ,$$

where V_μ is the appropriate vector current. I leave it as an exercise to show that as $q \to 0$

$$f_+ + f_- \to \frac{f_\pi}{f_K}$$

in the chiral limit. This is known as the Callan-Treiman (C-T) relation. Experimenters customarily use

---

*The validity of this procedure is discussed (e.g.) in ref. 25. If, alternatively, we keep $M_\pi = 0$ and let $q \to 0$, the nucleon pole terms in $q^\mu q^\nu R_{\mu\nu}$ give contributions of $O(q)$ which must be kept (they do not contribute to the $M_\pi \neq 0$ amplitude at threshold by parity conservation); but, carefully interpreted, the final result is the same (see e.g. ref. 24).

$$f(t) \equiv (M_K^2 - M_\pi^2)^{-1}(k-q)^\mu \langle\pi|V_\mu|K\rangle$$

$$= f_+(t) + \frac{t}{M_K^2 - M_\pi^2} f_-(t) \ ,$$

which they parameterize as

$$f_+(0) \left(1 + \frac{\lambda_o t}{M_\pi^2}\right) \ .$$

The C-T relation gives $f(M_K^2) = f_\pi/f_K$ or

$$\lambda_o = \frac{M_\pi^2}{M_K^2}\left(\frac{f_K}{f_\pi} - 1\right) = 0.020 \ ,$$

to be compared to the experimental value of 0.020 ± 0.003.

We see that the predictions of chiral SU(2) × SU(2) seem to hold to O(10%). (Chiral SU(3) × SU(3) — reached in the limit $m_s \to 0$ — holds to O(30%).) No unambiguous exceptions to this statement are now known and the departures from chiral symmetry are also fairly well understood.[26] In the past, however, there were three glaring failures (in addition to a serious disagreement between some early data and the C-T relation): a) The U(1) problem (i.e. the prediction $M_\eta < \sqrt{3}M_\pi$ alluded to in Section 1) which may be solved in QCD.[8] b) The prediction that $\eta \to 3\pi$ is forbidden in the chiral symmetry; this difficulty may also be resolved by instantons in QCD (although, like the resolution of the U(1) problem, this is controversial — see R. Crewther, ref. 8). c) The prediction of Sutherland and Veltman (S-V) that $\Gamma(\pi^o \to \gamma\gamma)$ is forbidden in the chiral limit.[27]

We now know that the S-V disaster is triumphantly averted by the Bell-Jackiw-Adler anomaly.[28,29] Some kinematics must preceed a detailed discussion. Consider an amplitude involving two photons and an axial current

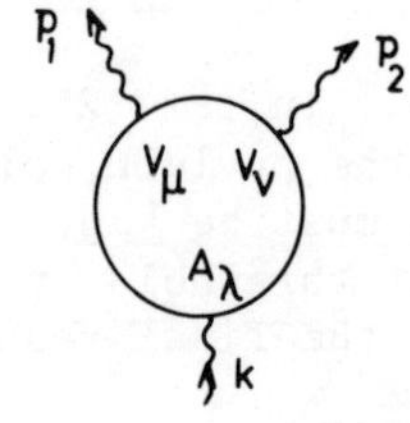

which is described by a tensor $T_{\mu\nu\lambda}(p_1,p_2)$ which satisfies Bose statistics

$$T_{\mu\nu\lambda}(p_1,p_2) = T_{\mu\nu\lambda}(p_2,p_1)$$

and gauge invariance

$$p_1^\mu T_{\mu\nu\lambda} = p_2^\nu T_{\mu\nu\lambda} = 0 \ .$$

The most general tensor with these properties can be written*

$$T_{\mu\nu\lambda} = AT^1_{\mu\nu\lambda} + BT^2_{\mu\nu\lambda} \ ,$$

where

$$T^1_{\mu\nu\lambda} = p_1^\alpha p_2^\beta \varepsilon_{\alpha\beta\mu\nu} k_\lambda$$

$$T^2_{\mu\nu\lambda} = p_1^\alpha p_2^\beta [\varepsilon_{\alpha\beta\nu\lambda}(p_1-p_2)_\mu + \varepsilon_{\alpha\beta\mu\lambda}(p_1-p_2)_\nu]$$

$$+ [p_1^2 p_2^\alpha - p_2^2 p_1^\alpha + (p_1\cdot p_2)(p_1-p_2)^\alpha]\varepsilon_{\alpha\mu\nu\lambda} \ .$$

The "form factors" A and B are free from kinematic singularities. The pion pole

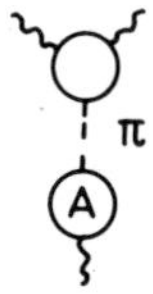

contributes

$$\frac{k_\lambda f_\pi}{k^2 - M_\pi^2} C_\pi \ p_1^\alpha p_2^\beta \ \varepsilon_{\alpha\beta\mu\nu}$$

---

*The identity

$$a\cdot f|bcde| + b\cdot f|cdea| + c\cdot f|deab| + d\cdot f|eabc| + e\cdot f|abcd| = 0,$$

where $|bcde| = b^\mu c^\nu d^\alpha e^\beta \ \varepsilon_{\mu\nu\alpha\beta}$ is needed to show this.

to A, where C_π describes $\pi^0 \to \gamma\gamma$. We now consider two "proofs" of the S-V theorem: 1) Consider*

$$<0|A_\lambda(0)|\gamma(p_1,\varepsilon_1)\gamma(p_2,\varepsilon_2)> = \varepsilon_1^\mu \varepsilon_2^\nu T_{\mu\nu\lambda} .$$

Using $\partial^\lambda A_\lambda = 0$ (which will turn out to be false) we find

$$k^\lambda \varepsilon_1^\mu \varepsilon_1^\nu T_{\mu\nu\lambda} = [Ak^2 + B(2p_1 \cdot p_2 - p_1^2 - p_2^2)]\varepsilon_1^\mu \varepsilon_2^\nu p_1^\alpha p_2^\beta \varepsilon_{\mu\nu\alpha\beta}$$
$$= 0 . \quad \text{[false]}$$

As $k^2, p_1^2, p_2^2 \to 0$, the coefficients of A and B vanish and they only contribute if they have (dynamical) poles at this point. Thus only the pion pole term survives (since $M_\pi = 0$ in the chiral limit) and the false theorem implies

$$C_\pi = 0 \to \Gamma_\pi = 0 .$$

We shall see that in fact

$$i\partial^\lambda A_\lambda = 2M_0\bar{\psi}\gamma_5\psi + \frac{e^2}{16\pi^2}\varepsilon^{\mu\nu\alpha\beta}F_{\mu\nu}F_{\alpha\beta} ,$$

which does not vanish in the chiral limit $M_0 \to 0$, in which the anomalous term gives the result quoted in Section 1 for $\pi^0 \to \gamma\gamma$ in the GMZ quark model. 2) Consider

$$T^5_{\mu\nu\lambda} = \int d^4x d^4y e^{ip_1 \cdot x + ip_2 \cdot y} <0|T(V_\mu(x)V_\nu(y)A_\lambda(0))|0> .$$

Imposing the vector Ward identifies

$$p_1^\mu T_{\mu\nu\lambda} = p_2^\nu T_{\mu\nu\lambda} = 0$$

and Bose symmetry we get

---

*I give results here and below for the axial current in QED. In QCD the anomaly connected with the third component of the isovector axial current (relevant to π^0 decay) is entirely due to electromagnetism (gluons being isoscalar) and is easily constructed from the QED result (QCD contributes an anomaly to the U(1) current which is intimately connected with the proposed resolution of the U(1) problem).

$$T^5_{\mu\nu\lambda} = a\, T^1_{\mu\nu\lambda} + bT^2_{\mu\nu\lambda}$$

where T^1 and T^2 were given above. In the chiral limit

$$k^\lambda T^5_{\mu\nu\lambda} \xrightarrow[k^2, p_i^2 \to 0]{} C_\pi f_\pi\, p_1^\alpha p_2^\beta \varepsilon_{\alpha\beta\mu\nu}$$

as before. Now consider

$$T^5_{\mu\nu} = \int d^4y d^4x\, e^{i(p_1 \cdot x + p_2 \cdot y)} \langle 0|T(V_\mu(x)V_\nu(y)\bar{\psi}(0)\gamma_5\psi(0)|0\rangle\ .$$

This seems to be related to $T^5_{\mu\nu\lambda}$ by the Ward identity (WI):

$$k^\lambda T^5_{\mu\nu\lambda} = 2M_o T^5_{\mu\nu} \qquad \text{[false]}$$

where I have used $i\partial^\lambda A_\lambda = 2M_o\bar{\psi}\gamma_5\psi$ (which is correct to lowest order in e, which is all we need here). This would imply that in the chiral limit ($M_o \to 0$)

$$k^\lambda T_{\mu\nu\lambda} = 0 \to C_\pi = 0\ .$$

We first consider the failure of the false WI in perturbation theory. To lowest order $T^5_{\mu\nu\lambda}$ and $T^5_{\mu\nu}$ are given by triangles

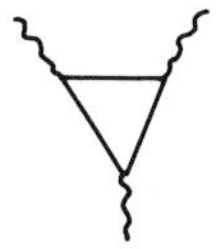

The WI could be verified without doing the integrals if we could shift integration variables in the expression for $T^5_{\mu\nu\lambda}$ in a cavalier manner. However, if we were careful we would realize that the integral is linearly divergent and the shift of variables introduces an extra surface term which spoils the false WI. The "anomaly" introduced by the linear divergence is associated with the short distance/high energy behaviour of the theory; for an exposition which stresses this aspect see ref. 30.

We proceed differently. Doing the spinor trace, the triangle gives

$$T^5_{\mu\nu} = M_o p_1^\alpha p_2^\beta \varepsilon_{\alpha\beta\mu\nu} \int d^4k F(k,p_1,p_2,M_o) \ .$$

F has dimensions M^{-6} so this integral is ultra-violet convergent.

Furthermore necessarily

$$\int d^4k F \xrightarrow[k^2, p_i^2 \to 0]{} \frac{\text{const.}}{M_o^2} \ .$$

Hence

$$2M_o T^5_{\mu\nu} \xrightarrow[k^2, p_i^2 \to 0]{} \text{const. (indep of } M_o) \ ,$$

whereas in perturbation theory (in which there is no pion and no pole in A or B):

$$k^\lambda T^5_{\mu\nu\lambda} \xrightarrow[k^2, p_i \to 0]{} 0 \ ,$$

and the Ward identity fails.

We can easily correct the Ward identity using the calculated value of $\int d^4kF$: the result is

$$k^\lambda T^5_{\mu\nu\lambda} = 2M_o T^5_{\mu\nu} + 8\pi^2 \varepsilon_{\mu\nu\alpha\beta} p_1^\alpha p_2^\beta \qquad \text{(true)} \ .$$

This is a useful procedure because it turns out that this relation, derived in lowest order, is actually true to all orders.[31] (Only triangle diagrams are mass divergent ($\to M_o^{-1}$ for $k^2, p_1^2, p_2^2 \to 0$) and radiative corrections to vertices and propagators in the triangle cancel, essentially due to Ward identities; alternatively in diagrams other than the bare triangle, the fermion loop integration is convergent if carried out first and the shift of integration variables needed to verify the naive WI is allowed.) As in the derivation of the GT relation, the Ward identity can be satisfied in two ways. Manifest symmetry: as $M_o \to 0$, $M_N \to 0$, $T^5_{\mu\nu}$ diverges and $2M_o T^5_{\mu\nu} \xrightarrow[k^2, p_i^2 \to 0]{} -8\pi^2 \varepsilon_{\mu\nu\alpha\beta} p_1^\alpha p_2^\beta$ while $k^\lambda T^5_{\mu\nu\lambda} \to 0$.

Hidden symmetry: as $M_o \to 0$, $M_\pi \to 0$ but $M_N \neq 0$. The anomalous term is now cancelled by the pion pole and $M_o T^5_{\mu\nu} \to 0$ since $T^5_{\mu\nu}$ is not divergent in the absence of fermions with zero physical mass.

Returning to the first "derivation" of the S-V theorem, the false divergence $\partial^\lambda A_\lambda = 2M_o\bar{\psi}\gamma_5\psi$ can be corrected by defining A_λ as the limit as $\varepsilon \to 0$ of

$$A_\lambda(x,\varepsilon) = \bar{\psi}(x+\varepsilon)\gamma_\lambda\gamma_5\psi(x-\varepsilon)\ \exp(-\ ie\int_{x-\varepsilon}^{x+\varepsilon} A(y)\cdot dy)\ ,$$

where the line integral ensures gauge invariance. $\partial^\lambda A_\lambda$ is then the limit of

$$2iM_o\bar{\psi}(x+\varepsilon)\gamma_5\psi(x-\varepsilon) + ie\ \varepsilon_\mu A_\lambda(x,\varepsilon)F^{\lambda\mu}(x) + \ldots .$$

The matrix elements of $A_\lambda(x,\varepsilon)$ diverge like ε^{-1} (A has dimensions x^{-3} but the axial vector character of A_λ reduces the divergence by two powers); the second term survives as $\varepsilon \to 0$ and gives the "anomalous" contribution.[32]

In conclusion: 1) The singularities of field theory may vitiate naive "canonical" manipulations. In particular they give rise to "anomalies" in $\partial_\lambda A^\lambda$ and in Ward identities involving the axial current. (For a complete classification of anomalous Ward identities see ref. 33.) 2) Throughout we have insisted that the vector Ward identities/gauge invariance be respected. As far as the currents are concerned, this is a matter of choice; complete canonical behaviour is impossible but we could have put the anomaly in the vector identities and kept the axial identity canonical or made both anomalous. (This is discussed in detail in ref. 29; it is instructive to see how it arises in dimensional regularization.[34]) However, since in fact the vector current is coupled to the photon this choice is not open to us if we wish to maintain electromagnetic gauge invariance (and keep $M_\gamma = 0$). 3) For currents coupled to non-Abelian gauge fields gauge invariance is essential for renormalizability and unitarity and no anomaly is allowed. Thus in the Weinberg-Salam model the total axial current

$$A_\lambda^{tot.} = A_\lambda^{quark} + A_\lambda^{lepton}$$

must be made anomaly free (See item 4.) and the "naive" Ward identity of the form

$$k^\lambda T^5_{\mu\nu\lambda} = 2M_o T^5_{\mu\nu}$$

is correct. As k^2, p_1^2, $p_2^2 \to 0$, the I=1, I_3=0 quark current gives a piece proportional to C_π on the left and zero on the right in the chiral limit $M_o^q \to 0$, while the lepton current gives zero on the left hand and a constant (M_0 independent) piece on the right, which determines C_π. The Ward identities for the quark and lepton currents are separately anomalous and $C_\pi \propto$ (quark anomaly) = -(lepton anomaly). 4) Since the anomalous terms are mass independent it is possible to arrange cancellations between the contributions of different species.[35] I leave it as an exercise to show that in SU(2) × U(1) the condition is

$$\sum_i Q_i (I_3{}^i)^2 = 0$$

where i runs over all fermion species and $\vec{I}^i$ is the weak isospin. This condition is satisfied by each generation separately (e.g. by e, ν, u and d) for GMZ quarks provided they have three colours.

4. QUANTIZATION OF GAUGE THEORIES

Every textbook on QED explains that canonical quantization fails due to gauge invariance and that it is necessary either to quantize in a particular non-covariant gauge (e.g. Coulomb or axial gauge) or to break gauge invariance in a way which does not change the physics (leading to a covariant formulation). Before turning to the solution of these difficulties in QCD, let us review one covariant method of quantizing QED.

4.1 QED

We first add a field ϕ to the usual Lagrangian in the following way

$$L = -\frac{1}{4} F_{\mu\nu} F^{\mu\nu} + \partial \cdot A\phi + \frac{\alpha\phi^2}{2} + g A_\mu J^\mu + L(\psi) \ ,$$

where $J_\mu = \bar{\psi}\gamma_\mu\psi$, thus breaking gauge invariance. The equations of motion are

$$\partial \cdot A = -\alpha\phi$$

$$\partial^\nu F_{\mu\nu} + \partial_\mu \phi = J_\mu \ .$$

The derivative of the second equation gives

$$\Box \, \phi = 0 \ ,$$

i.e. ϕ and hence $\partial \cdot A$ are free fields — which implies that the addition of ϕ does not change the physics and the S-matrix is unitary. Note that 1) the propagator is

$$\left[-g_{\mu\nu} + (1-\alpha)\frac{k_\mu k_\nu}{k^2}\right](k^2)^{-1} .$$

(Without the ϕ field, the operator $\Box g_{\mu\nu} - \partial_\mu\partial_\nu$ which appears in the equation of motion is singular and the propagator is not defined.) The special cases $\alpha = 1$ and $\alpha = 0$ are known as Feynman and Landau gauges respectively. 2) The same rules are obtained from

$$L = -\frac{1}{4}F_{\mu\nu}F^{\mu\nu} - \frac{1}{2\alpha}(\partial \cdot A)^2 + g\, J_\mu A^\mu + L(\psi) .$$

3) Because $\partial \cdot A$ is free,

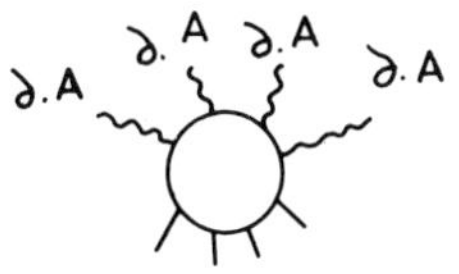

is zero when the external fermions are on shell. Consequently if the amplitude to emit photons with polarizations ε_1, ε_2 ... and momenta k_1, k_2 ... is $(\varepsilon_1^\mu \varepsilon_2^\nu \ldots)\, A_{\mu\nu\ldots}$, then $k_1^\mu A_{\mu\ldots} = 0$ etc. This ensures that unitarity is satisfied, e.g.

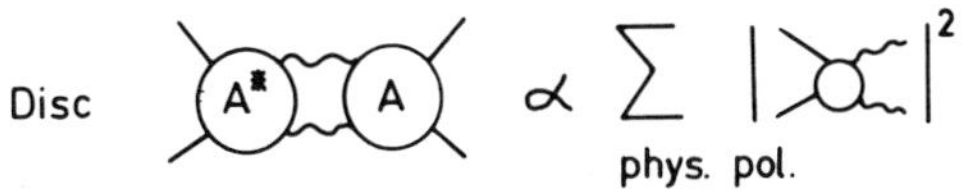

Summing over one polarization index, holding the other fixed, we have

$$-g^{\mu\nu} A^*_\mu A_\nu = A^*_1 A_1 + A^*_2 A_2 + A^*_3 A_3 - A^*_0 A_0$$

on the left, whereas on the right

$$\sum \varepsilon^{*\mu}\varepsilon^\nu A^*_\mu A_\nu = A^*_1 A_1 + A^*_2 A_2$$

if k is along the 3-axis. However, since $k \cdot A = 0$, $A_0 = A_3$ and these expressions are equal. 4) The fact that $\partial \cdot A$ is free does <u>not</u> mean that we can drop all $k_\mu k_\nu$ terms in propagators in

calculating Green functions: a) Clearly

is only zero when both fermion legs are on shell. b) Although

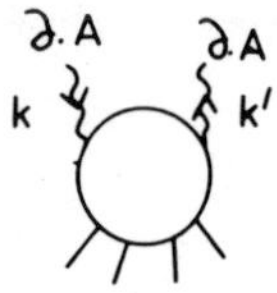

is zero for $k \neq k'$ when all the fermions legs are on shell, the contribution of the $k_\mu k_\nu$ part of the propagator in the diagram

is not zero. Such terms make non-zero contributions to Green functions and to wave function renormalization, although these cancel when we construct the S-matrix. (This is shown later for QCD; beware — Bjorken and Drell mislead about this on p. 199-200 of vol. 2.)

4.2 QCD

An obvious first guess is to try the trick which works for QED and put

$$L_{QCD} \to L_{QCD} + \partial^\mu \vec{A}_\mu \cdot \vec{\phi} + \frac{\alpha \vec{\phi}^2}{2} .$$

The equations of motion are

$$\partial \cdot \vec{A} = -\alpha \vec{\phi}$$

$$D^\nu \vec{G}_{\mu\nu} + \partial_\mu \vec{\phi} = \vec{J}_\mu$$

where $G_{\mu\nu}$ is the covariant field tensor

$$\vec{G}_{\mu\nu} = \partial_\nu \vec{A}_\mu - \partial_\nu \vec{A}_\mu + g\, \vec{A}_\mu \times \vec{A}_\nu$$

and D^ν the covariant derivative

$$D^\nu = \partial^\nu + g\,\vec{A}^\nu \times \,.$$

Contracting the second equation of motion with D^μ (using $D^\mu J_\mu = 0$ and $D^\mu D^\nu G_{\mu\nu} = 0$, which is an example of Poincare's lemma discussed in Wess' lectures) we obtain

$$D \cdot \partial\phi = 0 \,,$$

i.e. in contrast to QED, ϕ is not a free field. Consequently with the Feynman rules constructed from this Lagrangian, longitudinal gluons are coupled and unitarity fails, e.g.

Disc 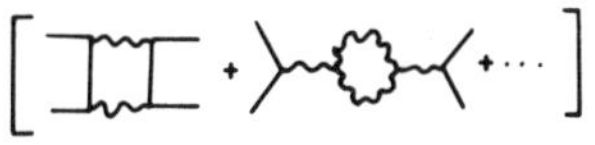 $\neq \sigma_{q\bar{q}\to gg}$

To restore unitarity additional "ghost" contributions are needed on the left hand side to cancel the contribution of longitudinal and time like gluons.[36]

Following Fradkin and Tyutin,[37] we can construct a satisfactory theory using the Lagrangian above if we let ϕ be the following functional of $\vec{A}_\mu$ and a new independent field $\vec{\theta}$:

$$\phi(x) = \int dy\, [D \cdot \partial]^{-1}(x,y)\, \Box_y \theta(y) \,.$$

The equation $D \cdot \partial\phi = 0$ now becomes

$$\Box\, \theta = 0 \,,$$

i.e. the auxiliary field θ is free. Hence the matrix elements of θ and $\partial \cdot A$ between physical states are zero and we would expect the theory to be unitary. (This is verified later.)

The S-matrix can be constructed from the functional integral representation of the generating functional

$$Z(J) = \int DA\; D\theta\; e^{i\int d^4x (L + J \cdot A)}$$

in the standard way, which is reviewed briefly below. (We omit fermions for simplicity; here $J_\mu(x)$ is a c-number function.) To find the Feynman rules we proceed as follows. 1) Change variables from θ to ϕ using

$$D\theta = D\phi \,\det\, (D \cdot \partial\Box^{-1}) \,.$$

2) Shift from ϕ to ϕ', where

$$\phi' = \sqrt{\alpha}\,\phi + \frac{1}{\sqrt{\alpha}}\,\partial \cdot A \ ,$$

so that

$$L = L_{QCD} - \frac{1}{2\alpha}(\partial \cdot A)^2 + \frac{1}{2}\phi'^2 \equiv L' + \frac{1}{2}\phi'^2 \ .$$

The ϕ' integral gives an irrelevant constant and we have

$$Z(J) = \int DA \ \det\,(D \cdot \partial\square^{-1})e^{i\int dx(L' + J \cdot A)}$$

$$= \int DA \ e^{iS_{eff}} \quad ,$$

where the effective action is

$$S_{eff} = \int dx(L' + J \cdot A) - i\mathrm{Tr}\,\ell n(D \cdot \partial\square^{-1}) \ .$$

We can proceed from S_{eff} to the Feynman rules in the standard way. The last term gives

$$\mathrm{Tr}\ \ell n\ (1 + g\ \vec{A}_\mu \times \partial^\mu\square^{-1}) = \mathrm{Tr} \sum_n \frac{1}{n}(g\ \vec{A}_\mu \times \partial^\mu\square^{-1})^n \ .$$

The $n = 0$ term contributes an irrelevant constant and the $n = 1$ term vanishes. Written explicitly, the $n = 2$ term is

$$\frac{1}{2}\int dx\,dy\ f_{abc}A^b_\mu(x)\frac{\partial}{\partial x_\mu}\square^{-1}(x,y)f_{cea}A^e_\nu(y)\frac{\partial}{\partial y_\nu}\square^{-1}\ (y,x) \ ,$$

which corresponds to the contribution of the diagram

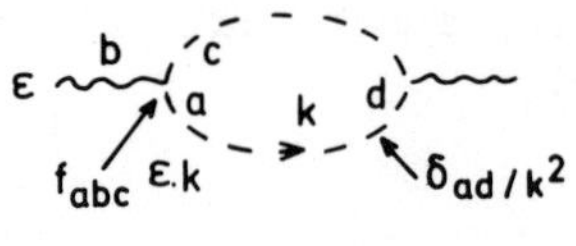

where the dashed line represents a spin zero field quantized with Fermi statistics, accounting for the minus sign relative to the contribution of a genuine spin zero particle. It is the discontinuity of this contribution which restores unitarity in the example above. In fact the whole of the Tr ℓn contribution can be generated by replacing L' by

$$L' - \partial^{\mu}\vec{\eta} \cdot (\partial_{\mu} + g\,\vec{A}_{\mu}\times)\vec{\omega}\ ,$$

where $\vec{\eta}$ and $\vec{\omega}$ are fictitious spin zero fermions, known as Faddeev-Popov ghosts,[38] which only appear in closed loops.

4.3 Path Integrals

We now briefly review some properties of path integrals (which we used above and will use again below). It is very instructive to derive the path integral representation of the S-matrix from the path integral formulation of quantum mechanics/field theory.[39] Instead we shall assume that the Feynman rules are already familiar (e.g. from the canonical approach) and demonstrate that the path integral representation reproduces them. It is therefore a succinct representation of the whole perturbation series.

We start from the elementary integrals

$$\int dx\ \exp(-\tfrac{1}{2}Kx^2 + Jx) = \sqrt{\frac{2\pi}{K}}\ \exp(\tfrac{1}{2}J^2/K)\ ,$$

$$\int dx_1 \ldots . dx_n \exp(-\tfrac{1}{2}x_i K_{ij} x_j + J_i x_i) = \frac{(2\pi)^{n/2}}{\sqrt{\det K}} \exp(\tfrac{1}{2}J_i K^{-1}_{ij} J_j)\ ,$$

where K is a real symmetric matrix. A slightly more complicated case is

$$\int \prod_{i=1}^{N} d\phi_i \exp\left(-\frac{i}{2}\phi_i\left(K_{ij} - i\epsilon\delta_{ij}\right)\phi_j - iJ_j\phi_j\right)$$

$$= \text{const.}\ \exp\left(\frac{i}{2}J_i(K - i\epsilon I)^{-1}_{ij}J_j\right)\ ,$$

where the constant is independent of J and the $i\epsilon$ provides for convergence. If we suppose that ϕ_i is a field defined on a discrete set of points in space-time, this suggests that in the case of a field $\phi(x)$ defined on a continuum we write

$$\int D\phi(x)\ \exp\left[-\frac{i}{2}\int \phi(x)A(x,y)\phi(y)d^4xd^4y - i\int J(x)\phi(x)d^4x\right]$$
$$\propto \exp\left[-\frac{i}{2}\int d^4xd^4yJ(x)\Delta(x,y)J(y)\right]\ , \qquad (2)$$

for the case

$$A = K(x,y) - i\varepsilon\delta^4(x-y) \ ,$$

with K a real symmetric operator, where

$$\int \Delta(x,z)A(z,y)d^4z = -\delta^4(x-y) \ .$$

We take equation (2) to be the definition of the "integral" over $D\phi$;[48] it is therefore incontestable. (In all the expressions below suitable $i\varepsilon$ factors should be understood, with the limit $\varepsilon \to 0$ to be taken after calculating amplitudes.)

We now show that, with this definition, if

$$Z(J) \equiv \int D\phi e^{i\int (L+J\cdot A)dx}$$

then the n point Green function is given by

$$(i)^n G(x_1,\ldots,x_n) = \frac{1}{Z(0)} \frac{\delta^n Z(J)}{\delta J(x_n)\ldots\delta J(x_n)}\bigg|_{J=0} \ .$$

(The functional derivative is defined by the obvious generalization from the discrete case; if in doubt about functional manipulations the reader is advised to return to the discrete case.) Suppose that

$$L = L_0 + V(\phi) \ ,$$

where $L_0 = \int \phi(x)\ K(x,y)\phi(y)dy$. Then

$$Z(J) = e^{i\int dxV(-i\delta/\delta J(x))}\int D\phi e^{i\int (L_0+J\phi)dx}$$

$$= e^{i\int dxV(-i\delta/\delta J(x))} e^{-\frac{i}{2}\int J(x)\Delta(x,y)J(y)dx\,dy}$$

using our definition of the functional integral. As an example we take $V = \lambda\phi^4/4!$ and check that Z(J) generates G(x,y) correctly to $O(\lambda)$. The relevant part of Z(J) is

$$1 + F(J) + \frac{i\lambda}{4!}\int dt\left(\frac{-i\delta}{\delta J(t)}\right)^4\left[\frac{1}{2}F(J)^2 + \frac{1}{3!}F(J)^3\right] + \ldots$$

where

$$F(J) = -\frac{i}{2}\int J(x)\Delta(x,y)J(y)\,dx\,dy \quad .$$

Differentiating with respect to J(x) and J(y), recognizing Δ(x,y) as the Feynman propagator and remembering to divide by Z(0), we obtain the correct expression for G(x,y) which has the diagrammatic representation

Exercise: convince yourself that Z(J) generates other Green functions correctly and that if Z(J) = exp(i X(J)) then functional derivatives of X generate connected Green functions.

We refer to the literature for a proper discussion of the functional method, the treatment of fermions etc.[39]

4.4 Path Integral Formulation of Gauge Theories[41]

Because of gauge invariance, an infinite set of fields A_μ corresponds to each physical situation, i.e. we can write

$$A_\mu = \bar{A}_\mu + \Lambda_\mu$$

where $L = L(\bar{A})$ is independent of the gauge variables Λ but changes when $\bar{A}$ changes. Faddeev and Popov pointed out[38] that consequently

$$\int DA e^{i\int L dx} = \int D\Lambda \int D\bar{A} e^{i\int L(\bar{A})dx}$$

diverges; this is the origin of the failure of canonical quantization for gauge theories from the path integral point of view. Their suggested remedy is to fix the gauge and drop the factor $\int D\Lambda$, i.e. to integrate over each physical configuration — or history — once. 1) For example, if we choose axial gauge, n•A = 0, we should use

$$Z(J) = \int DA\,\delta(n\cdot A)e^{i\int (L+J\cdot A)dx}$$

according to Faddeev and Popov. We derive the Feynman rules corresponding to this representation of Z(J) below. They agree with the rules obtained by quantizing canonically in this gauge, which shows that the F-P ansatz is correct.[42] 2) If we impose the gauge condition F(A) = f we should solve for the gauge variables $\Lambda = \Lambda(\bar{A},f)$ and introduce $\delta(\Lambda-\Lambda(\bar{A},f))$ in the functional integral. We then write

$$\delta(\Lambda - \Lambda(\bar{A},f)) = \delta(F(A)-f) \det \frac{\partial F}{\partial \Lambda}$$

so that

$$Z(J,f) = \int DA\delta(F(A) - f) \det \frac{\partial F}{\partial \Lambda} e^{i\int (L+J\cdot A)dx} .$$

Although the Green functions obtained from this representation will depend on f, the S-matrix will not because it is gauge invariant. We can therefore multiply Z by a functional $\psi(f)$ and integrate over f without changing the S-matrix, provided ψ is chosen so that the integral converges. With the standard choice $\psi = \exp(-\frac{i}{2}\int (f(x))^2 dx)$ we obtain

$$Z(J) = \int Df\ \psi(f) Z(J,f) = \int DA \det \frac{\partial F}{\partial \Lambda} e^{i\int (L - \frac{1}{2}F^2(A))dx} .$$

We now discuss the derivation of the Feynman rules and some of their properties in the two cases discussed here.

<u>Axial Gauge</u>

If we set

$$Z(J) = \int DA\delta(n\cdot A)e^{i\int (L+J\cdot A)dx}$$

$$= \int DAD\phi e^{i\int (L+n\cdot A\phi+J\cdot A)dx}$$

we can derive the Feynman rules from the effective Lagrangian

$$L_{eff} = L + n\cdot A\phi$$

in the standard way. The equations of motion are

$$n\cdot A = 0$$

$$\Box A_\mu - \partial_\mu(\partial\cdot A) - n_\mu\phi = J_\mu$$

where the current J_μ includes vector meson and also fermion contributions if the gauge field is coupled to fermions. Contracting the second equation with n_μ gives

$$-n^2\phi - n\cdot\partial\partial\cdot A = n\cdot J ,$$

which can be used to eliminate ϕ. It is then a simple exercise to

Fourier transform and obtain the propagator

$$\Delta_{\mu\nu} = \left[-g_{\mu\nu} + \frac{n^2 k_\mu k_\nu}{(n\cdot k)^2} - \frac{n_\mu k_\nu + n_\nu k_\mu}{n\cdot k}\right](k^2)^{-1}$$

which satisfies

$$\tilde{A}_\mu(k) = \Delta_{\mu\nu}\tilde{J}^\nu(k) \ .$$

Note the following.

1) $n^\mu \Delta_{\mu\nu} = 0$,

$$\lim_{k^2\to 0} k^2 k^\mu \Delta_{\mu\nu} = 0 \ ,$$

$$\lim_{k^2\to 0} k^2 g^{\mu\nu}\Delta_{\mu\nu} = 2 \ ,$$

i.e. only two components of $\Delta_{\mu\nu}$ have poles, corresponding to two physical degrees of freedom which propagate finite distances. We therefore expect the Feynman rules to lead to a unitary S-matrix. 2) The correct prescription for treating the poles at $n\cdot k = 0$ is to take the principal value — this makes their contribution real and ensures that they do not contribute to the unitarity equation or give rise to poles in S-matrix elements.[43] 3) Because only physical degrees of freedom propagate (in contrast to covariant gauges — where unphysical particles propagate but make cancelling contributions to S-matrix elements), the diagrammatic structure of some amplitudes is especially simple in axial gauge (see Section 7); however, explicit calculation is hard due to the singularities at $n\cdot k = 0$. (Light cone gauge, $n^2 = 0$, leads to some simplifications.[44])

Covariant Gauges

If we take the gauge-fixing function F to be $\alpha^{-1/2}\partial\cdot A$, the F-P determinant, obtained by subjecting F to an infinitessimal gauge transformation, is

$$\det \frac{\partial F^a}{\partial \Lambda_b} = \det(\Box\delta_{ab} + g\, f_{acb}\partial^\mu[A^c_\mu])$$

$$= \det(1 + g\,\partial^\mu(\vec{A}_\mu \times)\Box^{-1})\det\Box \ .$$

The factor det$\Box$ is an irrelevant (field independent) constant. We have already encountered the factor $\det(D \cdot \partial \Box^{-1})$ and seen that its effects are given by adding fictitious spin 0 fermion fields ($\vec{\eta}$ and $\vec{\omega}$) to the effective Lagrangian. We therefore find the same result as before

$$L_{eff} = L_{QCD} - \frac{1}{2\alpha}(\partial \cdot A)^2 - \partial^\mu \eta^a D_\mu^{ab} \omega^b ,$$

from which the Feynman rules are easily read off. (Note that

$$\int D\theta D\phi e^{i\int \theta M \phi dx} \propto (\det M)^{-1}$$

if θ and ϕ commute; we have just shown that

$$\int D\eta D\omega e^{i\int \eta M \omega dx} \propto \det M$$

for anticommuting variables.)

It is not at all obvious that the S-matrix is unitarity in covariant gauges or that it is gauge invariant. To demonstrate these properties requires Ward identities, which we shall now derive using the transformation introduced by Becchi, Rouet and Stora (BRS).[45]

4.5 BRS Transformations in QED

In QED ghosts are not needed in standard covariant gauges. (Exercise — show this using the F-P ansatz; see ref. 46 for an instructive example of a gauge in which ghosts are needed in QED.) However, no harm is done by introducing an uncoupled ghost (which we treat as a boson) and using

$$L_{eff} = L_{QED} - \frac{1}{2\alpha} (\partial \cdot A)^2 + \eta \Box \omega .$$

This Lagrangian is invariant under the BRS transformation

$$\delta A_\mu = \varepsilon \partial_\mu \omega$$

$$\delta \psi = ig\varepsilon \delta \psi$$

$$\delta \eta = \frac{\varepsilon}{\alpha} \partial \cdot A$$

$$\delta \omega = 0 ,$$

the role of the ghost Lagrangian being to cancel the variation of the gauge fixing term under a gauge transformation with parameter

ω. The invariance of Green functions under this exact symmetry of the theory leads directly to Ward identities, the three most important cases being:

1) $\varepsilon^{-1}\delta\langle 0|T(\eta A_\nu)|0\rangle = 0$

$$= \alpha^{-1}\langle 0|T(\partial\cdot AA_\nu)|0\rangle + \langle 0|T(\eta\partial_\nu\omega)|0\rangle \quad .$$

Since η and ω are free fields the second term can be calculated exactly, giving

$$q^\mu D_{\mu\nu} = -\frac{\alpha q_\nu}{q^2}$$

for the photon propagator. (The derivative in $\partial\cdot A$ can be taken outside the T operator since $[A_o(\vec{X},0),A_\nu(0)] = 0$.) This equation implies that the longitudinal part of the photon propagator is trivially renormalized. (It is equivalent to the statement that the vacuum polarization tensor $\pi_{\mu\nu}$ is transverse.)

2) $\varepsilon^{-1}\delta\langle 0|T(\eta\bar{\psi}\psi)|0\rangle = 0$

$$= \alpha^{-1}\langle 0|T(\partial\cdot A\bar{\psi}\psi)|0\rangle - ig\langle 0|T(\eta[\omega\bar{\psi}]\psi)|0\rangle + ig\langle 0|T(\eta\bar{\psi}[\omega\psi]|0\rangle,$$

or, in terms of diagrams

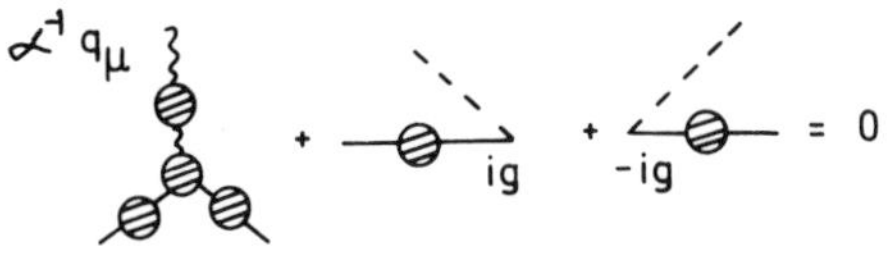

where the dashed line is the ghost propagator. As $p^2 \to m^2$ (where m is the physical mass), the fermion propagator behaves (by definition) as $Z_2/(\not{p}-m)$ while for $p^2,(p')^2 \to m^2$, $q^2 \to 0$ the one particle irreducible vertex function behaves as $Z_1^{-1}i\gamma_\mu g$. Using these definitions and the previous identity for $q^\mu D_{\mu\nu}$ to evaluate the Ward identity in the on-shell limit we obtain

$$\frac{-\not{q}(Z_2)^2}{Z_1}\frac{1}{\not{p}-m}\frac{1}{\not{p}'-m} - \frac{Z_2}{\not{p}-m} + \frac{Z_2}{\not{p}'-m} = 0$$

$$\to Z_1 = Z_2$$

3) $\delta\langle 0|T(\eta\partial\cdot A..\partial\cdot A\bar{\psi}..\bar{\psi}\psi..\psi)|0\rangle = 0$

Noting that $\delta(\partial\cdot A) = \varepsilon\Box\omega = 0$ by the equation of motion, we can represent this relation by

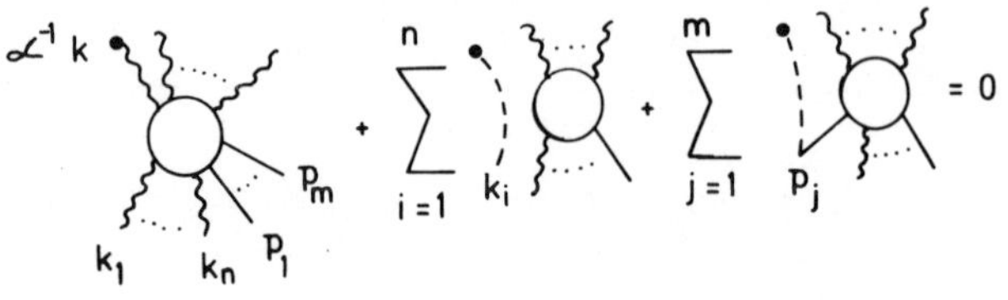

The middle term vanishes (unless $p_i = -k$; even then it is zero if the ith gluon is "physically" polarized with $\varepsilon^i\cdot k^i = 0$). The last term has no pole as $p_j^2 \to m^2$ since the fermion carries momentum (p_j-k). Therefore if we multiply by $\prod_i k_i^2 \prod_j (p_j^2-m^2)$ and let $k_i^2 \to 0$, $p_j^2 \to m^2$ only the first term survives — but it is zero according to the Ward identity, i.e. $\partial\cdot A$ decouples from on-shell particles (as asserted above).

4.6 BRS Transformations for QCD

In standard covariant gauges L_{eff} for QCD is invariant under the BRS transformation

$$\delta\vec{A}_\mu = \varepsilon\partial_\mu\vec{\omega}$$

$$\delta\psi = ig\varepsilon\frac{\vec{\Lambda}\cdot\vec{\omega}}{2}\psi$$

$$\delta\vec{\eta} = \varepsilon\partial\cdot\vec{A}$$

$$\delta\vec{\omega} = -\frac{\varepsilon}{2}g\,\vec{\omega}\times\vec{\omega},$$

where the infinitessimal parameter ε must now be an anticommuting object in order that δA_μ and $\delta\psi$ commute but $\delta\eta$ and $\delta\omega$ anticommute i.e. $\varepsilon^2 = 0$, $\varepsilon\vec{\omega} = -\vec{\omega}\varepsilon$, $\varepsilon\vec{\eta} = -\vec{\eta}\varepsilon$ but $\varepsilon A_\mu = A_\mu\varepsilon$ etc. (Exercise: show that L_{eff} is indeed invariant under this transformation.) Note that if $\vec{\omega}$ and $\vec{\eta}$ were bosons, $\vec{\omega}\times\vec{\omega}$ would be zero and this transformation would not be a symmetry of L_{eff}.

The Ward identities are a direct expression of the invariance of Green functions under BRS transformations. We shall now use them to prove a) unitarity in standard covariant gauges and b) that the S-matrix is independent of the gauge-fixing function F(A). In addition the Ward identities for one particle irreducible vertex functions can be used to give an extremely compact proof of renormalizability using algebraic methods, for which we refer to the literature.[47]

The Ward identity we need is

$$\varepsilon^{-1}\delta\langle 0|T(\eta A_{\mu_1}A_{\mu_2}\ldots.)|0\rangle = 0$$

$$= \alpha^{-1}\langle 0|T(\partial\cdot AA_{\mu_1}A_{\mu_2}\ldots.)|0\rangle - \sum_i \langle 0|T(\eta A_{\mu_1}\ldots D_{\mu_i}\omega\ldots)|0\rangle ,$$

where all quantities are unrenormalized and here and below we omit fermions for simplicity, leaving it as an exercise to show that they do not change the conclusions. Attaching sources $\hat{J}_\mu = iJ\alpha^{-1}k_\mu$ and $J^i_{\mu_i}$ to the external particles, we have diagrammatically[48]

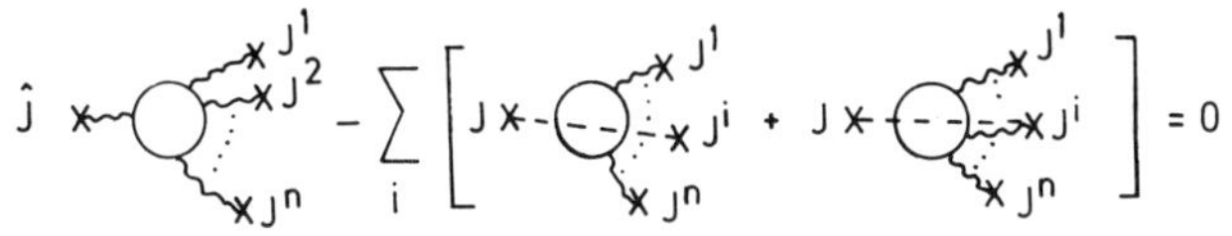

where J^i couples to the two parts of $\delta\vec{A}_\mu = D_\mu\vec{\omega} = \partial_\mu\vec{\omega} + g\,\vec{A}_\mu\times\omega$ in the second and third terms. Note that a) in passing through the blob the dotted ghost line interacts in all possible ways, b) the propagators are fully dressed. We let the sources J^i create on-shell particles, i.e. take $J^i \propto k_i^2$ and let $k_i^2 \to 0$. Only the pole contributions

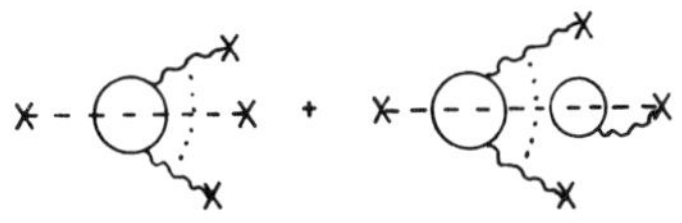

survive in the second and third terms. If, however, we take "physical" sources, defined so that

$$J^i_\mu\langle\text{ghost}|D_\mu\omega|0\rangle = 0 ,$$

these pole contributions are zero. (In covariant gauges, this condition reads $k^\mu\tilde{J}^i_\mu(k) = 0$ for the Fourier transform $\tilde{J}$ of the sources and ensures that the gluons they produce satisfy $\varepsilon\cdot k = 0$.)

It follows that $\partial \cdot A$ decouples from physically polarized on-shell particles.

We now proceed to prove <u>unitarity</u> for two particle discontinuities.[48] (Generalization to other contributions is straightforward.) We choose renormalization conditions such that the gluon propagator is $-Z_3 g_{\mu\nu}(k^2)^{-1}$ for $k^2 \to 0$. The Ward identity $q^\mu D_{\mu\nu} = -\alpha q_\nu$, which is valid in QCD as well as QED, therefore requires that $\alpha = Z_3$*. Symbolically, then, two-gluon discontinuities are given by

$$\text{Disc} \quad \underset{}{\overset{}{\bigcirc}} \propto \left[\underset{k}{\overset{k'}{\bigcirc}} \begin{array}{c} -g_{\alpha\beta} \\ -g_{\mu\nu} \end{array} \bigcirc \right] (Z_3)^{-2}$$

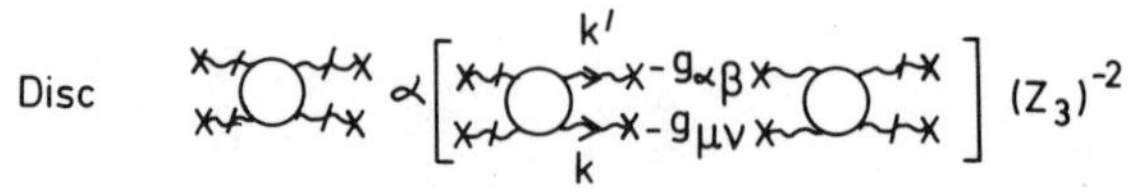

where each line attached to a source (×) is on-shell. (I.e. each of the two blobs is multiplied by k^2, $(k')^2$ etc. and the limit $k^2 \to 0$ taken.) The slash on a propagator indicates physical polarization. (The factor $(Z_3)^{-2}$ arises because there are two propagators carrying k[k'] on the right but only one on the left.) With k along the 3 axis ($k = |k|(1,0,0,1)$) we define

$$\delta^T_{\mu\nu} \equiv \begin{pmatrix} 0 & & & \\ & 1 & & \\ & & 1 & \\ & & & 0 \end{pmatrix}$$

$$\bar{k} \equiv |k|(-1,0,0,1)$$

and write

$$-g_{\mu\nu} = \delta^T_{\mu\nu} - \frac{1}{2|k|^2}(k_\mu \bar{k}_\nu + \bar{k}_\mu k_\nu) .$$

To establish unitarity we must show that the contribution of the $k\bar{k}$ terms cancels the two-ghost discontinuity, leaving just a δ^T term. Using the fact that $\partial \cdot A$ decouples from amplitudes involving on-shell particles with physical polarizations we can write

---

*Note: 1) With this choice the renormalized propagator is $-g_{\mu\nu}(k^2)^{-1}$ for $k^2 \to 0$. 2) The Ward identity for $q^\mu D_{\mu\nu}$ can be derived from

$$\varepsilon^{-1}\delta <0|T(\eta \partial \cdot A)|0> = \alpha^{-1} <0|T(\partial \cdot A \partial \cdot A)|0> = 0 .$$

Taking the derivatives outside the T operator, the divergence of the propagator is related to a canonical equal time commutator.

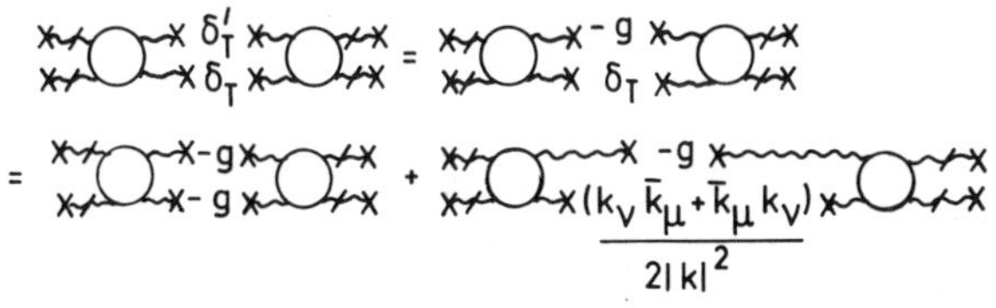

Then, using the Ward identity,

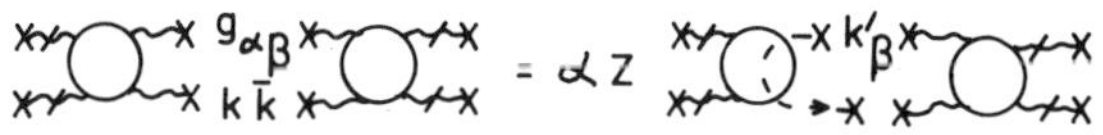

where Z is defined by

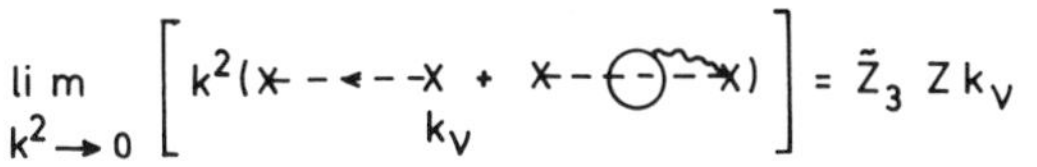

(By definition the ghost propagator behaves as $\tilde{Z}_3(k^2)^{-1}$ as $k^2 \to 0$.) However, the Ward identity $\delta\langle 0|T(\eta A_\nu)|0\rangle = 0$ implies that $\tilde{Z}_3 Z = 1$. (Exercise: verify this.) Hence we find that the $k\bar{k} + k\bar{k}$ contribution to the two-gluon discontinuity is

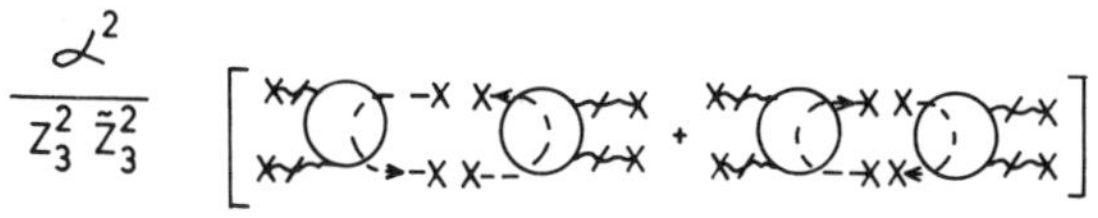

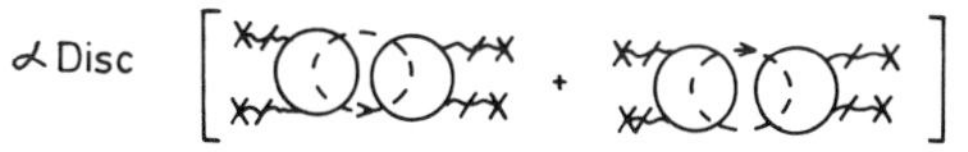

using $k\cdot\bar{k} = -2$ and $\alpha = Z_3$, i.e. the unphysical contributions obtained using $g_{\mu\nu}$ are exactly cancelled by the ghosts, leaving the contribution due to $\delta^T_{\mu\nu}$.

Next we consider <u>gauge invariance</u> for a general gauge-fixing function F which gives rise to

$$L_{eff} = L_{QCD} - \frac{1}{2} F^2 + \eta \frac{\delta F}{\delta \Lambda} \omega ,$$

suppressing indices.[50] The effective action is invariant under a BRS transformation with

$$\delta A_\mu = \varepsilon D_\mu \omega$$

$$\delta \eta = \varepsilon F ,$$

and $\delta\omega$ satisfying

$$\delta\left(\frac{\delta F}{\delta \Lambda}\right)\omega + \frac{\delta F}{\delta \Lambda}\, \delta\omega = 0 .$$

Suppose we change $F \to F + f$ (for f infinitesimal); we have

$$L_{eff} \to L_{eff} + \Delta L$$

with

$$\Delta L = -Ff + \eta \frac{\delta f}{\delta \Lambda} \omega ,$$

dropping terms of order f^2. Note that

$$- \varepsilon^{-1} \delta(\eta f) = \Delta L .$$

Hence

$$\begin{aligned} - \varepsilon^{-1} \delta \langle 0|T([\eta f]A^1 \ldots A^n)|0\rangle &= 0 \\ &= \langle 0|T(\Delta L A^1 \ldots A^n)|0\rangle + \sum_i \langle 0|T([\eta f]A^1 \ldots D_{\mu_i}\omega \ldots A^n)|0\rangle , \end{aligned}$$

or, graphically:

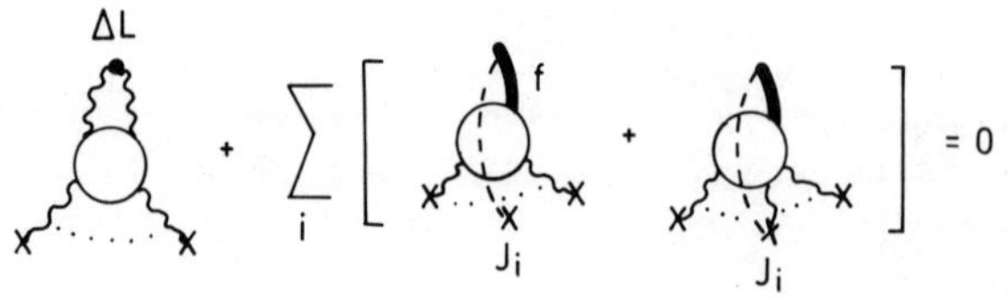

Taking physical sources and letting the external legs go on-shell, the first term gives ΔG_n- the change of the physical on-shell n-point function G_n due to ΔL. Recalling that ghost poles decouple from physical sources, the only other terms which have poles in the external momenta and survive on-shell are

The part of this diagram which connects the source J_i to the gluon propagator gives a constant, which we call X, as $k^2 \to 0$, i.e. each of these diagrams give XG_n, and the identity reads

$$G_n + nXG_n = 0 .$$

Applying this to the two-point function we have

$$\Delta Z_3 + 2XZ_3 = 0 .$$

Since the n-point S-matrix element is given by

$$S_n = \frac{1}{(Z_3)^{n/2}} G_n ,$$

it follows that $\Delta S_n = 0$, i.e. that the S-matrix is gauge invariant, although Green functions and wave function renormalization are of course gauge dependent.

5. RENORMALIZATION AND THE RENORMALIZATION GROUP EQUATIONS (RGE)

The literature contains excellent expositions of renormalization.[51] Nevertheless I include (without proof) a brief explanation of the procedure in order to allow a discussion of the important question of prescription dependence.

5.1 Renormalization Prescriptions

As an example we consider the Lagrangian

$$L = \frac{1}{2}(\partial_\mu \phi)^2 - \frac{1}{2}m^2\phi^2 - \frac{\lambda\phi^4}{4!} .$$

If quantum field theory did not give rise to infinities, we could calculate all Green functions in terms of the two parameters m and λ. (We could then eliminate m and λ to obtain relations between measurable quantities; note that it is only in lowest order that m and λ are equal to the mass and the value of the four-point function and are themselves directly observable.) In fact, of course, m and λ diverge and cannot be used to parameterize the theory. In order to calculate we must choose a "renormalization prescription", i.e. decide how to parameterize the theory. Of course relations between observables will not depend on this choice — this gives rise to the RGE.

In QED it is natural to relate all other observables to m_e and the fine structure constant α. (If "on-shell" renormalization is used the theory is parameterized directly in terms of these quantities.) In QCD, the quark mass is (presumably) not measurable and furthermore, there is no analogue of the Thomson limit and no natural definition of α_s in terms of observables. Therefore the definitions of the parameters α_s and m_q are necessarily prescription dependent and so is the way that observables depend on them. I stress this point because it is of vital importance in understanding calculations beyond leading order. (One consequence is that the convergence of the perturbation series for a particular process is scheme dependent.[52])

Returning to ϕ^4 theory, if we start with a Lagrangian L of the form above we soon encounter divergences e.g. in

The remedy is (I hope) well known to the reader. We mutilate/cut off the theory in some way designed to make the divergent integral finite (e.g. by keeping $k^2 < \Lambda^2$ in the Wick rotated integral, going to $n-\varepsilon$ dimensions — as discussed below — or modifying the propagators). We then use

$$L = L_o + L_c$$

where $L_c = \lambda f(\Lambda)\phi^2$ generates the diagram

and f is chosen so that it cancels the divergent part of the previous diagram when the cut-off is removed. We then continue to calculate using L, modifying $f(\Lambda)$ and adding other terms to cancel new divergences as they arise. In a renormalizable theory L_c and $\hat{L}$ have the same form, i.e. we obtain

$$L_c = A(\partial_\mu \phi)^2 + B\, m^2 \phi^2 + C\lambda\phi^4 \ ,$$

where A, B and C diverge when the cut-off is removed. The divergent parts of A, B and C are well defined, in a given cut-off scheme, but the finite parts are completely arbitrary; they must be specified by a renormalization prescription. For a given prescription, the finite quantities m and λ parameterize the theory.

Note that we can write

$$L = \frac{Z_3}{2} (\partial_\mu \phi)^2 - \frac{m^2 Z_2}{2} \phi^2 - \frac{\lambda Z_1 \phi^4}{4!}$$

$$= \frac{1}{2} (\partial_\mu \phi_B)^2 - \frac{m_B^2 \phi_B^2}{2} - \frac{\lambda_B \phi_B^4}{4!} \ ,$$

where the Z's and the bare quantities (ϕ_B, M_B and λ_B) are divergent but Green functions $\langle 0|T(\phi..\phi)|0\rangle$ calculated using renormalized fields are finite functions of the finite parameters m and λ.

It is evident that changing the renormalization prescription, corresponding to changing the Z's, amounts to a rearrangement of the perturbation series and cannot affect relations between measurable quantities. (We could calculate S-matrix elements as functions $S(g_B, m_B, \Lambda)$, eliminate g_B and m_B in favour of two measurables and remove Λ — obtaining relations without making reference to a prescription.)

We now consider two particular prescriptions.

5.1.i Momentum Subtraction

A set of conditions which specifies the counter terms uniquely in $\lambda\phi^4$ theory is

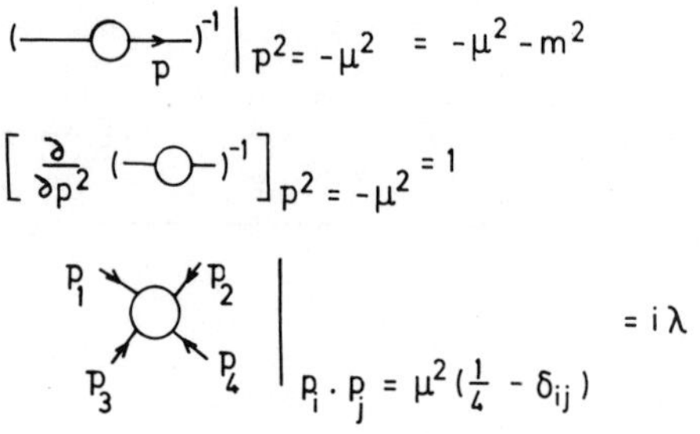

$$\left(-\!\!\bigcirc\!\!\!\rightarrow_p\!\!-\right)^{-1}\Big|_{p^2=-\mu^2} = -\mu^2 - m^2$$

$$\left[\frac{\partial}{\partial p^2}(-\!\!\bigcirc\!\!-)^{-1}\right]_{p^2=-\mu^2} = 1$$

$$\left.\begin{array}{c}\vphantom{X}\\ \includegraphics{} \end{array}\right|_{p_i\cdot p_j = \mu^2(\frac{1}{4}-\delta_{ij})} = i\lambda$$

Relations between physical quantities do not depend on μ, which is completely arbitrary (although the values of λ and g needed to obtain given values of physical quantities will of course depend on μ). Hence S-matrix elements satisfy

$$\mu \frac{dS}{d\mu}\Big| = \left(\mu\frac{\partial}{\partial\mu} + \beta\frac{\partial}{\partial\lambda} + \gamma_m\, m\frac{\partial}{\partial m}\right) S = 0 \ ,$$

where $\beta = \mu\frac{\partial\lambda}{\partial\mu}\Big|$, $\gamma_m = \mu\frac{\partial \ell n m}{\partial\mu}\Big|$ and $\frac{\partial}{\partial x}\Big|$ means that the derivative is carried out for fixed measurable quantities or, equivalently, fixed bare parameters and cut off. ($S = S(m_B, \lambda_B, \Lambda)$ is evidently independent of μ.) Since

$$\lambda = \frac{\lambda_B Z_3{}^2}{Z_1}, \quad m = m_B\sqrt{\frac{Z_3}{Z_2}} \ ,$$

we find

$$\beta = \mu\lambda \frac{\partial \ell n (Z_3{}^2 Z_1{}^{-1})}{\partial\mu}\Bigg| \ ,$$

$$\gamma_m = \frac{\mu}{2} \frac{\partial \ell n (Z_3 Z_2^{-1})}{\partial\mu}\Bigg| \ .$$

The n-point Green function G_n is related to the S-matrix element S_n by

$$S_n = (Z)^{-n/2} G_n \ ,$$

where

$$G_2 \xrightarrow[p^2 \to M^2]{} Z(p^2 - M^2)^{-1} ,$$

where M is the physical mass of the ϕ particle. Hence

$$\left(\mu \frac{\partial}{\partial\mu} + \beta \frac{\partial}{\partial\lambda} + (m\gamma_m) \frac{\partial}{\partial m} + n\gamma\right) G_n = 0 ,$$

where $\gamma = (-1/2)(\partial \ell n Z/\partial\lambda)$. (Note that if G_n is defined to have external legs removed, $n \to -n$ in this equation, which is sometimes a source of confusion.) Alternatively let $G_n(G_n^B)$ be the vacuum expectation value of the T product of n fields $\phi(\phi_B)$. Using

$$G_n^B = (Z_3)^{n/2} G_n$$

and the fact that $G_n^B(m_B, \lambda_B, \Lambda)$ is independent of μ, we recover the RGE above for G_n with $\gamma = \frac{1}{2} \frac{\partial \ell n Z_3}{\partial \mu}\Big|$. (The previous expression for γ is recovered by applying the RGE to G_2.)

5.1.ii Minimal Subtraction

Momentum subtraction works for any cut-off procedure whereas minimal subtraction is only defined for dimensional regularization,[53] which we now review very briefly. The basic idea is to replace $\int d^4k f(k)$ by $\int d^n k f(k)$. Writing

$$d^n k = k^{n-1} dk \, d\Omega_n$$

the angular integral can be done for any integral n and then continued while the radial integral is defined for all n. Integrals which diverge as $k \to \infty$ for $n = 4$ can obviously be made convergent by going to sufficiently small n. (The same method is very convenient for treating infrared divergences by going to $n > 4$.[54]) The divergences show up as poles as $n \to 4$; counter terms are added to cancel these poles in the usual way.

As an example we consider the one-loop correction to the propagator. First, it is convenient to let $\lambda \to \mu^\varepsilon \lambda$ for $n = 4 + \varepsilon$, where μ is an arbitrary parameter, in order that λ remains dimensionless. We then obtain

$$i\lambda\mu^{\varepsilon}\int\frac{d^nk}{(2\pi)^n}\frac{i}{k^2-m^2}=\frac{i\lambda\pi^{n/2}}{(2\pi)^{n/2}}\frac{\Gamma(1-n/2)}{\Gamma(1)}\frac{\mu^{\varepsilon}}{(m^2)^{1-n/2}}$$

$$=i\lambda m^2\left(\frac{1}{8\pi\varepsilon}+a+b\,\ell n\,m/\mu+0(\varepsilon)\right)\ .$$

To remove the divergence as $\varepsilon \to 0$, we must add a counter term $m^2\lambda\phi^2/8\pi\varepsilon$. In minimal subtraction this is the entire counter term. (In contrast, we would add $m^2\lambda\phi^2(1/8\pi\varepsilon + a + b\,\ell n\,m/\mu)$) in momentum subtraction.*)

In fact the minimal prescription is that L_c should only contain terms which diverge as $n \to 4$ if it is written as a Laurent series in n-4. (The μ^{ε} in $\lambda\mu^{\varepsilon}\phi^4$ should not be expanded when this is done — otherwise the dimensions would come out wrong.) Thus the finite parameters λ and m and the field ϕ are related to the corresponding bare quantities by

$$\lambda_B=\mu^{\varepsilon}\left[\lambda+\sum_{r=1}^{\infty}\frac{a_r(\lambda)}{\varepsilon^r}\right]\ ,$$

$$m_B=m\left[1+\sum\frac{b_r(\lambda)}{\varepsilon^r}\right]\ ,$$

$$\phi_B=\phi\left[1+\sum\frac{c_r(\lambda)}{\varepsilon^r}\right]\ .$$

Note that: 1) the coefficients a_r, b_r and c_r are unique; 2) $G_n^B(\lambda_B,m_B,\varepsilon)$ is manifestly independent of μ and so our previous derivation of the RGE still works. In terms of the functions in L_c,

$$\beta=-a_1+\lambda\frac{\partial a_1}{\partial\lambda}-\varepsilon\lambda$$

$$\gamma=\lambda\frac{\partial c_1}{\partial\lambda}$$

---

*The parameter μ introduced to keep λ dimensionless (not to be confused with the subtraction point, also called μ above) is irrelevant in this scheme since renormalized Green functions do not depend on it in the limit $\varepsilon \to 0$.

$$\gamma_m = \lambda \frac{\partial b_1}{\partial \lambda} \quad .$$

Exercise: verify that $\mu \to \mu + \delta\mu$ with $\mu\delta\lambda = \beta\delta\mu$, $\mu\delta m = m\gamma_m\delta\mu$, $\mu\delta\phi = \gamma\phi\delta\mu$ leaves the bare quantities invariant and is indeed an exact symmetry of the theory. Note that invariance need only be checked to order ε^0 since the divergent terms are unique. Their necessary invariance implies relations between the a_r, b_r and c_r which allow the terms with $r > 1$ to be calculated in terms of $r = 1$ terms. (Exercises a) check this statement. b) Convince yourself that in a given order of perturbation theory the only divergences which cannot be expressed in terms of integrals already evaluated in lower order are of order ε^{-1}, i.e. the $r = 1$ coefficients determine the $r > 1$ coefficients; this is the fundamental reason why leading, next to leading etc. logs can be calculated to all orders in terms of low order calculations using the RGE.)
3) The coefficients a_r, b_r and c_r are independent of m and μ.[53] (This follows from the — I hope plausible — facts that a) they could only depend on $\ell n\, m/\mu$ and b) they are finite as $m \to 0$.) Consequently in minimal subtraction β, γ and γ_m only depend on λ. This is not true in momentum subtraction. Conceptually this makes the RGE much simpler but it is probably no advantage in practice since perturbation theory is probably more slowly convergent in minimal than momentum subtraction in most cases of interest.[55]

5.2 Gauge Theories

After the introduction of counter terms the QCD Lagrangian will read (dropping quark masses)

$$L_{QCD} = -\frac{Z_3}{4}\vec{F}^2_{\mu\nu} - \frac{Z_1 g}{2}\vec{F}^{\mu\nu}\cdot(\vec{A}_\mu\times\vec{A}_\nu)$$

$$-\frac{1}{4}Z_4 g^2(\vec{A}_\mu\times\vec{A}_\nu)^2 - \frac{1}{2\alpha}(\partial\cdot\vec{A})^2 - \tilde{Z}_3\partial^\mu\vec{\eta}\cdot\partial_\mu\vec{\omega}$$

$$-\tilde{Z}_1 g\partial^\mu\vec{\eta}\cdot(\vec{A}_\mu\times\vec{\omega}) + iZ_3^F\bar{\psi}\not{\partial}\psi + gZ_1^F\,\overline{\psi\vec{T}}\cdot\vec{\not{A}}\psi$$

where

$$\vec{F}_{\mu\nu} = \partial_\mu\vec{A}_\nu - \partial_\nu\vec{A}_\mu \quad .$$

BRS invariance of L (on which we must insist since unitarity depends on it) requires

$$\frac{Z_1^{\ F}}{Z_3^{\ F}} = \frac{Z_1}{Z_3} = \frac{\tilde{Z}_1}{Z_3} , \frac{Z_4}{Z_3} = \left(\frac{Z_1}{Z_3}\right)^2 ,$$

which we shall refer to as Slavnov-Taylor (S-T) identities.[56] This is most easily seen by equating L to the expression given previously in terms of bare quantities (we should now understand $A \rightarrow A^B$, $\eta \rightarrow \eta^B$, $g \rightarrow g^B$, $\alpha \rightarrow \alpha^B$ etc. in this expression), which also gives

$$g = \frac{Z_3^{\ 3/2}}{Z_1} \ g_B, \ \alpha_B = Z_3\alpha .$$

(The Ward identity $q^\mu D^B_{\mu\nu} = -\alpha_B q_\nu$ then ensures that α is finite.)

The S-T identities are usually derived in a more complicated way whose purpose we now explain. We require unitarity and hence BRS invariance order-by-order. Assuming that L is invariant up to a given order, we must be sure that it remains invariant when we add the counter terms generated in the next order. BRS invariance of L implies that Green functions satisfy Ward identities (given a cut-off procedure which respects the invariance). We must show that these Ward identities imply the BRS invariance of the counter terms needed to cancel the divergences. This very plausible result is implied by the general results of Kluberg-Stern and Zuber,[57] and Joglekar and Lee.[58] It was first shown by deriving S-T identities explicitly.[56]*

In minimal subtraction the S-T identities are automatically satisfied. In momentum subtraction the value at the subtraction point of a subset of superficially divergent amplitudes can be used to parameterize the theory; the value of the others is then determined by the S-T identities. Relations between observables do not depend on which subset is chosen, of course, but the form of the perturbation series does depend on the prescription.[52]

We proved that S is independent of the gauge fixing term** (i.e. $S = S(g_B, m_B, \Lambda)$, independent of α_B), but it is important to

---

*The relation for $Z_1^F (Z_3^F)^{-1}$ has only been derived explicitly in Landau gauge[56] to my knowledge, which is a pity in view of the formal nature of the general results.

**S is actually infra-red divergent but we can (and should really) replace S everywhere by infra-red finite measurable quantities.

realize that $S(g,m,\alpha,\mu)$ will depend on α in general (although clearly relations between observables will be α- as well as μ-independent). In minimal subtraction, however, S is α independent.[59] Consider

$$\left.\frac{dS}{d\alpha}\right| = \left(\frac{\partial}{\partial\alpha} + \left.\frac{\partial g}{\partial\alpha}\right| \frac{\partial}{\partial g} + \left.\frac{\partial m}{\partial\alpha}\right| \frac{\partial}{\partial m}\right) S = 0 \ .$$

It follows from this equation (evaluated for two different elements of S) that $\left.\frac{\partial g}{\partial\alpha}\right|$ and $\left.\frac{\partial m}{\partial\alpha}\right|$ are finite as $\varepsilon \to 0$, i.e. they have the form $\sum_{n=0}^{\infty} C_n \varepsilon^n$. Using this information in the equation

$$\left.\frac{\partial g}{\partial\alpha}\right| + \sum_{r=1}^{\infty} \frac{1}{\varepsilon^r} \left[\frac{\partial a_r}{\partial\alpha} + \frac{\partial a_r}{\partial g} \left.\frac{\partial g}{\partial\alpha}\right|\right] = 0$$

(obtained by differentiating the equation which relates g_B to g), which must be satisfied order-by-order in g and ε^{-1}, we find that $\left.\frac{\partial g}{\partial\alpha}\right|$ is zero and the a_r are independent of α. Similarly it is easy to show that $\left.\frac{\partial m}{\partial\alpha}\right| = 0$ and $b_r = b_r(g)$. Hence

$$\frac{\partial S}{\partial\alpha} = 0 \ ,$$

and β and γ_m are independent of α in minimal subtraction. None of these statements is true in general. (An example is given in ref. 57; γ's for fields are of course gauge dependent in <u>all</u> schemes.)

The RGE for gauge theories contain the extra term

$$\mu\left.\frac{\partial\alpha}{\partial\mu}\right| \frac{\partial}{\partial\alpha} = -\frac{\alpha\mu}{Z_3} \left.\frac{\partial Z_3}{\partial\mu}\right| \frac{\partial}{\partial\alpha} \ ,$$

because we must change α when we change μ if we wish L to be invariant. This term is clearly absent in Landau gauge ($\alpha = 0$), in which the tensor structure of the propagator cannot change with k^2 (why?).[60] For this reason Landau gauge is frequently used in discussions of deep inelastic scattering.

5.2.i Properties of β and γ

We begin by considering "mass independent schemes", in which the Z's do not depend on masses. (One example is minimal subtraction — for others see ref. 61.), or other schemes with all masses neglected. Let the coupling constant be g with one prescription and g' with another prescription or gauge. If $\partial g'/\partial g \neq 0$ (true in perturbation theory for small g) then[62]: 1) Since

$$\beta'(g') = \mu \frac{\partial g'}{\partial \mu}\bigg| = \frac{\partial g'}{\partial g}\beta(g) ,$$

the existence of a zero of β implies a zero of β' and vice-versa (although $g \neq g'$ at the zero). 2) Since

$$\frac{\partial \beta'}{\partial g'} = \frac{\partial^2 g'}{\partial g' \partial g}\beta(g) + \frac{\partial g'}{\partial g}\frac{\partial g}{\partial g'}\frac{\partial \beta}{\partial g} ,$$

it follows that

$$\frac{\partial \beta'}{\partial g'} = \frac{\partial \beta}{\partial g}$$

at a zero. 3) Assuming $Z'(g') = F(g)Z(g)$ with $F \neq 0$, it is easy to show that $\gamma'(g') = \gamma(g)$ at a zero of β. 4) If

$$\beta(g) = b_o g^3 + b_1 g^5 + 0(g^7)$$

and

$$g' = g + Ag^3 + \ldots$$

$$g = g' - Ag'^3 + \ldots$$

then

$$\beta'(g') = \frac{\partial g'}{\partial g}\beta(g)$$

$$= b_o g'^3 + b_1 g'^5 + 0(g'^7) .$$

These properties must be true since (as should become clear below) the existence of zeros, $\gamma|_{\beta=0}$ and b_o and b_1 are directly related to observables. It is straightforward to generalize properties 1 and 3 to the case of many coupling constants (for points $\beta_i(g_j) = 0$) and/or mass dependent schemes (for $\beta_i(g,m) = 0$, $\gamma_{m_i}(g,m) = 0$; formally masses and additional coupling constants play the same role). Property 2 also generalizes to

$$\frac{\partial \beta_i''}{\partial g_j''} = \frac{\partial \beta_i}{\partial g_j}$$

at a zero if we change from g' to $g''_i = \left.\frac{\partial g_i}{\partial g'_j}\right|_{\beta=0} g'_j$. The fourth property does not generalize (and indeed the $O(g^3)$ term of β is α dependent for $m \neq 0$ with momentum subtraction[63]).

5.2.ii Solutions of the RGE

With one coupling, no masses and $\alpha = 0$ (Landau gauge) in the case of a gauge theory, a dimensionless function of dynamical variables Q_i with dimensions of a mass satisfies the RGE

$$\left[-\frac{\partial}{\partial t} + \beta\frac{\partial}{\partial g} + \gamma\right] G(g,t,Q_i/Q_j) = 0 ,$$

where $t = -\frac{1}{2}\ln \mu^2/Q^2$. To solve, define $\bar{g}$ by

$$\frac{\partial \bar{g}(g,t)}{\partial t} = \beta(\bar{g}),\ \bar{g}(g,0) = 0 ,$$

or

$$\int_g^{\bar{g}(g,t)} \frac{dx}{\beta(x)} = t$$

from which it follows that DG = 0 where

$$D \equiv -\frac{\partial}{\partial t} + \beta\frac{\partial}{\partial g} .$$

This equation can be "understood" by realizing that $\bar{g}(g,t)$ and $\bar{g}(g+\delta g,t)$ must follow "parallel" trajectories as functions of t. (This visualization is especially useful in the case of multiple coupling constants.) Hence

$$D\psi(\bar{g}) = 0$$

and

$$G(g,t) = G(\bar{g}(g,t),0)\ \exp\left(\int_0^t \gamma(\bar{g}(g,t'))dt'\right)$$

$$= G(\bar{g}(g,t),0)\ \exp\left(\int_0^{\bar{g}} \frac{\gamma(g')}{\beta(g')}\,dg'\right) .$$

The asymptotic behaviour is controlled by β, with $\bar{g}$ increasing (decreasing) with increasing t for $\beta(\bar{g}) > 0(<0)$ reaching "fixed points" $g = g^*$ at the zeros of β as shown in Fig. 2.

With many couplings

$$\left[-\frac{\partial}{\partial t} + \Sigma\beta_i \frac{\partial}{\partial g_i} + \gamma\right] G = 0$$

is solved by putting

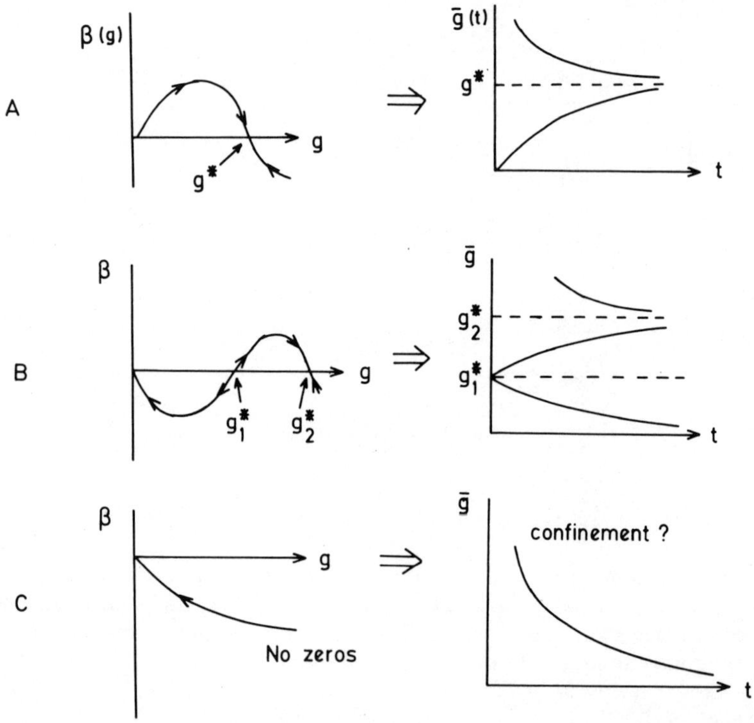

Fig. 2. Three examples of possible behaviour of the β function (the arrows indicate the change of $\bar{g}$ with increasing t) and the corresponding behaviour of $\bar{g}$. Case B is asymptotically free if $\bar{g}(t) < g_1^*$; C is always asymptotically free and presumably leads to confinement because of infra-red slavery ($\bar{g} \to 0$ as $t \to 0$).

$$\frac{\partial \bar{g}_i(g,t)}{\partial t} = \beta_i(\bar{g},t), \quad \bar{g}_i(g,0) = g_i ,$$

from which it follows that

$$\left[-\frac{\partial}{\partial t} + \sum_k \beta_k \frac{\partial}{\partial g_K} \right] \bar{g}_i = 0$$

so that

$$G(g_i,t) = G(\bar{g}_i,0) \exp(-\int_0^t (\bar{g}(g,t'))dt') .$$

Two examples follow. 1. In gauge theories (with masses still zero)

$$\left[-\frac{\partial}{\partial t} + \beta\frac{\partial}{\partial g} + \delta\frac{\partial}{\partial \alpha} + \gamma \right] G = 0 ,$$

and so α plays exactly the same role as a coupling constant. With minimal subtraction β is independent of α,

$$\frac{\partial \bar{g}}{\partial t} = \beta(\bar{g}) ,$$

as usual and

$$\frac{\partial \bar{\alpha}}{\partial t} = \delta(\bar{g},\bar{\alpha}) = \frac{\bar{\alpha}\bar{g}^2}{8\pi^2} (\frac{13}{3} - \bar{\alpha}) C_2(G) + 0(\bar{g}^4)$$

for a pure gauge theory (no fermions), where $C_2(G)$ is the quadratic Casimir operator for the adjoint representation. Hence $\bar{\alpha}$ has a uv(ir) fixed point at $\bar{\alpha} = \frac{13}{3}$ ($\bar{\alpha} = 0$). 2) With a mass and in Landau gauge ($\delta = 0$), or in a non-gauge theory:

$$\left[\mu\frac{\partial}{\partial \mu} + \beta \frac{\partial}{\partial g} + m\gamma_m \frac{\partial}{\partial m} + \gamma \right] G = 0 ,$$

which implies

$$\left[-\frac{\partial}{\partial t} + \beta \frac{\partial}{\partial g} + (\gamma_m - 1)M\frac{\partial}{\partial M} \right] G(g,t,M=\frac{m}{\mu}) = 0 ,$$

where M plays the role of a second coupling constant. Hence $\bar{M}$, which satisfies

$$\frac{\partial \bar{M}}{\partial t} = \bar{M}(\gamma_m(\bar{g})-1)$$

$$\bar{M}(g,0) = M$$

in minimal subtraction (γ_m independent of M — the physical conclusion is scheme independent of course), behaves as

$$\bar{M} = Me^{-t} \exp(\int_0^t \gamma_m(\bar{g}(t'))dt')$$

and vanishes as the energy $Q \to \infty$ like $Q^{-1} \times (\ln Q)^a$ in asymptotically free theories. (With a fixed point at $g^* \neq 0$, $\bar{M}$ behaves as $(Q)^{-1+\gamma(g^*)}$.) This justifies the neglect of masses at large Q^2 in QCD.

5.2.iii <u>RGE and Deep Inelastic Scattering</u>

We already derived the RGE which controls the behaviour of non-singlet structure functions in Section 2 assuming factorization. (We now see that for gauge theories we had implicitly assumed minimal subtraction or the use of Landau gauge, since otherwise there should have been a $\delta(\partial/\partial\alpha)$ term.) We indicated how the relevant γ's can be calculated perturbatively. In the singlet case the moments have the form

$$C_n^q(\mu,Q^2)M_n^q(\mu,m) + C_n^g(\mu,Q^2)M_n^g(\mu,m) + O(m^2/Q^2)$$

where the C's are independent of the target (but the M's are not) and q(g) stands for a quark (gluon) contribution. (This should become clear in subsequent sections.) By simultaneously considering scattering from quarks and gluons, we can proceed as in the non-singlet case to derive coupled RGE with a matrix of anomalous dimensions γ_{ij} which can be found perturbatively. Although this is simple conceptually, in practice it is easier to use the traditional operator product expansion (OPE) method. This involves relating moments to matrix elements of products of currents which are expanded using the OPE

$$J(x)J(0) \underset{x\to 0}{\to} \sum_i C^i(x)O_i(0)$$

where C^i are c-numbers and O_i operators. (This is essentially a statement of the completeness of operators.) It can then be shown that

$$\int x^n F(x,Q^2)dx = C_n^q(Q^2)\langle p|O_n^q|p\rangle + C_n^g(Q^2)\langle p|O_n^g|p\rangle + O(m^2/Q^2) ,$$

where the operators $O_n^{q,g}$ are the unique flavour singlet operators made of quark and gluon fields with spin n and "twist" (= dimension-spin) two. The C's satisfy the RGE, the γ's being given by the Z's which renormalize the $O_n^{q,g}$. This is discussed in detail in the literature.[64]

6. DEEP INELASTIC PROCESSES IN LEADING LOG APPROXIMATION

We now consider the derivation of the factorization property discussed in Section 2 and show that deep inelastic processes are described by a parton-like picture in QCD. In this section we work to leading order in α_s; higher orders are discussed in Section 7. Many authors have discussed the leading order.[65] Here I shall follow the approach developed by myself[66] and Ash Carter, indicating the thrust of the argument and referring to our recent paper[67] for details.

We consider first deep inelastic lepton scattering as a paradigm. In perturbation theory

$$\begin{aligned}\int x^n F(x,Q^2)dx = A + Bg^2\ell nQ^2 &+ Cg^4(\ell nQ^2)^2 + \ldots \\ + B'g^2 \quad &+ C'g^4\ell nQ^2 + \ldots \\ &+ C''g^4 + \ldots\end{aligned}$$

The strategy is to identify the origin of the terms in the top line (the leading logs), calculate and sum them; we shall see that they give the leading order in α_s. In order g^2 there are two diagrams

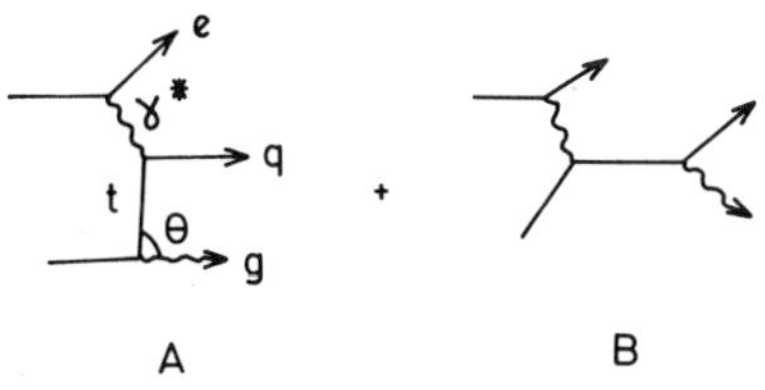

which give a contribution of order $g^2\ell nQ^2$. Provided we only sum over the physical transverse polarization states of the produced gluon, the log comes entirely from the "parton" diagram A. (Including the mutually cancelling contributions of time-like and longitudinal gluons by using $\Sigma\varepsilon^*_\mu\varepsilon_\nu = -g_{\mu\nu}$ the A-B interference term also gives a log.)

Setting m = 0 (temporarily), quark helicity is conserved in the vertex

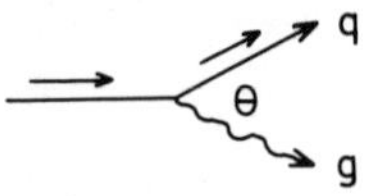

which therefore vanishes for physical gluons (helicity ± 1) as $\theta \to 0$, $t \to 0$; in fact it behaves like $\sqrt{t}$. Thus the amplitude A behaves as $\sqrt{t}/t$ for small t and

$$F \sim \int dt \left[\left(\frac{\sqrt{t}}{t}\right)^2 + O\left(\frac{\sqrt{t}}{t}\right)\right]$$

Only the first term, due to $|A|^2$, is log divergent and therefore only it gives $\ell nQ^2/m^2$ for $m \neq 0$. Note that the log is contributed by the small t/small θ region in which the gluon is emitted parallel to the incoming quark; this is the beginning of a jet.

Going to higher orders and using the optical theorem

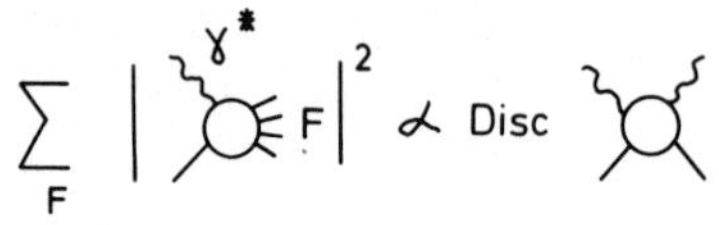

we find that while ladder diagrams like

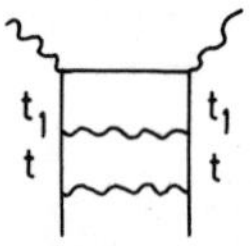

give leading logs ($O(g^4(\ell nQ^2)^2)$ in this case), crossed ladders such as

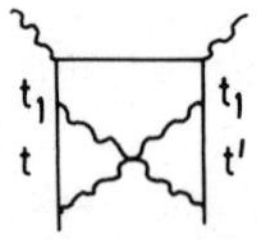

do not if we stick to physical polarizations. The reason is again that each vertex × propagator gives $\sqrt{t_i}/t_i$ and we need a symmetric ladder diagram in which these factors are all squared to get a leading log.

Using only physical polarizations for real gluons henceforth, we need only consider ladders dressed with virtual corrections in order to get the leading logs. Logs from virtual contributions are of two kinds: "parallel" and ultraviolet.

A) "Parallel Logs"

Amplitudes diverge when virtual processes become real[68] (giving poles or causing divergences in coordinate space loop integrals by allowing different parts of diagrams to separate to arbitrarily large distances). This can almost happen when virtual particles become parallel to the particle which emits them, e.g. in the diagram

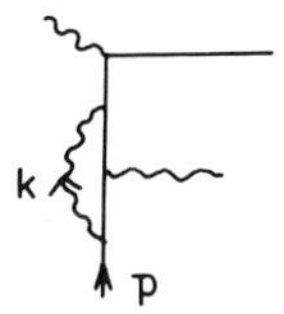

there is a (logarithmic) divergence if $p^2 = m^2 = 0$ from the point $k^2 = (p-k)^2 = 0$, $k_\mu \propto p_\mu$. For $p^2 = m^2 \neq 0$, the virtual processes cannot quite become real and the region where k^2 and $(p-k)^2$ are small yields $\ln Q^2/m^2$. In calculating the coefficient of this "parallel log" we can safely set $(p-k)^2 = k^2 = p^2 = m^2 = 0$, $p_\mu \propto k_\mu$ in the numerator and write

$$(\not{p}-\not{k})\gamma_\mu u(p) \propto k_\mu u(p) ,$$

i.e. the gluon propagator is contracted with k_μ. This is always true for parallel emission, and parallel logs are therefore due to longitudinal gluons. We can now: either choose an axial gauge in which $k^2 k^\mu D_{\mu\nu} \xrightarrow[k^2\to 0]{} 0$ and there are no "parallel logs" diagram-by diagram (see the next section); or use a covariant gauge and sum up the parallel logs due to sets of diagrams into simpler effective diagrams using Ward identities of the type discussed in Section 4, e.g. in the Abelian case

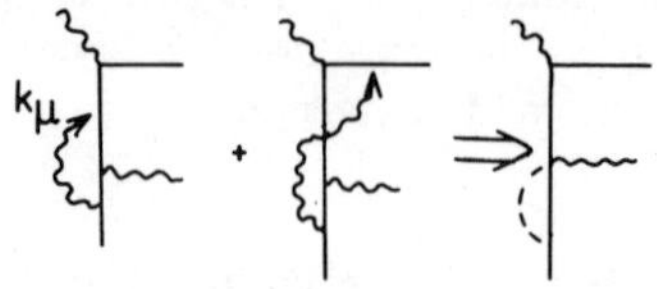

The non-Abelian case is slightly more complicated,[67] more diagrams being involved. This simplification is hardly surprising since the singularities as $m \to 0$ are associated with the propagation of real intermediate states, to which longitudinal gluons cannot contribute.

It is convenient to combine the effects of real and virtual parallel logs, which are individually infra-red divergent, into a single effective diagram e.g.

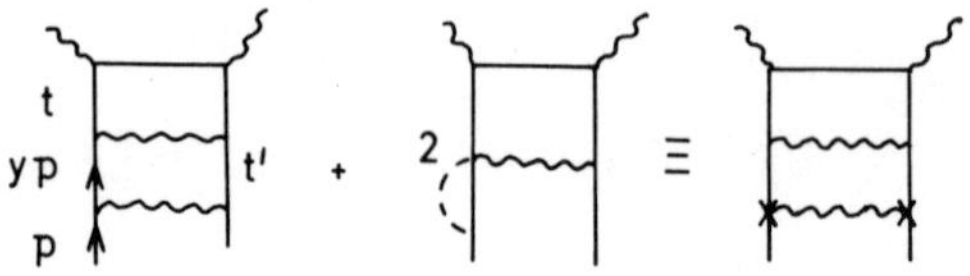

$$\ell n t/m^2 \left[\frac{1+y^2}{1-y} - \delta(1-y) \int \frac{1+z^2}{1-z} dz \right] \equiv \ell n t/m^2 g(y)$$

(where the t' integral has been done). The y-dependent part is defined as a distribution (i.e. its moments are well defined); their form should be extremely familiar from other lectures at this school.* Real parallel logs must always combine with virtual contributions in this way into an i.r. finite result, i.e. the net effect of virtual logs is that effectively

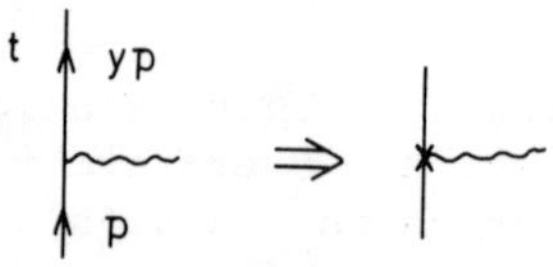

$$\frac{\sqrt{t}}{t}\left[\frac{1+y^2}{1-y}\right] \to \frac{\sqrt{t}}{t} g(y)$$

in the ladders.

*Actually the finite part of the coefficient of $\delta(1-y)$ is gauge-dependent. The result here is for Landau gauge ($\alpha = 0$).

B. Ultra-violet Logs

U.V. divergences give logs scaled by μ^2, e.g.

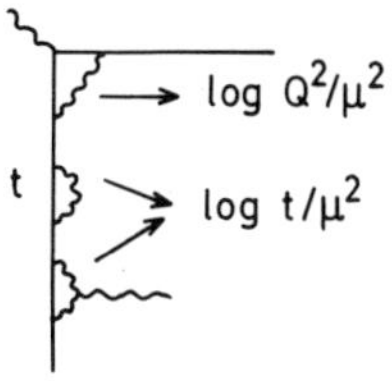

The net effect of all u.v. corrections is that*

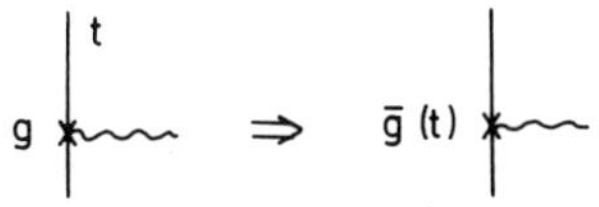

in the ladder. For t = t' = s (all spacelike)

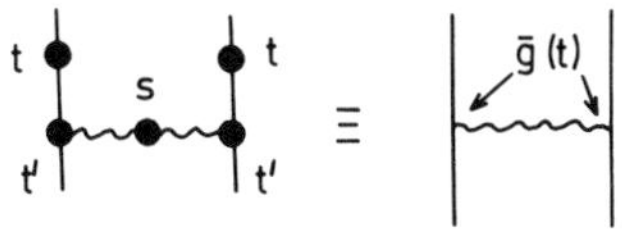

(Exercise: verify this.) Having removed all potential (parallel) logs of t'/t, s/t, whose contribution we already considered, we can safely expand in $\bar{g}(t)$ getting just $\bar{g}(t)$ to leading order for $t \neq t' \neq s$. The use of this expression for time-like s is justified since we anticipate integration over s washing out threshold effects (a similar smearing is really needed before expanding $\sigma_{e^+e^-}(s)$ in powers of $\bar{g}(s)$).

We have now shown that the parts of diagrams through which large momenta flow have a ladder structure with vertices

---

*For $\alpha \neq 0$ there are corrections to the e.m. vertex and to the quark propagator at the top of the ladder, which compensate the terms proportional to α contributed by virtual parallel gluons.

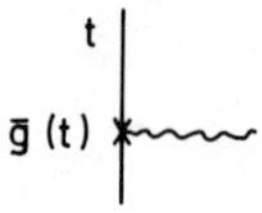

to leading order in $\alpha_s(t)$. This does not mean that only ladders matter since t can become small making α_s large. However, parts of diagrams through which large momenta flow must have a ladder structure. Thus all diagrams begin as ladders to leading order in $\alpha_s(Q^2)$:

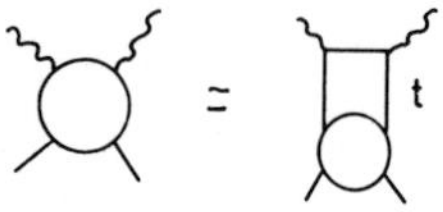

Splitting the t range into $0 < t < \Lambda^2$ and $\Lambda^2 < t < Q^2$ with Λ large enough so that we can neglect $\alpha_s(\Lambda^2)$, the next element of the diagram must be a ladder for $t > \Lambda^2$. For $t < \Lambda^2$ it is an incalculable "black box", i.e.

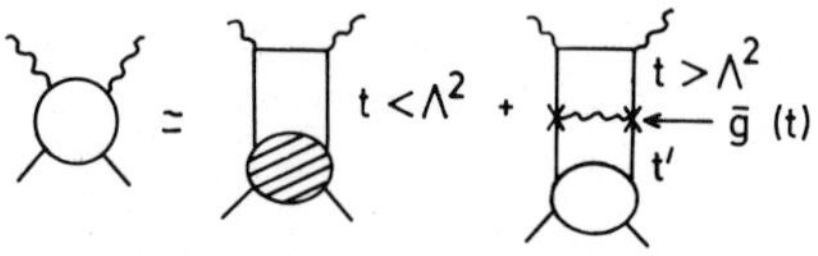

Proceeding to split up the t' and subsequent integrals in the same way we obtain the picture discussed in Secion 2:

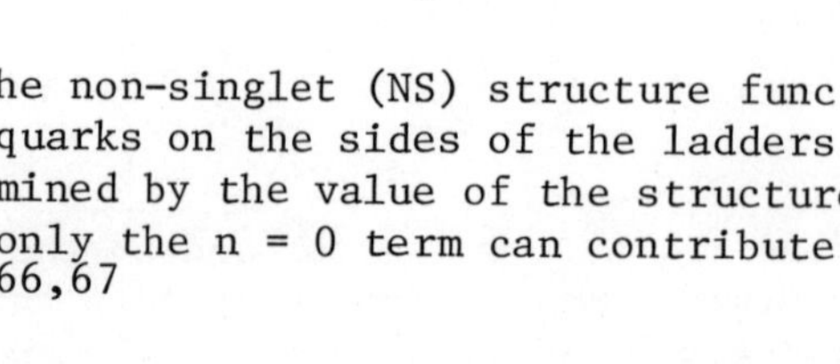

for the non-singlet (NS) structure function in which there are only quarks on the sides of the ladders. The "black box" is determined by the value of the structure function for $Q^2 = \Lambda^2$, when only the n = 0 term can contribute. Summing the ladder then gives[66,67]

$$M_n(Q^2) \equiv \int x^n F_{NS}(x,Q^2)\,dx$$

$$= \int^{\Lambda^2} dt \left(\frac{\ell n Q^2/\lambda^2}{\ell n t/\lambda^2}\right)^{-bB_n} X_n(t)[1 + 0(\alpha_s(Q^2))]$$

$$= \left(\frac{\ell n Q^2/\lambda^2}{\ell n \Lambda^2/\lambda^2}\right)^{-bB_n} M_n(\Lambda^2)$$

in the notation of Section 2 (where some implications of this result were discussed) and we can identify the function f(y) introduced there with $(6\pi^2)^{-1}g(y)$, where g(y) was defined explicitly above.

In general there are singlet contributions from diagrams like

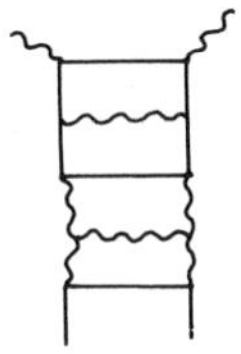

which are absent by definition in the NS case (which is isolated by considering quantities like $F_2^{ep} - F_2^{en}$ or F_3; exercise — explain why). The singlet case therefore involves a black box amplitude for emitting gluons as well as quarks. Consequently the predictions are more complicated than in the N-S case and we will not discuss them in detail. One interesting result is that QCD predicts the fraction of the proton's momentum carried by quarks to become 3/7 as $Q^2 \to \infty$ for four flavours (as $Q^2 \to \infty$ the number of rungs of the ladder $\to \infty$ so we would expect the ratio of the momentum in quarks and gluons to be a calculable number independent of the black boxes). This is consistent with the value of about 50% slowly decreasing with Q^2 which has been observed for $Q^2<100$ GeV^2. Gluck and Reya[69] have made the interesting observation that in many other theories the asymptotic fraction would be expected to be considerably greater than 1/2 and an increase with Q^2 should have been seen. Their quantitative results for other theories are not necessarily credible since they calculate $\gamma(g^*)$ perturbatively assuming a fixed point at $g^* << 1$ and, as discussed in Section 2, this may be invalid. Qualitatively, however, their conclusion is very plausible. We would expect the momentum partition to be (very roughly) proportional to the number of degrees of freedom, which is greater for the gluons in QCD than in the other models considered by Gluck and Reya (Abelian glue, scalar glue etc.). Thus the observed value and behaviour of the momentum fraction is circumstantial evidence for multicoloured vector glue.

Having recovered the classic results for the moments of structure functions, let us see what the diagrammatic analysis tells us about the structure of the final states. Consider a typical diagram, omitting vertex and propagator corrections but including pieces which contribute to the propagator corrections to the cross bars of the ladder formed by squaring the amplitude

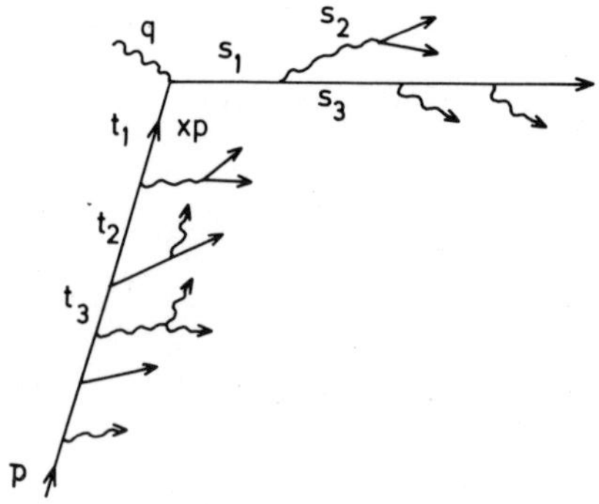

In the leading log configuration all the t_i's and s_i's are small (the log divergence coming from t_i, $s_i \to 0$ for $m = 0$), all radiation is almost parallel and the dominant contribution consists of two jets of particles parallel to $\vec{p}$ and $x\vec{p}+\vec{q}$ respectively. In order for particles to be produced at large $(O(\sqrt{Q^2}))$ p_T relative to these directions, at least one s_i or t_i must become large. However clearly $s_1 > s_2 > s_3$ ... and (less obviously) $t_1 > t_2 > t_3$ .... Since a log (and hence a power of α_s) is lost for each s_i or t_i which is of order $\sqrt{Q^2}$, the "cheapest" way to produce particles at large p_T is for t_1 or s_1 to be large. The rest of the ladder remains the same and gives the same quark distribution $q(x,Q^2)$. The effect of radiation from the quark or gluon produced at large p_T can be absorbed by changing $g \to \bar{g}$, i.e. the dominant large p_T contribution is given by

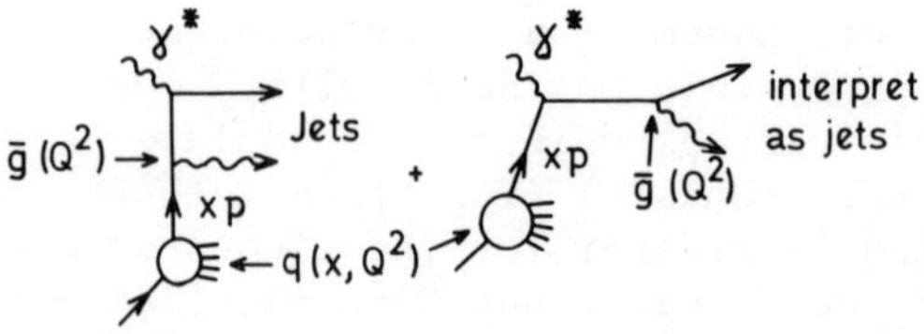

and a similar contribution for hard scattering off a gluon. This prescription is discussed in detail in other courses at this school.

The cross section for longitudinal virtual photons can be calculated in a similar way. With $\vec{p}\,||\,\vec{q}$ we can write $\varepsilon^L_\mu = Ap_\mu + Bq_\mu$. The second term gives zero since $q^\mu J_\mu = 0$. We must therefore calculate

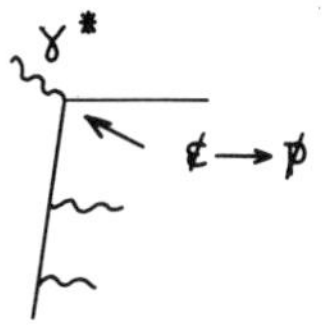

etc.

The $\not p$ kills the potential t^{-1} singularity and hence one log, so that σ_L is of order α_s. With no singularity $\langle t\rangle \sim O(Q^2)$ the dominant final state for longitudinal photons contains 3 jets. Quantitatively σ_L is given by

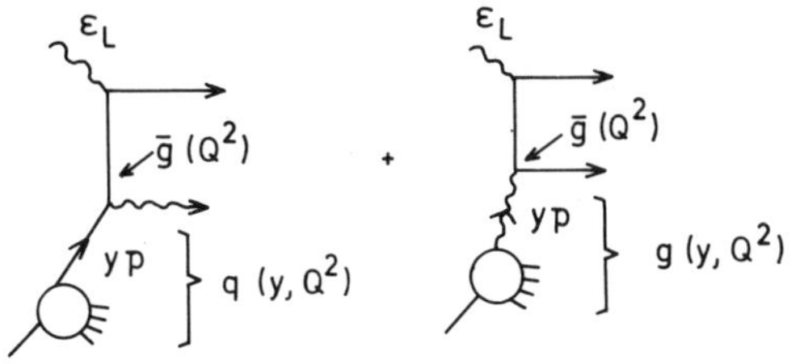

Turning to e^+e^- annihilation the dominant diagrams have the form

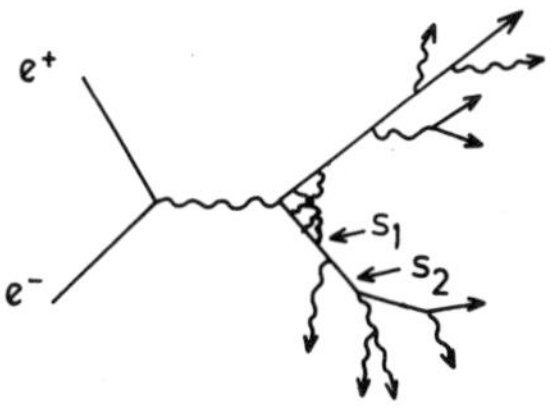

with s_i small i.e. the final state consists of two jets. The total effect of real radiation cancels that of virtual corrections to the electromagnetic vertex, to leading order in α_s, because of a Ward identity. The total cross section (to $O((\alpha_s)^0)$) is therefore given by

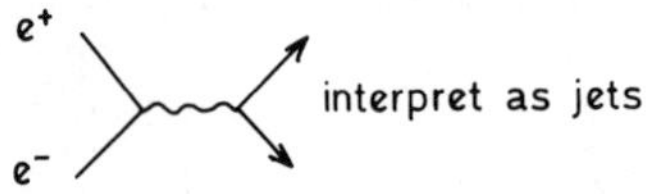

which we already knew (Section 2) because it is mass finite. Likewise diagrammatic or mass finiteness arguments show that three jet cross-sections are given by the O(g) diagrams with $g \to \bar{g}(Q^2)$ and the quarks and gluons interpreted as jets.

The techniques described here can also be used to justify a QCD corrected picture of other processes (e.g. $pp \to \mu\bar{\mu}x$ or $pp \to$ large p_T).

7. FACTORIZATION, INCLUDING NON-LEADING LOGS

The operator product expansion (OPE), which guarantees the factorization of the moments of structure functions, is very plausible since it amounts to a statement of the completeness of operators, but it has only been proved in perturbation theory. In the last year several groups of authors[70] have given perturbative proofs of factorization for processes to which the OPE does not apply (not surprisingly most of the proofs are similar to proofs of the OPE). Here we outline the idea of the proof due to R.K. Ellis, Georgi, Machacek, Politzer and G.G. Ross.

Consider an inclusive "hard" process involving a large momentum scale Q in which certain final state particles may be singled out but the rest are not observed, which we represent by

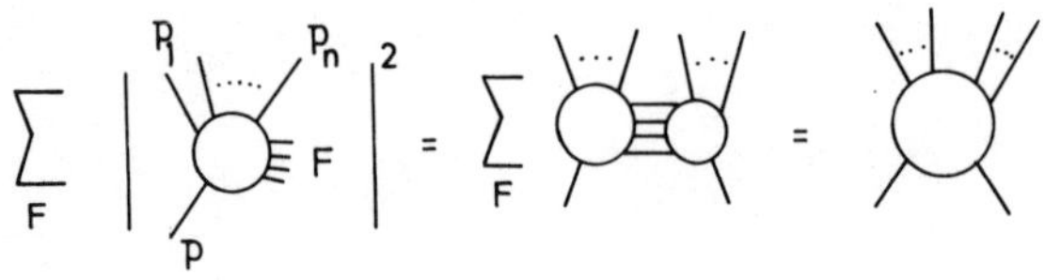

As in the leading order discussion in Section 6, there are two steps. 1) Set $m = 0$ and identify diagrams which give rise to singularities as $p_i^2 \to 0$. Concentrating on $p^2 \to 0$, the result (explained below) is that, provided $|p \cdot p_i|$ is large, singularities come only from

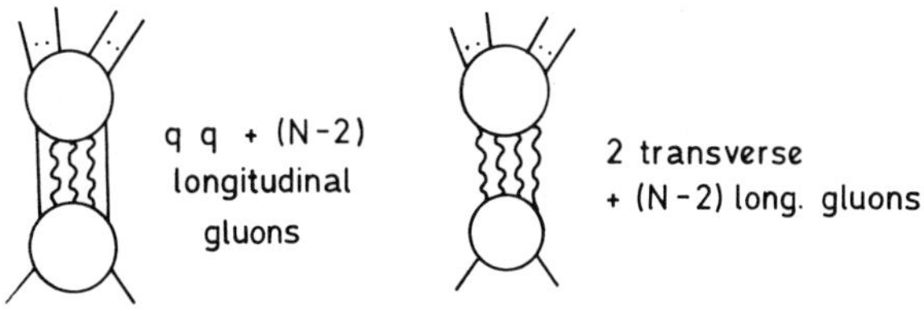

in the configuration in which $k_i^2 \to 0$, $\vec{k}_i \parallel \vec{p}$ for all the exposed internal particles. In axial gauge, the propagator for longitudinal gluons does not have a pole as $k^2 \to 0$ and the only diagrams which are singular as $p^2 \to 0$ are

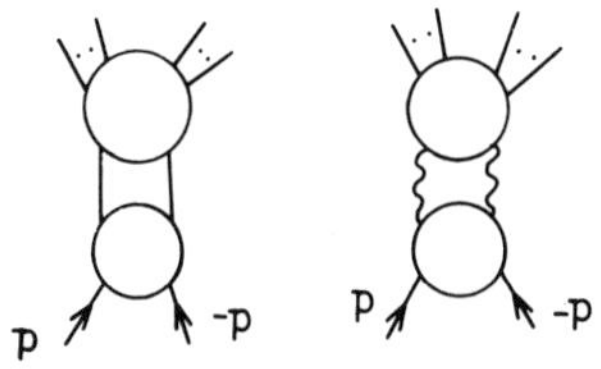

i.e. if we write the cross-section as a ladder sum of two particle irreducible diagrams in the (p,-p) channel, all p^2 singularities are due to the two particle intermediate states. 2) Show that to order p^2/Q^2, the ladder sum can be reduced to a one-dimensional parton-like convolution of pieces which are/are not singular as $p^2 \to 0$ (which correspond to parton distributions/hard scattering cross-sections).

In step one, the KLN theorem[71] guarantees that there are no singularities as $(p_i^F)^2 \to 0$ for the unobserved final state particles. The danger occurs when virtual processes can become real, as discussed in Section 6. As $p^2 \to 0$ this occurs in

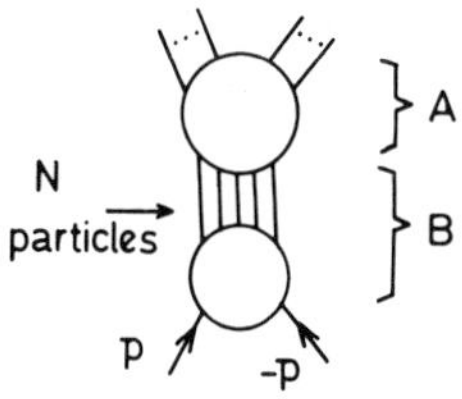

when $k_i^2 = 0$, $k_\mu^i \propto p_\mu$ for all particles in the exposed intermediate state; provided $|p \cdot p_i|$ is large we are not close to any other potential singularities. Consider the case of spinless particles. We multiply the whole diagram by a factor $(Q^2)^D$ to make it dimensionless. Using the fact that $1/P^2$ singularities never occur, $(Q^2)^D$ times the A part of the diagram, which has dimensions M^{-N+2}, behaves as

$$A \sim (Q^2)^{1-N/2}(1 + 0(p^2/Q^2))$$

up to logs, while $B \sim M^{N-2}$ behaves as

$$B \sim (p^2)^{N/2-1} \quad .$$

Dropping p^2/Q^2 terms, we have the desired result that only $N = 2$ matters.

Turning to QCD, suppose the N particle intermediate state contains P gluons and N-P quarks. The A and B factors are now P rank tensors and (up to logs)

$$B \sim p_{\mu_1} \cdots p_{\mu_R} g_{\alpha\beta} \cdots g_{\mu\nu} (p^2)^{N/2 - 1 - R/2} \quad .$$

Negative powers of p^2 being impossible, the leading contribution occurs for $R = N-2$. Since $N > P > R$ we have P=N or N-2.
<u>P = N-2</u>: The best case occurs when $A \sim q_{\mu_1} \cdots q_{\mu_P} (Q^2)^{-P}$ giving $(q \cdot p/Q^2)^P$, when contracted with B, which scales in the Bjorken limit (up to logs). Note that every gluon emerging from B produces $p_{\mu_i} \propto k_{\mu_i}$ in the singular configuration i.e. it is longitudinal. This is the $(q\bar{q})$ + (N-2) longitudinal gluon contribution.
<u>P = N</u>: The best case occurs when $A \sim (q_{\mu_1} q_{\mu_2} \text{ or } Q^2 g_{\mu_1\mu_2}) q_{\mu_3} \cdots q_{\mu_N} (Q^2)^{N-1}$ and also scales (up to logs). Since $B \sim g_{\mu_1\mu_2} p_{\mu_3} \cdots p_{\mu_N}$, two gluons are transverse and the remaining N-2 are longitudinal.

This completes step 1. The argument is easy to understand but a real proof is difficult because of possible additional "soft" divergences which may occur as gluon momenta go to zero. It must be shown that non-factorizing soft divergences cancel.[72] Step 2 consists of straightforward but lengthy algebra. We shall only indicate the idea.

We can express the cross-section as a sum of graphs which are two particle irreducible [2PI] (indicated by a cross) in the (p,-p) channel

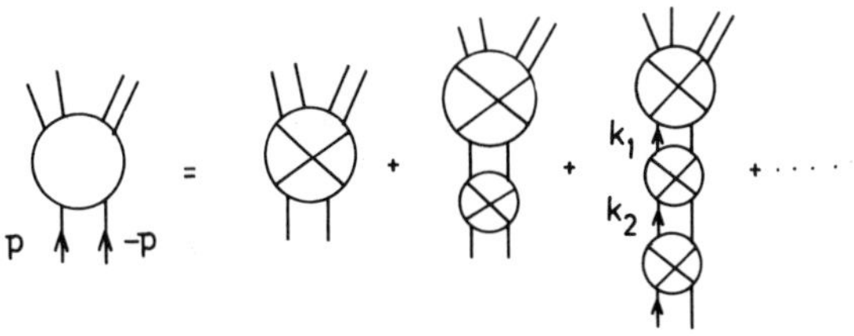

As a preliminary step we sum up all sub-ladders in which adjacent two-particle intermediate states contain longitudinal gluons i.e. a cross now means 2PI except for states containing a longitudinal gluon and the exposed two-particle states contain quarks or transverse gluons. As $p^2 \to 0$ these two particle states (and only these) give rise to singularities when $k_i^2 = 0$. Writing $k_i = xp + \tilde{k}$ it is a kinematic fact that $\tilde{k} \to 0$ as p^2, $k_i^2 \to 0$ provided $x \neq 0$. (The subsequent analysis therefore only applies to "hard" processes in which the kinematics require $x > 0$, e.g. it applies to $pp \to$ large p_T jets but not to the total pp cross-section.) It is also a fact that if $k_i^2 \neq 0$, k_j^2 cannot vanish for $j < i$. Now suppose that $k_{i-1}^2 > \Lambda^2$ but $k_i^2 < \Lambda^2$, where Λ is a factorization scale which plays essentially the same role as it did in Section 6:

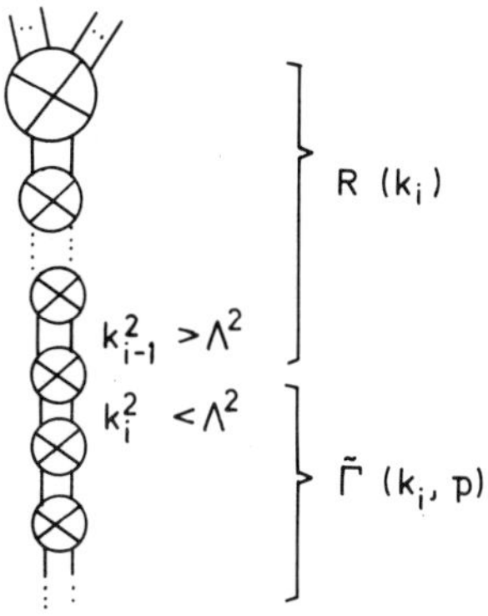

Since $k_j^2 \neq 0$ for $j < i$, R is free from mass singularities, i.e. $R(k_i)$ is finite for $k_i^2 = 0$. We can therefore write

$$R(k_i) = \Delta R + R(xp)\Big|_{p^2=0} \ .$$

The contribution of the second term can be written in the desired parton form

$$\int R(xp)\Big|_{p^2=0} \tilde{\Gamma}(k,p)d^4k = \int R(xp)\Big|_{p^2=0} \Gamma(x,p^2)dx \ .$$

The contribution of the first term does not contain singularities from the ith two-particle state since $\Delta R \to 0$ as $k_i^2 \to 0$. Therefore if we combine ΔR with the next element of the ladder we can define a new function $R'(k_{i+1})$ which is finite for $k_{i+1}^2 = 0$:

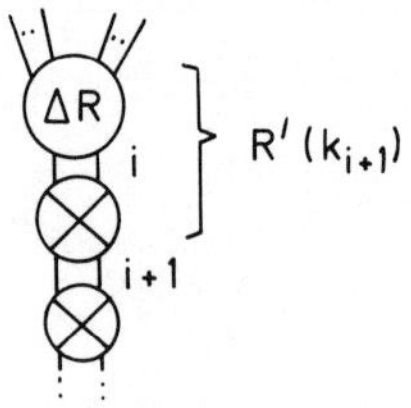

We then write $R' = \Delta R' + R'(xp)\Big|_{p^2=0}$ and repeat the argument.

We end up with an infinite sum of terms all of which have the parton form:

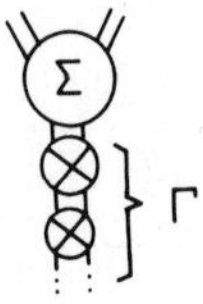

which corresponds to

$$\int \Sigma(xp)\Big|_{p^2=0} \Gamma(x,p^2)dx \quad .$$

All p^2 dependence resides in Γ which is manifestly independent of the hard process ("universal"). In electroproduction, for example, we have a diagrammatic structure

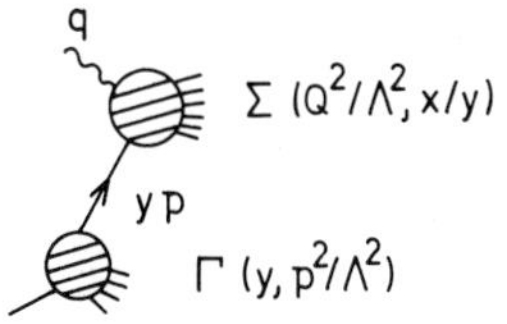

or

$$F(x) = \int_x^1 \Sigma(Q^2/\Lambda^2, x/y)\Gamma(y, p^2/\Lambda^2)\,dy + O(p^2/Q^2) \ ,$$

where $x = Q^2/2q\cdot p$ and the moments of F factorize. In lowest order Σ is just σ_{γ^*q} but this is not true in higher orders. In fact Σ is related to the two-particle irreducible diagrams in a complicated way involving ladders containing projections onto momenta greater than Λ^2 and unphysical polarizations.[73]

Obviously the analysis also applies to singularities as $(p')^2 \to 0$ for particles in the final state and shows that they can be isolated in a universal fragmentation function Γ'. As a final illustration consider $eN \to e\pi(p_\pi)X$. After integrating over the transverse momentum of the pion (relative to the γ^*-nucleon collision axis) there are three variables Q^2, $x = Q^2/2q\cdot p$ and $z = -2p_\pi\cdot q/Q^2$, which is a measure of the longitudinal momentum of the pion (in the frame where $q_0 = 0$, $z = 2p_\pi^L/\sqrt{Q^2}$). The differential cross-section has the form

$$d\sigma = \int_x^1 dy \int_z^1 d\eta\,\Gamma'(\eta, p_\pi^2/\Lambda^2)\Gamma(y, p^2/\Lambda^2)A\left(\frac{z}{\eta}, \frac{x}{y}, \frac{Q^2}{\Lambda^2}\right) + O(p^2/Q^2)$$

with the mass singularities "factorized" into Γ and Γ', corresponding to the picture

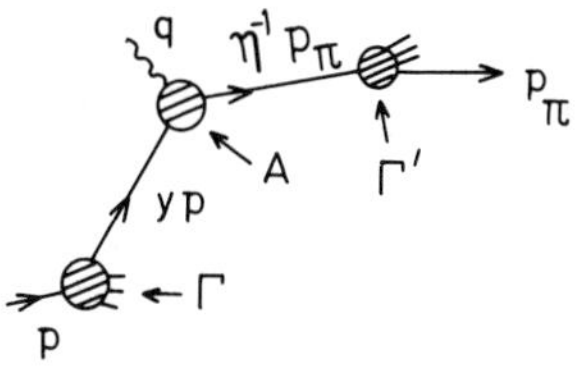

Taking moments

$$d\sigma_{nm} = \int_0^1 x^n dx \int_0^1 z^m dz\, d\sigma = \Gamma_n(p^2/\Lambda^2)\Gamma'_m(p_\pi^2/\Lambda^2)A_{nm}(Q^2/\Lambda^2) \;,$$

where

$$\Gamma_n = \int_0^1 y^{n+1}\Gamma dy$$

$$\Gamma'_m = \int_0^1 \eta^{n+1}\Gamma' d\eta$$

$$A_{nm} = \int_0^1 r^n dr \int_0^1 s^m ds\, A(r,s,Q^2/\Lambda^2) \;.$$

Note that the statement that the mass singularities factorize should not be taken to mean that $A_{nm} \propto \delta_{nm}$. In fact A_{nm} is diagonal to leading order in α_s, and "naive" factorization occurs, but in higher orders it is not diagonal and we expect $d\sigma_{nm}\Gamma_n^{-1}$ to be n-dependent. (The $O(\alpha_s)$ contributions to A_{nm} have been calculated by Sakai, and Altarelli et al.[74])

REFERENCES

1. For recent reviews see: W. Marciano and H. Pagels, Physics Reports 36C, 137 (1978); A. Peterman, CERN TH2581 (1978), to be published in Physics Reports; A.J. Buras, Fermilab-Pub-79/17-Thy (1979), to be published in Rev. Mod. Phys.

2. This feature of non-Abelian forces was pointed out by Y. Nambu, in "Preludes in Theoretical Physics," ed. A. De-Shalit, H. Feshbach and L. Van Hove (North Holland, 1966). It was rediscovered by H. Lipkin, Phys. Lett. 45B, 267 (1973); and R.P. Feynman, Proc. Fifth Hawaii Topical Conference (1973), ed. P.N. Dobson, V.Z. Peterson and S.F. Tuan (University of Hawaii Press, 1974), and in "Weak and Electromagnetic Interactions at High Energy," ed. R. Balian and C.H. Llewellyn Smith (North Holland, 1977) whose treatment I follow.

3. A. De Rujula, H. Georgi and S.L. Glashow, Phys. Rev. D12, 147 (1975).

4. H.J. Lipkin, Weizmann preprint WIS-79/20-Ph (1979).

5. N. Isgur and G. Karl, Phys. Rev. D19, 2653 (1979) and refs. therein and N. Isgur, G. Karl and R. Koniuk, Phys. Rev. Lett. 41, 1269 (1978).

6. This is discussed by C. De Tar, these proceedings.

7. S. Weinberg, Phys. Rev. D11, 3583 (1975).

8. G. 't Hooft, Phys. Rev. Lett. 37, 8 (1976) and Phys. Rev. D14, 3432 (1976). The conclusions are still controversial; see R. Crewther, Rivista del Nuovo Cimento 2, 63 (1969), for a critical review. The mechanism has recently been discussed in terms of the $1/N_c$ expansion by E. Witten, Nucl. Phys. B149, 285 (1979), and Harvard preprint HUTP-79/A014; and G. Veneziano, CERN TH 2651 (1979).

9. D.J. Gross and F. Wilczek, Phys. Rev. Lett. 30, 1343 (1973); H.D. Politzer, Phys. Rev. Lett. 30, 1346 (1973).

10. R.D. Field, these proceedings.

11. Higher order processes are discussed by R.D. Field and by A.J. Buras (these proceedings and ref. 1) who give further references.

12. J.S. Bell, Nuovo Cimento 41A, 495 (1977); P.M. Fishbane and M.T. Grisaru, Nucl. Phys. B114, 462 (1978).

13. S. Brodsky and P. Lepage, SLAC-PUB 2294 (1979) and others (e.g. G.R. Farrar and D.R. Jackson, Phys. Rev. Lett. 43, 264 (1979)) have shown that "dimensional counting rules" may be justified as a zeroth order approximation in QCD for form factors and perhaps also for exclusive processes, such as $pp \to pp$, and that QCD perturbation theory may be used to calculate logarithmic modifications of these rules.

14. For a review of the parton model see (e.g.) S.D. Ellis, these proceedings; C.H. Llewellyn Smith in "Hadron Structures and Lepton-Hadron Interactions - Cargèse 1977," ed. M. Lévy et al. (Plenum, 1979); or F.E. Close, "An Introduction to Quarks and Partons" (Academic Press, 1979).

15. G. Altarelli and G. Parisi, Nucl. Phys. B126, 298 (1977).

16. J. Kogut and L. Susskind, Phys. Rev. D9, 697 and 3391 (1974).

17. P.C. Bosetti et al., Nucl. Phys. B142, 1 (1978).

18. J.G.H. de Groot et al., Phys. Lett. 82B, 292 (1979).

19. For a recent discussion of this point and references to the original literature see J. Ellis, CERN TH 2701 (1979).

20. The next order corrections to the predicted slopes are relatively small (and scheme independent); M. Pennington and G.G. Ross, Oxford preprint 23/79 (Phys. Lett. - in press).

21. H. Harari, SLAC-PUB-2254 (1979).

22. A.J. Buras and K.J.K. Gaemers, Nucl. Phys. B132, 249 (1978).

23. G.S. Guralnik et al., Advances in Particle Physics 2, 567 (1968).

24. J.D. Bjorken and M. Nauenberg, Ann. Rev. Nucl. Sci. 18, 229 (1968).

25. V. De Alfaro et al., "Currents in Hadron Physics" (North Holland, 1973).

26. H. Pagels, Phys. Rep. 5, 219 (1975).

27. M. Veltman, Proc. Roy. Soc. A301, 107 (1967); D.G. Sutherland, Nucl. Phys. B2, 433 (1967).

28. J.S. Bell and R. Jackiw, Nuovo Cimento 60A, 47 (1969); S.L. Adler, Phys. Rev. 177, 2426 (1969).

29. Two excellent reviews are S.L. Adler in "Lectures on Elementary Particles and Quantum Field Theory," ed. S. Deser, M. Grisaru and H. Pendleton (MIT press, 1970); R. Jackiw in "Lectures on Current Algebra and its Applications," by S. Treiman, R. Jackiw and D. Gross (Princeton, N.J., 1972).

30. J. Ellis in "Weak and Electromagnetic Interactions at High Energy," cited above (ref. 2).

31. S.L. Adler and W. Bardeen, Phys. Rev. 182, 1517 (1969).

32. This was discovered by J. Schwinger, Phys. Rev. 82, 664 (1951), but subsequently forgotten. See Jackiw (ref. 29) or B. Zumino, in CERN Yellow Report 69-7 (1969) p. 361, for a clear discussion.

33. W.A. Bardeen, Phys. Rev. 184, 1848 (1969); J. Wess and B. Zumino, Phys. Lett. 37B, 95 (1971); R. Aviv and A. Zee, Phys. Rev. D5, 2372 (1972).

34. M. Chanowitz, M. Furman and I. Hinchliffe, LBL preprint - LBL 8855 (1979). The conclusions of this paper are disputed by S. Gottlieb and J.T. Donohue, ANL preprint ANL-HEP-PR-79-08 (1979).

35. C. Bouchiat et al., Phys. Lett. 38B, 519 (1972); D.J. Gross and R. Jackiw, Phys. Rev. D6, 477 (1972); H. Georgi and S.L. Glashow, Phys. Rev. D6, 429 (1972).

36. R.P. Feynman, Acta. Phys. Polon. 24, 697 (1963) and in paper cited above (ref. 2).

37. E.S. Fradkin and I.V. Tyutin, Phys. Rev. D2, 2841 (1970).

38. L.D. Faddeev and V.N. Popov, Phys. Lett. 25B, 29 (1967).

39. See for example: R.P. Feynman and A.R. Hibbs, Quantum Mechanics and Path Integrals (McGraw-Hill, 1975); L.D. Faddeev in "Methods in Field Theory," ed. R. Balian and J. Zinn-Justin (North Holland, 1976); E.S. Abers and B.W. Lee, Phys. Rep. 9C, No. 1 (1973).

40. This approach follows J. Zinn-Justin in Lecture Notes in Physics 37, ed. J. Ehlers et al. (Springer-Verlag, 1975).

41. This is the standard approach - see for example Abers and Lee (ref. 39); B.W. Lee in "Methods in Field Theory" (ref. 39); J.C. Taylor, Gauge Theories of Weak Interactions (Cambridge University Press, 1976).

42. This argument, due to Faddeev, is discussed in detail by S. Coleman in "Laws of Hadronic Matter," part A, ed. A. Zichichi (Academic Press, 1975).

43. W. Kummer, Acta. Phys. Austriaca 41, 315 (1975); W. Konetschny and W. Kummer, Nucl. Phys. B100, 106 (1975) and B108, 397 (1976).

44. D.J. Pritchard and W.J. Stirling, Cambridge preprint 79/1 (1979).

45. C. Becchi, A. Rouet and R. Stora, Annals of Physics 98, 287 (1976). The discussion here is derived from R.A. Brandt, Nucl. Phys. B116, 414 (1976).

46. G. 't Hooft and M. Veltman, CERN Yellow Report 73-9 (1973).

47. J. Zinn-Justin (ref. 40). The method is reviewed by Lee and Taylor (ref. 41).

48. This identity was proved by G. 't Hooft and M. Veltman, Nucl. Phys. B50, 318 (1972) using much more cumbersome diagrammatic techniques.

49. The lines of the proof here follow G. 't Hooft, Nucl. Phys. B33, 173 (1971).

50. The proof here follows ref. 48.

51. Good reviews of renormalization and the RGE and further references may be found in the lectures by C. Callan and D.J. Gross in Methods in Field Theory (cited in ref. 39). For an excellent introduction to renormalization see S. Coleman in "Properties of the Fundamental Interactions" Part C, ed. A. Zichichi (Editrice Compositori Bologna, 1973).

52. Explicit examples are given by W. Celmaster and R.J. Gonsalvos, U.C. San Diego preprint UCSD-10P10-202 (1979).

53. G. 't Hooft and M. Veltman, Nucl. Phys. B44, 189 (1972) and ref. 46. See also ref. 54 for a convenient summary. Note that trace of γ matrices etc. are also n-dependent. Minimal Subtraction was introduced by G. 't Hooft, Nucl. Phys. B61, 455 (1973).

54. W.J. Marciano, Phys. Rev. D12, 3861 (1975), and refs. therein.

55. Subtracting near the energy scale of physical interest makes higher order corrections to 1PI Green functions small by definition, which tends to improve the convergence for all Green functions. For explicit examples see ref. 52.

56. J.C. Taylor, Nucl. Phys. B33, 436 (1971). A.A. Slavnov, Theor. Math. Phys. 10, 99 (1972).

57. H. Kluberg-Stern and J.B. Zuber, Phys. Rev. D12, 467, 482 and 3159 (1975).

58. S.D. Joglekar and B.W. Lee, Annals of Physics 97, 100 (1976).

59. We follow arguments due to W. Caswell and F. Wilczek, Phys. Lett. 49B, 291 (1974) (see also D.J. Gross (ref. 51)). A refined version is given in ref. 57.

60. It is possible to let $\alpha_B = \alpha_B(k^2)$ chosen in such a way that $D_{\mu\nu}$ has other fixed tensor structures for all k^2 (e.g. $g_{\mu\nu}$) and there is no $\partial/\partial\alpha$ term in the RGE for these stagnant gauges (C.H. Llewellyn Smith, in preparation); this is convenient for some purposes (A. Carter and C.H. Llewellyn Smith - ref. 67).

61. S. Weinberg, Phys. Rev. D8, 3497 (1973).

62. This listing follows D.J. Gross (ref. 51) where the properties are discussed in more detail.

63. This is shown explicitly by the results of O. Nachtmann and W. Wetzel, Nucl. Phys. B146, 273 (1978).

64. See A.J. Buras (ref. 1) and references therein.

65. Yu.L. Dokshitser, Sov. Phys. JETP46, 641 (1977); Yu.L. Dokshitser, D.I. D'Yakonov and S.I. Troyan "Inelastic Processes in QCD," Proc. 13th Leningrad Winter School (1978), available as SLAC Trans.-183; D. Amati et al., Nucl. Phys. B140, 54 (1978); A.V. Efremov and A.V. Radyushkin, Dubna preprints E2-11535, 11725, 11726 and 11849 (1978); W.R. Frazer and J.F. Gunion, Phys. Rev. D19, 2447 (1979); J. Frenkel, M.J. Shailer and J.C. Taylor, Nucl. Phys. B148, 228 (1979); Y. Kazama and Y.-P. Yao, Phys. Rev. Lett. 41, 611 (1978) and University of Michigan preprints UM HE 78-53, 54; A.V. Radyushkin, Phys. Lett. 69B, 245 (1977); W.J. Stirling, Nucl. Phys. B141, 311 (1978) and B145, 477 (1978); C.T. Sachrajda, Phys. Lett. 76B, 100 (1978) and 73B, 185 (1978).

66. C.H. Llewellyn Smith, Acta Physica Austriaca Suppl. XIX, 331 (1978).

67. A.B. Carter and C.H. Llewellyn Smith, Oxford preprint 34/79.

68. S. Coleman and R.E. Norton, Nuovo Cimento 38, 438 (1965).

69. M. Gluck and E. Reya, DESY preprint 79/13.

70. D. Amati et al., Nucl. Phys. B146, 29 (1978); R.K. Ellis et al., Phys. Lett. 78B, 281 (1978) and Nucl. Phys. B152, 285 (1979); S. Gupta and A.H. Mueller, Phys. Rev. D20, 118 (1979); G. Sterman and S. Libby, Phys. Rev. D18, 3252 and 4737 (1978) and Phys. Lett. 78B, 618 (1978).

71. J. Kinoshita, J. Math. Phys. 3, 650 (1962); T.D. Lee and M. Nauenberg, Phys. Rev. 133, 1549 (1964). For an introductory discussion see R. Field, these proceedings.

72. For a recent treatment and references see G.G. Ross (Oxford preprint in preparation).

73. R.K. Ellis et al., ref. 70.

74. N. Sakai, CERN TH 2641 (1979); G. Altarelli et al., MIT preprint CTP 793 (1979).

JETS AND QUANTUM CHROMODYNAMICS

Stephen D. Ellis

University of Washington

Seattle, Washington

1. INTRODUCTION: JETS IN HADRON PHYSICS

In these lectures I will attempt to provide a fairly complete overview of our understanding of the rather nebulous concept of hadronic jets. The first few lectures will constitute a review of the simple concepts which were in vogue in the "old days," i.e., prior to the advent of quantum chromodynamics (QCD), and which are loosely referred to as the parton model.[1-3] This procedure is relevant, even though I personally believe QCD is probably the correct theory, for several reasons. First and foremost, the naive, pre-QCD picture of jets still provides an adequate description of essentially all of the present data. Secondly, it will provide a simple framework within which to define the concepts which will be discussed later within the context of QCD. Thirdly, it is important to realize that at the current level of our (in)ability to actually calculate with QCD, the theory serves primarily to legitimatize the parton model and provide a well-defined procedure for calculating corrections (some of which are sizable) to the parton model.[4] It is important to recall that in the late 1960's and early 1970's no field theory was understood which could lead both to the essentially free parton model and simultaneously to the confinement of quarks. It has only been with our increasing appreciation of non-Abelian gauge theories that this has become a distinct possibility. (Finally, I have a bias against rejecting old concepts simply because they seem somewhat dated—as a student at Caltech I was encouraged to believe that quantum field theory was no longer a useful tool for studying weak and strong interactions.)

Subsequent lectures will review the evidence for these general properties of hadronic jets in nature and the problems inherent in detecting them directly. We will then outline what QCD can say about jets. Of course the central tool is perturbative QCD which will provide only a partial description of jets. We will discuss various techniques for avoiding these difficulties in order to achieve precise predictions by focusing on quantities other than the theorists' concept of a jet. The final lectures will cover methods (i.e. proposals) for extending the kinematic regions where QCD is a useful calculational tool and, if there is time, cover some other even more speculative and less central issues. These lectures are admittedly incomplete, slanted as they are to cover topics which I at least partially understand. It is my fond hope that items which appear as question marks in these lectures will be fully elucidated in those of my fellow lecturers. In particular I anticipate some overlap with Chris Llewellyn Smith[5] and considerable convergence with Rick Field.[6]

Let us begin with a brief (and biased) overview of "all" of hadron physics[7] circa 1970. For our purposes the important features of hadronic interactions are the properties of particle production defined in terms of the single particle distribution

$$\frac{dN}{d^3p/E} = \frac{1}{\sigma_{TOT}} \frac{d\sigma}{d^3p/E} . \tag{1.1}$$

We first note that in all high energy hadronic interactions there is a unique spatial direction, the longitudinal direction, defined by the incoming beam (or beams in clashing ring machines) which dominates the structure of the final state. With respect to this direction, defined by unit vector $\hat{b}$, we can define three variables. The first is rapidity, suitable for discussing particles which are relatively slow in the overall CM and given by

$$y = \frac{1}{2} \ln \left[\frac{E + \vec{p}\cdot\hat{b}}{E - \vec{p}\cdot\hat{b}}\right] \underset{E>>m_\perp}{\approx} \pm \ln \frac{2E}{m_\perp} , \quad m_\perp \equiv \sqrt{m^2 + p_\perp^2} \tag{1.2}$$

where $\vec{p}_T$ is the component of momentum transverse to $\hat{b}$ ($|\vec{p}_T| = |\vec{p}\times\hat{b}|$) and is the second relevant variable. Definition (1.2) is chosen to give $y > 0$ for particles in the beam direction, $y < 0$ for particles in the "target" direction ($-\hat{b}$) and $y \approx 0$ for particles at rest in the CM. Boosting to another frame (e.g. the laboratory frame) corresponds to adding a constant to y (e.g., adding $Y/2 = \ln \sqrt{s}/m$, where s is the total CM energy squared, in the CM frame $-Y/2 < y < Y/2$, in the lab $0 < y < Y$). Finally, there is the variable appropriate to the energetic particles (the Feynman scaling variable)

$$x_F \equiv \frac{\vec{p}\cdot\hat{b}}{p_{max}} \underset{CM}{=} \frac{2\vec{p}\cdot\hat{b}}{\sqrt{s}} \underset{E >> m}{\approx} \pm \frac{2E}{\sqrt{s}} \ . \tag{1.3}$$

Note the range $0 < |x_F| \leq 1$ corresponds to a finite range in y near the edge of phase space, independent of s. For example, in the CM

$$\Delta y \equiv \frac{Y}{2} - y \approx \ell n \frac{\sqrt{s}}{2E} = \ell n \frac{1}{x_F} \ ,$$

while the range of y which grows as ℓns corresponds to $x_F \approx 0$.

In terms of these variables the single particle distribution exhibits the following (approximate) properties (variables not appearing explicitly have been integrated over): 1) p_T cutoff

$$\frac{dN}{dp_T^2 \, dy} \propto e^{-bp_T} \tag{1.4}$$

where b is approximately constant in y for $y < Y/2$ and about equal to 6/(GeV/c). (Actually a better form, applicable to all particle species, is $\exp(-b\sqrt{m^2+p_T^2})$ where m is the appropriate mass.)
2) (Feynman) Scaling:

$$\frac{dN_{ij}}{dx_f} \approx f_{ij}(x_F) \ , \tag{1.5}$$

independent of s. Here j = T or B and i is the produced particle type. For $x_F > 0$ (the beam fragmentation region) the functional form of $f_{ij=B}(x_F)$ depends only on the particle type i being detected and the beam particle type. While for $x_F < 0$ (target fragmentation) $f_{ij=T}$ depends only on the target type and i. Thus we have (approximate) factorization of the fragmentation regions. Experimentally it turns out that, for $|x_F|$ near 1 and $i \neq j$, $f(x_F)$ is well described by the functional form[8]

$$f_{ij}(x_F) \underset{|x_F| \to 1}{\approx} \text{const.} \times (1-|x_F|)^{\alpha_{ij}} \tag{1.6}$$

where α_{ij} is approximately a positive integer depending on i and j (for i=j the effective α is negative due to diffractive effects), and essentially measures how difficult it is to produce i from j.

3) Rapidity plateau

$$\left.\frac{dN_i}{dy}\right|_{y<<Y} \approx C_i \tag{1.7}$$

where C_i is (approximately) a constant dependent only on the particle type i being detected and not on the target or beam particle types.

4) Short-range correlations. With the two particle distributions d^2N/dy_1dy_2, we can define a correlation function describing how the probability to find a particle at one rapidity is affected by the presence of another particle at a second rapidity

$$C(y_1,y_2) \equiv \frac{\frac{d^2N}{dy_1dy_2} - \frac{dN}{dy_1}\frac{dN}{dy_2}}{\frac{dN}{dy_1}\frac{dN}{dy_2}} \propto e^{-\gamma|y_1-y_2|} \quad \text{for } |y_1-y_2| > 1. \tag{1.8}$$

The (approximate) exponential behavior for large $\Delta y = |y_1-y_2|$ indicates the short range nature of the correlation. Experimentally γ has a value near 1/2 (correlation length $\approx$ 2 units of rapidity). The short range (in rapidity) nature of hadronic interactions is also indicated by the fact that exclusive cross sections (except elastic) always fall rapidly with energy ($\sim s^{-\beta}$).

This general longitudinal picture of the final state is illustrated in Fig. 1. We see immediately that the average total multiplicity must behave essentially as

$$\langle n_{TOT}\rangle \simeq Y \sum_i C_i + \text{const.} \tag{1.9}$$

where the sum is over particle types. Note also that since $dy \simeq \frac{dp}{E} \approx \frac{dp_{||}}{E} \approx \frac{dx_F}{x_F}$ (where $p_{||} = |\vec{p}\cdot\hat{b}|$ and a p_T cutoff is

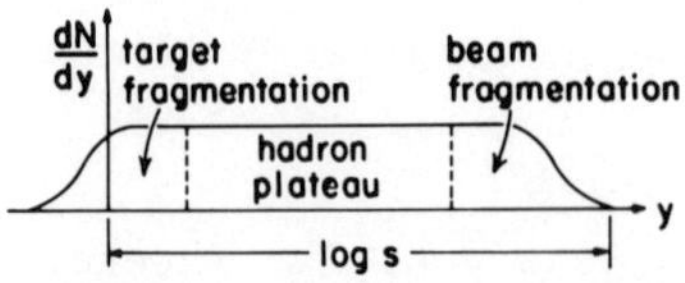

Fig. 1. Idealized final particle distribution in hadronic collisions plotted versus rapidity.

understood), Eq. (1.7) implies

$$\left.\frac{dN_i}{dx_F}\right|_{x_F \simeq 0} \approx \frac{C_i}{x_F} \quad . \tag{1.10}$$

We will soon see how these properties relate to the parton model and QCD. It is important to understand first how one's intuition was led by the observations. For contrast the interested student should consider how these properties, particularly the constants α_{ij} and γ, were understood originally in the context of the Multi-peripheral Model and Regge Theory. These concepts must eventually find a home in QCD. A partial list of useful references is given in Refs. 9 and 7.

2. THE PARTON MODEL

Let us now turn to a brief review of the simple parton model.[1-3] We wish to focus on processes where there exists a large invariant momentum transfer, Q^2. The classic examples as illustrated in Fig. 2 are deep inelastic lepton-hadron scattering(a) and large p_T production in purely hadronic reactions(d), corresponding to spacelike momentum transfer, and e^+e^- annihilation into hadrons(b) and lepton pair production(c) in hadronic reactions ("Drell-Yan"), corresponding to time-like momentum transfer. In this limit one can motivate what is essentially an impulse approximation to describe the hadronic process in terms of a more fundamental "hard" process involving the underlying constituents of the hadrons, the partons. (We will not bother to motivate that hadrons must contain underlying "gearwheels" but assume that it is obvious.) Via the impulse approximation we ignore the interactions between the partons in a single hadron during the time of the interaction, i.e. we implicitly assume that there is a fixed momentum scale Λ_s characterizing the strong confining forces such that they can be ignored on time scales characterized by $|Q^2| \gg \Lambda_s^2$. Thus we have complete, <u>incoherent</u> factorization of the above processes into effectively three factors.

The first factor describes <u>all</u> the strong (presumably confining) interactions between the partons within a single hadron <u>prior</u> to the hard interaction. It is a product (one for each incoming hadron) of <u>probability</u> distributions (recall the incoherent factorization) describing the probability to find partons of certain momenta within each of the incoming hadrons. The absence, in the old days, of any true quantum field theory which both confined partons and simultaneously allowed the impulse approximation discussed above led to a certain skepticism on the part of a large

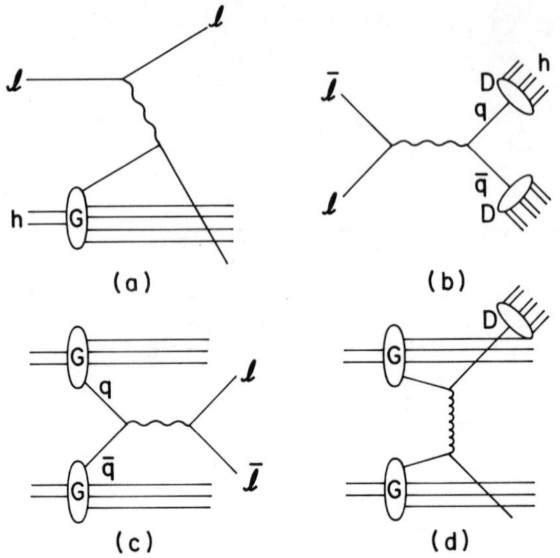

Fig. 2. Processes treated in the parton model: a) deep inelastic lepton-hadron scattering, b) e^+e^- annihilation into hadrons, c) hadronic production of lepton pairs ("Drell-Yan"), d) large p_T production in hadronic reactions.

fraction of the physics community concerning the validity of the parton model and also required the partisans of this approach to simply conjecture a form for this distribution.[1-3] Fortunately considerable quantities of intuition had been derived from the study of, admittedly incomplete, field theory models and from the phenomenological observations noted earlier along with those arising from the first measurements of deep inelastic lepton-proton scattering. There was general agreement (amongst the partisans) that the relevant variables for describing the parton distributions in a frame where the hadron is highly relativistic are x, the longitudinal momentum fraction carried by the parton $p_{L,parton} = x\, p_{hadron}$, and the transverse momentum $\vec{k}_T$ of the quark relative to the hadron direction. It is important to note that, since we are discussing (essentially) a wave function squared, this distribution is not Lorentz invariant (the frame choice is important if the explicit Hamiltonian is not known) and that, since we treat the partons as free during the hard interaction, they have effectively a fixed mass ("on-shell" partons). Thus we can define the probability to find a parton of type a in hadron A in momentum range dx d^2k_T as

$$dP_a^A = G_a^A(x,k_T^2)\,dx\ d^2k_T\ . \tag{2.1}$$

The observations in hadron physics discussed earlier led to the expectation that a) the k_T distribution is rapidly cut off; b) the x distribution behaves as $(1-x)^{\alpha_{Aa}}$ as $x \to 1$ (dependent on A and a); c) the x distribution behaves as $\bar{C}_a(dx/x)$ as $x \to 0$ (independent of A). In the earlier work the k_T distribution was taken to be a δ function.

The second factor in the total scattering process is the cross section describing the hard process itself which involves the scattering of essentially free, "on-shell", point-like partons. We label this quantity generically $d\hat{\sigma}/d\hat{t}(Q^2)$. It is a function of the large momentum transfer Q^2 and various dimensionless ratios (angles). For the processes involving leptons and hadrons (Figs. 2a-c) this cross section is determined in lowest order by the weak and electromagnetic interactions. For the purely hadronic process (Fig. 2d) it remains a point of debate (see the Field Lectures[6]) although one gluon exchange was the early choice of parton model purists.

Since $d\hat{\sigma}/d\hat{t}$ was known for deep inelastic lepton-hadron scattering, data on this process were a source of information about $G_a^A(x)$. By the factorization hypothesis this G_a^A can then be used to calculate the other processes. Note that in the case of 2 hadrons in the initial state (Figs. 2c,d) it was necessary to "show" that possible initial state hadronic interactions did not substantially alter the parton distributions. A modern treatment of this question is the domain of the lectures of C. Llewellyn Smith.[5]

<u>Finally</u>, we come to the third factor which is the central issue of these lectures. The result of the hard scattering process (except in Fig. 2c) is to place one or more partons at an isolated position in momentum space. Recall that in an incoming hadron all the energetic (finite x), essentially free, partons are nearby each other in rapidity space and in k_T space (presumably the relevant variables) and also, of course, in configuration space. After the hard scattering one parton is now going the opposite direction (Fig. 2a in the Breit frame), a parton pair exists at equal and opposite momenta where none existed before (Fig. 2b) or 2 partons emerge at large k_T (Fig. 2d). Since their momenta are very different, these partons will start to separate in configuration space. Since partons are not observed as asymptotic states in the laboratory, as this separation develops the isolated parton must evolve (and interact) so as to produce the observed hadrons. This "fragmentation" process was postulated to exhibit the by now standard features of scaling in the momentum fraction z along the original quark direction and of a sharp cutoff in the momentum components transverse to the parton direction. Thus the distribution of hadrons of type B resulting from the fragmentation of parton of type b is described by a function D_b^B

$$dN_b^B = D_b^B(z, q_T^2)\,dz\; d^2q_T \tag{2.2}$$

which is presumed to a) be cut off in q_T; b) behave as $(1-z)^{\alpha_{bB}}$ as $z \to 1$ (dependent on B and b); c) behave as $\bar{\bar{C}}_B(dz/z)$ for $z \to 0$ (independent of b). Again the transverse momentum distribution was ignored in the earliest applications.

The general result for a hadronic process $d\sigma$ is a convolution over the above parton model factors. For the case $A+B \to C+x$ where, for example, C is produced with transverse momentum p_T at angle θ in the CM and A and B have total energy s we have (ignoring internal transverse momenta for now)

$$E_C \frac{d\sigma}{d^3p_C}(s, p_T, \theta) \underset{\sim}{\sim} \sum_{a,b,c} \int dx_a\, dx_b\, dz_C\, G_a^A(x_a) G_b^B(x_b) \frac{d\hat{\sigma}}{\pi d\hat{t}}(\hat{s}, \hat{t}) \times D_c^C(z_C)\, \delta[\quad] \tag{2.3}$$

where $\hat{s}, \hat{t}$ are the internal variables of the hard scattering process and all constraints of energy-momentum conservation (which serve to define $\hat{s}, \hat{t}$ in terms of s, p_T, θ, x_a, x_b, and z) and quantum number conservation (relating a,b and c) are in the final factor. The implementation of this complete parton model structure is left to other lecturers.

In the next lecture we shall discuss the general (pre-QCD) properties ascribed to the distribution D, i.e. to the <u>jet</u> of hadrons which result from the fragmentation from an essentially isolated (by assumption) parton. The term jet arose from the image of a narrow, in q_T, spray of hadrons looking much like half of Fig. 1. The thinking student will note that by this reasoning the partons which are spectators with respect to the hard scattering (see Fig. 2) can also lead to a jet of hadrons along the beam lines as in ordinary hadronic reactions. Are these two types of jets similar? identical? Are the $\bar{\bar{C}}_B$ equal to the C_B? Are the C's really constants? [There are hints of $\ell n^2 s$ behavior.] We return to these questions below. We will also shortly consider real data and check whether jets do appear and where.

3. PARTON MODEL JETS AND DATA

We are now at the conceptual situation of the mid-1970's. By this point the partons were generally considered to be quarks (good old u, d, s) except for certain other possibilities advocated[11] in the purely hadronic process (Fig. 2d). The quarks were furthermore thought to form a $\underset{\sim}{3}$ under a new symmetry group for the strong interactions, SU_3 of color. While it was perceived that this implied that the quarks interacted via the exchange of an $\underset{\sim}{8}$ of colored gluons, it was not yet understood how to translate this into quantitative calculational techniques to treat the problem of jets. Actually there was progress being made in understanding the distribution function G in the context of this colored theory but not in a form useful to the parton model. Hence the fragmentation process was still treated largely by theoretical conjecture and phenomenological construct.[12]

Consider first the conceptually simplest process, e^+e^- annihilate into a virtual photon which in turn produces a quark-anti-quark pair. As the quark and the anti-quark (a color $\underset{\sim}{3}$ and a color $\underset{\sim}{\bar{3}}$) separate in space something happens. Typical words and images (even today) are that a color flux tube forms between them (many gluons?) and this in turn leads to the production of many quark-anti-quark pairs as idealized in Fig. 3. Some time later these pairs regroup themselves into the observed hadrons, all color singlets. The parton model says nothing in detail about these intermediate steps. (We will pursue this topic in the context of perturbative QCD later in these lectures.) The parton model, however, does include (postulate) considerable detail about the distribution of hadrons as discussed earlier. There should be limited transverse momentum with respect to the quark (jet) direction. There should be scaling in the variable

$$z = \frac{\vec{p}_h \cdot \vec{p}_q}{|\vec{p}_q|^2} \simeq \frac{2\vec{p}_h \cdot \hat{p}_q}{E_{cm}}$$

where $\hat{p}_q$ is the unit vector in the quark (jet) direction and $E_{cm}^2 = Q^2 (= W^2)$ is the total energy squared of the e^+e^- system. A plateau should form in the central (low momentum) region (often called the current plateau). Finally by the factorization assumption, corresponding to having only short range correlations (except for energy conservation and those long range correlations suggested by having always a quark and its anti-quark), the event-

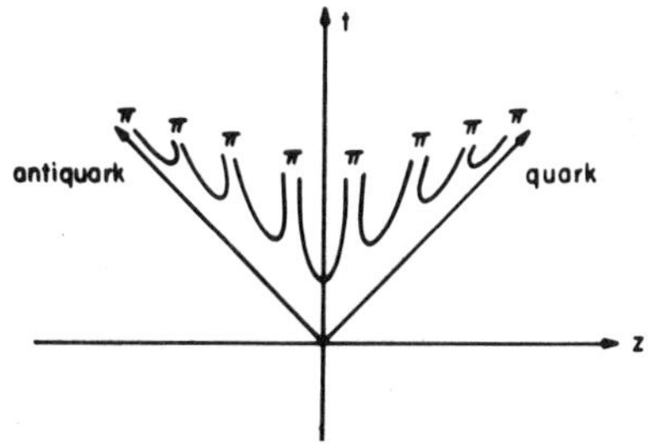

Fig. 3. Idealized view of how a quark pair evolves into hadrons.

by-event structure of the two jets should be independent. This general situation is illustrated in Fig. 4, the analogue of Fig. 1.

The fact that these properties are very nearly exactly true in the data from SPEAR[13] was a great boost to this whole scheme. To analyze the data a jet axis was defined in each event by finding that direction which minimizes the sum of squares of transverse momenta with respect to that direction (the "sphericity" axis).* The distributions of particles as functions of the variables p_T^2, $x_{||} = 2p_{||}/E_{cm}$ and $y = (1/2)\ell n((E+p_{||})/(E-p_{||}))$, defined with respect to this jet axis, are indicated in Fig. 5a,b,c. In Fig. 5a,c the events were required to have at least one particle with $x_{||} \geq 0.3$ to help define the jet axis. Clearly p_T is cut off, there is scaling in $x_{||}$ and there is a plateau. Actually $\langle p_T \rangle$ varies with $E_{cm} = W \equiv \sqrt{Q^2}$ as shown in Fig. 6 but appears to asymptote near 365 MeV/c at the highest energy. (These data contain some contamination from charmed meson production which affects the large p_T behavior.[13]) By trying various cuts of the form of requiring at least one hadron with $x_{||} > x_{max}$ for various x_{max}, kinetmatic factorization was tested. As indicated in Fig. 7, the result is that the jet from which the fast meson comes is highly distorted but the other jet is essentially unaffected.

In order to pass on to a new generation of physicists an "olde time" confusion, I note that $\langle p_T \rangle$ in the above data also depends on $x_{||}$ as indicated in Fig. 8. At least some of this effect can be understood as the classic "seagull" effect seen also in hadron physics and lepton hadron physics.[14] The effect is easily

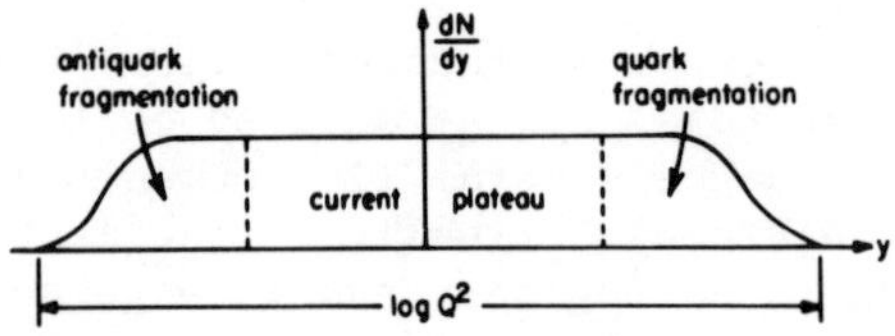

Fig. 4. Idealized final particle distribution in e^+e^- annihilation plotted versus rapidity.

---

*For completeness I include the definitions of the three standard variables used to define jet axes in multiparticle final states: 1) sphericity $\equiv 3/2 \min \{\sum_i |\vec{p}_{Ti}|^2 / \sum_i |\vec{p}_i|^2\}$,

2) thrust $\equiv 2 \max \{\tilde{\sum_i} p_{||i} / \sum_i |\vec{p}_i|\}$ where $\tilde{\Sigma}$ is a sum over a single hemisphere defined by the axis, 3) spherocity $\equiv (4/\pi)^2 \min \{\sum_i |\vec{p}_{Ti}| / \sum_i |\vec{p}_i|\}^2$.

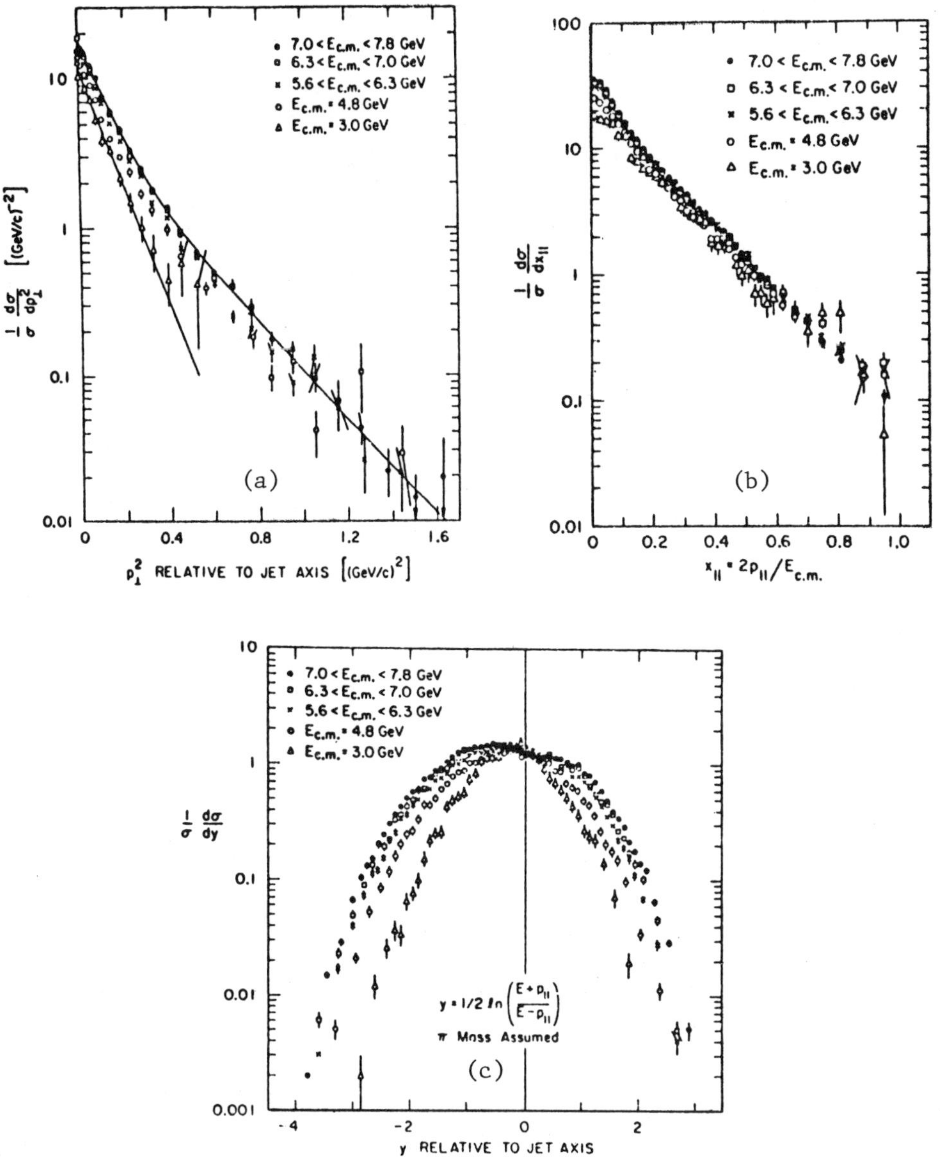

Fig. 5. Particle distributions observed in e^+e^- annihilation plotted versus: a) p_T^2, b) $x_{||}$, c) y, for various values of E_{cm}. In a) and c) there is a requirement that the fastest particle has $x_{max} > 0.3$ and the distributions are normalized to the total rate of such events. In a) the particles are in the hemisphere opposite the fastest one and in c) the x_{max} particle is not plotted but would be at positive y (Ref. 13).

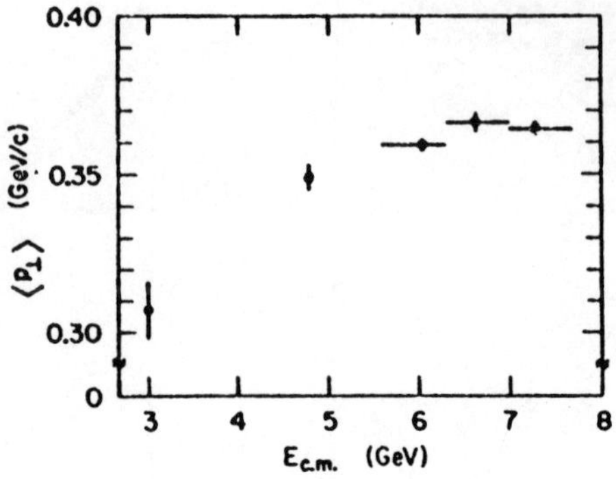

Fig. 6. Average p_T relative to the jet axis in events with $x_{max} > 0.3$ versus E_{cm} (Ref. 13).

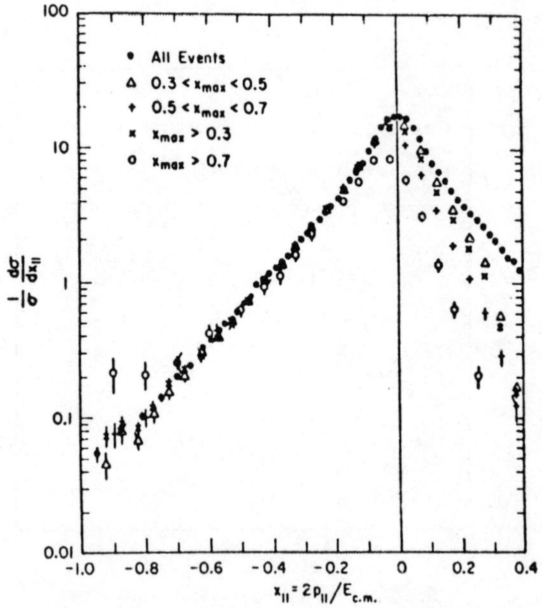

Fig. 7. Observed particle distributions in e^+e^- annihilations versus $x_{\|}$ with $7.0 < E_{cm} < 7.8$ GeV. The fastest particle in one hemisphere has $x_{\|}$ in the specified x_{max} range, defines the positive $x_{\|}$ direction and is not plotted. Each curve is normalized to the total rate for the specified x_{max} bin (Ref. 13).

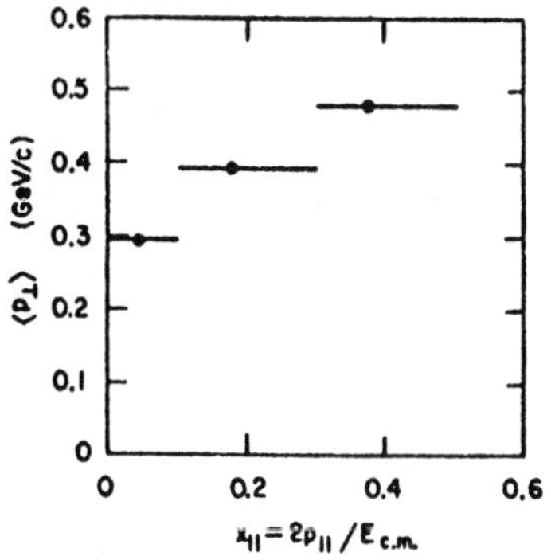

Fig. 8. Average p_T relative to the jet direction in e^+e^- annihilation versus $x_{\parallel}$ for $7.0 < E_{cm} < 7.8$ GeV and for particles in the jet opposite a particle with $x_{max} > 0.3$ (Ref. 13).

"understood" if one presumes that the p_T distribution is constant in rapidity. Then when variables are changed from y to $x_{\parallel}$ there is an extra factor of $E_{cm}/2E$ introduced where $E = \sqrt{p_{\parallel}^2 + m^2 + p_T^2}$ ($dy = dx_{\parallel} \cdot E_{cm}/2E$) and thus the p_T distribution is $x_{\parallel}$ ($p_{\parallel}$)-dependent. The questions of whether there is something else afoot and why, in fact, the p_T distribution is constant in y not $x_{\parallel}$ remain to be resolved. Note that, in any case, large $x_{\parallel}$ hadrons have $\langle p_T \rangle$ well above the average. Another feature of the data worth noting at this point is that the height of the current plateau is in fact experimentally the same as the hadronic plateau. The $\bar{C}$'s do apparently equal the C's (i.e., the coefficient of ℓnW^2 in $\langle n \rangle$ is equal in e^+e^- and in hadronic events). While this possibility was favored for various reasons,[15] it is probably not a trivial observation even at our present naive level. For example, if we postulate[16] that, whatever the details of the mechanism which ultimately produces the final hadrons, their multiplicity behaves as the appropriate bremsstrahlung spectrum of gluons for separating color charges, then we expect 4/9 the multiplicity for separating $\underset{\sim}{3}$-$\underset{\sim}{\bar{3}}$ (as in e^+e^- annihilation) as for separating $\underset{\sim}{8}$-$\underset{\sim}{8}$ (as would be the case in processes involving gluon exchange between hadrons). The factor 4/9 is just the ratio of the "charges" squared for a $\underset{\sim}{3}$ versus an $\underset{\sim}{8}$ (more later). In this context the universality of the plateau suggests that all processes are "$\underset{\sim}{3}\underset{\sim}{\bar{3}}$-like."

Turning to the case of deep inelastic lepton-hadron scattering we have only a slightly more complex situation. In the naive model we view the interaction as the collision of a nucleon and a virtual γ or W as illustrated in Fig. 9 for a W^+ on a proton (νp scattering). The relevant variables are the x of the struck quark, $x = Q^2/2m\nu$, and the mass squared and lab energy, Q^2 and ν, of the

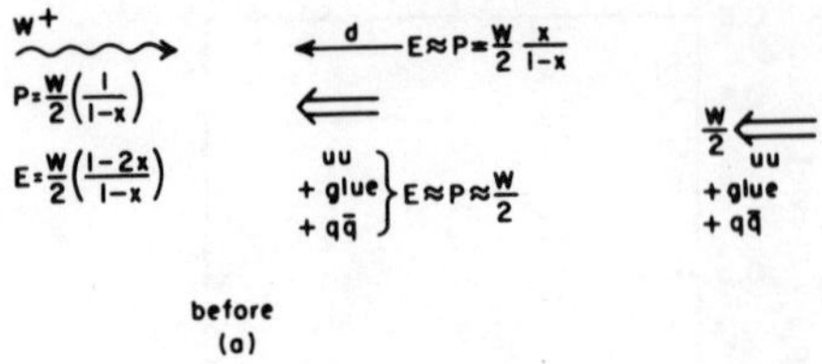

Fig. 9. Naive view of lepton-hadron scattering as the collision of a virtual W^+ (for νp) and a d-quark as seen in the proton-W^+ CM system.

γ or W. Just after the collision a single quark moves right while the spectator quarks (and glue and pairs) move left. Potentially this will result in a right moving quark or current jet, as in e^+e^-, and a left moving hadron jet, as in πp, with a transition region in between which was itself once thought to be complicated.[2] However this situation is pictorially and apparently physically a separating $\underset{\sim}{3}\,\underset{\sim}{\bar{3}}$ just as in e^+e^-. Data[14] indicate only a single smooth plateau (but who knows what will happen at large energies) and the picture is approximately as in Fig. 10 where the quark jet is defined[2] to have length $\ell n\ Q^2$ in rapidity and the hadron jet $\ell n\ \omega$ ($\omega = 2m\nu/Q^2 = 2q\cdot p/Q^2$). Data suggesting the single plateau at Fermilab[14] are shown in Fig. 11. We note for fun that the same data show the "seagull" effect as illustrated in Fig. 12 where the scaling variable $x_F\big|_{exp} = E_h/E_{vis} \simeq p_h/\nu = z$.

Perhaps most importantly for the argument of a universal plateau the data show almost no non-kinematic Q^2 dependence in interesting variables for $|Q^2| > 4\ \text{GeV}^2$.

I will not discuss the observation of jets in large p_T physics in any detail as that will be covered by R. Field.[6] I note only that there is ample evidence[17] that large p_T particles tend to appear in fairly well collimated groups with typical tranverse sizes characterized by 500 MeV/c. Whether this is significantly different from the e^+e^- result of 360 MeV/c is unclear as the biases of the sampling procedure are quite different, with the large p_T experiments being strongly biased toward larger z values where $\langle p_T \rangle$ is larger. There also exists a fundamental (theoretical) uncertainty for those experiments which trigger on jets at large p_T in associating a specific plateau (slow) particle with a specific jet. Since the trigger rate falls rapidly with total jet p_T, the corresponding uncertainty in this quantity ($\pm$500 MeV/c $\underset{\sim}{\approx}$ $\pm$1 slow particle) leads to a large uncertainty in the magnitude of the jet cross section. It also appears that the "size" (in angle or p_T) of the observed jets is strongly influenced by the size of the detector (calorimeter) used.[18]

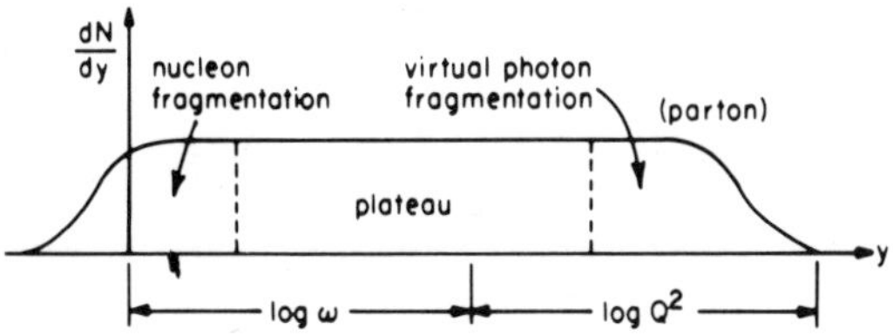

Fig. 10. Idealized final particle distribution in deep inelastic lepton-hadron scattering plotted versus rapidity.

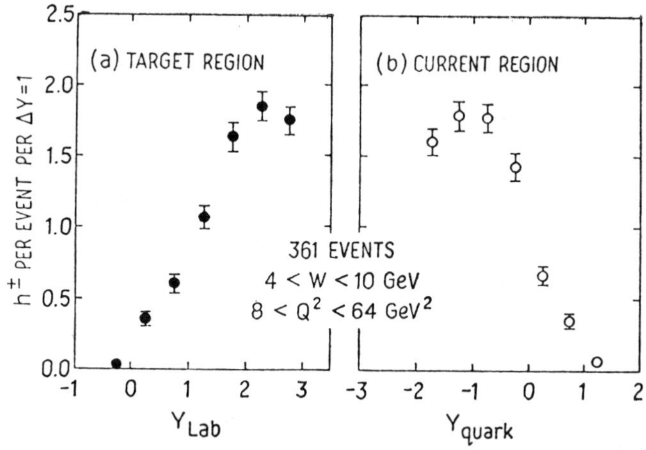

Fig. 11. Observed charged hadron distributions in neutrino-nucleon scattering plotted versus a) Y_{lab}, and b) $Y_{quark} = Y_{lab} - \ell n(2m\nu/1\ \mathrm{GeV}^2)$ and exhibiting a similar plateau in the two regions (Ref. 14).

In summary, the naive parton model of jets works quite nicely (for the carefully chosen data I've shown you) and before confronting QCD proper I would like to include a little more discussion of simple intuitive ideas about how jets (quarks) fragment. I remind you that we still can't calculate everything (in fact, very little) from first QCD principles.

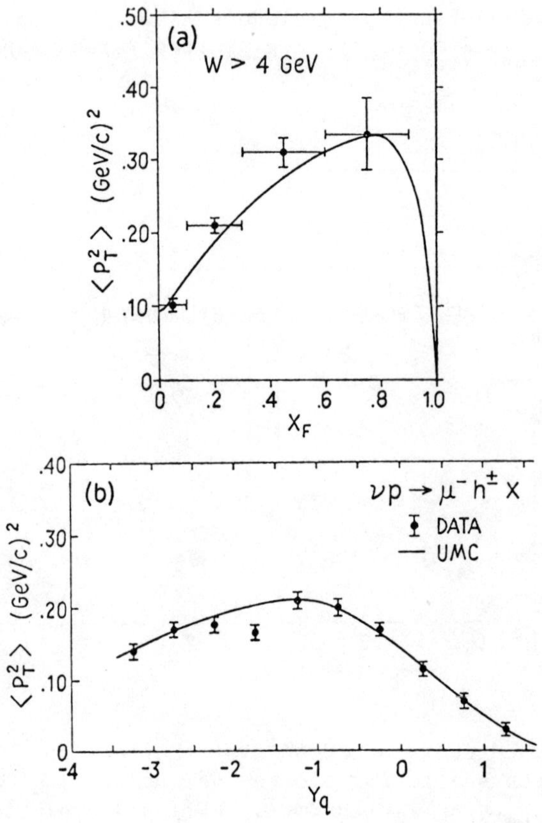

Fig. 12. Data from neutrino-nucleon scattering on the average p_T^2 versus a) x_F for positive x_F (current fragmentation), b) $Y_q = Y_{lab} - \ln(2m\nu/1\ GeV^2)$ where positive x_F corresponds to $Y_q > -2$ (Ref. 14).

4. A SPECIFIC JET MODEL

We have seen that, within the context of the naive quark-parton model plus a few general properties of jets abstracted from hadronic data, it is possible to understand the qualitative features of the final states in deep inelastic lepton-hadron and hadron-hadron processes. Can we be more specific without invoking the dynamical details of QCD? The answer is, of course, yes and it serves us well to understand the features of the data in a more dynamical, if phenomenological, fashion before turning to QCD to see if these features are reproduced. In fact few of the details to be discussed here are derivable in perturbative QCD, i.e. we are still focusing on "soft" physics, but connections can be made!

The basic motivating picture in the following is discussed by Field and Feynman[12] and others.[19] In this picture we imagine that as a quark of type a, which is isolated in momentum space, starts to become isolated in configuration space it simply fragments by emitting a meson (only pseudoscalars for the moment) leaving a quark of type b as illustrated in Fig. 13a. Note that this "first rank" meson contains the initial quark (it is an $a\bar{b}$ meson). The probability that this process leaves a quark b with faction η of the original quark's momentum is assumed to be given by

$$dP \equiv f(\eta)d\eta \ . \tag{4.1}$$

This function is a priori unknown (and will remain essentially so in these lectures) being related to the unsolved confinement question and is the major input to this problem (along with a p_T cutoff, and the treatments of resonances and flavor symmetry breaking). Since there can be only one such meson containing the original quark, the normalization is

$$\int_0^1 d\eta \ f(\eta) = 1 \ . \tag{4.2}$$

Having treated the first quark as fragmenting in isolation, we proceed in the same fashion with quark b. It will fragment into a meson and a quark c where the fragmentation is again described by $f(\eta')$ where η' is the fraction of b's momentum carried by c. This sequence is to proceed as in Fig. 13b until the remaining momentum is some fixed amount $p_0 \gtrsim$ few hundred MeV/c. Thus the probability to find a meson at a given z (to z + dz) in a quark fragmentation jet, D(z)dz, has two pieces. The meson can be the first rank meson and be described by f(1-z) or it can be a product of the subsequent fragmentation. An integral equation results

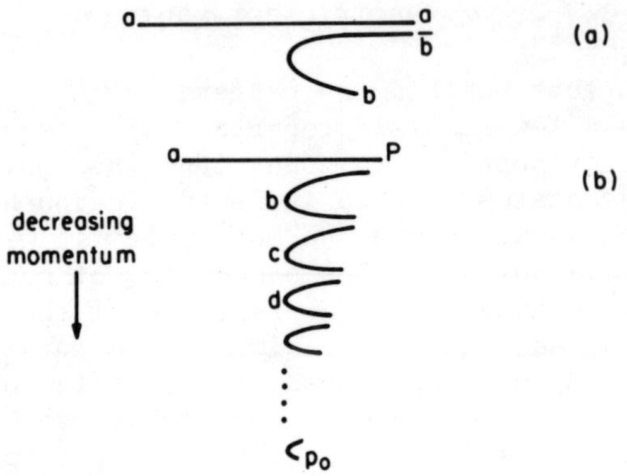

Fig. 13. Naive picture of simple hierarchal cascade development of jet with the first step in a) and the result of several steps (until a small momentum p_0 is reached) displayed in b).

from this picture

$$D(z) = f(1-z) + \int_z^1 \frac{d\eta}{\eta} f(\eta)\ D(z/\eta) \tag{4.3}$$

where we have implicitly set f(z) and D(z) to zero for z outside the interval $0 \le z \le 1$, and we are ignoring flavor quantum numbers for the moment. Such (Volterra) integral equations (so important also in the QCD analysis) are easily treated by taking moments. Following the notation of Ref. 12, define

$$M(r) \equiv \int_0^1 z^r\ D(z)\ dz\ , \tag{4.4a}$$

$$C(r) \equiv \int_0^1 \eta^r\ f(\eta)\ d\eta\ , \tag{4.4b}$$

and

$$A(r) \equiv \int_0^1 z^r\ f(1-z)\ dz\ . \tag{4.4c}$$

Then the rth moment of Eq. (4.3) becomes

$$M(r) = A(r) + C(r)\ M(r) \tag{4.5a}$$

or

$$M(r) = \frac{A(r)}{1-C(r)} = A(r) \times \left(1 + \frac{C(r)}{1-C(r)}\right)\ . \tag{4.5b}$$

Thus, introducing a new function by

$$\int_0^1 \eta^r \; g(\eta) \; d\eta \equiv \frac{C(r)}{1-C(r)} \quad , \tag{4.6}$$

we have

$$D(z) = f(1-z) + \int_z^1 g(\eta) \; f(1-z/\eta) \; \frac{d\eta}{\eta} \; . \tag{4.7}$$

The function $g(\eta)$ can be interpreted as the probability that all the mesons of lower rank than a given meson (that is all those mesons which were emitted "earlier" in the cascade) have left behind a fraction η of the total jet momentum. Several observations are immediately possible. From Eqs. (4.2) and (4.4b,c) it follows that

$$A(1) = C(0) - C(1) = 1 - C(1) \tag{4.8}$$

and thus that

$$M(1) = 1 \; . \tag{4.9}$$

Hence a distribution of the form of Eq. (4.3) automatically conserves momentum, a necessary feature. Also for small r, when C(r) and A(r) approach 1, M(r) diverges as 1/r. In detail we have

$$M(r) \xrightarrow[r\to 0]{} \frac{R}{r} \; , \tag{4.10a}$$

where

$$\frac{1}{R} = - \left. \frac{dC(r)}{dr} \right|_{r=0} = - \int_0^1 \ell n \; \eta \; f(\eta) \; d\eta \; . \tag{4.10b}$$

This in turn implies that, for small z, D(z) behaves as R/z and the fragmentation function indeed has a rapidity plateau as desired. More specifically the multiplicity of mesons (one type for the moment) in a jet of total momentum P with individual momenta greater than p_0 is given by

$$\langle \eta \rangle = \int_{p_0/P}^1 dz \; D(z) = R \; \ell n \frac{P}{p_o} + O[(p_o/P)^0] \; . \tag{4.11}$$

Hence R gives the height of the plateau.

Turning now to the probability of finding two mesons, we can proceed in an exactly analogous fashion. Consider first the case where the meson at z_1 is of "lower rank," i.e. in following along the quark lines it is "closer" to the original quark, than the meson at z_2. (Note that there is no constraint on the relative size of z_1 and z_2.) If the probability to find this situation is $\bar{D}_2(z_1,z_2)$, then there follows immediately the integral equation

$$\bar{D}_2(z_1,z_2) = \frac{f(1-z_1)}{1-z_1} D(z_2/(1-z_1)) + \int \frac{d\eta}{\eta^2} f(\eta)\ D_2(z_1/\eta, z_2/\eta) \ , \tag{4.12}$$

where the interpretation of the two terms should be immediate. In precisely the same fashion as above we solve for $\bar{D}_2$ as

$$\begin{aligned} \bar{D}_2(z_1,z_2) = \frac{f(1-z_1)}{1-z_1} D(z_2/(1-z_1)) + \int_{z_1+z_2}^{1} g(\eta)\ f(1-z_1/\eta) \\ \times D(z_2/(\eta-z_1)) \frac{d\eta}{(\eta-z_1)} \end{aligned} \tag{4.13}$$

with the same function $g(\eta)$. Now in general the two-meson distribution includes both "rank orderings" and the full distribution is given by

$$D_2(z_1,z_2) = \bar{D}_2(z_1,z_2) + \bar{D}_2(z_2,z_1) \ . \tag{4.14}$$

With the explicit form of Eq. (4.13) it is straightforward to again show generalized momentum conservation as it appears in the two-particle distribution. Define

$$\begin{aligned} M(r,s) &\equiv \int_0^1 dz_1 z_1^r \int_0^{1-z_1} dz_2 z_2^s\ D_2(z_1,z_2) \\ &= \int_0^1 dz_2 z_2^s \int_0^{1-z_2} dz_1 z_1^r\ D_2(z_1,z_2) \ , \end{aligned} \tag{4.15a}$$

and

$$C(r,s) \equiv \int_0^1 dz\ z^r (1-z)^s\ f(1-z) \ . \tag{4.15b}$$

Then using the definitions of Eqs. (4.4), we find

$$M(r,s) = \frac{C(r,s)M(s) + C(s,r)M(r)}{1 - C(r+s)} . \tag{4.16}$$

In particular the average momentum of particle 2 is found from

$$M(r,1) = \frac{C(r,1)M(1) + C(1,r)M(r)}{1 - C(r+1)} \tag{4.17a}$$

or

$$\begin{aligned} M(r,1) &= \frac{A(r) - A(r+1) + M(r)\,[C(r) - C(r+1)]}{1 - C(r+1)} \\ &= M(r) - \frac{A(r+1)}{1 - C(r+1)} - \frac{M(r)[1 - C(r)] - A(r)}{1 - C(r+1)} \end{aligned} \tag{4.17b}$$

where the special cases of Eq. (4.15b) were evaluated using Eq. (4.4) and single particle momentum conservation, Eq. (4.9), was inserted. Now using the result (4.5) one easily finds

$$M(r,1) = M(r) - M(r+1) . \tag{4.18a}$$

This implies the desired result

$$\int dz_2\; z_2\; D_2(z_1,z_2) = (1-z_1)\; D(z_1) . \tag{4.18b}$$

The simplest and, for many reasons, the most appealing solutions to the above equations arise for $f(\eta)$ being power behaved

$$f(\eta) = (\alpha+1)\; \eta^{\alpha} . \tag{4.19}$$

This leads immediately to

$$C(r) = \frac{\alpha + 1}{r + \alpha + 1} , \tag{4.20a}$$

$$C(r)/(1-C(r)) = \frac{\alpha + 1}{r} , \tag{4.20b}$$

and thus

$$g(\eta) = \frac{\alpha + 1}{\eta} . \tag{4.20c}$$

Using this result in Eq. (4.7) yields

$$D(z) = \frac{1}{z}\, f(1-z) = \frac{\alpha + 1}{z}\, (1-z)^{\alpha} . \tag{4.21}$$

As discussed earlier this form exhibits the correct qualitative behavior both for $z \to 0$ and $z \to 1$. Actually the first expression for $D(z)$ in Eq.(4.21) (i.e., $f(1-z)/z$) is a reasonable first approximation for more general forms for $f(\eta)$.

The two-particle distribution corresponding to Eq. (4.19) has the analogue simple form

$$D_2(z_1,z_2) = \frac{(\alpha+1)f(1-z_1-z_2)}{z_1z_2} = \frac{(\alpha+1)^2}{z_1z_2}(1-z_1-z_2)^\alpha . \tag{4.22}$$

This generalizes immediately to the N-particle distribution which is obviously of the form

$$D_N(z_1,z_2\ldots z_N) = (\alpha+1)^N \quad [1 - \sum_{i=1}^{N} z_i]^\alpha \prod_{i=1}^{N} \frac{1}{z_i} . \tag{4.23}$$

This clearly satisfies the generalized momentum sum rule

$$\int dz_N\, z_N\, D_N(z_1,z_2\ldots z_{N-1},z_N) = (1 - \sum_{i=1}^{N-1} z_i) D_{N-1}(z_1,z_2\ldots z_{N-1}) . \tag{4.24}$$

While our derivation of Eq. (4.23) seemed fairly special, it is in fact a rather general (approximate) result in the limit where we ignore correlations and transverse momenta and assume a rapidity plateau. It is found to arise naturally in 1+1 dimensional field theories[20] where particles are produced uniformly in rapidity with no correlations except momentum conservation. For various phenomenological and theoretical reasons Field and Feynman originally preferred[21] a $D(z)$ function such that $D(1) = f(1) \neq 0$. They chose

$$f^{FF}(\eta) = 1 - a + 3a\eta^2 \tag{4.25}$$

such that

$$D^{FF}(z) = \frac{1}{z}\left[\frac{3}{(3-2a)} + \frac{3a\, z^2}{2a-1} + \frac{2a(2a^2-3a-2)}{(3-2a)(2a-1)}\, z^{3-2a}\right] . \tag{4.26}$$

This form is, however, no longer in theoretical favor and the case $D(1) = 0$ as above seems more likely correct.

We can now go back and keep track of flavor, i.e. isospin and strangeness, still producing only pseudoscalar mesons. From the "physical" picture of the process expressed in Fig. 13, it is clear that the question of flavor propagation rests on the flavor of the secondarily produced quark pairs. We are, of course, ignoring baryon production and other subtleties which are discussed, for example, in the last two papers of Ref. 19. Assuming that these pairs (the "sea") respect isospin but not SU_3, we label the probability of u or d quarks by γ and strange quarks $1-2\gamma$, i.e., $\gamma_u = \gamma_d = \gamma$, $\gamma_s = 1-2\gamma$. Guessing that a 2 to 1 (u/s, d/s) ratio is a reasonable description of the observed SU_3 breaking Field and Feynman chose $\gamma = 0.4$. Thus, using quark labels a,b,c = u,d,s (...), the flavorful analogue of Eq. (4.3) for a quark q to fragment into a meson with valence quarks $a\bar{b}$ is given by

$$D_q^{a\bar{b}}(z) = \delta_{qa}\gamma_b f(1-z) + \int \frac{d\eta}{\eta} f(\eta) \sum_c \gamma_c \, D_c^{a\bar{b}}(z/\eta) \qquad (4.27)$$

where as usual the first term corresponds to the observed meson containing the initial quark while the second term arises from the produced pairs, with uncorrelated flavor. Thus mesons which arise from the "sea" of pairs are described by

$$D_{sea}^{a\bar{b}}(z) \equiv \sum_c \gamma_c D_c^{a\bar{b}}(z) \; . \qquad (4.28)$$

Using Eq. (4.27) on the right-hand side yields

$$D_{sea}^{a\bar{b}}(z) = \gamma_a\gamma_b \; f(1-z) + \int \frac{d\eta}{\eta} f(\eta) \; D_{sea}^{a\bar{b}} \qquad (4.29a)$$

or

$$D_{sea}^{a\bar{b}} = \gamma_a\gamma_b \; D(z) \qquad (4.29b)$$

by our previous techniques (see Eq. (4.7)). Thus only the "first rank" meson exhibits flavor correlations with the initial quark

$$\begin{aligned} D_q^{a\bar{b}} &= \delta_{qa} \; \gamma_b \; f(1-z) + \gamma_a\gamma_b \; (D(z) - f(1-z)) \\ &\equiv \delta_{qa} \; \gamma_b \; f(1-z) + \gamma_a\gamma_b \; \bar{F}(z) \end{aligned} \qquad (4.30)$$

in the notation of Ref. 12. The first term is the "first rank" contribution and the second is from all higher ranks. The interested student can derive all sorts of interesting (and, generally speaking, true) results from these relations. We note a few here. Since we put in isospin invariance

$$D_u^{\pi^+}(z) = D_{\bar{u}}^{\pi^-}(z)$$
$$D_u^{K^+}(z) = D_{\bar{u}}^{K^-}(z)\ , \quad \text{etc.} \tag{4.31a}$$

Also

$$D_u^{K^+}(z) = \frac{\gamma_s}{\gamma_d} D_u^{\pi^+}(z) = 0.5\ D_u^{\pi^+}(z)$$
$$D_u^{K^-}(z) = \frac{\gamma_s}{\gamma_d} D_u^{\pi^-}(z) = 0.5\ D_u^{\pi^-}(z) \tag{4.31b}$$

and thus

$$D_u^{K^-}(z)/D_u^{K^+}(z) = D_u^{\pi^-}(z)/D_u^{\pi^+}(z) \tag{4.31c}$$

which is the statement that the fractional effect of the first rank meson is the same for π's and K's. Finally we also have

$$D_u^{\pi^+}(z)-D_u^{\pi^-}(z) = D_s^{K^-}(z)-D_s^{K^+}(z) = \frac{\gamma_u}{\gamma_s}(D_u^{K^+}-D_u^{K^-})$$
$$= \gamma\ f(1-z) = 0.4\ f(1-z)\ . \tag{4.31d}$$

The charge correlation due to the first rank particle can be seen in the ratio

$$D_u^{\pi^+}(z)/D_u^{\pi^-}(z) = \frac{\gamma\ f(1-z) + \gamma^2\ \bar{F}(z)}{\gamma^2\ \bar{F}(z)} \tag{4.32}$$

which diverges as $z \to 1$ and $\bar{F}$ vanishes. This can be tested (as we shall see) in neutrino scattering where a ν (W^+) scatters primarily from d quarks producing u quarks while a $\bar{\nu}$ (W^-) produces d quarks (ignoring scattering from antiquarks).

Finally we can ask about the electric charge distribution in a jet initiated by a quark q. Since a meson $a\bar{b}$ has charge $e_a - e_b$ (quark a has charge e_a), we have

$$\langle Q_q(z)\rangle = \sum_{a,b} (e_a - e_b) D_q^{a\bar{b}}(z) = (e_q - \sum_b \gamma_b e_b)\ f(1-z) \tag{4.33}$$

where, due to the symmetry of quarks and antiquarks in the sea, the $\bar{F}$ term makes no contribution. The quantity

$$e_{sea} \equiv \sum_a \gamma_a e_a = \gamma - 1/3 \tag{4.34}$$

is the average charge of a quark in the sea and clearly is zero in the SU_3 limit. Summing over all z (the z-dependence will be highly model dependent and be affected by resonance decays, etc.) we find the average charge on a u quark jet is

$$\langle Q_u \rangle = 2/3 - (\gamma - 1/3) = 1 - \gamma$$

and for d and s

$$\langle Q_d \rangle = \langle Q_s \rangle = -\gamma \quad . \tag{4.35}$$

This is a fairly general result and independent of the limitation discussed below.[22] In order to be composed finally of only (color singlet) hadrons, a quark-initiated jet must pick up an anti-quark (or reject a quark). Assuming this anti-quark is an (uncorrelated) anti-quark from the sea, we have the charge on the jet as equal to that on the initial quark minus that on an average quark in the sea as above. Note, however, for an "average" jet (e.g. initiated by a quark from the sea) the average charge vanishes $\sum_a \gamma_a \langle Q_a \rangle \equiv 0$.

5. A SPECIFIC JET MODEL REVISITED

To make connection with data using the pre-"perturbative QCD" jet model discussed in the previous lecture, we must account for two more real effects — resonances and internal q_T of hadrons with respect to the jet. In the previous discussion we produced only pseudoscalar mesons whereas we expect vector (and perhaps more rapidly spinning) mesons to be produced in the primary process at a comparable rate. These will subsequently decay into the observed pseudoscalars and change both the shape of the distribution and the presence of correlations. In fact it appears[23] that the bulk of the short range correlations arise from the presence of resonances or, more generally, "clusters". A careful study of these effects requires a good Monte Carlo program,[12] here we make only a simple observation. If a "parent" meson decays into two massless child mesons with an isotropic decay spectrum, the fraction of the parent's momentum carried by either child is uniformly distributed between 0 and 1. Hence, if the parent's distribution is $D_P(y)$, the child's distribution $D_{CH}(z)$ is given by (a factor 2 for 2 children)

$$D_{CH}(z) = \int_z^1 \frac{2\,dy}{y} D_P(z/y) \ . \qquad (5.1)$$

Hence, if $D_P(y)$ behaves as $(\alpha+1)(1-y)^{\alpha}$ for y near 1, it follows from Eq. (5.1) that $D_{CH}(z)$ behaves as $2(1-z)^{\alpha+1}$ for z near 1. In general the children will exhibit a steeper spectrum as the decays produce more slow final mesons. Thus if we are looking for final mesons near $z = 1$ we will not see the decay products of resonances. This is seen in large p_T single particle production[6] where the trigger biases favor large z. It is observed[17] that the primary $\rho^{\pm}$ production rate is comparable to the π rate ($\rho^+ + \rho^- \approx \pi^0$) but only a few percent of the π triggers result from ρ decay.

Transverse momentum within the jet can be added in various (necessarily ad hoc) ways. The procedure of Ref. 12 is to give each pair in the sea a zero net transverse momentum but a gaussian distribution for the back-to-back transverse momentum of each member of the pair. A primary meson is then assigned the total transverse momentum carried by the quark and anti-quark which compose it. This appears to describe most of the existing jet data if $\langle k_T \rangle_{primary} = 439$ MeV/c. As this pertains to the size of the final hadrons we shall not be able to calculate this number from QCD here. Note that due to the decays of some (typically half) of the primaries, this parameter choice gives final pions with average $\langle k_T \rangle_{\pi} \simeq 323$ MeV/c.

As a brief illustration that all of this does a reasonable job of describing the bulk of existing jet data (mostly at low Q^2) I show the graph from Ref. 12 showing the fit to deep inelastic data on distributions in jets in Fig. 14. Note that it is a good fit for $a \approx 0.77$ ($D(1) \approx .07$) if half of the produced particles are vector mesons and decay. The quantum number structure is illustrated in ν data[24] in Fig. 15 where the enhancement of positive over negative mesons from u quark jets as $z \to 1$ is shown

$$\left(\frac{\nu \to \pi^-}{\bar{\nu} \to \pi^-} = \frac{d \to \pi^-}{u \to \pi^-} = \frac{u \to \pi^+}{u \to \pi^-} \right) \ .$$

Finally, from other recent neutrino data[25] we have measurements of the average charge on quark jets. These results depend to some extent on how the plateau (i.e. sea) is divided between quark and target fragmentation but generally the results are

$$\langle Q_u \rangle \simeq 0.61 \pm .09 \ ,$$

and (5.2)

$$\langle Q_d \rangle = -.15 \pm .21$$

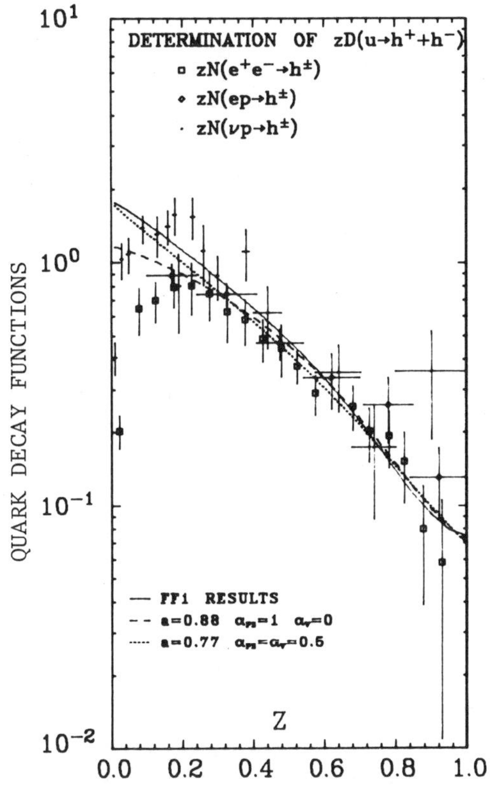

Fig. 14. Charged particle distribution $D_u^{h^+}(z) + D_u^{h^-}(z)$ arising from the jet model of Ref. 12 for $\alpha_{ps} = 1$, $\alpha_V = 0$ (only pseudoscalars), a = 0.88 (dashed curve), and for $\alpha_{ps} = \alpha_V = 0.5$ (equal numbers of primary pseudoscalars and vectors), a = 0.77 (dotted curve) compared to various kinds of data. The solid curve is from an earlier, less complete Field and Feynman treatment.

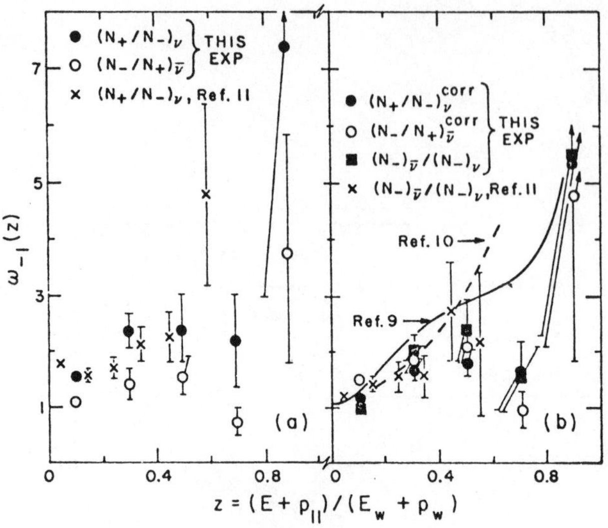

Fig. 15. Data from neutrino reactions on a heavy liquid target on the ratios of produced charged hadrons $(N_+/N_-)_\nu$ and $(N_-/N_+)_{\bar{\nu}}$ for total hadronic energy in the range 1 to 10 GeV. Part a) includes some proton contamination while b) has been corrected for this. The curve labeled Ref. 9 is from Ref. 21 of the present paper (Ref. 24).

where I have simply added the error of various techniques of definition in quadrature and averaged the means. This is in excellent agreement with Eqs. (4.34) for $\gamma = 0.4$ given the uncertainty of $\langle Q_d \rangle$.

Along with these upbeat conclusions comes the observation from the calculations of Ref. 12 that "inelastic" quark jets, including all the above effects, especially resonance decay, are really not very quark like. As an example Fig. 16, from Ref. 12, shows how the excess charge (Eq. (4.34)), isospin and strangeness of a u quark jet are distributed in rapidity $Y_z = \ell n\, 1/z$. They are clearly spread out (over a range of 4 units) and hence not distinctively concentrated at the large z end. These calculations are a telling reminder that there is "no free lunch," it will be difficult to experimentally associate a specific quark flavor with an observed jet structure. Likewise the sum of the energies and momenta of the hadrons in a jet model calculation[12] never exactly equal those of the quark.

Before turning to perturbative QCD itself it is useful to evaluate the extent to which the above naive picture can be close to the truth. Besides lacking the explicit QCD corrections to be

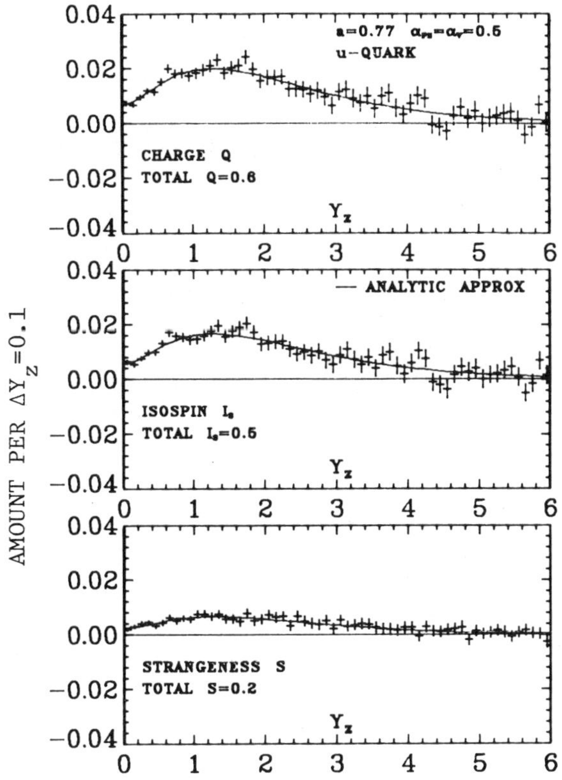

Fig. 16. Distribution of the charge Q, isospin component I_3, and strangeness S in $Y_z = \ell n(1/z)$ for a u quark jet in the model of Ref. 12. The solid curves are analytic approximations to the full Monte Carlo calculations.

discussed below, the model has one major conceptual flaw. We know that the cascade concept of a quark sequentially emitting a meson and becoming a slower quark of different flavor which in its turn emits a meson and a slower quark yet and so on until the final product is a quark with momentum $p_0 \simeq 300$ MeV/c cannot be literally true.[26] Consider the case of $e^+e^- \to q\bar{q}$. If the quark and anti-quark fragment independently in the manner just described they will each travel some average distance d in opposite directions before the final low momenta q and $\bar{q}$ appear with low enough invariant mass to form a final state meson. Now the distance d will increase with the number of steps in the cascade and hence with the total energy W ($d \propto \ell n\ W$) and hence I can imagine an experiment where the final $q\bar{q}$ are a FOOT apart! This surely is not the case of confined quarks. Hence the actual temporal sequence of appearance of final hadrons must be in the reverse order (inside-out cascade[2]) with the slowest hadrons appearing first. In words the actual picture[2] must correspond approximately to, for the e^+e^- example, 1) separating $q\bar{q}$ pair; 2) formation of gluon cloud (string) between them; 3) for $q\bar{q}$ separation of order 1 Fermi the energy in the gluon cloud is sufficient to lead to substantial quark pair formation, some of which become associated to form color singlet, low mass pairs (final mesons), and some are carried along with (accelerated by) the original quark or anti-quark, partially screening its charge, and generating polarization currents; 4) the polarization currents are slowly accelerated toward the corresponding leading quark and anti-quark drawing energy from them (at a rate near 1 GeV/fermi traveled) and producing final hadrons at ever increasing rapidity; 5) the polarization current catches the leading quark and the final hadrons are produced.

We note that, while some form of long range correlation (vector gluons) is implied by the screening of the fractional charges of the original quark and anti-quark, the final hadron distribution may not show strong correlations as they appear from the screened region "behind" the polarization current. Since the actual interactions implied here are basically soft (low momentum and energy transfers) they are consistent with the impulse approximation of the parton model and the observed p_T cutoff. It is my view that, while the explicit sequential picture of the above jet model is incorrect, the actual distributions calculated are a reasonable first approximation to the real world at some large but fixed momentum scale. The actual (if unfortunate) effects of resonances are adequately expressed. The general shape in z is a good first approximation, being essentially the form of any model with only small, short-range correlations and a p_T cutoff. Thus I believe this model to give a reasonable description of the soft physics part of the fragmentation process — including the general z shape, a rapidity plateau with short-range correlations, and a p_T cutoff.

We turn next to what perturbative QCD can say about this. We will see that perturbative QCD leads to what we might call "hard" corrections to the above soft skeleton — giving a Q^2 dependence to the z shape, hard tail to the p_T distribution, and a further smearing of the quantum number correlations. However, it will not be possible to motivate the soft physics results purely from perturbative QCD — they remain part of the nonperturbative confinement question. Hence, if we wish to work only with perturbative QCD, we will have to learn to ask different questions — in particular the concept of jet, as described above, must be modified somewhat. This later conclusion is a nice complement to the experimental situation that it is, in principle, impossible to define a jet precisely. For a given final state which particles are to be associated with a specific jet? Since a quark can never fragment in total isolation (without leaving some quarky footprints), the wee (sea) hadrons "associated" with that quark must always correspond to some interactions with other quarks and hence have ill-defined parentage. (Again there are long-range correlations to screen the quark quantum numbers.) So to make matters more precise, we will eventually turn to new experimental measures.

6. PERTURBATIVE QCD AND THE LEADING LOG

We turn finally to the question of what QCD tells us about jets. I will leave to Chris Llewellyn Smith[5] the demonstration that the general factorized structure of the parton model survives the inclusion of perturbative corrections and assume that the student is already familiar with the technical details of QCD.[5,6] I will give here only a brief review of these points.

The property of QCD (or non-Abelian gauge theories in general) which suggests the possibility of a parton model like approximation is, of course, asymptotic freedom, i.e., the effective coupling constant specifying interactions at a momentum scale Q^2 is given by

$$\alpha_s(Q^2) = \frac{\bar{g}^2(Q^2)}{4\pi} \approx \frac{g_\mu^2}{4\pi} \frac{1}{1 + \frac{g_\mu^2}{\pi^2}(11N_c - 2N_F)\ell n(Q^2/\mu^2)} , \qquad (6.1)$$

where g_μ^2 is the coupling at the (arbitrary) reference scale μ, N_c is the number of colors ($SU(N_c)$) and N_F is the number of flavors. This result is the correctly renormalized coupling resulting from keeping the leading log correction at each order in g_μ^2. Clearly for an appropriate choice of μ^2, call it Λ^2, we have (for $N_c = 3$)

$$\alpha_s(Q^2) \approx \frac{4\pi}{(11 - \frac{2}{3}N_F)\ \ell n(Q^2/\Lambda^2)} \qquad (6.2)$$

$$\equiv \frac{4\pi}{\beta_o\ \ell n(Q^2/\Lambda^2)} ,$$

where this choice, in effect, corresponds to minimizing the next order corrections and may be process dependent. Assuming that we are studying processes where $Q^2/\Lambda^2 >> 1$ (it turns out that realistic Λ are roughly of order the hadron scale Λ_s discussed earlier $\sim$100 MeV) as appropriate to the parton model, we are encouraged to proceed by calculating perturbative ($\alpha(Q^2) << 1$) corrections to the basic model. Typical contributions are illustrated in Fig. 17. Note that there are three basic varieties arising from real gluon emission, virtual gluon emission and reabsorption (including propagator corrections not shown), and fundamentally new terms where one of the incident partons is a gluon. For our purposes this last contribution means that we must also consider the properties of gluon initiated jets.

When one proceeds to calculate perturbative corrections such as in Fig. 17, one finds immediately that they exhibit logarithmic mass divergences (and ultraviolet divergences which cancel against the wave function renormalization). These logarithms tend to cancel the advantages of a coupling which is vanishing as in Eq. (6-2) and must be treated to all orders. If we control the divergences by giving all quanta a mass Q_0^2, then diagram-by-diagram there are terms both like $(\alpha_s \ln Q^2/Q_0^2)^n$ and $(\alpha_s \ln^2 Q^2/Q_0^2)^n$. In momentum space these arise both from configurations where two or more quanta are parallel-moving (collinear divergences) and from the emission (or absorption) of soft quanta (true infrared divergences). The overlap of these two conditions results in the $\ln^2$ terms which cancel when real and virtual emissions are added coherently. In configuration space these divergences correspond to the properties of the quarks and gluons over long time scales, both

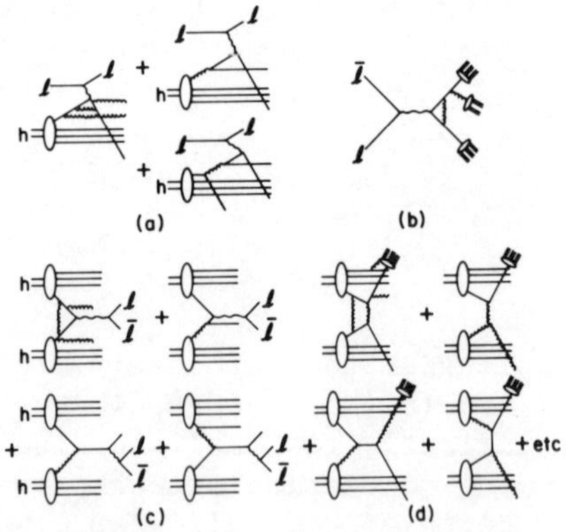

Fig. 17. Typical perturbative QCD corrections to the parton model processes: a) deep inelastic lepton-hadron scattering, b) e^+e^- annihilation, c) hadronic lepton pair production, d) large p_T production. Wavy lines are W's or γ's while curly lines are gluons.

before and after the hard scattering process. Thus the factorization property[27] of these singularities discussed in detail in the Llewellyn Smith lectures[5] is perhaps not surprising. The singularities do require a renormalization of the various distributions but leave the basic structure of the parton model intact. In particular, those logarithms having to do with long times after the hard collision affect the structure of the final state and thus the structure of jets. We will treat this topic first and return later to the subject of the mass finite corrections to the final state in perturbative QCD.

We note first that, while the renormalization properties of the structure functions, the G's, can be treated formally (and completely) by short distance operator product expansion techniques,[28] there is at present no analog powerful tool for the D's. However, just as in the case of the structure functions, the effects of the leading (single) logarithmic singularities ($\sim[\alpha_s \ell n\ Q^2/Q_0^2]^n$) can be summarized, to all orders n, in terms of an integral-differential equation.[29,30,31] Furthermore, due to the factorization properties of the infrared singularities just mentioned, one can formally show[32] that the effects of the leading logarithms (anomalous dimensions) for the fragmentation functions should be the same as for the structure functions.

Focusing on the fragmentation processes in Fig. 17 we have the perturbative corrections illustrated in Fig. 18. The structure of these diagrams immediately suggests an integral equation structure very similar to Eq. (4.3), describing how momentum fractions are shared. The difference here is that we are simultaneously focusing on the dependence of D on the momentum scale Q^2. We are isolating the part of the diagram which yields the "leading log" behavior in $\ell n\ Q^2/Q_0^2$. (The ℓn^2 terms have already been cancelled against the virtual contributions.) Now Q_0^2 is a cutoff supplied, for example, by giving the outgoing quark/gluon a mass Q_0^2, which is chosen to be a scale at which perturbation theory is good ($\alpha_s(Q_0^2) << 1$) and power corrections ($\sim\Lambda^2/Q_0^2$) can be ignored (i.e.

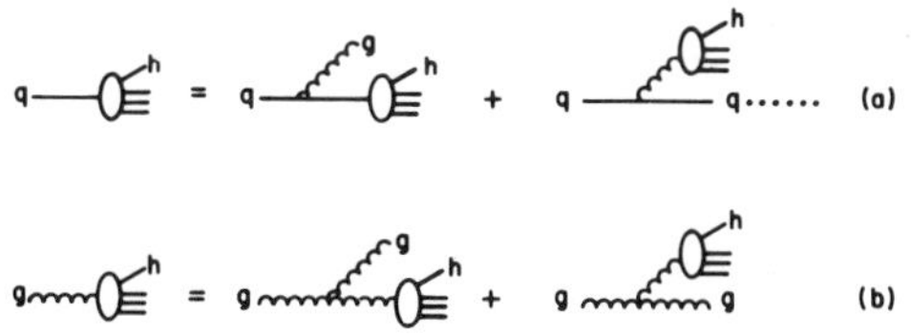

Fig. 18. "Perturbative" corrections to the fragmentation process: a) quark fragmentation with the final hadron coming either from the further fragmentation of the original quark after gluon emission or from the fragmentation of the emitted gluon, b) the corresponding situation for gluon fragmentation.

$Q_0^2 \gg \Lambda^2$). This connection to the "leading log approximation" is accomplished by equating the content of Fig. 18 to the derivative of D with respect to the variable $t = \ell n\, Q^2/\Lambda^2$. Thus in QCD the D functions <u>cannot</u> scale but are rather functions of some scale like Q^2. Of course the precise value of Q^2 in the D's will depend on how the factorization process (factoring the jet from the hard scattering) is carried out but it is, in principle, well defined by the perturbation theory since it should be chosen to minimize the higher order (mass non-divergent) perturbative corrections. For our purposes we take Q^2 as given. Thus, in the leading log approximation, Fig. 18 becomes

$$\frac{\partial}{\partial t} D_q^h(z,t) = \frac{\alpha(t)}{2\pi} \int_z^1 \frac{d\eta}{\eta} \left[P_{qq}(\eta) D_q^h(\tfrac{z}{\eta},t) + P_{q\to G}(\eta) D_G^h(\tfrac{z}{\eta},t)\right] \tag{6.3a}$$

for each type of quark q and

$$\frac{\partial}{\partial t} D_G^h(z,t) = \frac{\alpha(t)}{2\pi} \int_z^1 \frac{d\eta}{\eta} \left[P_{GG}(\eta) D_G^h(\tfrac{z}{\eta},t) + P_{G\to q}(\eta) \sum_{q_i,\bar{q}_i}^{N_f} D_q^h(\tfrac{z}{\eta},t)\right] \tag{6.3b}$$

for gluons where there is an explicit sum over quark flavors ($P_{G\to q,\bar{q}}$ is flavor independent). The functions P_{qq}, $P_{q\to G}$, $P_{G\to q}$, P_{GG} are the analogues of the function $f(\eta)$ in Eq. (4.3) describing how the momentum is shared amongst the quarks and gluons in the infrared divergent parts of Fig. 18. Note that the role of the inhomogeneous term $f(1-z)$ in Eq. (4.3) is played by the appropriate "off diagonal" term. The P functions are given by[29]

$$P_{qq}(x) = P_{\bar{q}\bar{q}}(x) = C_F\left(\frac{1+x^2}{1-x}\right)_+ \tag{6.4a}$$

$$P_{q\to G} = P_{\bar{q}\to G} = C_F\left(\frac{1+(1-x)^2}{x}\right) \tag{6.4b}$$

$$P_{G\to q} = P_{G\to\bar{q}} = \frac{1}{2}(x^2 + (1-x)^2) \tag{6.4c}$$

$$P_{GG} = 2C_A\left(\frac{1-x}{x} + x(1-x) + \left(\frac{x}{1-x}\right)_+ - \frac{\delta(1-x)}{12}\right) - \frac{N_f}{3}\delta(1-x) , \tag{6.4d}$$

with the definition

$$\int_0^1 dx\, f_+(x) g(x) \equiv \int_0^1 dx\, f(x) \big(g(x) - g(1)\big) , \tag{6.4e}$$

where $C_F = (N_C^2 - 1)/2N_C = 4/3$ and $C_A = N_C = 3$ are the effective "charges" (color traces) of the quarks and gluons.

I remind you again that these functions are the coefficients of the single logarithmic divergent pieces, with the double logarithms removed, explaining the presence of the $(\ )_+$ functions and the δ functions. Also note that since the number of quarks minus anti-quarks in a jet is fixed and since momentum is conserved, these functions, as order α_S corrections to a zeroth order distribution, satisfy

$$\int_0^1 dx\, P_{q\to q}(x) = 0\ , \tag{6.5a}$$

$$\int_0^1 dx\, x\, [P_{q\to q}(x) + P_{q\to G}(x)] = 0\ , \tag{6.5b}$$

$$\int_0^1 dx\, x\, [P_{G\to G}(x) + 2N_f\, P_{G\to q}(x)] = 0\ . \tag{6.5c}$$

The equations (6.3) describe the evolution of a q or G at scale Q into a q or G at scale Q_0 via the emission of essentially collinear quarks and gluons. Again, since this is not really a perturbative correction, due to the logarithms, these bits must be summed and treated completely. This is possible because, if factorization is both true to all orders (in logs) and universal (see Refs. 5, 6 and 27), then in the above equations the $P_{a\to b}$ have a systematic (calculable) expansion in $\alpha_s\,(Q^2)$ (i.e. in $1/\ell n\, Q^2$), derivable for the G function from the OPE, including the expansion of $\alpha_s(Q^2)$ itself.

Again we have equations of the Volterra type (coupled now instead of inhomogeneous) which we may solve by taking moments as before. Thus we define

$$M_a^h(r,t) \equiv \int_0^1 dz\, z^r\, D_a^h(z,t) \tag{6.6a}$$

and the "anomalous dimensions"

$$A_{a\to b}(s) \equiv \int_0^1 dz\, z^s\, P_{a\to b}(z)\ . \tag{6.6b}$$

Taking the moment of Eq. (6.3) we have

$$\frac{\partial}{\partial t} M_q^h(r,t) = \frac{\alpha(t)}{2\pi}\left[A_{q\to q}(r)M_q^h(r,t) + A_{q\to G}(r)M_G^h(r,t)\right]\ , \tag{6.7a}$$

$$\frac{\partial}{\partial t} M_G^h(r,t) = \frac{\alpha(t)}{2\pi}\left[A_{G\to q}(r)\sum_{q,\bar{q}}^{N_f} M_q^h(r,t) + A_{G\to G}(r)M_G^h(r,t)\right] . \quad (6.7b)$$

Now, with $\alpha(t)/2\pi = 2/\beta_0 t$, I define the variable

$$\kappa = \frac{2}{\beta_0}\,\ell n(t/t_0) \quad (6.8)$$

where $t_0 = \ell n\, Q_0^2/\Lambda^2 >> 1$ by assumption of the relevance of perturbation theory above. Using an obvious matrix notation we solve the above as (very similar to a result in Ref. 6)

$$M_a^h(r,t) = [\exp \kappa A(r)]_{a\to b}\, M_b^h(r,t_0) \quad (6.9)$$

where $M_b^h(r,t_0)$ is typically a $2N_f + 1$ dimensional vector and it is understood that $A_{a\to b}$ is diagonal in flavor for $q \to q$ but flavor and charge conjugation independent for $G \to q(\bar{q})$. More conventionally (and simply) we can consider the flavor nonsinglet contribution by looking at $D_q^h - D_q^{\bar{h}}$ or $D_{q_i}^h - D_{q_j}^h$ in which case the gluon contributions drop out. Focusing on the former we have (e.g., for all π's in the final state, $D_G^{\pi^+} - D_G^{\pi^-}$

$$\frac{\partial}{\partial t}\left[M_q^h(r,t) - M_q^{\bar{h}}(r,t)\right] = \frac{\alpha(t)}{2\pi} A_{q\to q}(r)\left[M_q^h(r,t) - M_q^{\bar{h}}(r,t)\right] \quad (6.10a)$$

or

$$\left[M_q^h(r,t) - M_q^{\bar{h}}(r,t)\right] = \left[\exp \kappa A_{q\to q}(r)\right]\cdot\left[M_q^h(r,t_0) - M_q^{\bar{h}}(r,t_0)\right] \quad (6.10b)$$

$$= \left[\frac{\ell n\ Q^2/\Lambda^2}{\ell n\ Q_0^2/\Lambda^2}\right]^{\frac{2A_{q\to q}(r)}{\beta_0}}\cdot\left[M_q^h(r,t_0) - M_q^{\bar{h}}(r,t_0)\right]. \quad (6.10c)$$

Evaluating the A(r) we have $\left(C_F = \frac{N_C^2 - 1}{2N_C} = \frac{4}{3},\ C_A = N_C = 3\right)$

$$A_{q\to q}(r) = C_F\left[-\frac{1}{2} + \frac{1}{(r+1)(r+2)} - 2\sum_{j=2}^{r+1}\frac{1}{j}\right] = A_{NS}(r)\ , \quad (6.11a)$$

$$A_{q\to G}(r) = A_{\bar{q}\to G}(r) = C_F\,\frac{r^2 + 3r + 4}{r(r+1)(r+2)}\ , \quad (6.11b)$$

$$A_{G\to q}(r) = A_{G\to\bar{q}}(r) = \frac{1}{2}\,\frac{r^2+3r+4}{(r+1)(r+2)(r+3)}\ , \qquad (6.11c)$$

$$A_{G\to G}(r) = C_A\left[-\frac{1}{6} + \frac{2}{r(r+1)} + \frac{2}{(r+2)(r+3)} - 2\sum_{j=2}^{r+1}\frac{1}{j}\right] - \frac{N_f}{3}\ . \qquad (6.11d)$$

This can be compared to data[25] from deep inelastic neutrino scattering (where the flavor of the scattered quark can be controlled) once the factorization of x dependence ($x = Q^2/2M\nu$) from z dependence ($z = 2\vec{h}\cdot\vec{q}/Q^2$) has been established. (If it does not factor as it shouldn't if higher order corrections are important one can use simultaneous moments in x and z instead.[33]) In Fig. 19 it is indicated that the data of Ref. 25 are consistent (only) with factorization, at least for $Q^2 > 10\ \mathrm{GeV}^2$. Various moments are plotted in Fig. 20 a) and b) versus Q^2 where lowest order (LO) QCD predictions (those above) with $\Lambda^2_{LO} = .5\ \mathrm{GeV}^2$ are shown as solid lines for comparison. In Fig. 20 c) and d) are shown the "Perkins Plots" for D moments (slopes are ratios of A(r)) and again the solid lines are for Eqs.(6.10) and (6.11a). Finally Fig. 21 indicates that if the ν data are used to find the value of Λ^2_{LO} the results are consistent with the structure function (xF_3) results.

To study the corresponding singlet function we define an average over q and $\bar{q}$ (the analog of D_{sea})

$$D^h_{q_s} \equiv \frac{1}{2N_f}\sum_{i=1}^{2N_f} D^h_{q_i}(z,t) \qquad (6.12)$$

so that

$$\frac{\partial}{\partial t} D^h_{q_s}(z,t) = \frac{\alpha(t)}{2\pi}\int_z^1 \frac{d\eta}{\eta}\left[P_{q\to q}(\eta)\,D^h_{q_s}(\tfrac{z}{\eta},t) + P_{q\to G}(\eta)\,D^h_G(\tfrac{z}{\eta},t)\right]\ , \qquad (6.13)$$

or in terms of moments we have the 2x2 equation (with (6.3b))

$$\frac{\partial}{\partial t}\begin{pmatrix} M^h_{q_s}(r,t) \\ M^h_G(r,t)\end{pmatrix} = \frac{\alpha(t)}{2\pi}\begin{pmatrix} A_{qq}(r) & A_{q\to G}(r) \\ 2N_f A_{G\to q}(r) & A_{G\to G}(r)\end{pmatrix}\begin{pmatrix} M^h_{q_s}(r,t) \\ M^h_G(r,t)\end{pmatrix}. \qquad (6.14)$$

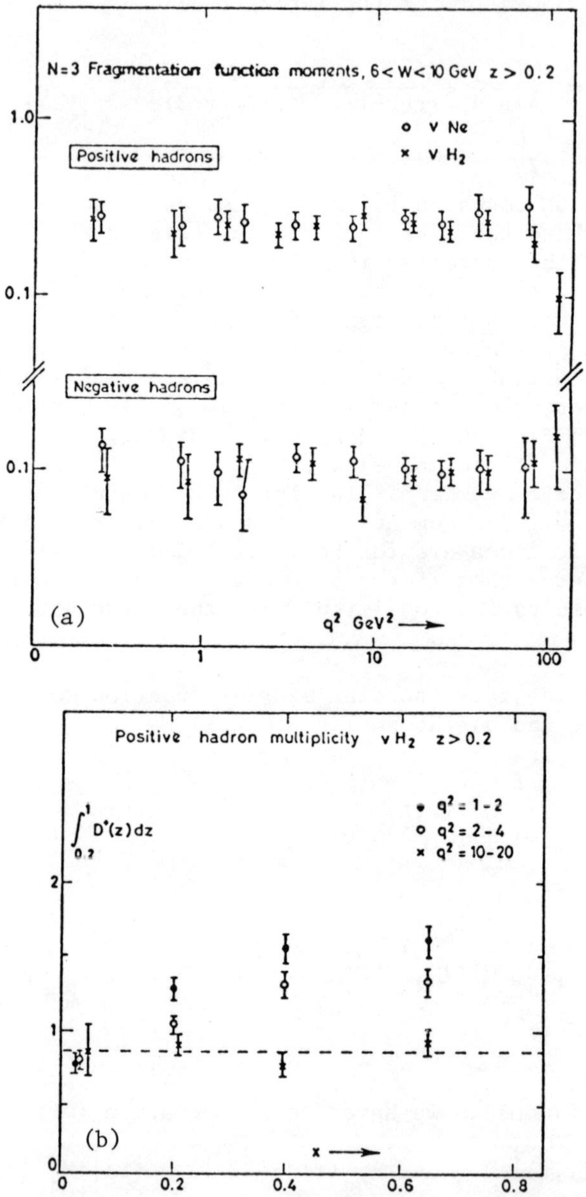

Fig. 19. Hadron distributions in neutrino scattering experiments; a) r = 2 (N = 3) moment versus q^2 at fixed W, b) r = 0 (N = 1) moment versus x for various q^2 (Ref. 25).

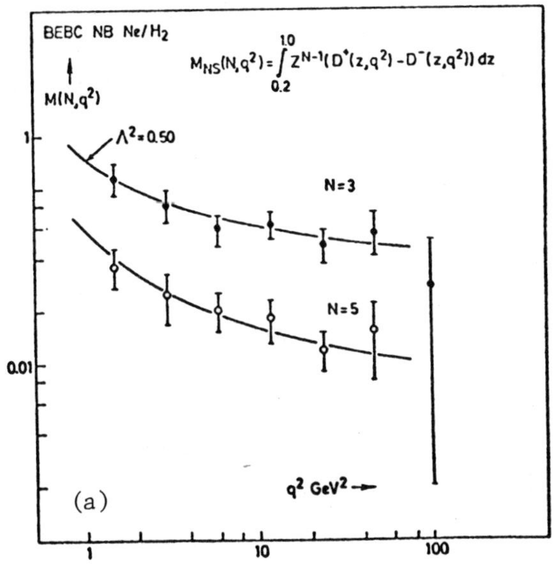

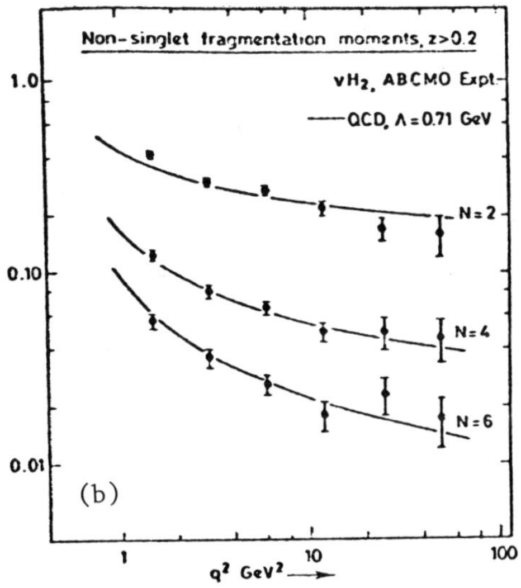

Fig. 20a,b. Figure caption follows Fig. 20c,d.

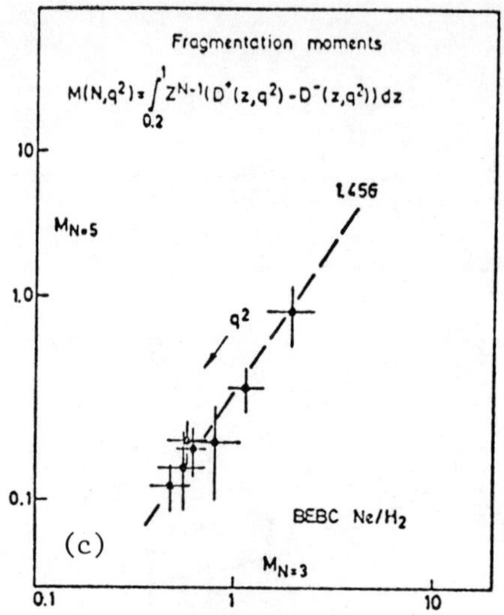

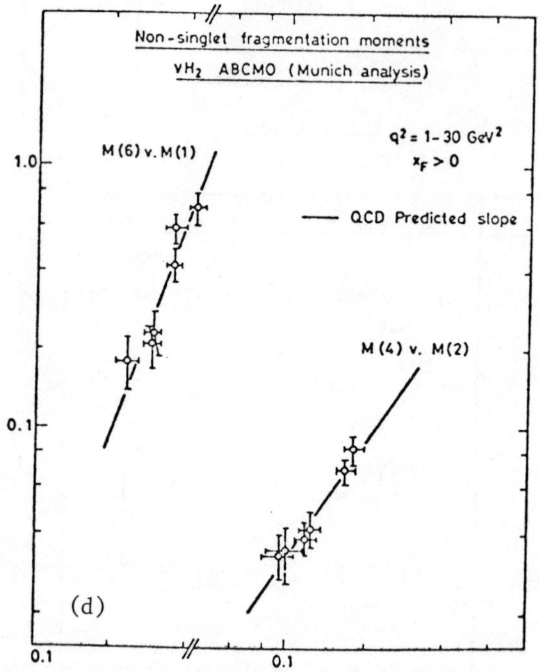

Fig. 20c,d. Moments of observed particle distributions in neutrino reactions for various experiments, plotted in various ways: a) nonsinglet r=2, r=4 moments versus q^2, b) same for nonsinglet r=1,3,5 moments from different experiments, c) logarithm of r=4 moment versus that of the r=2 moment ("Perkins Plot"), d) same for r=5 versus r=0 and r=3 versus r=1. The solid lines are lowest order (LO) QCD predictions with $\Lambda^2_{LO} = 0.5\ GeV^2$. N.b. N = r + 1. (Ref. 25)

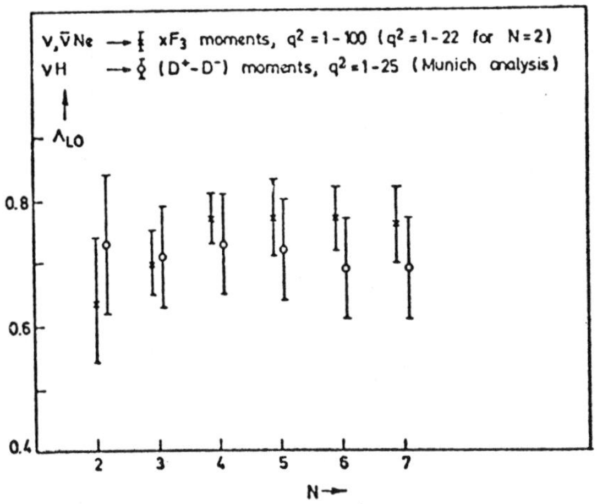

Fig. 21. Indication that analyses of the various moments of the structure function (xF_3) and fragmentation function (D^+-D^-) yield similar values for Λ_{LO} (Ref. 25). Again the parameter N corresponds to $r+1$ in the text.

Just as in the structure function case we can diagonalize and define

$$A_{\pm}(r) \equiv \frac{1}{2}\left[A_{q\to q}(r)+A_{G\to G}(r)\pm\sqrt{\left(A_{q\to q}(r)-A_{G\to G}(r)\right)^2+8N_f A_{G\to q}(r)A_{q\to G}(r)}\right], \tag{6.15a}$$

$$\alpha(r) \equiv \left(A_+(r) - A_{qq}(r)\right)/\left(A_+(r) - A_-(r)\right) \quad , \tag{6.15b}$$

$$\beta(r) \equiv A_{q\to G}(r)/\left(A_-(r) - A_+(r)\right) \quad , \tag{6.15c}$$

$$\gamma(r) \equiv \left(A_+(r) - A_{q\to q}(r)\right)\left(A_-(r) - A_{q\to q}(r)\right)/ \left(A_{q\to G}(r)\cdot\left(A_+(r) - A_-(r)\right)\right) \quad . \tag{6.15d}$$

Then,

$$\begin{pmatrix} M^h_{q_s}(r,t) \\ M^h_G(r,t) \end{pmatrix} = \begin{bmatrix} (1-\alpha(r))e^{\kappa A_+(r)} + \alpha(r)e^{\kappa A_-(r)} , & \beta(r)\left[e^{\kappa A_-(r)} - e^{\kappa A_+(r)}\right] \\ \gamma(r)\left[e^{\kappa A_-(r)} - e^{\kappa A_+(r)}\right], & \alpha(r)e^{\kappa A_+(r)} + (1-\alpha(r))e^{\kappa A_-(r)} \end{bmatrix}$$

$$\times \begin{bmatrix} M^h_{q_s}(r,t_o) \\ M^h_G(r,t_0) \end{bmatrix} \equiv \overline{A}(\kappa,r) \begin{bmatrix} M^h_{q_s}(r,t_o) \\ M^h_G(r,t_o) \end{bmatrix} \tag{6.16}$$

For our later purposes a special result is the case for $r = 1$, the momentum sum rule. We have

$$\begin{aligned} A_{q\to q}(1) &= -\frac{4}{3} C_F , \\ A_{q\to G}(1) &= +\frac{4}{3} C_F , \\ A_{G\to q}(1) &= \frac{1}{6} , \\ A_{G\to G}(1) &= -\frac{N_f}{3} . \end{aligned} \tag{6.17}$$

Thus $\left(\text{note that } \sum_h \sum_{q,\bar{q}}^{N_f} M^h_q = 2N_f \sum_h M^h_{q_s}\right)$

$$\frac{\partial}{\partial t} \sum_h M^h_{q_s}(1,t) = \frac{\alpha(t)}{2\pi} \frac{4}{3} C_F \left[-\sum_h M^h_{q_s}(1,t) + \sum_h M^h_G(1,t) \right] , \tag{6.18a}$$

$$\frac{\partial}{\partial t} \sum_h M^h_G(1,t) = \frac{\alpha(t)}{2\pi} \frac{N_f}{3} \left[\sum_h M^h_{q_s}(1,t) - \sum_h M^h_G(1,t) \right] , \tag{6.18b}$$

so that if, at any scale t, momentum is conserved

$$\sum_h M^h_{q_s}(1,t) = \sum_h M^h_G(1,t) = 1 \text{ (or any constant)} ,$$

then conservation remains true for all t.

To match the corresponding discussions in Refs. 5 and 6, we also consider the Q^2 evolution of the individual $r = 1$ moments. In this case the elements of the matrix $\overline{A}_{ij}(\kappa, r=1)$ can be interpreted as the average momentum fraction carried by partons of type j and scale Q_0 in a jet initiated by parton of type j and scale Q. The explicit forms of the $\overline{A}(\kappa,1)$ are

$$\overline{A}_{qq}(\kappa,1) = \frac{N_f + 4C_F \exp[-\frac{\kappa}{3}(4C_F + N_f)]}{4C_F + N_f} , \tag{6.19a}$$

$$\overline{A}_{q\to G}(\kappa,1) = \frac{4C_F}{4C_F + N_f}(1 - \exp[-\frac{\kappa}{3}(4C_F + N_f)]) , \tag{6.19b}$$

$$\overline{A}_{G\to q}(\kappa,1) = \frac{N_f}{4C_F + N_f}(1 - \exp[-\frac{\kappa}{3}(4C_F + N_f)]) , \tag{6.19c}$$

and

$$\overline{A}_{GG}(\kappa,1) = \frac{4C_F + N_f \exp[-\frac{\kappa}{3}(4C_F + N_f)]}{4C_F + N_f} , \tag{6.19d}$$

where for $N_f = 4$ and $C_F = 4/3$ the exponential factor is $(t_0/t)^{56/75}$. In the limit of large κ (truly huge Q^2), the ratio of energy in quarks to energy in glue is the same for both types of jets and is given by

$$\frac{\overline{A}_{iq}(\kappa,1)}{\overline{A}_{iG}(\kappa,1)} \xrightarrow[\kappa\to\infty]{} \frac{N_f}{4C_F} = \frac{3}{4} . \tag{6.20}$$

In the opposite limit ($\kappa \to 0$) quarks are mostly quarks and glue is mostly glue, as expected. The approach to the asymptotic limit of Eq. (6.20) is remarkably slow. This ratio, which is infinite for $\kappa = 0$, for $i = q$, doesn't reach unity until $Q^2 \approx 10^9$ GeV^2 for $\Lambda = .5$ GeV and $Q_0 = 1$ GeV.

We can, as we did in Lecture 4, study the multiplicity in jets by considering the limit $r \to 0$. In this case $A_{qq} \to 0$, $A_{q\to G} \to 2C_F/r$, $A_{G\to q} \to 1/3$ and $A_{GG} \to 2C_A/r$. Substituting into Eqs. (6.15 and 6.16) yields $A_+ \to 2C_A/r$ and $A_- \to -2/3\, C_f\, N_f/C_A$ and elements in $\overline{A}(K, r\to 0)$ which are singular like $\exp[2\kappa C_A/r]$ as $r \to 0$. While this behavior is difficult to interpret directly, we can take the ratio of multiplicities in quark and gluon jets as

$$\frac{M^h_{q_s}(t,r\to 0)}{M^h_G(t,r\to 0)} \equiv \frac{\langle n_h\rangle_{q_s}}{\langle n_h\rangle_G} \to \frac{\bar{A}_{qG}(\kappa,r\to 0)}{\bar{A}_{GG}(\kappa,r\to 0)} = \frac{C_F}{C_A} \tag{6.21}$$

a result which was suggested earlier.[16] Thus a glue jet has more (slow) hadrons than a quark jet.

We can also study the $z\to 1$ behavior of the D function by considering the $r\to\infty$ limit of the moments. In particular a behavior like r^{-p-1}, when inverted, gives $(1-z)^p$ while $r^{-p-1}/\ln r$ means $(1-z)^p / \ln(1/1-z)$ and so on. Note that, as $r\to\infty$, A_{qq} and A_{GG} both grow as $\ln r$ (a specifically vector theory property). With sufficient care to keep nonleading terms we find in the limit $r\to\infty$ $(C_F < C_A)$

$$\bar{A}_{aa}(\kappa,r) \to r^{-2C_F\kappa}\, e^{-2C_F(\gamma-3/4)\kappa} , \tag{6.22a}$$

$$\bar{A}_{qG}(\kappa,r) \to \frac{C_F}{2(C_A-C_F)r\,\ln r}\, r^{-2C_F\kappa}\, e^{-2C_F(\gamma-3/4)\kappa} \tag{6.22b}$$

$$\bar{A}_{Gq}(\kappa,r) \to \frac{N_f}{2(C_A-C_F)r\,\ln r}\, r^{-2C_F\kappa}\, e^{-2C_F(\gamma-3/4)\kappa} \tag{6.22c}$$

$$\bar{A}_{GG}(\kappa,r) \quad \frac{C_F\,N_f}{4(C_A-C_F)^2\, r^2\, \ln^2 r}\, r^{-2C_F\kappa}\, e^{-2C_F(\gamma-3/4)\kappa} \tag{6.22d}$$

where γ is Euler's constant (0.5772...). Inverting the $\bar{A}$'s to find the distribution of quarks and gluons in quark jets and gluon jets leads to the conclusion[30,31] that for both kinds of jets the ratios of gluons to quarks as $z\to 1$ is

$$\frac{D^G_i(z,t)}{D^q_i(z,t)} \xrightarrow[z\to 1]{} \frac{(1-z)}{4(C_A-C_F)\kappa\,\ln\frac{1}{1-z}} , \tag{6.23}$$

i.e., at large t (Q^2), the distribution of gluons in a jet is always softer than that of quarks. While if one compares the quark distribution in a gluon jet to that in a quark jet, one finds as $z\to 1$

$$\frac{D_G^q(z,t)}{D_q^q(z,t)} \rightarrow \frac{N_f(1-z)}{4C_F(C_A - C_F)\kappa \, \ell n \, \frac{1}{1-z}} \tag{6.24}$$

implying (along with Eq. (6.23)) that the hadron distribution in gluon jets is softer than in quarks by approximately a factor $(1-z)/\ell n(1/1-z)$.

7. "PURE PERTURBATION": ENERGY PATTERNS

We have noted in the previous lectures that the distribution function, D_j^h, describing the distribution of hadrons within a jet resulting from the fragmentation of a quark or a gluon remains a viable theoretical concept even when perturbative corrections are added to the parton model, at least in the leading log approximation. However, it is a quantity which requires renormalization (its moments have, in general, nonzero anomalous dimensions) and therefore some experimental (nonperturbative) input (its value at some reference Q^2) is needed in order to fully evaluate D. The moments satisfy a renormalization group equation of the generic form (ignoring corrections which vanish as powers of $1/Q^2$.)

$$(\mu \frac{\partial}{\partial\mu} + \beta(g)\frac{\partial}{\partial g} + \gamma_n) \, M(n) = 0 \quad , \tag{7.1}$$

expressing the "independence" of M(n) of the choice of renormalization point μ where g is defined, $g \equiv g(\mu)$, and where γ_n is the anomalous dimension. This was illustrated explicitly in perturbation theory by the appearance of logarithms containing an infrared cutoff (e.g., a gluon mass) which are associated with $\gamma_n \neq 0$. These logarithms arise (as explained earlier) both from the emission of soft quanta (true infrared singularities) and from the emission of hard but collinear quanta and from the corresponding regions of phase space for the virtual corrections. Furthermore, they negate the advantages of having an asymptotically vanishing coupling.

Thus there are several reasons for seeking an alternative approach for studying "jetty" final states to that which we have thus far discussed. From a theoretical standpoint we should like to discuss measures (characterizations) of these final states whose dependence on perturbative QCD effects is large (i.e., measurable) but at the same time fully calculable. Thus we would like these measures to be insensitive to the full renormalization of the D function and hence not be highly sensitive to input data for both normalization and Q^2 dependence. For example, if we discuss quantities (call them $\tilde{D}$) with vanishing anomalous dimension, Eq. (7.1) becomes

$$(\mu \frac{\partial}{\partial \mu} + \beta(g) \frac{\partial}{\partial g}) \tilde{D} = 0 \tag{7.2}$$

which means that all Q^2 dependence (except for corrections decreasing as powers of Q^2) can be expressed in terms of the running coupling $\bar{g}^2(Q^2)$. Furthermore, since $g^2(Q^2)$ vanishes as $Q^2 \to \infty$, $\tilde{D}$ is reliably calculable in asymptotically free perturbation theory. (Of course, we will still require experimental input in the form of the parameter Λ, but this we know how to handle in perturbation theory.) In practice this means that when $\tilde{D}$ is calculated in perturbation theory (for a coupling defined at scale μ and with cutoff mass m), the function $\tilde{D}(g_\mu, Q^2/\mu^2, m^2/\mu^2)$ has no logarithmic dependence on the cutoff m^2, order by order in g_μ^2, and the limit $m^2 \to 0$ exists (only positive powers of m^2/μ^2 or m^2/Q^2 appear). In modern language $\tilde{D}$ is mass finite. This also implies that the calculation of $\tilde{D}$ is regularization scheme independent. Furthermore we would like the power behaved corrections to $\tilde{D}$, including those arising from nonperturbative effects, to be both reliably estimateable and small.

From an experimental standpoint we desire a quantity which is insensitive to questions of with which jet to associate a given "slow" hadron since this association is necessarily ambiguous, both experimentally and theoretically. Examples of quantities which are sensitive are the already mentioned large normalization uncertainties in large p_T jet physics due to the trigger bias and the problem of specifying a "3 jet" event in e^+e^- annihilation which is illustrated in Fig. 22 where the same perturbative graph leads not only to collinear nonscaling corrections as in Fig. 22a, but also to a fat jet (large p_T within a jet) as in Fig. 22b, and to two "separate" jets as in Fig. 22c. In principle the theory makes a smooth transition, and the separation of the effects is clearly not unambiguous (we shall return to this point). We should also like to be insensitive to typical incompleteness in the data. For example schemes which involve event-by-event determination of "jet axes" are sensitive to differences in the efficiency of detecting charged and neutral hadrons. Finally we should like to study effects which are sizable compared to the relevant QCD-independent background.

It turns out that there are experimental measures which, to a large degree, satisfy all of the above restrictions. Consider first the theoretical requirement. We will discuss here the specific example of e^+e^- annihilation into hadrons which is conceptually the simplest (where we are assuming we are far from any new particle thresholds). The techniques to be discussed are

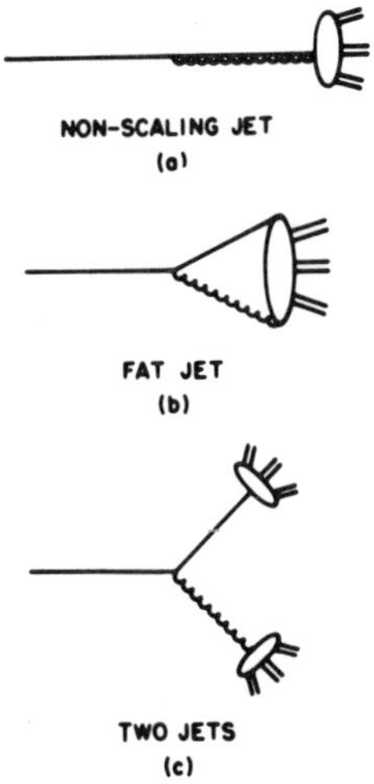

Fig. 22. Single perturbative graph for gluon emission leading to a) collinear nonscaling corrections (zero opening angle, b) a "fat" jet (small opening angle), c) two jets (large opening angle).

applicable to other situations but the situation is typically much more complicated due to hadrons in the initial state which necessarily contain nonperturbative distributions in x and k_T. As discussed by R. Field[6] in the context of the KLN theorem[34], one way to avoid the explicit logarithms of m^2 is to sum over states with any number of quanta within a cone of fixed angular size δ carrying at least some finite fraction, $1 - \varepsilon$, of the total energy. This will supply new, dimensionless cutoffs and replace logarithms of m^2/Q^2 by logarithms of δ and ε. This is exactly what was proposed by Sterman and Weinberg[35] and further developed by others.[36]

A second approach, which I will discuss in detail here (and which I prefer), involves weighting the final states in the perturbative diagrams with some quantity, i.e., measuring some quantity, which is insensitive to both soft emissions and collinear bifurcations. There are countless possibilities[37] but one[38-41] stands out for its theoretical and experimental simplicity. We imagine doing experiments with calorimeters and thus in the perturbative calculation we weight each graph by the energy being "broadcast" into a fixed angular region and coherently sum over all configurations which send nonzero energy in that direction. By direct application[42] of the methods used to demonstrate the factorization of the mass singularity logarithms, it can be shown that, when these same graphs are weighted by energies, there are no logarithms at any order in perturbation theory. I do not intend to go through this proof here as it would be only a slight modification of a large fraction of C.H. Llewellyn Smith's lectures.[5] The energy weighting serves to remove sensitivity to soft emissions while the sum on collinear configurations removes the collinear singularities.

Another way to motivate the finiteness of energy weighted cross sections is to note what is essentially the conservation of energy-momentum. Recall that in the leading log approximation (LLA) the first moment of D_a^h summed over h, i.e., $\sum_h M_a^h(1,t)$, in fact, does not change with t. Thus, if our reference jets, $D_a^h(z,t_0)$ conserve energy, all jets (in LLA) do. This, I believe, continues to be true to all order in the logarithms because such contributions, even though they are no longer diagonal (even in an axial gauge) and hence not interpretable as probabilities, still correspond to collinear configurations which serve only to redistribute the energy in the variable z. When all final quanta (or hadrons) are summed over, the result of $\int_0^1 zdz$ remains the same. These are the contributions of Fig. 22a. There are, of course, truly perturbative (no logarithms and regularization-scheme independent) corrections which I wish to interpret "as multijet" events as in Fig. 22c. I propose to define jets so that there are two jets in Fig. 22c. Each jet continues to conserve energy itself but such distinct final states do generate perturbative corrections to the basic hard scattering process. I have not here proved (nor do I have a rigorous proof at the moment) that this interpretation (i.e., definition of jets) is true and, in particular, that this sort of factorization is process-independent. Rick Field, for example, assumes a different approach. Furthermore I have clearly not treated Fig. 22b adequately.

This discussion, among other things, serves to point up the value of energy weighted cross sections. Since my result will be mass finite, I can simply proceed to calculate physical observables in perturbation theory and needn't attempt to give a separate interpretation to each part of Fig. 22. Also, as I will explicitly demonstrate later, the nonperturbative corrections due to the finite size and, therefore, finite internal q_T of hadrons can be reliably estimated and are power behaved.

From an experimental standpoint, let us first consider the situation at very large energy where $\langle n \rangle_h \to \infty$. Like an electrical engineer studying an antenna, who does not ask about individual photons but studies the energy flow versus angle, the "antenna pattern," we use a calorimeter to study where hadronic energy goes in the final state, the hadronic antenna pattern. Such inclusive-inclusive cross sections (i.e., measure rate of everything in some direction plus anything else in all other directions) result in higher rates than simple single particle inclusive measurements. Also such measurements involve sums over events, not event-by-event analyses. Thus systematics are less likely to be important.

In an e^+e^- machine with luminosity L, total energy W, and using a calorimeter of angular acceptance $d\Omega$ which measures a total energy $\Delta E = \Sigma \Delta E_i$ in time T we have the energy pattern

$$\frac{d\Sigma}{d\Omega} \equiv \frac{\Delta E}{W L T d\Omega} \quad . \tag{7.3}$$

In terms of the N hadron (or QCD quanta) cross section the definition is

$$\frac{d\Sigma}{d\Omega} = \sum_{N=2}^{\infty} \prod_{a=1}^{N} \frac{d^3p_a}{E_a} \frac{d^{3N}\sigma}{\frac{d^3p_1}{E_1} \cdots \frac{d^3p_N}{E_N}} S_N \sum_{b=1}^{N} \frac{E_b}{W} \delta(d\hat{\Omega} - d\hat{\Omega}_b) \tag{7.4}$$

where S_N is a statistical factor to avoid any multiple counting of identical particles. With this normalization we have

$$\int d\Omega \frac{d\Sigma}{d\Omega} = \sigma_{tot} \quad . \tag{7.5}$$

As an aside we note that since the e^+e^- annihilate, in lowest order, into a single photon, polarized and unpolarized beams yield the same information. The spin density matrix into which the hadronic density matrix is multiplied is

$$L_{jk} = \frac{1}{2}(1-P^2)(\delta_{jk} - \hat{\ell}_j\hat{\ell}_k) + P^2\hat{b}_j\hat{b}_k \quad , \tag{7.6}$$

where P is the polarization in the magnetic field direction $\hat{b}$ and $\hat{\ell}$ is the beam direction $(\hat{\ell}\cdot\hat{b} = 0)$ and we define angles as in Fig. 23.

To calculate $d\Sigma/d\Omega$ to order α_s we evaluate the diagrams in Fig. 24 which have already been discussed in detail.[6] I will simply quote some of the results so that they can be compared. I will use a gluon mass m to regulate singularities in individual graphs and use $\Sigma\, \varepsilon_\mu \varepsilon_\nu = g_{\mu\nu}$ gauge. Thus the matrix element is just as in Ref.6 except that here I fix the direction of one of the quarks or the gluon to be $\hat{r}$ (the direction of the calorimeter), weight by its energy fraction $(x_i = 2E_i/W)$, and integrate over x_1 and x_2 $(x_3 = x_G = 2 - x_1 - x_2)$.

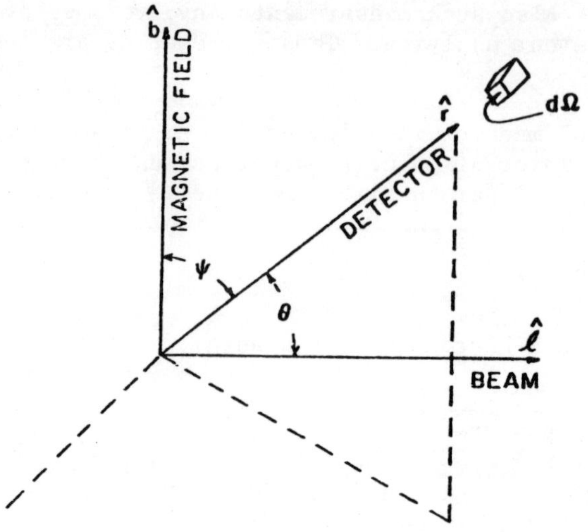

Fig. 23. Geometry for the energy pattern experiment.

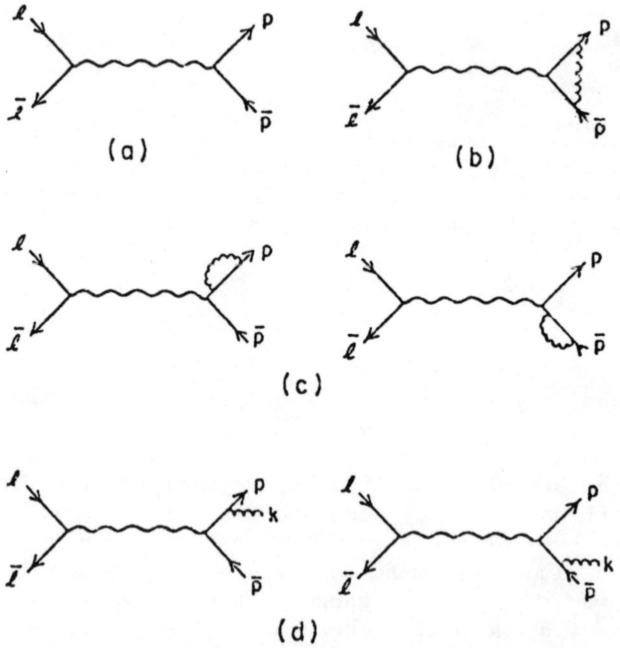

Fig. 24. Lowest order diagrams relevant for e^+e^- annihilation: a) zeroth order, b) vertex modification, c) self-energy insertion, d) gluon emission. (Wavy lines are photons and curly lines are gluons.)

In lowest order, Fig. 24a, we have for general polarization

$$\frac{d\Sigma_o}{d\Omega} = \frac{d\sigma}{d\Omega} = \frac{\alpha^2}{2W^2}\sum_f 3Q_f^2[P^2\sin^2\psi + \tfrac{1}{2}(1-P^2)(1+\cos^2\theta)]$$

$$\equiv \frac{\sigma_o}{(8\pi/3)}[P^2\sin^2\psi + \tfrac{1}{2}(1-P^2)(1+\cos^2\theta)] \qquad (7.7)$$

where I have factored out the usual quark total cross section. The contributions from the virtual emissions, Fig. 23c, and the interference of (23a) and (23b) give

$$\frac{d\Sigma^{(virt)}}{d\Omega} = \frac{d\sigma^{(virt)}}{d\Omega} = \frac{-\sigma_o}{(8\pi/3)}\frac{C_F\alpha_s}{2\pi}\left[\ln^2\frac{W^2}{m^2} - 3\ln\frac{W^2}{m^2} - \frac{\pi^2}{3} + \frac{7}{2}\right]\left[P^2\sin^2\psi + \tfrac{1}{2}(1-P^2)(1+\cos^2\theta)\right] . \qquad (7.8)$$

"Real" gluon emission in Fig. 23d gives the following result for a detected quark or anti-quark

$$\frac{d\Sigma^{(real)}_{q+\bar{q}}}{d\Omega} = \frac{\sigma_o}{(8\pi/3)}\frac{C_F\alpha_s}{2\pi}\left\{\left[\ln^2\frac{W^2}{m^2} - \frac{13}{3}\ln\frac{W^2}{m^2} - \frac{\pi^2}{3} + \frac{80}{9}\right]\right.$$

$$\times\left[P^2\sin^2\psi + \tfrac{1}{2}(1-P^2)(1+\cos^2\theta)\right] + \tfrac{1}{2}[P^2(3\cos^2\psi-1) \qquad (7.9)$$

$$\left. + \tfrac{1}{2}(1-P^2)(1-3\cos^2\theta)]\right\} .$$

It is amusing to note that at order α_s the $q+\bar{q}$ cross section with real gluon emission is

$$\frac{d\sigma^{(real)}_{q+\bar{q}}}{d\Omega} = \frac{\sigma_o}{(8\pi/3)}\frac{C_F\alpha_s}{2\pi}\left\{\left[\ln^2\frac{W^2}{m^2} - 3\ln\frac{W^2}{m^2} - \frac{\pi^2}{3} + 5\right]\right.$$

$$\times\left[P^2\sin^2\psi + \tfrac{1}{2}(1-P^2)(1+\cos^2\theta)\right] + \left[P^2(3\cos^2\psi-1)\right. \qquad (7.10)$$

$$\left.\left. + \tfrac{1}{2}(1-P^2)(1-3\cos^2\theta)\right]\right\}$$

and when added to Eqs. (7.7 - 7.9) yields a finite result. This is to be expected since with quark and anti-quark multiplicity 1, these contributions come in with exactly the same coefficients in

$d\sigma/d\Omega$ as in σ_{TOT} which we know to be finite and which differs by only a finite angular integral. In order α_s^2, when quark pairs are produced, this is not the case.

The final contribution to the antenna pattern arises when the gluon is "detected." The result of evaluating this contribution is

$$\frac{d\Sigma_G^{(real)}}{d\Omega} = \frac{\sigma_o}{(8\pi/3)} \frac{C_F\alpha_s}{2\pi} \left\{ \left[\frac{4}{3} \ln \frac{W^2}{m^2} - \frac{35}{9}\right] \times \left[P^2 \sin^2\psi + \frac{1}{2}(1-P^2)(1+\cos^2\theta)\right] + \left[P^2(3\cos^2\psi-1) + \frac{1}{2}(1-P^2)(1-3\cos^2\theta)\right]\right\} . \tag{7.11}$$

Now the whole point of defining the energy weighted cross section is that we now sum these contributions which gives the total energy in the direction $\hat{r}$, a mass finite quantity. The sum of Eqs. (7.7 - 7.9, 7.11) is, with $C_F = 4/3$ and $\alpha_s = \alpha_s(W^2)$,

$$\frac{d\Sigma^{QCD}}{d\Omega} = \frac{\sigma_o}{(8\pi/3)} \left\{ \left[1 + \frac{\alpha_s}{\pi}\right] \left[P^2 \sin^2\psi + \frac{1}{2}(1-P^2)(1+\cos^2\theta)\right] + \frac{\alpha_s}{\pi}\left[P^2(3\cos^2\psi-1) + \frac{1}{2}(1-P^2)(1-3\cos^2\theta)\right]\right\} . \tag{7.12}$$

Note that when integrated over $d\Omega$, Eq. (7.12) gives σ_{TOT} to order α_s as required by Eq. (7.5), and there are no residual logarithms, even of finite ratios. The antenna pattern is plotted in Fig. 25 for W=5, 30, and ∞ GeV with the parameter Λ in α_s equal to 500 MeV and P=1.

Our analysis is not yet complete. While summing energies has removed dependence on the shape and Q^2 variation of the z distribution in $D_a^h(z,Q^2)$ and we have included perturbative smearing of the energy pattern, we must still include the nonperturbative smearing (the "postremordial" smerge in R. Field's language).

We estimate these effects by using a parameterization of the nonperturbative structure of the jet such as that discussed in Lectures 4 and 5. The analysis[38] will, in fact, require only a q_T cutoff and a plateau. We write

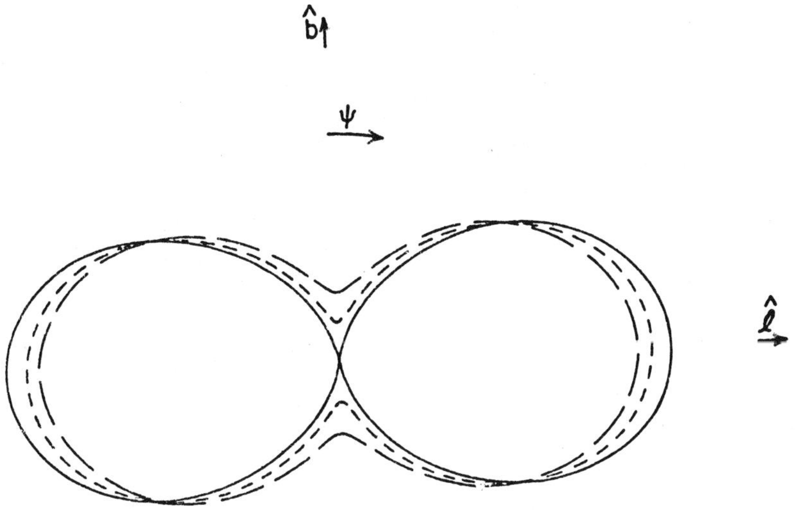

Fig. 25. QCD predictions for normalized antenna pattern $(1/\sigma_{TOT})d\Sigma/d\Omega$ with perfectly polarized e^+e^- beams for W=5 GeV (long dashes), 30 GeV (short dashes), infinity (solid line) (Ref. 38).

$$dN = dz\, d^2h_T\, D(z,h_T) = \frac{dz}{z}\, d^2h_T\, f(z,h_T) \approx \frac{d^3h}{h_o}\, f(z,h_T) \qquad (7.13)$$

and, with the expectation that this will be a small correction, we smear only the lowest order cross section Eq. (7.7). If we define the quark direction by $\hat{p}$ as in Fig. 26 and specialize to P=1, we find for a calorimeter $\Delta\Omega$ (2 for $q+\bar{q}$ and $\cos\chi \equiv \hat{b}\cdot\hat{p}$)

$$\Delta\Sigma^{qf} = 2\int d\Omega_{\hat{p}}\, \frac{\sigma_o}{(8\pi/3)}\, \sin^2\chi \int_{\Delta\Omega} \frac{d^3h}{h_o} \left(\frac{h_o}{W}\right) f(z,h_T) \;, \qquad (7.14)$$

where qf symbolizes the quark fragmentation contribution. If we define ($\cos\eta \equiv \hat{p}\cdot\hat{h}$, $h_T = \frac{W}{2} z \tan\eta$)

$$F_1(\eta) \equiv \frac{2}{W}\int h^2dh\; f_1(z,h_T) = \frac{1}{\pi W \sin^3\eta}\int d^2h_T\, h_T\, f\left(\frac{2h_T}{W}\cot\eta,\, h_T\right) , \qquad (7.15)$$

we have for $\eta \gg \langle h_T\rangle/W$ that $z \to 0$ as $W \to \infty$ so that in this limit

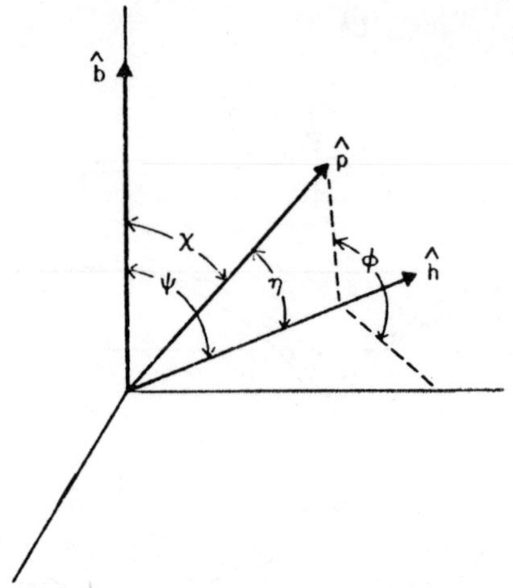

Fig. 26. Geometry for fragmentation of a quark moving in direction $\hat{p}$ into a hadron in the direction $\hat{h}$ with the polarization direction $\hat{b}$ shown.

$$F_1(\eta) \approx \frac{1}{\pi W \sin^3\eta} \int d^2h_T\, h_T\, f(0,h_T) = \frac{C}{2\pi} \frac{\langle h_T \rangle}{W} \frac{1}{\sin^3\eta} \quad (\eta \le \pi/2)$$

$$= 0 \qquad (\eta > \pi/2) \quad , \tag{7.16}$$

where

$$\langle h_T \rangle \equiv \frac{\int d^2h_T\, h_T\, f(0,h_T)}{\int d^2h_T f(0,h_T)} \tag{7.17a}$$

and

$$C \equiv 2 \int d^2h_T f(0,h_T) \tag{7.17b}$$

is just the height of the e^+e^- plateau (2 jets)

$$\langle n \rangle_{e^+e^-} \approx C\, \ell n\, W + \text{const.} \quad . \tag{7.17c}$$

Thus $F_1(\eta)$ decreases as $1/W$ for $\eta > 0$ but for $\eta \approx 0$

$$F_1(0) = \frac{2}{W} \int h^2 dh\, f(\frac{2h}{W}, 0) = \frac{W^2}{4} \int_0^1 z^2 dz\, f(z,0) \quad , \tag{7.18}$$

increasing as W^2 with a model dependent coefficient. Since

$D(z,h_T)$ conserves energy by assumption and construction, we have

$$\int d\Omega F_1(\eta) = \int dz\, d^2h_T\, z\, D(z,h_T) = 1 \tag{7.19}$$

and thus the peak at $\eta \gtrsim 0$ must have width $1/W$ ($d\Omega \gtrsim \eta d\eta \cdot 2\pi$ for $\eta \gtrsim 0$) and $F_1(\eta)$ looks as in Fig. 27.

Finally, by the law of cosines and averaging over the azimuthal angle in Fig. 26, we have

$$\sin^2\chi \rightarrow \sin^2\psi + \frac{1}{2}\sin^2\eta(3\cos^2\psi - 1)$$

and hence, changing the order of integration in Eq. (7.14),

$$\begin{aligned} \frac{d\Sigma^{qf}}{d\Omega} &= \frac{\sigma_o}{(8\pi/3)} \int d\Omega_{\hat{p}} [\sin^2\psi + \frac{1}{2}\sin^2\eta(3\cos^2\psi-1)]F_1(\eta) \\ &\equiv \frac{\sigma_o}{(8\pi/3)} [\sin^2\psi + \frac{1}{2} \langle\sin^2\eta\rangle^{qf} (3\cos^2\psi - 1)] \end{aligned} \tag{7.20}$$

with the definition

$$\langle\sin^2\eta\rangle^{qf} \equiv \int_{\eta < \pi/2} d\Omega \sin^2\eta\, F_1(\eta) \quad .$$

This is the energy weighted average opening angle (squared) of a jet due to nonperturbative effects. It is insensitive to small η and we can use the expression in Eq. (7.16) to find

$$\langle\sin^2\eta\rangle^{qf} = \frac{\pi C}{2} \frac{\langle h_T\rangle}{W} \quad . \tag{7.21}$$

Comparing Eq. (7.12) to (7.20b) suggests that perturbative QCD leads to an average opening angle given by

$$\langle\sin^2\eta\rangle^{QCD} \gtrsim \frac{2\alpha_s(W^2)}{\pi} \tag{7.22}$$

which is an expression of the statement that $\langle q_\perp^2\rangle \sim \alpha_s(Q^2) \cdot Q^2$ in perturbative QCD. We shall give a more explicit description below. Data ($W < 10$ GeV) suggest $C \simeq 2.5$, $\langle h_\perp\rangle \simeq 350$ MeV/c, which lead to comparable numbers in Eqs. (7.21) and (7.22) for $W = 10$ GeV and a three times larger QCD result for $W = 30$ GeV. [What is happening experimentally to $\langle n\rangle$ at larger W is less clear.] We now simply add Eqs. (7.12) and (7.20) (note that

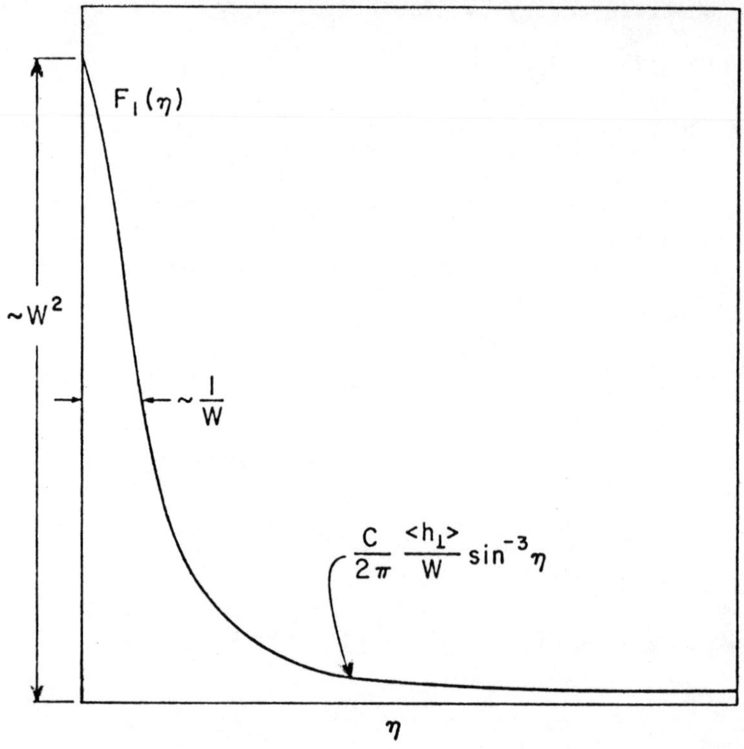

Fig. 27. Schematic illustration of the hadronic angular distribution $F_1(\eta)$ for the fragmentation of a quark resulting from nonperturbative effects.

$$\int d\Omega \frac{d\Sigma^{qf}}{d\Omega} = \sigma_o$$

as required by energy-momentum conservation) so that

$$\frac{d\Sigma}{d\Omega} \simeq \frac{\sigma_o}{(8\pi/3)} \left\{ \left[1 + \frac{\alpha_s}{\pi}\right] \cdot \left[P^2\sin^2\psi + \frac{1}{2}(1-P^2)(1+\cos^2\theta)\right] + \left[\frac{\alpha_s}{\pi} + \frac{1}{2}<\sin^2\eta>^{qf}\right] \cdot \left[P^2(3\cos^2\psi-1) + \frac{1}{2}(1-P^2)(1-3\cos^2\theta)\right]\right\}. \tag{7.23}$$

This form is illustrated in Fig. 28 where the effects of heavy lepton decays are also estimated as discussed in Ref. 28. These fall as $1/W^2$ and are essentially always negligible. Assuming an

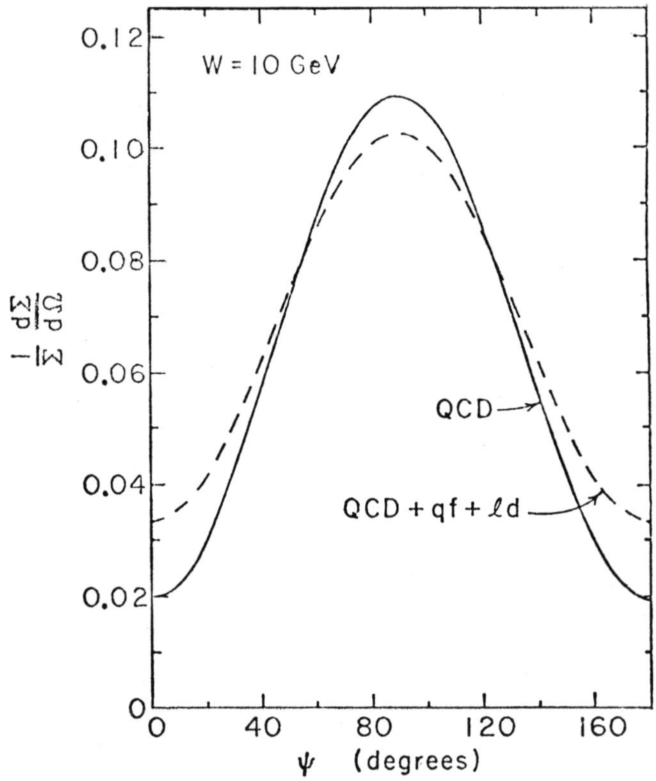

Fig. 28a. Figure caption follows Fig. 28d.

iso

isotropic decay in the lepton rest frame, a branching ratio β into semi-leptonic decays, and an average energy fraction into hadrons of γ (for P = 1)

$$\frac{d\Sigma^{\ell}}{d\Omega} \simeq \frac{\alpha^2}{2W^2}\, \beta^2\gamma \left[\sin^2\psi + \frac{4M^2}{W^2}(3\cos^2\psi - 1)\right] \quad . \tag{7.24}$$

If we assume that the angular distribution of energy in neutrals is the same as in charged particles then we can evaluate the correction to the angular pattern in the charged-particle inclusive cross section data[43] (unpolarized data)

$$\frac{d}{dz d\Omega} \sim (1 + \alpha(z)\cos^2\theta) \quad . \tag{7.25}$$

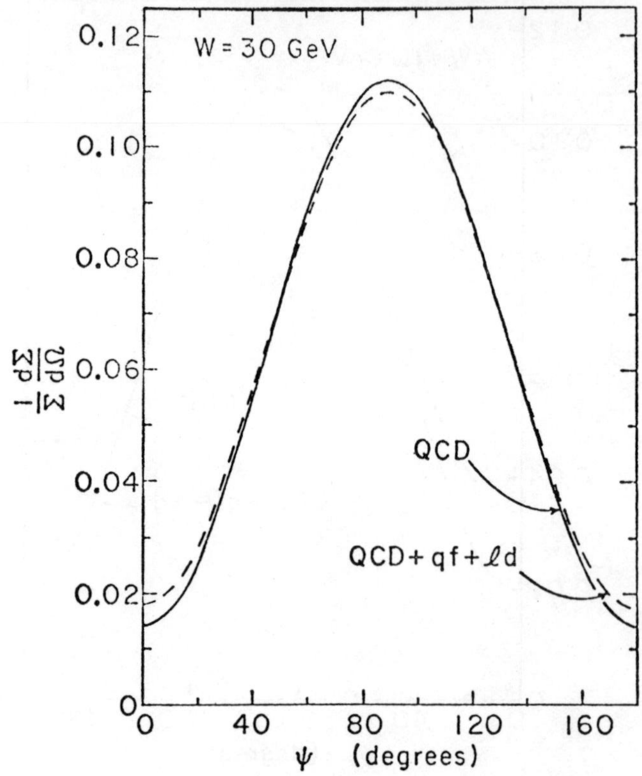

Fig. 28b. Figure caption follows Fig. 28d.

Thus

Thus

$$\frac{1}{2}\langle\sin\eta\rangle_{TOT} \equiv \frac{1}{2}\langle\sin^2\eta\rangle^{qf} + \frac{\alpha_s}{\pi} = \frac{\int_0^1 dz\, z\frac{d\sigma}{dz}\left(\frac{1-\alpha(z)}{3+\alpha(z)}\right)}{\int_0^1 dz\, z\frac{d\sigma}{dz}} \quad . \qquad (7.26)$$

At W = 7.4 GeV this yields $\langle\sin^2\rangle_{TOT} \simeq .34 \pm .06$ in remarkably (embarrassingly) good agreement with Eq. (7.22) plus (7.21) evaluated at such low energies. Undoubtedly corrections are needed to account for heavy lepton decays (small) and charm decays (probably sizable at 7.4 GeV). Very recent results[44] from DESY also show (too) remarkable agreement.

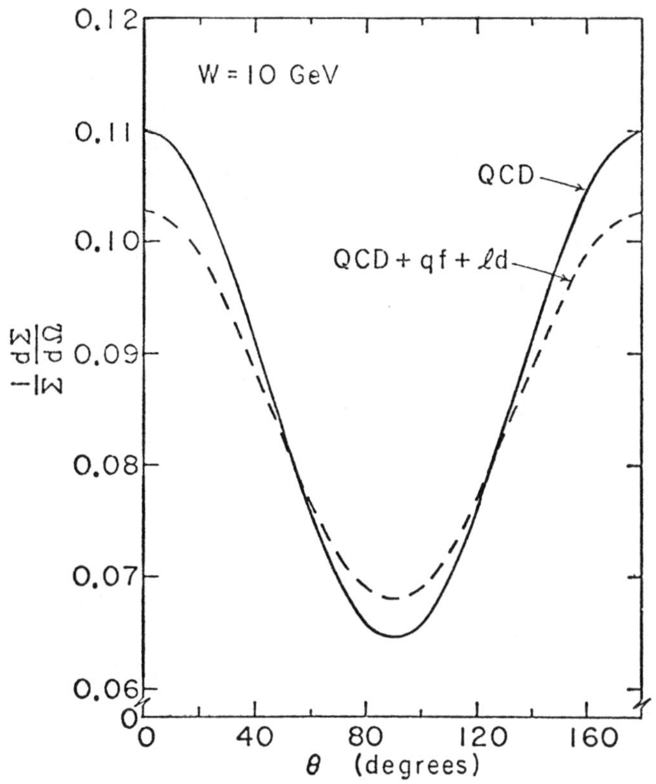

(c)

Fig. 28c. Figure caption follows Fig. 28d.

As Fig. 25 illustrates, the energy pattern does not show really large perturbative effects, even though the nonperturbative corrections (Fig. 28) are small. In the next lecture we shall discuss a more complicated energy weighted measure, in what is presumed to be a complete hierarchy of such measures, the energy-energy correlation function. This measure has the advantages of being nonzero only at order α_s (no order 1 bit) and of allowing us to study more directly the "jetty" nature of final states.

8. ENERGY CORRELATIONS

To obtain a finer measure of the structure of the final state in e^+e^- annihilation which more clearly exhibits perturbative QCD effects and the "jetty" nature of the state, we consider now experiments with two calorimeters. The two calorimeters have

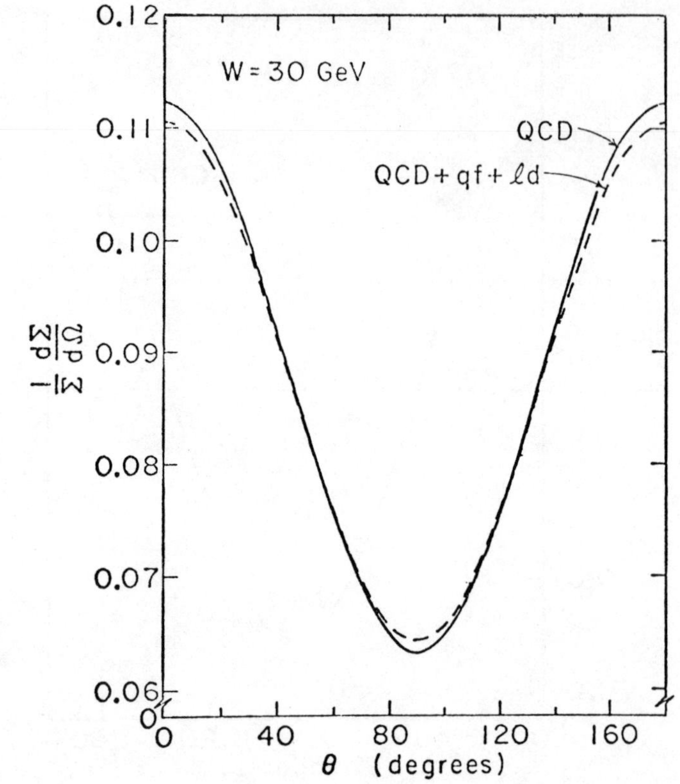

Fig. 28d. Antenna pattern with and without nonperturbative quark fragmentation and heavy lepton contributions for perfect polarization versus ψ for a) W = 10 GeV, b) W = 30 GeV; and for zero polarization versus θ for c) W = 10 GeV, d) W = 30 GeV (Ref. 38).

acceptances $d\Omega$ and $d\Omega'$ and angular directions $\hat{r}$ and $\hat{r}'$ with respect to the interaction region as in Fig. 29. The polar angles with respect to the beam ($\hat{\ell}$) and polarization ($\hat{b}$) directions are defined in the obvious way as is the relative angle $\cos\chi = \hat{r} \cdot \hat{r}'$. In this situation the analogue equations to Eq. (7.3) and (7.4) are

$$\frac{d^2\Sigma}{d\Omega d\Omega'} = \frac{\Sigma_{\text{events}}(dE \cdot dE')}{TW^2 d\Omega dr'} , \tag{8.1}$$

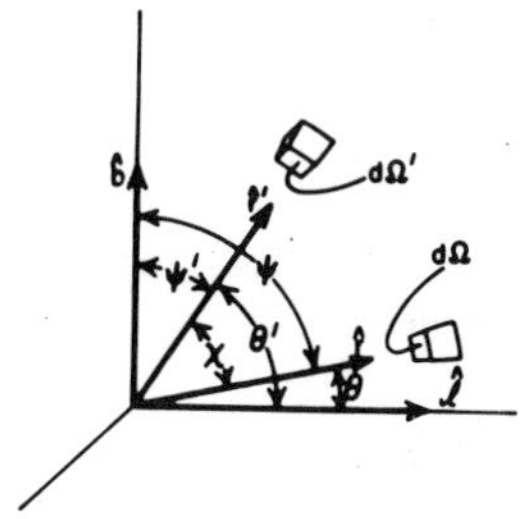

Fig. 29. Geometry for the two-calorimeter experiment showing both the beam direction $\hat{\ell}$ and polarization direction $\hat{b}$.

where dE and dE' are the energies observed in the two calorimeters in a given event, and

$$\frac{d^2\Sigma}{d\Omega d\Omega'} = \sum_{N=2}^{\infty} \int \prod_{a=1}^{\infty} \left(\frac{d^3p_a}{E_a}\right) \frac{d^{3N}\sigma}{\frac{d^3p_1}{E_1}\cdots\frac{d^3p_N}{E_N}} S_N \left[\sum_{b,c=1}^{N} \frac{E_b E_c}{W^2} \delta(\Omega-\Omega_b)\,\delta(\Omega'-\Omega_c)\right]. \tag{8.2}$$

Note in particular that we have considered "transparent" calorimeters and kept the b = c terms. Thus two coincident ($\hat{r} = \hat{r}'$) calorimeters detect the same energy. This insures the mass finiteness of Eq. (8.2) and the normalization

$$\int d\Omega' \frac{d^2\Sigma}{d\Omega d\Omega'} = \frac{d\Sigma}{d\Omega} \quad . \tag{8.3}$$

This last result is not precisely true if heavy leptons are present and energy is lost to undetected neutrinos. However, as we shall see, these effects are totally negligible (away from threshold).

Applying Eq. (8.2) to perturbative final states (quarks and gluons) the terms with $b = c$ lead to $\delta(\Omega-\Omega')$ contributions with (collinearly) divergent coefficients. As we shall see in detail,[38,39] this divergence cancels against the singular contribution of other terms, order beyond order in perturbation theory, when an integration is performed over a small angular region about $\Omega = \Omega'$ (collinear detectors). The perturbation theory is not uniformly convergent in the limit $\Omega' \to \Omega$. This is precisely the residue of the ambiguity about the structure of the final state illustrated in Fig. 22. Finite (fixed) order perturbation theory is not adequate as the "two jets" merge into "one jet." This is the smile of the collinearly logarithmic Cheshire Cat which we caused to vanish by energy weighting. A similar situation arises for the back-to-back configuration ($\Omega' \to -\Omega$) where there is a $\delta(\Omega + \Omega')$ contribution from the pure quark + antiquark state plus virtual corrections. The coefficient of this term is $(\log)^2$ divergent and again, in fixed order perturbative theory, is only finite when integrated with the other contributions over a small angular region near $\Omega = -\Omega'$. Since in both of these kinematic regions, the problem arises due to using finite order perturbation theory, the ordinary (i.e. nonsingular) but strongly peaked, true behavior can be obtained by summing by some technique the most singular pieces to all orders in perturbation theory. This procedure will lead to an extension[45-49] of the kinematic region accessible to perturbation theory techniques and the discussion of it will constitute the last topic of these lectures.

Returning to the energy-energy correlation we note that the hadronic factor must be characterized by a tensor H_{jk} which is to be contracted with the lepton spin density matrix L_{jk} of Eq. (7.6). Since L_{jk} is symmetric we can treat H_{jk} as symmetric and also note that H is even under $\hat{r} \leftrightarrow \hat{r}'$ and so involves only functions of $\cos\chi = \hat{r} \cdot \hat{r}'$ times the three possible matrix forms. In terms of the positive, symmetric, mutually orthogonal projection matrices

$$P^{(1)}_{jk} = \frac{(r_j + r_j')(r_k + r_k')}{2(1 + \hat{r}\cdot\hat{r}')} , \tag{8.4a}$$

$$P^{(2)}_{jk} = \frac{(r_j - r_j')(r_k - r_k')}{2(1 - \hat{r}\cdot\hat{r}')} \tag{8.4b}$$

and

$$P^{(3)}_{jk} = \delta_{jk} - P^{(1)}_{jk} - P^{(2)}_{jk} , \tag{8.4c}$$

we have

$$\frac{d^2\Sigma}{d\Omega d\Omega'} = \sum_{a=1}^{3}{}' A^{(a)}(\hat{r}\cdot\hat{r}')P^{(a)}_{jk} L_{jk} . \tag{8.5}$$

Since the $P^{(a)}$ are orthogonal and positive, the three numbers $P^{(a)}_{jk} L_{jk}$ can take on independent positive values. Here positivity of the cross section requires positivity of the A's. While this explicitly positive form is amusing, it is cumbersome for displaying the results of perturbative calculations. We use instead $(2\delta_{jk} - r_j r_k - r_j' r_k')$, $[\delta_{jk}(r;r') - \frac{1}{2} r_j r_k' - \frac{1}{2} r_k' r_j]$ and δ_{jk}. Thus we define, for general polarization,

$$\frac{d^2\Sigma}{d\Omega d\Omega'} \equiv \frac{\sigma_o}{(8\pi/3)}\{A(\chi)[P^2(\sin^2\psi + \sin^2\psi') + \frac{1}{2}(1-P^2)$$

$$\times (1+\cos^2\theta + 1+\cos^2\theta')] + \mathcal{B}(\chi)[P^2(\cos\chi - \cos\psi\cos\psi')$$

$$+ \frac{1}{2}(1-P^2)(\cos\chi + \cos\theta\cos\theta')] + \mathcal{C}(\chi)\} \quad , \tag{8.6}$$

where one can now write various constraints on A, $\mathcal{B}$, and $\mathcal{C}$ arising from the positivity of the $A^{(a)}$. Note that only four of the five angles θ,θ', ψ,ψ', and χ are actually independent.

To order α_s for $|\cos\chi| < 1$ (i.e. neither collinear nor back-to-back calorimeters) the entire contribution to the energy-energy correlation is from the graphs of Fig. 24d. There are only two cases. The first is where the q and $\bar{q}$ enter the calorimeters and the second for the q($\bar{q}$) and G into the calorimeters. Both contributions are individually nonsingular for $|\cos\chi| < 1$ and, while singular as $|\cos\chi| \to 1$, are uniformly behaved in these limits and can be evaluated easily. [In Ref. 38 these terms are actually calculated in $\nu > 4$ dimensions explicitly so as to facilitate the study of the singular limit.] For the angular coefficient functions defined above we have, in terms of the variable

$$\zeta = \frac{1}{2}(1 - \cos\chi) \quad , \tag{8.7}$$

$$A^{QCD}(\chi) = \frac{\alpha_s}{\pi}\frac{1}{12\pi}\frac{1}{1-\zeta}[(\frac{3}{\zeta^5} - \frac{4}{\zeta^4})\ln(1-\zeta) + \frac{3}{\zeta^4} - \frac{5}{2\zeta^3} - \frac{1}{\zeta^2}] \quad , \tag{8.8a}$$

$$\mathcal{B}^{QCD}(\chi) = \frac{\alpha_s}{\pi}\frac{1}{12\pi}\frac{1}{1-\zeta}[(\frac{12}{\zeta^5} - \frac{16}{\zeta^4} + \frac{4}{\zeta^3})\ln(1-\zeta) + \frac{12}{\zeta^4} - \frac{10}{\zeta^3}] \quad , \tag{8.8b}$$

$$\mathcal{C}^{QCD}(\chi) = 0 \quad . \tag{8.8c}$$

As predicted earlier these functions are singular in the limits $\chi \to 0$, $\zeta \to 0$

$$A^{QCD}(\chi) \to \frac{\alpha_s}{\pi}\frac{1}{12\pi}\frac{7}{12\zeta} \tag{8.9a}$$

$$\mathcal{B}^{QCD}(\chi) \to \frac{\alpha_s}{\pi}\frac{1}{12\pi}\frac{1}{3\zeta} \quad , \tag{8.9b}$$

and $\chi \to \pi$, $\zeta \to 1$

$$A(\chi) \xrightarrow{QCD} \frac{\alpha_s}{12\pi^2} \frac{1}{1-\zeta} [\ell n \frac{1}{1-\zeta} - \frac{1}{2}] \tag{8.10a}$$

$$\mathcal{B}(\chi) \xrightarrow{QCD} \frac{\alpha_s}{12\pi^2} \frac{2}{1-\zeta} \quad . \tag{8.10b}$$

We will return shortly to discuss this in detail.

An analysis of the nonperturbative smearing can be carried out much as we did for the antenna pattern earlier. (See Ref. 38 for a complete description.) It turns out that for scaling nonperturbative corrections (recall we are explicitly dealing with the nonscaling perturbative corrections via our energy weighting prescription) the dominant [O(1/W)] correction[50] can be given simply in the terms of the function $F_1(\eta)$ defined in Eq. (7.15). The configurations correspond to a quark or antiquark going directly into one of the calorimeters and the other calorimeter catching fragments of that jet or the opposite jet. The functional form is

$$\frac{d\Sigma^{qf}}{d\Omega d\Omega'} = \frac{1}{2}[F_1(\chi) + F_1(\pi-\chi)] \frac{\sigma_o}{(\frac{8\pi}{3})} [P^2(\sin^2\psi + \sin^2\psi') + \frac{1}{2}(1-P^2)(1 + \cos^2\theta + 1 + \cos^2\theta')] \quad , \tag{8.11}$$

or, from Eq. (7.16),

$$A^{qf}(\chi) = \frac{C}{4\pi} \frac{\langle h_T \rangle}{W} \sin^{-3}\chi + O(1/W^2) \tag{8.12}$$

for $|\cos\chi| < 1$. Note that, ignoring corrections of order $1/W^2$, there are no fragmentation corrections to $\mathcal{B}$ or $\mathcal{C}$ and that A^{qf} is symmetric under $\chi \leftrightarrow \pi-\chi$ ($\zeta \leftrightarrow 1-\zeta$). This symmetry corresponds to comparing two particles from the same "jet" to two particles from different "jets." Asymmetries will arise from the correlations between the particles in the same jet which we argued above are generally small and, in any case, will be of order $1/W^2$. This symmetry is illustrated by evaluating $\langle\sin^2\chi\rangle$ separately in the forward ($0 \le \chi \le \pi/2$) and backward ($\pi/2 \le \chi \le \pi$) hemispheres,

$$\langle\sin^2\chi\rangle \equiv \frac{1}{\sigma_{TOT}} \int d\Omega d\Omega' \sin^2\chi \frac{d\Sigma}{d\Omega d\Omega'}$$

$$= 2\pi \int d\chi \sin^3\chi\, [2A(\chi) + \cos\chi\mathcal{B}(\chi) + \frac{3}{2}\mathcal{C}(\chi)] \quad , \tag{8.13a}$$

giving

$$\langle \sin^2 \chi \rangle_F^{qf} = \langle \sin^2 \chi \rangle_B^{qf} = \langle \sin^2 \eta \rangle^{qf} \tag{8.13b}$$

where the last quantity is that in Eq. (7.21). This is to be contrasted with the perturbative QCD contribution which is clearly asymmetric at the Feynman graph level and yields two distinct measures of "jet opening angle,"

$$\langle \sin^2 \rangle_F^{QCD} = (.55)\,\frac{2\alpha_s(W^2)}{\pi}\ , \tag{8.14a}$$

$$\langle \sin^2 \rangle_B^{QCD} = (.89)\,\frac{2\alpha_s(W^2)}{\pi}\ . \tag{8.14b}$$

This is to be contrasted with the antenna pattern result of Eq. (7.22).

These forms are explicitly plotted in Fig. 30 where again the (negligible $1/W^2$) corrections due to heavy lepton decay are also indicated. Since the nonperturbative corrections are of order $1/W^2$, the coefficient $\mathcal{B}$ and asymmetry $\mathcal{D}$, defined by

$$\mathcal{D}(\chi) \equiv A(\pi-\chi) - A(\chi) \tag{8.15}$$

and illustrated in Fig. 31, should be good quantities to test perturbative QCD. They both have small nonperturbative corrections and no order 1 "background." Also they vary with $W^2 = Q^2$ as $\alpha_s(Q^2)$, characteristic of QCD and should exhibit a shape indicative of <u>vector</u> gluon emission. For comparison the rather different shapes of the same quantities in an (admittedly unrealistic) scalar gluon model[51] are exhibited in Fig. 32. All these quantities should be straightforward, even with reduced rates, to measure with calorimeters or with magnetic detectors under the assumption of identical energy <u>patterns</u> for charged and neutral hadrons.

Let us now return to the limiting behavior for $\chi \to \pi$, 0 where, as we have already seen, perturbation theory breaks down if W^2 is fixed while the angular limit is taken. Consider first $\chi \to \pi$ which has been discussed to some extent by R. Field.[6] The kinematic situation for the two-calorimeter experiment is illustrated in Fig. 33. That the energy-energy correlation must behave as $1/(1-\zeta)\,\ell n\, 1/(1-\zeta)$ in this limit can be understood by considering the contributions to the energ-energy correlation from purely virtual terms (i.e. vertex corrections) as, for example, the interference of Figs. 24 a) and b) in order α_s. With only a quark and antiquark in the final state, the contribution is constrained to be proportional to $\delta(\Omega+\Omega')$ and contribute only for $\chi \equiv \pi$. Order-by-order

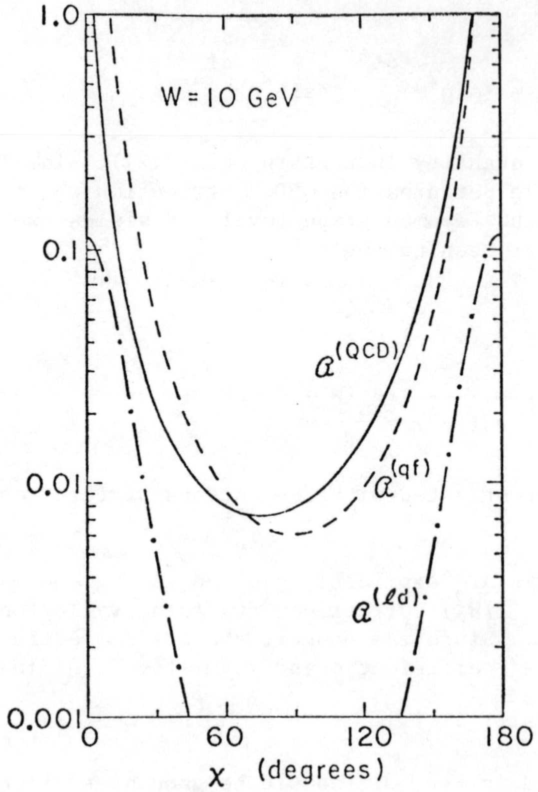

Fig. 30a. Figure caption follows Fig. 30c.

in perturbation theory (except for zeroth order) the coefficient is infrared singular. For example, in the dimensional regularization scheme (ν dimensions) for Ref. 38, the coefficient of $\delta(\Omega+\Omega')$ in order α_s is

$$\left.\frac{d^2\Sigma^{(virt)}}{d\Omega d\Omega'}\right|_{\chi \approx \pi} \simeq \frac{\sigma_o}{(\frac{8\pi}{3})} [P^2 \sin^2\chi + \frac{1}{2}(1-P^2)(1+\cos^2\theta)] \times \frac{\alpha_s C_F}{4\pi}\left[\frac{-2}{(\frac{\nu}{2}-2)^2} + 0\left(\frac{1}{(\frac{\nu}{2}-2)^2}\right)\right] \tag{8.16}$$

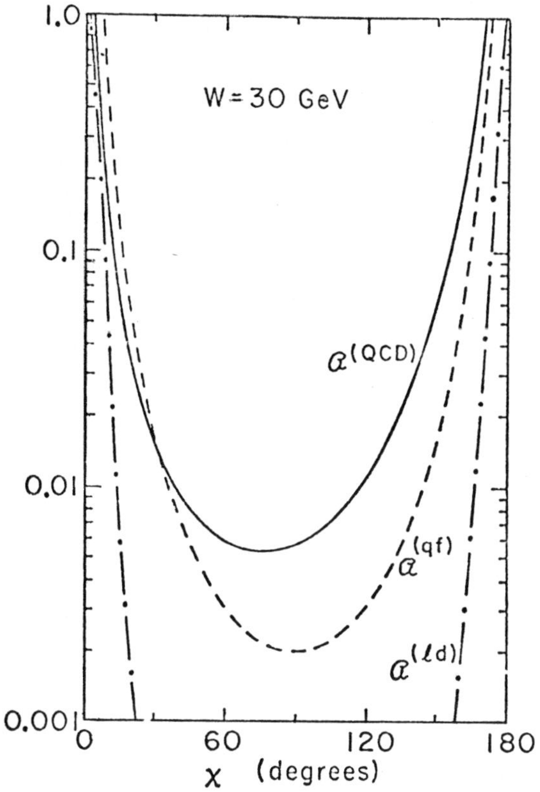

Fig. 30b. Figure caption follows Fig. 30c.

and in general is a power series in $\left[\frac{\alpha_s C_F}{(\frac{\nu}{2}-2)^2}\right]^n$ (beginning with $n = 0$ for Fig. 24a squared). Noting that in ν dimensions

$$d^{\nu-2}\Omega = \sin^{\nu-3}\chi d\chi d^{\nu-3}\Omega = 2[4\zeta(1-\zeta)]^{\frac{\nu}{2}-2} d\zeta d^{\nu-3}\Omega \ , \qquad (8.17a)$$

and

$$\int d^{\nu-3}\Omega = \frac{2\pi^{\frac{\nu}{2}-1}}{\Gamma(\frac{\nu}{2}-1)} \qquad (8.17b)$$

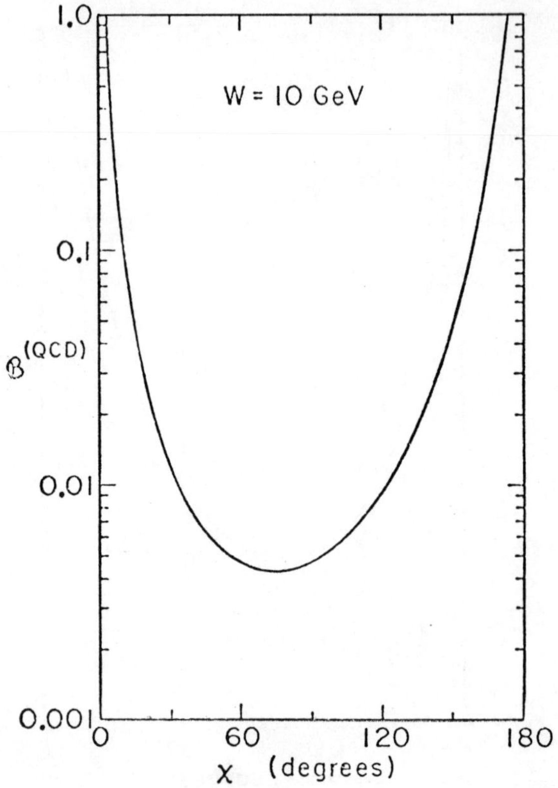

Fig. 30c. The coefficient $A(\chi)$ versus χ at a) W=10 GeV, b) W=30 GeV, indicating the relative energy dependence of the QCD contribution, the 1/W nonperturbative correction and that due to heavy leptons (Ref. 38). In c) is displayed the coefficient $\mathcal{B}(\chi)$ at 10 GeV. This term decreases as $1/\ln W$ with W.

for completeness, we see that a behavior like $\frac{1}{1-\zeta}\,\ln\frac{1}{1-\zeta}$, when integrated over a small region near $\zeta = 1$, yields a term with behavior $(\nu/2 - 2)^{-2}$. In fact, the singularities in the two contributions do exactly cancel when integrated over such a small angular patch as discussed in detail in Refs. 38 and 39. Furthermore one expects that the leading log part of these purely virtual corrections exponentiates to yield the "on-shell" form factor squared (familiar[52] from QED) $\sim \left[\exp - \frac{\alpha_s}{4\pi} C_F \ln^2 \frac{Q^2}{\mu^2}\right]^2$ where a gluon mass μ has been introduced as a cutoff. Here I have included the explicit Casimir factor C_F for the quark representation.

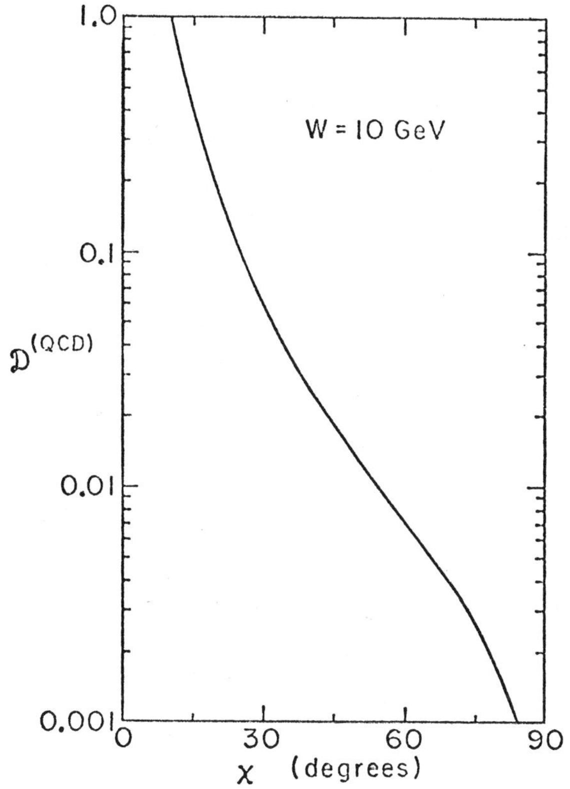

Fig. 31. The asymmetry $\mathcal{D}(\chi) = A(\pi - \chi) - A(\chi)$, which should have only small nonperturbative corrections (Ref. 38).

That this exponentiation should occur with the only change from QED being $1 \to C_F$ is nontrivial since it requires the exact cancellation (in leading log) of the purely non-Abelian terms proportional to C_A (crossed graphs and 3 gluon vertices). If this is the case, the coefficient of $\delta(\Omega+\Omega')$ then vanishes as $\mu^2 \to 0$. Since we still require an order-by-order cancellation with the smooth part of $d^2\Sigma/d\Omega d\Omega'$, integrated about $\zeta = 1$, it also must exponentiate. (The upper limit of ζ in this regularization scheme is now $1 - \mu^2/W^2$.) By explicit calculation[45,46] to order $\alpha_s{}^2$ we expect

$$\left.\frac{d^2\Sigma}{d\Omega d\Omega'}\right|_{\zeta \approx 1} \simeq \frac{d\sigma}{d\Omega} P(\chi) \equiv \frac{d\sigma}{d\Omega}\,\frac{\alpha_s}{\pi}\,\frac{C_F}{8\pi}\,\frac{\ln\left(\frac{1}{1-\zeta}\right)}{1-\zeta}\,\exp\left(-\frac{\alpha_s C_F}{2\pi}\ln^2\frac{1}{1-\zeta}\right)$$

$$= \frac{d\sigma}{d\Omega}\,\frac{1}{4\pi}\,\frac{d}{d\cos\chi}\,\exp\left[-\frac{\alpha_s C_F}{2\pi}\ln^2\left(\frac{2}{1+\cos\chi}\right)\right] \tag{8.18}$$

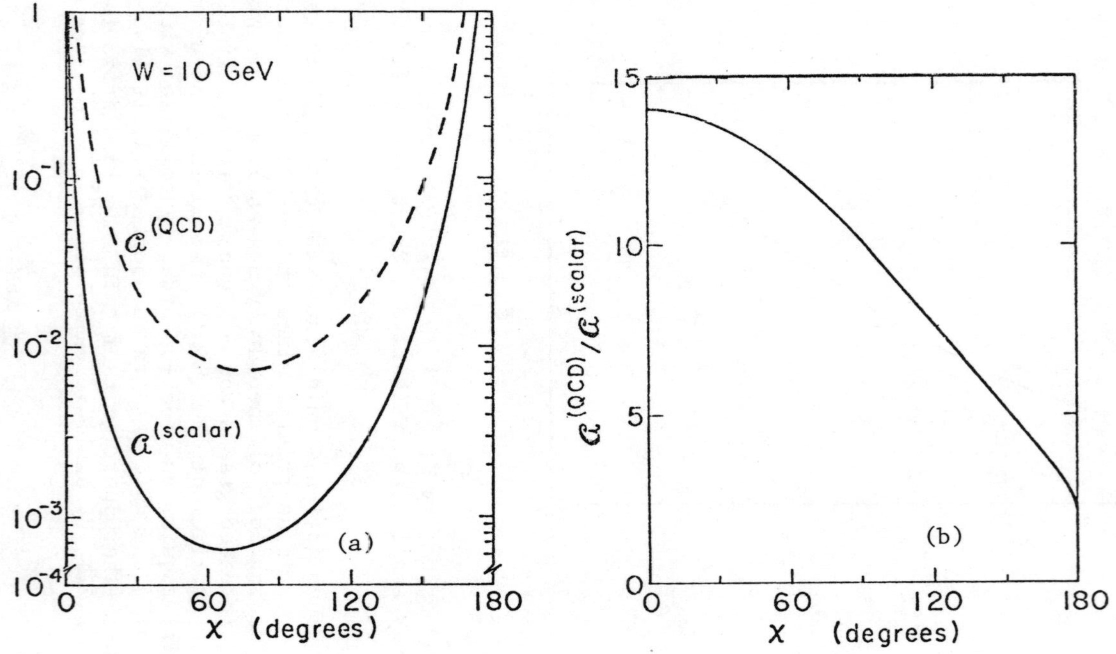

Fig. 32a,b. Figure caption follows Fig. 32c,d.

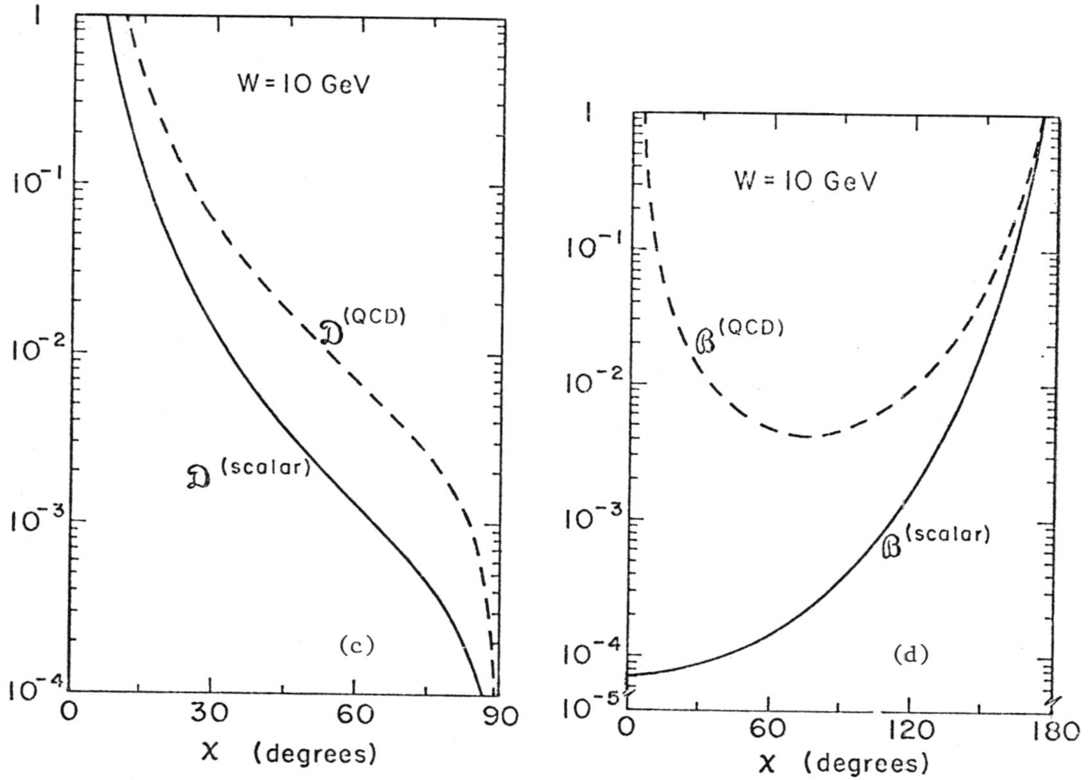

Fig. 32c,d. Comparison of the quantities a) and b) $A(\chi)$, c) $\mathcal{D}(\chi) = A(\pi - \chi) - A(\ )$, d) $\mathcal{B}(\chi)$ for scalar and vector (QCD) theories at $W = 10$ GeV (Ref. 46).

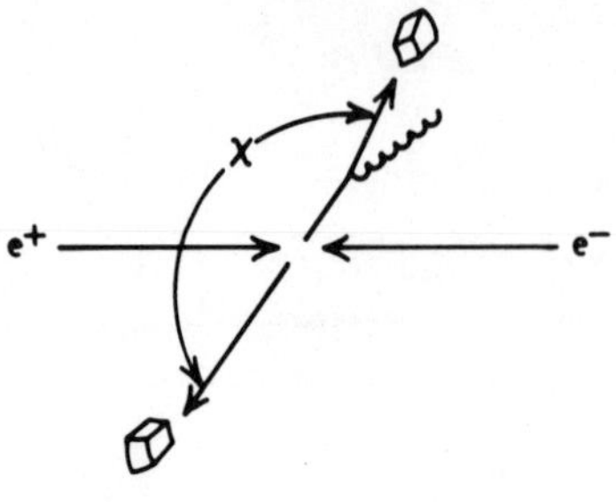

Fig. 33. Limit of geometry for two calorimeter experiment as $\chi \to \pi$.

where

$$\frac{d\sigma}{d\Omega} \equiv \frac{\sigma_o}{(\frac{8\pi}{3})} [P^2 \sin^2\psi + \frac{1}{2}(1-P^2)(1+\cos^2\theta)] \quad . \tag{8.19}$$

This form also arises in the more detailed analysis of Ref. 47 but not Ref. 48. Not only does this form have the correct perturbative behavior for χ fixed, $W \to \infty$ but it also satisfies the sum rule

$$\int_{\chi \approx \pi} d\Omega' \frac{d^2\Sigma}{d\Omega d\Omega'} = \frac{1}{2} \frac{d\sigma}{d\Omega}[1 + O(\alpha_s)] \quad . \tag{8.20}$$

[Note that $\int^{1-\mu^2/W^2} d\zeta$ of this expression gives precisely the correct term to cancel the form factor coefficient of $\delta(\Omega+\Omega')$.] The function $P(\chi)$ is plotted in Fig. 34 which indicates the expected peaking near $\chi = \pi$. As $W \to \infty$ (the perturbative limit) $P(\chi)$ should approximate a delta function at $\chi = \pi$ (the parton model limit). In the above approximation the peak in $P(\chi)$ occurs at

$$\pi - \chi_{max} \approx \exp\left[-\frac{2\pi}{C_F \alpha_s (W^2)}\right] \sim \left(\frac{\Lambda}{W}\right)^{25/16} \tag{8.21}$$

for $N_f = 4$. For W = 10 GeV and Λ = .5 GeV, this yields an angle of about 1°. Unfortunately this dramatic behavior will be obscured by nonperturbative smearing effects (the postremordial smerge) which, as discussed earlier, have an angular width behaving as 1/W. It is also the case, as noted by Field,[6] that, at forseeable energies, the leading logs do not stand out in a range of ζ where they are not already strongly damped by the exponentiation. It is important to note that in the above analysis the choice of the argument of α_s, i.e. W^2 rather than $W^2(1-\zeta)$

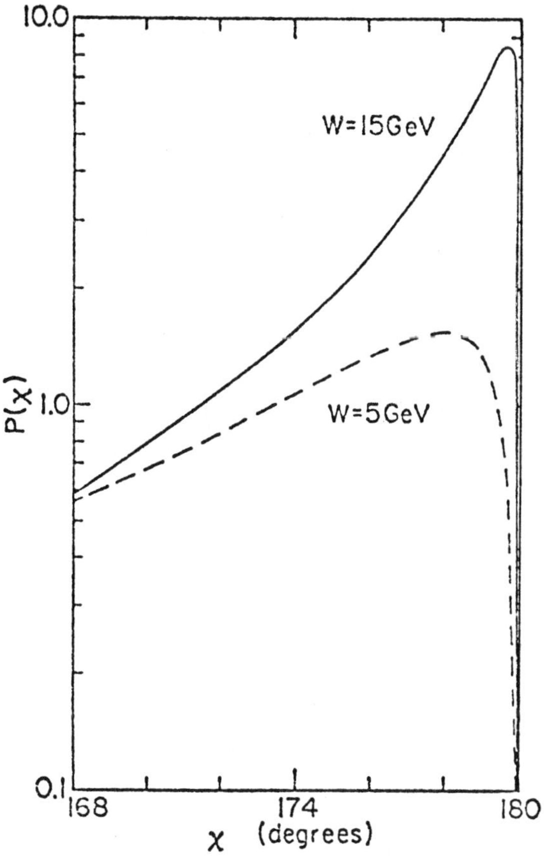

Fig. 34. The function $P(\chi)$ defined in Eq. (8.18) versus χ for W=5 and 15 GeV (Ref. 45).

for example, is only a correction to the next to leading log and is not relevant. This is not the case below.

Now consider $\chi \to 0$, as illustrated in Fig. 35. In this case we must not only consider the purely virtual contributions but also the singular contributions due to the "transparent" calorimeter definition. Both contribute to the coefficient of $\delta(\Omega-\Omega')$ and in this case the pure infrared singularities cancel and the coefficient is a power series in $[\alpha_s/(\nu/2-2)]^n$ or $[\alpha_s \ln W^2/\mu^2]^n$ depending on the regularization scheme. [The gluon is detected and can't be soft.] Thus we expect $d^2\Sigma/d\Omega d\Omega'$ to behave as $1/\zeta$ as $\zeta \to 0$ as illustrated in Eq. (8.9). Again to proceed to truly small χ we must sum to all orders in α_s. But these are exactly the (single) logarithmic divergences we summed to find the Q^2

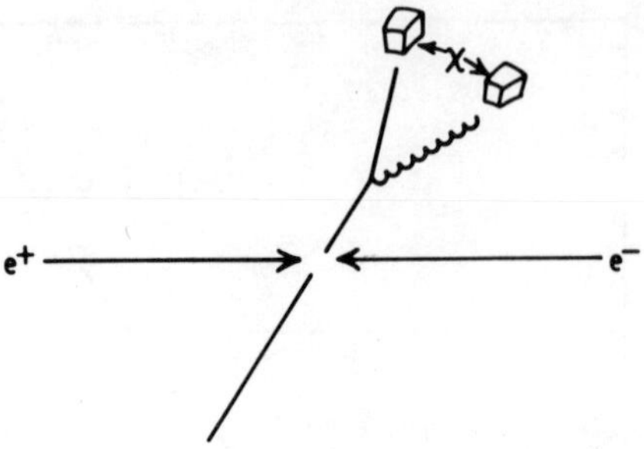

Fig. 35. Limit of geometry for two calorimeter experiment as $\chi \to 0$.

evolution of the D function. We must sum the logs due to the collinear emissions shown in Fig. 36. Emissions from the opposite quark or after the "hard" vertex are "summed" trivially by our energy weighting procedure. The cutoff for the logs being summed is provided by the angle χ. In this case the momentum dependence of the coupling cannot be ignored. In fact we must use the full glory of Eq. (6.16) where now the cutoff $W_o^{\ 2}$ is $\chi^2 W^2$ as $\chi^2 \to 0$ and we take r = 2 because we are taking the first (momentum) moment of <u>both</u> the final quark jet and the final gluon jet. I leave as an exercise the details[31,49] of this calculation and quote the results. Applying Eq. (6.16) for r = 2 actually results in the behavior of the integrated cross section from some $\chi_{min} \sim \Lambda/Q$ up to χ_{max}.

Define $\bar{\kappa} = \frac{2}{\beta_o} \ln \left(\frac{\ln W^2/\Lambda^2}{\ln \chi_{max}^{\ 2} W^2/\Lambda^2} \right)$, and we have

$$\int_{\chi_{min}}^{\chi_{max}} d\chi \frac{d^2\Sigma}{d\Omega d\chi} \simeq \frac{1}{2}\frac{d\sigma}{d\Omega}\left[\left(1-\alpha(2)-\beta(2)\right)e^{\bar{\kappa}A_+(2)} + \left(\alpha(2)+\beta(2)\right)e^{\bar{\kappa}A_-(2)}\right] \tag{8.22a}$$

$$\equiv \frac{1}{2}\frac{d\sigma}{d\Omega} Y(\chi_{max}) \quad . \tag{8.22b}$$

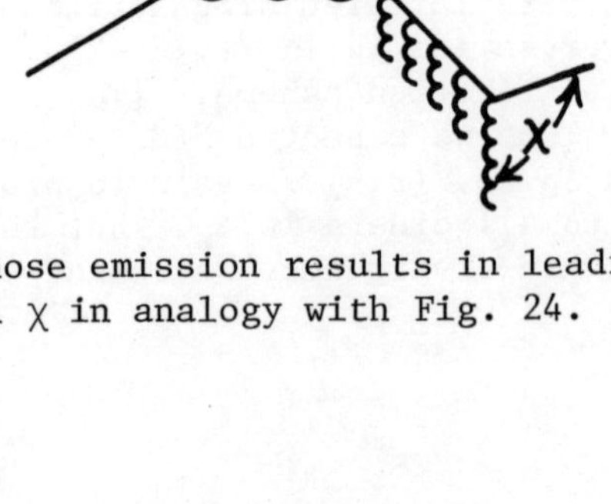

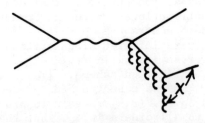

Fig. 36. Gluons whose emission results in leading log behavior for small χ in analogy with Fig. 24.

Here $Y(\chi_{max})$ is the analog of $P(\chi)$ above and is a measure of the fraction of the jet (energy) inside an opening angle of χ_{max}. Note that the required sum rule is satisfied as $Y \to 1$ for $\chi_{max} \sim O(1)$. The coefficients $M(r,t)$ in Eq. (6.16) are just 1 here because they are just the momentum sum rule for the quark jet and gluon jet (Fig. 35) separately. Note that for any fixed χ_{max} as $W \to \infty$ $\bar{\kappa} \to -2/\beta_0(\ln \chi^2_{max})/(\ln W^2/\Lambda^2) \to 0$. Consider now the limits $Y = 1-\varepsilon \lesssim 1$, $\bar{\kappa} \to 0$, where we can expand Eq. (8.22) using the definitions of Eq. (6.15). The result for a quark jet is that all but ε of the jet is contained inside an angle $\chi(\varepsilon)_{max}$ given by

$$\bar{\kappa}_q(\chi(\varepsilon)_{max}) \simeq \frac{-\varepsilon}{A(2)_{qq} + A(2)_{qG}} \quad . \tag{8.23a}$$

For comparison the corresponding result for a gluon jet is

$$\bar{\kappa}_G(\chi(\varepsilon)_{max}) \simeq \frac{-\varepsilon}{A(2)_{GG} + 2N_f A(2)_{Gq}} \quad . \tag{8.23b}$$

Thus the relative widths of a quark jet and glue jet are characterized by (for small ε)

$$\frac{\bar{\kappa}_G}{\bar{\kappa}_q} \simeq \frac{\ln \chi(\varepsilon)_{max,G}}{\ln \chi(\varepsilon)_{max,q}} = \frac{A(2)_{qq} + A(2)_{qG}}{A(2)_{GG} + 2N_f A(2)_{Gq}} = \frac{3/2\, C_F}{\frac{7}{5} C_A + \frac{N_f}{10}} = .435 \gtrsim C_F/C_A \quad . \tag{8.24}$$

Using somewhat different measures, essentially this result has been derived by many others.[31,36,49] The interpretation is simply that not only are glue jets softer and of higher multiplicity than quark jets, they are also fatter. In terms of W dependence we have

$$\chi_{max,q} \sim \left(\frac{\Lambda}{W}\right)^{\frac{\beta_0 \varepsilon}{3C_F}} \tag{8.25a}$$

and

$$\chi_{max,G} \sim \left(\frac{\Lambda}{W}\right)^{\frac{\beta_0 \varepsilon}{14/5\, C_A + N_f/5}} \tag{8.25b}$$

For $\varepsilon = .3$ and $N_f = 4$ the exponents are about 0.62 and 0.27 for quarks and gluons respectively.

Finally we can return to the differential form and evaluate numerically to find

$$\left.\frac{d^2\Sigma}{d\Omega d\chi}\right|_{\chi \gtrsim 0} \simeq \frac{d\sigma}{d\Omega}\frac{\alpha_s(\chi^2 W^2)}{\pi\,\chi}\left\{1.48\left[\frac{\ell n(\chi^2W^2/\Lambda^2)}{\ell n(W^2/\Lambda^2)}\right]^{0.61} - 0.48\left[\frac{\ell n(\chi^2W^2/\Lambda^2)}{\ell n(W^2/\Lambda^2)}\right]^{1.39}\right\} . \tag{8.26}$$

Note that in the perturbative limit with χ fixed, $W^2 \to \infty$, [ ] $\to 1$, we recover Eq. (8.9) with some algebra. This form [Eq. (8.26)] peaks at an angle χ_{peak} near $e^1 \times (\Lambda/W)$ which again is in a region controlled by nonperturbative effects. However, one may still hope that it will be possible (at large W^2) to detect deviations from the lowest order perturbative result in regions where non-perturbative effects are small. In any case the perturbation theory, after summing the leading logs, is by itself well behaved (integrable) as $\chi \to \Lambda/W$.

Let me close with a brief word concerning the statistics of energy measurements. This problem is under study by J. Sidles,[53] a graudate student at the University of Washington. The statistical uncertainty in, for example, the quantity $d\Sigma/d\Omega$ is given to a good approximation by the experimental quantities

$$\left|\frac{\Delta(d\Sigma/d\Omega)}{d\Sigma/d\Omega}\right| \simeq \sqrt{\frac{\sum_{\text{events}}(E_i^2)}{(\sum_{\text{events}} E_i)^2}} . \tag{8.27}$$

This quantity can in turn be estimated in the form

$$\left|\frac{\Delta(d\Sigma/d\Omega)}{d\Sigma/d\Omega}\right| \simeq \frac{\text{constant}}{\sqrt{L\cdot T\cdot \sigma_{TOT}}} , \tag{8.28}$$

where the constant can be evaluated in a jet model such as discussed earlier for specific experimental angular acceptances. It turns out that the constant is in the range 1-5 for reasonable values of the model parameter $\langle h_T \rangle$ and for reasonable experimental acceptances.

ACKNOWLEDGMENT

Warm thanks go to my colleagues at the University of Washington, especially Lou Basham, Lowell Brown and Sherwin Love, and to my fellow lecturers, especially Rick Field and Chris Llewellyn Smith, for many helpful discussions. I also acknowledge the prodigious efforts of K.T. Mahanthappa to make this Summer School possible.

REFERENCES

1. For a fairly complete review of the early parton model see R.P. Feynman, "Photon-Hadron Interactions" (Benjamin, Reading, Mass., 1972).

2. J.D. Bjorken, Lectures given at the International Summer Institute in Theoretical Physics, DESY, Hamburg, September, 1975, SLAC-PUB-1756(1976).

3. For early discussions of the role of jets in hadronic physics see J.D. Bjorken, Phys. Rev. D8, 4098 (1973); S.D. Ellis and M.B. Kislinger, Phys. Rev. D9, 2027 (1974).

4. For an early suggestion that this scenario would actually work see H.D. Politzer, Nucl. Phys. B129, 301 (1977).

5. Chris Llewellyn Smith, lectures at this Summer School.

6. Richard D. Field, lectures at this Summer School.

7. The classic papers of this era are D. Amati, S. Fubini and A. Stanghellini, Nuovo Cimento 26, 896 (1962); K. Wilson, Acta Phys. Austriaca 17, 37 (1963); and later J. Benecke, T.T. Chou, C.N. Yang and E. Yen, Phys. Rev. 188, 2159 (1969); R.P. Feynman, Phys. Rev. Letters 23, 1415 (1969). This last paper essentially marks the beginning of the parton model.

8. See, for example, Wolfgang Ochs, Nucl. Phys. B118, 397 (1977).

9. C.E. DeTar, Phys. Rev. D3, 128 (1971); C.E. DeTar, et. al., Phys. Rev. Letters 26, 675 (1971); H.D.I. Abarbanel, Phys. Rev. D3, 2227 (1971); A.H. Mueller, Phys. Rev. D4, 150 (1971); L.S. Brown, Phys. Rev. D5, 748 (1972); S.D. Ellis and A.I. Sanda, Phys. Rev. D6, 1347 (1972).

10. For a pre-QCD discussion see C.E. DeTar, S.D. Ellis and P.V. Landshoff, Nucl. Phys. B87, 176 (1975). This study remains relevant even to modern treatments.

11. For a discussion of the CIM (Constituent Interchange Model) see R. Blankenbecler, S.J. Brodsky and J.F. Gunion, Phys. Rev. D18, 900 (1978), where further references are given.

12. The standard reference for such studies is R.D. Field and R.P. Feynman, Nucl. Phys. B136, 1 (1978), which is the source for much of these lectures. See also Ref. 2.

13. Gail G. Hanson, Proceedings of the 13th Rencontre de Moriond, Les Arcs, France, March 12-24, 1978 and SLAC-PUB-2118 (1978).

14. J.C. VanderVelde, Talk at the Symposium on Jets in High Energy Collisions, Neils Bohr Institute, Copenhagen, Denmark, July, 1978 and Univ. of Michigan preprint UMBC 78-9 (1978).

15. See Refs. 1 and 2 and J. Bjorken and J. Kogut, Phys. Rev. D8, 1341 (1973).

16. S. Brodwky and J. Gunion, Phys. Rev. Letters 37, 402 (1976); S. Brodsky, T. DeGrand and R. Schwitters, Phys. Letters 79B, 255 (1978).

17. See, for example, W. Selove, Proceedings of the 14th Rencontre de Moriond, Les Arcs, France, March, 1979 and Univ. of Pennsylvania preprint UPR-70E (1979); M.J. Tannenbaum, ibid. and Rockefeller Univ. preprint COO-2232A-79 (1979).

18. See, for example, V. Cook et.al., preprint FERMILAB-PUB-79/68-EXP (1979).

19. U.P. Sukhatme, Phys. Letters 73B, 478 (1978) and ORSAY preprint LPTPE 78/36 (1978); A. Seiden, Phys. Letters 68B, 157 (1977). See also the earlier work of R.N. Cahn and E.W. Colglazier, Phys. Rev. D9, 2658 (1974); G.R. Farrar and J.L. Rosner, Phys. Rev. D9, 2226 (1974).

20. A. Casher, J. Kogut and L. Susskind, Phys. Rev. Letters 31, 792 (1973).

21. R.D. Field and R.P. Feynman, Phys. Rev. D15, 2590 (1977).

22. See, for example, S.J. Brodsky and N. Weiss, Phys. Rev. D16, 2325 (1977).

23. See, for example, C. Quigg, P. Pirila and G.H. Thomas, Phys. Rev. Letters 34, 290 (1975) and Phys. Rev. D12, 92 (1975).

24. H. Rudnicka et al., Paper presented at Neutrino-78 Conference, Purdue University, 1978 and Univ. of Washington-VTL preprint VTL-PUB-53 (1978).

25. D.H. Perkins, Paper presented at the Rutherford Laboratory Xmas Theoretical Physics Meeting, December, 1978 and Oxford University preprint 2/79 (1979).

26. J. Kogut, D. Sinclair and L. Susskind, Phys. Rev. D7, 3637 (1973).

27. A.H. Mueller, Phys. Rev. D18, 3705 (1978); S. Gupta and A.H. Mueller, Phys. Rev. D20, 118 (1979); C.H. Llewellyn Smith, Acta Physica Austriaca, Suppl. XIX, 331 (1978); W.R. Frazer and J.F. Gunion, Univ. of California preprints UCSD-10P10-194 (1978) and Phys. Rev. D19, 2447 (1979); G. Sterman, Phys. Ref. D17, 2773 and 2789 (1978); S. Libby and G. Sterman, Phys. Rev. D18, 3252 (1978); D. Amati, R. Petronzio and G. Veneziano, Nucl. Phys. B140, 54 (1978) and ibid., B146, 29 (1978); R.K. Ellis, H. Georgi, M. Machacek, H.D. Politzer and G.G. Ross, Phys. Letters 78B, 281 (1978) and Nucl. Phys. B152, 285 (1979).

28. K. Wilson, Phys. Rev. 179, 1499 (1969).

29. For the case of the G function see G. Altarelli and G. Parisi, Nucl. Phys. B126, 298 (1977).

30. For the D function see T. Uematsu, Phys. Letters 79B, 97 (1978); J.F. Owens, Phys. Letters 76B, 85 (1978).

31. K. Konishi, A. Ukawa and G. Veneziano, Phys. Letters 78B, 243 (1978); ibid. 80B, 259 (1979); Nucl. Phys. B157, 45 (1979).

32. A.H. Mueller, Phys. Rev. D9, 963 (1974); C.G. Callan and M.L. Goldberger, Phys. Rev. D11, 1542 (1975).

33. J. Ellis, M.K. Gaillard and W.J. Zakrzewski, CERN preprint TH. 2595 (1978).

34. T. Kinoshita, J. Math. Phys. 3, 650 (1962); T.D. Lee and M. Nauenberg, Phys. Rev. 133, 1549 (1964).

35. G. Sterman and S. Weinberg, Phys. Rev. Letters 39, 1436 (1977).

36. See, for example, M.B. Einhorn and B. Weeks, Nucl. Phys. B146, 445 (1978); B.G. Weeks, Univ. of Michigan preprint UM-HE 78-49 (1978); P.M. Stevenson, Nucl. Phys. B150, 357 (1979) and Imperial College preprint ICTP/78-79/16 (1979); K. Shizuya and S.-H.H. Tye, Phys. Rev. Letters 41, 787 (1978).

37. See, for example, H. Georgi and M. Machacek, Phys. Rev. Letters 39, 1237 (1977); E. Farhi, ibid. 1587 (1977); A. De Rújula, J. Ellis, E.G. Floratos, and M.K. Gaillard, Nucl. Phys. B138, 387 (1978); S.-Y. Pi, R.L. Jaffe and F.E. Low, Phys. Rev. Letters 41, 142 (1978).

38. C.L. Basham, L.S. Brown, S.D. Ellis and S.T. Love, Phys. Rev. D17, 2298 (1978); Phys. Rev. Letters 41, 1585 (1978); Phys. Rev. D19, 2018 (1979).

39. G.C. Fox and S. Wolfram, Phys. Rev. Letters 41, 1581 (1978); Nucl. Phys. B149, 413 (1979); Phys. Letters B82, 134 (1979); Caltech preprint CALT 68-723 (1979).

40. Yu.L. Dokshitzer, D.I. D'yakonov and S.I. Troyan, Proc. XIIIth Winter School of LNPI (Leningrad, 1978) part 1, pages 3-89 (Engl. translation SLAC-TRANS-183 (1978)).

41. For a discussion of the extension of the methods of Refs. 38 and 39 to processes other than e^+e^- annihilation into hadrons see, for example, R.D. Peccei and R. Rückl, Max Planck Institute preprints MPI-PAE/PTh 6/79 and 21/79 (1979); W. Furmanski and S. Pokorski, CERN preprint TH. 2665 (1979); M. Abud, M. Dacorogna, R. Gatto and C.A. Savoy, Phys. Letters 84B, 229 (1979).

42. G. Sterman, Phys. Rev. D19, 3135 (1979); see also G. Tiktopoulos, Nucl. Phys. B147, 371 (1979).

43. R.F. Schwitters, in Proc. of the 1975 International Symposium on Lepton Interactions at High Energies, Stanford, CA (SLAC, Stanford, 1975) p. 10; G.G. Hanson, talks at the 17th International Colloquium on Multiparticle Reactions, Munich, 1976, and in the Proc. of the XVIIIth International Conference on High Energy Physics, Tbilisi, 1976 (JINR, Dubna, USSR, 1976) and preprint SLAC-PUB-1814 (1976).

44. G. Zech, Proc. of the XIVth Rencontre de Moriond, March 11-23, 1979, Les Arcs, France and Seigen Univ. preprint SI 79-2 (1979).

45. C.L. Basham, L.S. Brown, S.D. Ellis and S.T. Love, Phys. Letters 85B, 297 (1979).

46. C.Y. Lo and J.D. Sullivan, Univ. of Illinois preprint ILL-(TH)-79-22 (1979).

47. See the last paper in Ref. 39 and G. Parisi and R. Petronzio, CERN preprint TH. 2627 (1979).

48. Yu.L. Dokshitzer, D.I. D'yakonov and S.I. Troyan, Phys. Letters 78B, 290 (1978); ibid. 79B, 269 (1978); and Ref. 40.

49. K. Konishi, A. Ukawa and G. Veneziano, Phys. Letters 80B, 259 (1979); R.K. Ellis and R. Petronzio, Phys. Letters 80B, 249 (1979); A.V. Smilga, preprint ITEP-78 (1978).

50. The numerical studies of these corrections found in Ref. 39 suggest that the nonleading (in powers of 1/W) corrections may, in fact, be quite large at foreseeable energies.

51. C. Louis Basham and Sherwin T. Love, Phys. Rev. D20, 340 (1979).

52. P.M. Fishbane and J.D. Sullivan, Phys. Rev. D4, 458 (1971).

53. J.A. Sidles, Univ. of Washington, Ph.D. thesis, in preparation.

PERTURBATIVE QUANTUM CHROMODYNAMICS AND APPLICATIONS TO LARGE MOMENTUM TRANSFER PROCESSES

R. D. Field

California Institute of Technology, Pasadena, Calif. 91125

I. INTRODUCTION

During the last several years, a new framework to describe strong interaction physics has emerged: quantum chromodynamics (QCD). It is the simplest field theory which incorporates a color-dependent force among quarks. The forces among the colored quarks are generated by the exchange of colored vector gluons which are coupled to the quarks in a gauge-invariant manner. The theory is closely related to the most successful quantum field theory: QED. The only (but very important) difference between QED and QCD is the gauge group involved. QED is an Abelian gauge theory (the photons do not couple to each other); QCD is a non-Abelian gauge theory [gauge group SU_3(color)]. (The gluons carry color and thus are coupled to each other.) The formal Lagrangian for QCD is given by [1,2]

$$\mathcal{L}_{QCD}(x) = -\frac{1}{4} F^a_{\mu\nu} F^{a\mu\nu} + i\bar{\psi}_{q_i}(x)\gamma_\mu D^\mu_{ij}\psi_{q_j}(x), \tag{1.1a}$$

where $F^a_{\mu\nu}$ is related to the vector potential A^a_μ according to

$$F^a_{\mu\nu} = \partial_\mu A^a_\nu - \partial_\nu A^a_\mu + g f_{abc} A^b_\mu A^c_\nu \tag{1.1b}$$

and

$$D^\mu_{ij} = \partial^\mu \delta_{ij} - igA^\mu_a T^a_{ij}. \tag{1.1c}$$

The structure constants (for the group SU(3) of color) are given by

$$[\lambda^a, \lambda^b] = 2if_{abc}\lambda^c, \tag{1.2a}$$

where λ^a_{ij} are the eight SU(3) matrices introduced by Gell-Mann[3] and

$$T^a = \frac{1}{2}\lambda^a. \tag{1.2b}$$

Quantizing in a renormalizable gauge leads to the Feynman rules summarized in Fig. 1.1[1]. Notice that all flavors of quarks as well as gluons couple with a universal strength g.

Although the theory is well defined, precisely what it predicts is not yet clearly known. For example, it is not known if the theory actually confines quarks and gluons within hadrons nor has the spectrum of hadronic states been calculated. At present, the mathematical complexities are still too great[4]. However, at very high energy or momentum transfer, Q, the theory is asymptotically free; the effective coupling between quarks and gluons decreases with increasing Q^2. As emphasized by Politzer, this permits calculation of those parts of a process involving high Q^2 by the use of perturbation theory and the Feynman rules in Fig. 1.1. Yet every real process involves both high <u>and</u> low Q^2 together and precisely how to separate these parts is just becoming clear.

In these lectures, we will examine some of the present day applications of QCD to processes involving large momentum transfer. Some of these lectures will be identical to those I gave at the La Jolla Summer Workshop last year[5]. I will, however, skip some of the topics I covered at La Jolla and I will update these lectures by including discussions of some of the progress made during the last year. In addition, I will take this opportunity to correct some mistakes I made in my La Jolla write-up and to improve some of the discussions. We will examine here more closely the nature of the mass singularities that arise in perturbative QCD. Also, we will investigate the significance of experimental data obtained within the last year at CERN and Fermilab. I suggest these lectures be read in conjunction with my La Jolla notes.

I will begin in Section II with a discussion of the QCD "effective" coupling $\alpha_s(Q^2)$. In Section III we will examine the process $e^+e^- \rightarrow$ hadron and discuss the quark and gluon fragmentation functions. We will then proceed to investigate how QCD affects the quark and

$$-i\delta^{ab}\left[\left(g_{\mu\nu}-\frac{k_\mu k_\nu}{k^2}\right)/k^2 + \alpha k_\mu k_\nu/k^4\right]$$

$$-i\delta^{ab}/k^2$$

$$i\delta^{ij} \rlap{/}k/k^2$$

$$-gf^{abc}\left[(p-q)_\nu g_{\lambda\mu} + (q-r)_\lambda g_{\mu\nu} + (r-p)_\mu g_{\nu\lambda}\right]$$

$$-ig^2 f^{abe}f^{cde}\left(g_{\lambda\nu}g_{\mu\sigma} - g_{\lambda\sigma}g_{\mu\nu}\right)$$
$$-ig^2 f^{ace}f^{bde}\left(g_{\lambda\mu}g_{\nu\sigma} - g_{\lambda\sigma}g_{\mu\nu}\right)$$
$$-ig^2 f^{ade}f^{cbe}\left(g_{\lambda\nu}g_{\mu\sigma} - g_{\lambda\mu}g_{\sigma\nu}\right)$$

$$gf^{abc}p^\mu$$

$$-ig\gamma^\mu T^a_{ij}$$

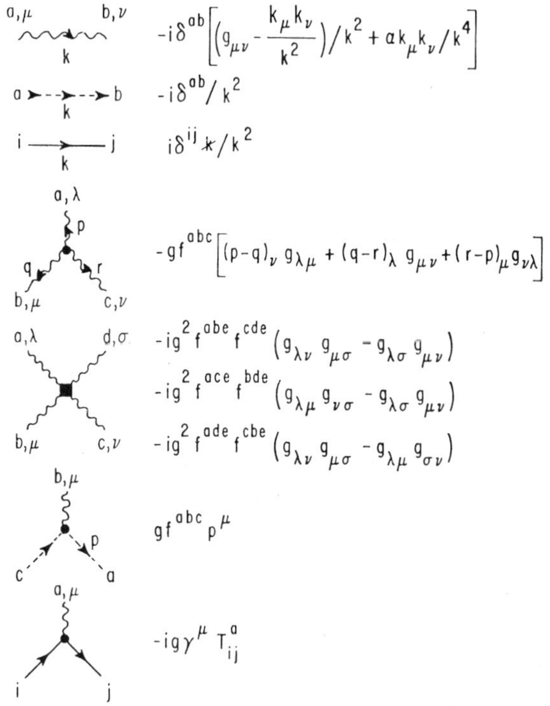

Fig. 1.1. Feynman rules for a gauge theory with fermions (from Ref. 1). The solid and wavy lines are fermions and vector gluons, respectively. The dotted lines are (non-existent) "ghost" particles which are introduced in the theory to aid in doing calculations that involve gluons.

gluon distributions within hadrons (Section IV) and to study large-mass muon-pair production in hadron-hadron collisions (Section V). In Section VI, I will write down the general QCD perturbative formalism and discuss the KLN theorem. The phenomenology of large $p_\perp$ meson and "jet" production in hadron-hadron collisions will be investigated in Section VII. Finally, Section VIII will be reserved for summary and conclusions. These lectures were given in conjunction with lectures by S. D. Ellis and C. H. Llewellyn Smith.

II. THE EFFECTIVE COUPLING $\alpha_s(Q^2)$

The theory of QCD does not produce inclusive cross sections that "scale". One cannot use dimensional counting arguments to determine the behavior of cross sections (at intermediate values of Q^2). This is because the theory has an intrinsic "scale" or mass parameter μ (the renormalization mass) that is generated as a result of the <u>interaction</u> between quarks and gluons. These interactions result in an effective strong interaction coupling $\alpha_s(Q^2)$ that decreases logarithmically with increasing Q^2, where Q is some characteristic momentum in a collision. I would like to discuss the origin of the Q^2 dependence of α_s, but first it is instructive to review the effective coupling in QED.

Effective Coupling in QED

In QED there is one parameter, the coupling, that must be determined experimentally. Suppose that we decide to experimentally measure the strength of electromagnetic interactions by scattering an electron off an electron and comparing the result with a theoretical calculation. The rate for electron-electron scattering is affected by the presence of vacuum polarization diagrams like that shown in Fig. 2.1a. One immediately runs into trouble, however, since the one electron loop correction to the photon propagator diverges like $\log\lambda^2$, where λ is some ultraviolet cutoff that can be arbitrarily large. In particular, the leading order bubble contribution is given (for $Q^2/m_e^2 >> 1$ and $Q^2/m_e^2 << 1$) by

$$\alpha_o B(Q^2) = -\frac{\alpha_o}{3\pi}[\log(\lambda^2/Q^2) + \frac{5}{3}]$$

$$\text{for } Q^2/m_e^2 >> 1, \quad (2.1a)$$

$$= -\frac{\alpha_o}{3\pi}[\log(\lambda^2/m_e^2) - \frac{1}{5}\frac{Q^2}{m_e^2}]$$

$$\text{for } Q^2/m_e^2 << 1, \quad (2.1b)$$

where $q^2 = -Q^2$ is the 4-momentum squared of the virtual (spacelike) photon and m_e is the electron mass. The coupling α_o is the "bare" electric charge ($\alpha_o = e_o^2/4\pi$). It is convenient to define an effective coupling, $\alpha_{eff}(Q^2)$, that incorporates all the vacuum polarization bubbles. Namely,

$$\alpha_{eff}(Q^2) = \alpha_o(1+\alpha_o B(q^2) + \alpha_o B(q^2)\alpha_o B(q^2)+\ldots.) \quad (2.2a)$$

$$\alpha_{eff}(Q^2) = \alpha_o/(1-\alpha_o B(q^2)) \quad (2.2b)$$

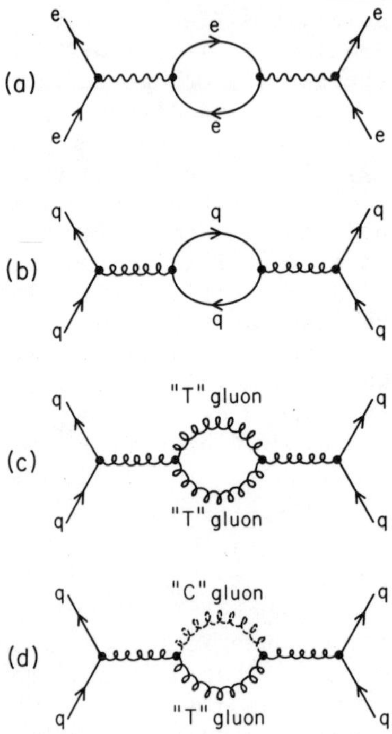

Fig. 2.1 - (a) Lowest order vacuum polarization correction to the electric charge.

(b) Lowest order correction to the quark-gluon coupling due to a virtual quark-antiquark pair.

(c) Lowest order correction to the quark-gluon coupling due to a virtual pair of transverse ("T") gluons in the coulomb gauge.

(d) Lowest order correction to the quark-gluon coupling due to a virtual pair of gluons, one transverse ("T") and one "coulomb" ("C") in the coulomb gauge.

or

$$1/\alpha_{eff}(Q^2) = 1/\alpha_o - B(q^2). \tag{2.2c}$$

The procedure for handling the ultraviolet divergences like those appearing in $B(q^2)$ in (2.1) is called renormalization. One defines an experimental electric charge, α, by the large distance behavior of the electric potential (Thompson limit)

$$\alpha = \alpha_{eff}(Q^2=0), \tag{2.3}$$

which experimentally is about 1/137. All results of calculations are now reexpressed in terms of the experimental coupling, α, rather than the unobservable bare coupling, α_o. In terms of α, the effective coupling is given by

$$1/\alpha_{eff}(Q^2) = 1/\alpha - (B(q^2)-B(0)), \tag{2.4}$$

where the quantity $B(Q^2) - B(0)$ is now independent of the artificial ultraviolet cutoff λ. The cutoff λ is now sent to infinity while holding α constant. From (2.1a) and (2.2b) we see that the large q^2 behavior of the effective coupling is given by

$$\alpha_{QED}(Q^2) = \alpha/[1-(\alpha/3\pi)\log(Q^2/m_c^2)]. \tag{2.5}$$

In QED as Q^2 increases, $\alpha_{eff}(Q^2)$ increases. No matter how small an α one has, one can always increase Q^2 to a point where $\alpha_{QED}(Q^2)$ becomes infinite. This means that perturbation theory breaks down at high Q^2 in QED. One needs to include higher and higher orders in α_{QED} as Q^2 increases. At low Q^2, on the other hand, $\alpha_{QED}(Q^2)$ is small ($\approx \alpha = 1/137$) and perturbation theory works well.

The physical reason for the rising effective charge with the increased Q^2 of the probing photon is illustrated in Fig. 2.2a. If Q^2 is small then the photon cannot resolve small distances and "sees" a "point" charge shielded by the vacuum polarization of the infinite sea of electron-positron pairs. As Q^2 increases, the photon "sees" a smaller and smaller spatial area and the shielding effect is less.

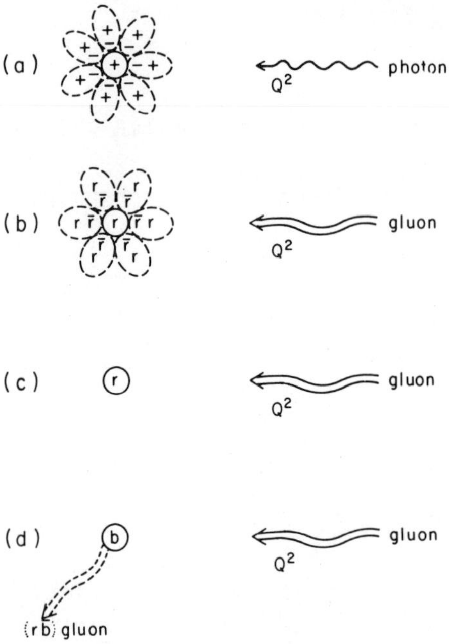

Fig. 2.2 - (a) Illustration of how vacuum polarization in QED will "shield" a bare positive charge when placed in a vacuum.

(b) The same shielding as in (a) but for a "red" charge in QCD.

(c) and (d) Show how a "red" charge can, in QCD, radiate away its red charge, r, and become a blue charge, b, via the emission of a virtual $r\bar{b}$ gluon.

Effective Coupling in QCD

In QCD the behavior of the effective coupling constant is strikingly different. The reason for this difference is the new feature of QCD, that the gluons carry charge (color) and interact with each other. The amount of the contributions of the various diagrams in Fig. 2.1b and Fig. 2.1c is gauge dependent. However, the situation is most clear in the coulomb gauge. In this gauge, the lowest order bubble contribution to the gluon propagator is given for large Q^2 by

$$\alpha_o B_{QCD}(Q^2) = -\alpha_o \tilde{a} \log(\lambda^2/Q^2), \tag{2.6}$$

where λ is the ultraviolet cutoff and α_o is the bare quark-gluon coupling and where

$$\tilde{a} = -\beta_o/4\pi \tag{2.7a}$$

with

$$\beta_o = -(\frac{2}{3} n_f + 5 - 16). \tag{2.7b}$$

The $+ \frac{2}{3} n_f$ and the +5 come from the quark loop and the transverse gluon loop in Fig. 2.1, respectively, and are of the same sign as the QED case. These contributions must be positive since the diagrams can be cut across the bubble and represent contributions to the physical rate for producing quark pairs or transverse gluon pairs which must be positive. The -16 in (2.7b) comes from the diagram with one transverse and one "coulomb" gluon in the bubble (Fig. 2.1d). This contribution need not be positive since the instantaneous "coulomb" gluon is not physical. If $\frac{2}{3} n_f < 11$ then β_o is positive and $\tilde{a}$ is negative in contrast to the QED case (2.1a).

As for the QED case, the ultraviolet divergences in (2.6) are handled by renormalization. Here, however, we cannot define the "experimental charge" by the $Q^2 \to 0$ limit of $\alpha_{eff}(Q^2)$ as we did in eq. (2.3). We instead choose some Q^2, say $Q^2 = \mu^2$, to define the coupling and express all predictions in terms of the coupling at this point (called the renormalization point or subtraction point). The effective coupling is then given by

$$1/\alpha_{eff}(Q^2) = 1/\alpha(\mu^2) - (B(Q^2) - B(\mu^2)). \tag{2.8}$$

As before, the quantity $B(Q^2) - B(\mu^2)$ is independent of the arbitrary cutoff λ. By the use of (2.6), we see that the leading order behavior of the coupling in QCD is

$$\alpha_s(Q^2) = \alpha(\mu^2)/[1+(\alpha(\mu^2)\beta_0/4\pi)\log(Q^2/\mu^2)], \tag{2.9}$$

which approaches zero as $Q^2 \to \infty$ (asymptotic freedom) as illustrated in Fig. 2.3. This means that for QCD, perturbation theory should work well at high Q^2 (short distances) but break down at small Q^2 (large distances) where $\alpha_s(Q^2)$ becomes large and hopefully (?) confines quarks within hadrons.

The nature of the QCD coupling constant $\alpha_s(Q^2)$ takes a bit of getting used to. In QED it is easy to define what one means by the charge of an electron e, by the large distance behavior of the electric potential (Thomson limit). One cannot do this for QCD since the $Q^2 \to 0$ limit of $\alpha_s(Q^2)$ cannot be calculated by perturbation theory. Instead we had to define some arbitrary point μ^2 at which the coupling is $\alpha(\mu^2)$. It, however, does not matter which point μ^2 one chooses. If one chooses instead the point M^2 then the two couplings are related (to lowest order) by

$$1/\alpha_s(\mu^2) - \tilde{a}\log(Q^2/\mu^2) = 1/\alpha_s(M^2) - \tilde{a}\log(Q^2/M^2) \tag{2.10a}$$

or

$$1/\alpha_s(\mu^2) + \tilde{a}\log(\mu^2) = 1/\alpha_s(M^2) + \tilde{a}\log(M^2). \tag{2.10b}$$

Thus, in some sense, the "real" parameter in the theory is not $\alpha(\mu^2)$

or μ^2 but rather a mass scale, Λ, that is independent of μ^2 and is given to this order by

$$\log\Lambda^2 = 1/(\tilde{a}\alpha(\mu^2)) + \log\mu^2. \qquad (2.11)$$

In terms of the mass scale Λ, the effective coupling is given by

$$\alpha_s(Q^2) \equiv \bar{g}^2(Q^2)/4\pi = \frac{4\pi}{\beta_o \log(Q^2/\Lambda^2)} \qquad (2.12)$$

with

$$\beta_o = 11 - \frac{2}{3} n_f. \qquad (2.13)$$

It is interesting to notice that eq. (2.12) becomes infinite at $Q^2 = \Lambda^2$ which corresponds to a distance of about 0.5 Fermi for Λ = 500 MeV. (This is extremely crude, however, since (2.12) is not applicable when $\alpha_s(Q^2)$ becomes large.)

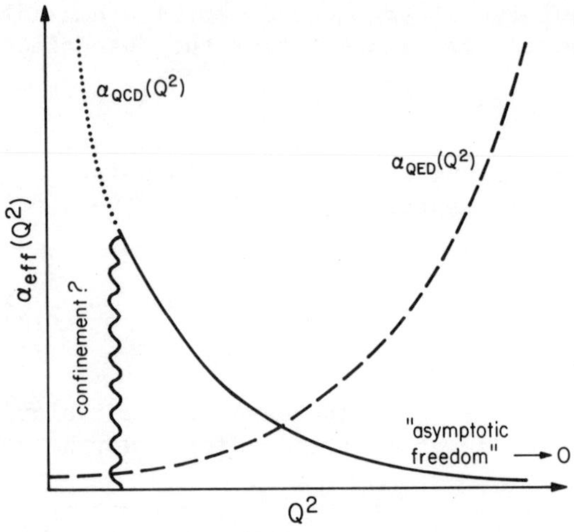

Fig. 2.3. Illustration of the behavior of the effective coupling calculated in perturbation theory in QED and QCD. In QED the effective coupling, $\alpha_{QED}(Q^2)$, is small at small Q^2, but becomes large at large Q^2 (small distances). In QCD, on the other hand, the effective coupling is large at small Q^2 (large distances) where confinement may occur, but decreases toward zero as Q^2 increases ("asymptotic freedom"). Perturbation theory should work well for QCD (QED) at large Q^2 (small Q^2).

A physical reason for the behavior of $\alpha_s(Q^2)$ is illustrated in Fig. 2.2. Quark-antiquark vacuum polarization shields the color charge as was the case in QED. However, now since the source can radiate charge (i.e., change from red to blue by emitting a red-blue-bar gluon), the charge is no longer located at a definite place in space. It is diffusely spread out due to gluon emission and absorption. As one increases the Q^2 of the incoming gluon probe, thereby looking at smaller and smaller spatial distances, it becomes less likely to find the "charge" (red in Fig. 2.2). This latter effect is stronger than the former and the effective charge thus appears weaker and weaker as the Q^2 of the probe increases.

One should exercise care when using eq. (2.12) for if one has calculated only to order $\bar{g}^2(Q^2)$, then one should write

$$\alpha_s(Q^2) = \frac{4\pi}{(11 - \frac{2}{3} n_f)(\log Q^2/\Lambda^2 + c)}, \tag{2.14}$$

where c is a constant that may, in general, differ from process to process, but which is, in principle, determinable by calculating to order $\bar{g}^4(Q^2)$ since

$$\frac{1}{\log Q^2/\Lambda^2 + c} \underset{\text{large } Q^2}{\approx} \frac{1}{\log Q^2/\Lambda^2} - \frac{c}{(\log Q^2/\Lambda^2)^2}, \tag{2.15}$$

where the coefficient of c in the second term is order $\bar{g}^4(Q^2)$. This means that to leading order there is no reason for Λ to be the same from process to process. The parameter Λ takes on a universal meaning only when calculations are made to order α_s^2.

In general, the interaction between quarks and gluons can be described by the effective coupling constant $\alpha_s(Q^2) = \bar{g}^2(Q^2)/4\pi$ which satisfies the following equation

$$\frac{d\bar{g}^2}{d\tau} = \bar{g}\beta(\bar{g}); \; \bar{g}(\tau=0) = g \tag{2.16a}$$

with

$$\tau = \log(Q^2/\mu^2) \tag{2.16b}$$

where g is the renormalized coupling and μ is the subtraction point at which the theory is renormalized. The function $\beta(g)$ determines how the effective coupling depends on Q^2 (i.e., τ) and can be calculated in perturbation theory yielding the series

$$\beta(g) = -\beta_o \frac{g^3}{16\pi^2} - \beta_1 \frac{g^5}{(16\pi^2)^2} + \ldots . \tag{2.17}$$

To lowest order (2.16) and (2.17), give $\frac{d\alpha_s(Q^2)}{d\tau} = -\frac{\beta_o}{4\pi}\alpha_s^2$, which is satisfied by eq. (2.12).

III. ELECTRON-POSITRON ANNIHILATIONS

Order α_s Corrections to the Total Rate for $e^+e^- \to$ hadrons

The naive quark parton model result from the amplitude A_o in Fig. 3.1 is

$$R \equiv \sigma(e^+e^-\to\text{hadrons})/\sigma(e^+e^-\to\mu^+\mu^-) = 3 \sum_{q_i=u,d,\ldots}^{n_f} e_{q_i}^2 , \tag{3.1}$$

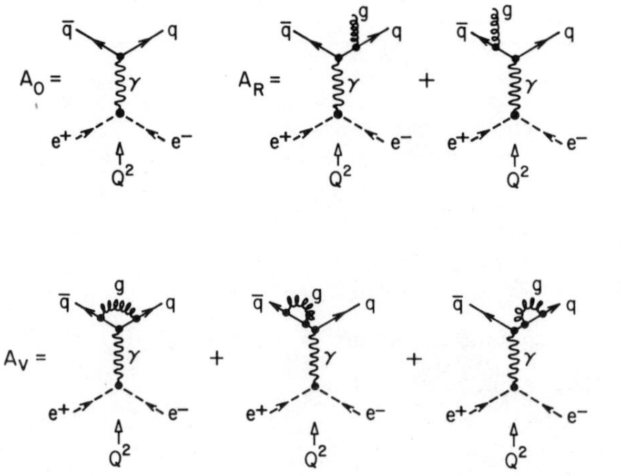

Fig. 3.1. The leading order "Born" amplitude, A_o, for the production of a quark and antiquark pair through the annihilation of an electron-positron pair together with the diagrams for the real, A_R, and virtual, A_V, gluon corrections.

where e_{q_i} are the charges of the quarks q_i and where

$$\sigma(e^+e^-\to\mu^+\mu^-) = \sigma_o = 4\pi\alpha^2/3Q^2 \tag{3.2}$$

with Q^2 the c.m. energy squared of the incoming e^+e^- pair.

Let us examine the order α_s correction to R in (3.1) due to the real, A_R, and virtual, A_V, gluon emissions in Fig. 3.1. The amplitude squared for the emission of a real gluon of 4-momentum p_3 in the process $\gamma^* \to q + \bar{q} + g$ is

$$|A_R|^2 = 8g_s^2\{t/s+s/t+2Q^4/st-2Q^2/s-2Q^2/t\}, \tag{3.3}$$

where s, t and u are invariants given by

$$s = (p_1+p_3)^2$$

$$t = (p_2+p_3)^2$$

$$u = (p_1+p_2)^2 \tag{3.4}$$

and

$$s + t + u = Q^2, \tag{3.5}$$

where p_1 and p_2 are the 4-momentum of the outgoing antiquark and quark, respectively, and $q^2 = Q^2$ is the momentum squared of the virtual photon. In calculating $|A_R|^2$ in (3.3), I have set the gluon and quark masses to zero (i.e., $p_i^2 = 0$; $i = 1,2,3$). Energy and momentum conservation require

$$\vec{q} = \vec{p}_1 + \vec{p}_2 + \vec{p}_3$$

$$Q = E_1 + E_2 + E_3. \tag{3.6}$$

It is convenient to define dimensionless variables

$$x_i = 2E_i/Q \qquad i = 1,2,3 \tag{3.7}$$

in which case (3.6) implies

$$x_1 + x_2 + x_3 = 2. \tag{3.8}$$

In the general case (with arbitrary masses), it is easy to show that

$$p_i \cdot p_j = \frac{1}{2} [Q^2 - 2q \cdot p_k + p_k^2 - p_i^2 - p_j^2] \tag{3.9}$$

which reduces to

$$p_i \cdot p_j = \frac{Q^2}{2} (1-x_k) \tag{3.10}$$

for the massless case. In this case, the invariants become

$$s = Q^2(1-x_2)$$

$$t = Q^2(1-x_1) \tag{3.11}$$

and (3.3) becomes

$$|A_R|^2 = 8g_s^2 \frac{x_1^2 + x_2^2}{(1-x_1)(1-x_2)} \tag{3.12}$$

resulting in a differential cross section for $e^+e^- \to q + \bar{q} + g$ given by

$$\frac{d\sigma}{dx_1 dx_2} = \frac{2\alpha_s}{3\pi} \sigma_o \frac{x_1^2 + x_2^2}{(1-x_1)(1-x_2)}, \tag{3.13}$$

where σ_o is given by (3.2).

In order to compute the order α_s correction to the total rate, we must integrate (3.13) over x_1 and x_2. The region of integration is determined by requiring that

$$-1 \le \cos\theta_{ij} \le 1, \tag{3.14}$$

which for the massless case results in the triangular region shown in Fig. 3.2a with

$$0 \le x_1 \le 1$$

$$1 - x_1 \le x_2 \le 1. \tag{3.15}$$

We are now faced with a problem that will occur over and over as we proceed to examine perturbative calculations. The differential cross section (3.13) diverges as x_1 or x_2 goes to 1. The origin of this divergence is clear. Consider the invariant t in (3.4). In the massless limit we have

$$t = 2p_2 \cdot p_3 = 2E_2\omega(1-\cos\theta_{23}), \tag{3.16}$$

where θ_{23} is the angle between $\vec{p}_2$ and $\vec{p}_3$ and where E_2 and ω are the energy of the outgoing quark and gluon, respectively. The

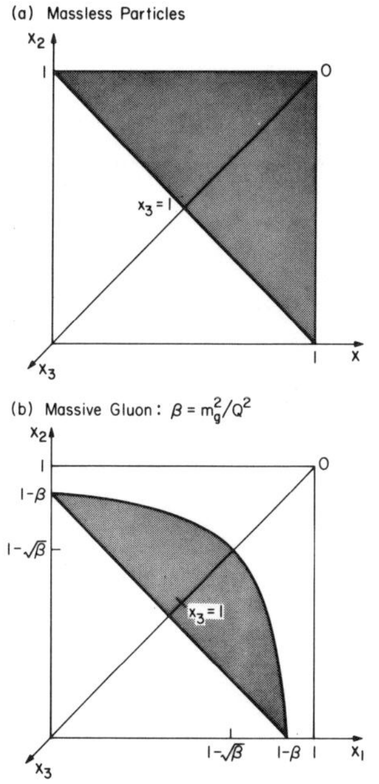

Fig. 3.2 - (a) Dalitz plot for the "decay" of a virtual photon with invariant mass Q into a massless quark, antiquark and gluon with fractional energies $x_i = 2E_i/Q$. The shaded region is the allowed kinematic range.

(b) Same as (a) except now the gluon is given a fictitious mass, m_g, and $\beta = m_g^2/Q^2$.

differential cross section diverges as $t \to 0$ which occurs when the energy of the gluon goes to zero ($\omega \to 0$, "soft divergence") or when the outgoing quark and gluon become parallel ($\cos\theta_{23} \to 1$, "parallel divergence"). The first type of divergence is usually referred to as an infrared divergence (it occurs in the limit $m_g \to 0$ and $m_q \neq 0$, where m_g and m_q are the gluon and quark masses, respectively), while the second is referred to as a mass singularity (it occurs as $m_q \to 0$ with $m_g = 0$).

In order to proceed, we must decide on some way of regularizing these infrared or mass singularities. In addition, it better be true that no experimentally observable prediction depends on the manner in which we perform the regularizing. The most eloquent and simple way to regularize is to perform the calculations in N rather than 4 dimensions (i.e., dimensional regularization). This is the method used by Altarelli, Ellis and Martinelli and is explained in Ref. 6. Here I will follow the work of Fox and Wolfram[7] and of Bashan, Brown, Ellis and Love[8] and regularize by giving the gluon a fictitious mass m_g. In this case, the differential cross section in (3.13) becomes

$$\frac{d\sigma}{dx_1 dx_2} = \sigma_o \frac{2\alpha_s}{3\pi} \frac{1}{(1-x_1)(1-x_2)} \{x_1^2+x_2^2+ \beta[2(x_1+x_2) - \frac{(1-x_1)^2+(1-x_2)^2}{(1-x_1)(1-x_2)}]+2\beta^2\}, \tag{3.17}$$

where $\beta = m_g^2/Q^2$ and the region of integration becomes

$$0 \le x_1 \le 1-\beta$$

$$1 - \beta - x_1 \le x_2 \le \left(\frac{1-x_1-\beta}{1-x_1}\right) \tag{3.18}$$

as shown in Fig. 3.2b. Integrating over x_2 results in

$$\frac{d\sigma}{dx_1} = \sigma_0 \frac{\alpha_s}{2\pi} \left(\frac{4}{3}\right)\left\{\frac{1+x_1^2}{1-x_1} \log\left(\frac{x_1(1-x_1)}{\beta}\right) - \frac{3}{2}\frac{1}{1-x_1}\right.$$

$$\left. + \frac{1}{2} x_1 + \frac{1}{2} + \frac{\beta(2-x_1)}{(1-x_1)^2} + \frac{1}{2}\frac{\beta^2}{(1-x_1)^3}\right\}, \tag{3.19}$$

where I drop terms that vanish in the limit $\beta \to 0$. The first term comes from

$$\int_{x_2^{min}}^{x_2^{max}} \frac{1}{1-x_2}\, dx_2 = \log\left(\frac{(x_1+\beta)(1-x_1)}{\beta}\right) \to \log\left(\frac{x_1(1-x_1)}{\beta}\right). \tag{3.20}$$

Care must be taken in not dropping the β and β^2 terms in (3.19) for these terms give finite contributions when integrating over x_1. For example,

$$\int_0^{1-\beta} \frac{\beta}{(1-x_1)^2}\, dx_1 - 1, \int_0^{1-\beta} \frac{\beta^2}{(1-x_1)^3}\, dx_1 = \frac{1}{2}. \tag{3.21}$$

We can thus replace the β and β^2 terms in (3.19) by a δ-function contribution as follows

$$\frac{d\sigma}{dx_1} = \sigma_0 \frac{\alpha_s}{2\pi} \left(\frac{4}{3}\right)\left\{\frac{1+x_1^2}{1-x_1} \log\left(\frac{x_1(1-x_1)}{\beta}\right) - \frac{3}{2}\frac{1}{1-x_1}\right.$$

$$\left. + \frac{1}{2} x_1 + \frac{1}{2} + \frac{5}{4}\,\delta(1-x_1)\right\}. \tag{3.22}$$

Integrating over x_1 and using

$$\int_0^{1-\beta} \frac{1}{1-x_1}\, dx_1 = -\log\beta$$

$$\int_0^{1-\beta} \frac{1}{1-x_1} \log(1-x_1) dx_1 = -\frac{1}{2}\log^2\beta$$

$$\int_0^{1-\beta} \frac{1+x_1^2}{1-x_1} \log x_1 = -\frac{\pi^2}{3} + \frac{5}{4} \tag{3.23}$$

yields

$$\sigma_{real} = \sigma_o \frac{\alpha_s}{2\pi} (\tfrac{4}{3})\{\log^2\beta+3\log\beta - \frac{\pi^2}{3} + 5\}. \tag{3.24}$$

As we expected, σ_{real} contains a $\log^2\beta$ and a $\log\beta$ divergence as $\beta \to 0$. However, we are not finished. We must include the order α_s contribution that arises from the virtual gluon contributions, A_V, in Fig. 3.1 interfering with the leading amplitude A_o (i.e., $\mathrm{Re}(A_o A_V^*)$). These contributions also diverge in the massless case and one must choose some regularization scheme. Allowing the gluon to have a mass yields

$$\sigma_{virtual} = \sigma_o \frac{\alpha_s}{2\pi} (\tfrac{4}{3})\{-\log^2\beta-3\log\beta + \frac{\pi^2}{3} - \frac{7}{2}\}, \tag{3.25}$$

which also contains a $\log^2\beta$ and a $\log\beta$ divergence. Adding the real and virtual corrections gives

$$\sigma_{tot} = \sigma_{real} + \sigma_{virtual} = \sigma_o\alpha_s/\pi, \tag{3.26}$$

which is finite in the limit $\beta \to 0$ and is independent of the particular scheme one uses to regularize σ_{real} and $\sigma_{virtual}$ (as long as one uses the same scheme for each). In higher orders there are ultraviolet divergences that are handled by renormalizing α_s leaving

$$R^{e^+e^-} = (1+\alpha_s(Q^2)/\pi)3\sum_{i=1}^{n_f} e_{q_i}^2 . \tag{3.27}$$

As Q^2 becomes large, $\alpha_s(Q^2) \to 0$ and R in (3.27) approach the naive parton model result in (3.1).

While calculating $R^{e^+e^-}$, we arrived at the differential cross section $d\sigma/dx_1$ in (3.22) that gave us the probability of finding the quark (or antiquark) carrying fractional energy x_1. Instead we could have started with $d\sigma/dx_3dx_1$ and integrated over x_1 leaving

$$\frac{d\sigma}{dx_3} = \sigma_o \frac{\alpha_s}{2\pi} \left(\frac{4}{3}\right)\{2\left(\frac{1+(1-x_3)^2}{x_3}\right)\log\left(\frac{x_3^2}{\beta}\right) - 4x_3$$

$$+ \left(4 \log^2 2 - \frac{\pi^2}{3}\right)\delta(x_3)\}, \tag{3.28}$$

where, in this case, the integration region is

$$\frac{1}{2}(2-x_3) - \frac{1}{2}\sqrt{x_3^2-4\beta} \le x_1 \le \frac{1}{2}(2-x_3) + \frac{1}{2}\sqrt{x_3^2-4\beta}$$

$$2\sqrt{\beta} \le x_3 \le 1 + \beta. \tag{3.29}$$

Equation (3.28) gives the probability (to order α_s) of finding a gluon carrying fractional energy x_3. As before, if we now integrate over x_3 and use (3.25), we get the result in (3.26).

Quark and Gluon Fragmentation Functions

In the naive parton model, the inclusive single hadron cross section defined by

$$\sigma^h(z,Q^2) = \frac{d\sigma}{dz}(e^+e^- \to h+x^-,z,Q^2), \tag{3.30}$$

which is given in terms of the quark fragmentation functions, $D^h_{o,q}(z)$, by

$$\sigma^h(z,Q^2) = \sigma^h_o(z) = 3\sigma_o \sum_{i=1}^{n_f} e^2_{q_i} (D^h_{o,q_i}(z)+D^h_{o,\bar{q}_i}(z)), \tag{3.31}$$

where $D^h_{o,q_i}(z)$ is the number of hadrons of type h with z between z and z + dz in a quark jet of type, q_i, and where z is the fractional momentum (or energy) carried by hadron h (i.e., $z = 2E_h/Q$). Energy conservation requires that

$$\sum_{\text{All } h} \int_o^1 zD^h_{o,q_i}(z)dz = 1 \tag{3.32}$$

and thus $\sigma^h(z,Q^2)$ is normalized so that

$$\sum_{\text{All } h} \int_o^1 \frac{z}{2}\,\sigma^h(z,Q^2)dz = \sigma_{tot}. \tag{3.33}$$

By the use of (3.22), (3.25) and (3.28), we can now compute the order α_s corrections to $\sigma^h(z,Q^2)$ in (3.31). Namely,

$$\sigma^h(z,Q^2) = 3\sigma_o \int_z^1 \frac{dy}{y} \{ \sum_{i=1}^{n_f} e^2_{q_i} [D^h_{o,q_i}(\tfrac{z}{y})+D^h_{o,\bar{q}_i}(\tfrac{z}{y})]$$

$$[\delta(1-y)+ \frac{\alpha_s}{2\pi} \log(Q^2/m_g^2)P_{q\leftarrow q}(y)+\alpha_s\tilde{f}_q(y)] \tag{3.34}$$

$$+ 2\sum_{i=1}^{n_f} e^2_{q_i} D^h_{o,g}(\tfrac{z}{y})[\frac{\alpha_s}{2\pi} \log(Q^2/m_g^2)P_{g\leftarrow q}(y)+\alpha_s f_g(y)]\}$$

with

$$P_{q\leftarrow q}(y) = \frac{4}{3}[\frac{1+y^2}{(1-y)_+} + \frac{3}{2}\,\delta(y-1)] \tag{3.35a}$$

$$P_{g\leftarrow q}(y) = \frac{4}{3}[\frac{1+(1-y)^2}{y}] \tag{3.35b}$$

$$\alpha_s\tilde{f}_q(y) = \frac{\alpha_s}{2\pi}\,(\frac{4}{3})[(1+y^2)(\frac{\log(1-y)}{1-y})_+ + \frac{1+y^2}{1-y}\log y - \frac{3}{2}\,\frac{1}{(1-y)_+}$$

$$+ \frac{1}{2}\,(y+1)+(\frac{\pi^2}{3} - \frac{9}{4})\delta(1-y)] \tag{3.35c}$$

$$\alpha_s f_g(y) = \frac{\alpha_s}{2\pi}\left(\frac{4}{3}\right)\left[2\left(\frac{1+(1-y)^2}{y}\right)\log y - 2y\right]. \tag{3.35d}$$

The two terms in (3.34) arise because, in general, the observed hadron h could have come from the quark jet, $D^h_{o,q}(z)$ or from the outgoing gluon jet, $D^h_{o,g}(z)$. The $\log^2\beta$ term in (3.25) has been absorbed into the definitions of the "+ functions"

$$(f(y))_+ = \lim_{\beta\to 0}\{f(y)\theta(1-y-\beta) - \delta(1-y-\beta)\int_0^{1-\beta} f(y)dy\}, \tag{3.36a}$$

which gives

$$\frac{1}{(1-y)_+} \equiv \frac{1}{1-y}\,\theta(1-y-\beta) + \log\beta\,\delta(1-y-\beta) \tag{3.36b}$$

$$\left(\frac{\log(1-y)}{1-y}\right)_+ \equiv \frac{1}{1-y}\log(1-y)\theta(1-y-\beta) + \frac{1}{2}\log^2\beta\,\delta(1-y-\beta). \tag{3.36c}$$

These functions are really only well defined when they are convoluted with a well behaved function. In particular

$$\int_z^1 \frac{D(\frac{z}{y})}{(1-y)_+}\frac{dy}{y} = D(z)\log(1-z) + \int_z^1 \frac{dy}{y}\,\frac{[D(\frac{z}{y}) - yD(z)]}{1-y} \tag{3.37a}$$

and

$$\int_z^1 D(z/y)\left(\frac{\log(1-y)}{1-y}\right)_+ dy = \frac{1}{2}D(z)\log^2(1-z)$$

$$+ \int_z^1 \frac{dy}{y}\,\frac{[D(\frac{z}{y}) - yD(z)]}{1-y}\log(1-y). \tag{3.37b}$$

If we now define Q^2 dependent fragmentation functions by

$$\sigma^h(z,Q^2) = 3\sigma_0 \sum_{i=1}^{n_f} e^2_{q_i}\left(1 + \frac{\alpha_s}{\pi}\right)\left(D^h_{q_i}(z,Q^2) + D^h_{\bar{q}_i}(z,Q^2)\right) \tag{3.38}$$

then to order α_s

$$D_q^h(z,Q^2) = \int_z^1 \frac{dy}{y}\{D_{o,q}^h(z/y)[\delta(1-y) + \frac{\alpha_s}{2\pi}\log(Q^2/m_g^2)P_{q\leftarrow q}(y)+\alpha_s f_q(y)]$$

$$+ D_{o,g}^h(z/y)[\frac{\alpha_s}{2\pi}\log(Q^2/m_g^2)P_{g\leftarrow q}(y)+\alpha_s f_g(y)]\}, \qquad (3.39)$$

where $f_g(y)$ is given by (3.35d) and

$$\alpha_s f_q(y) = \frac{\alpha_s}{2\pi}(\frac{4}{3})[(1+y^2)(\frac{\log(1-y)}{1-y})_+ + \frac{1+y^2}{1-y}\log y - \frac{3}{2}\frac{1}{(1-y)_+}$$

$$+ \frac{1}{2}(y+1) + (\frac{\pi^2}{3} - \frac{15}{4})\delta(y-1)]. \qquad (3.40)$$

We cannot hope to calculate $D_q^h(z,Q^2)$ at a given Q^2. The "bare" fragmentation functions $D_{o,q}^h(z)$ and $D_{o,g}^h(z)$ are unknowns. They represent the manner in which the outgoing quarks and gluons fragment into hadrons which takes place over a long time scale. It involves low momentum transfers and thus large effective couplings and, therefore, cannot be calculated by perturbation theory. Because of this, the functions $f_q(y)$ and $f_g(y)$ in (3.39) are not directly experimentally observable. This is fortunate since these "constant terms" depend, in general, on the particular renormalization scheme adopted. However, the coefficients of the $\log Q^2$ terms (i.e., $P_{q\leftarrow q}(y)$ and $P_{g\leftarrow q}(y)$ are not scheme dependent. They are unique. In addition,

$$\int_o^1 y[P_{q\leftarrow q}(y)+P_{g\leftarrow q}(y)]dy = 0 \qquad (3.41a)$$

and (in any regularization scheme)

$$\int_o^1 y[f_q(y)+f_g(y)]dy = 0, \qquad (3.41b)$$

which insures that if energy is conserved by $D_{o,q}^h$ and $D_{o,g}^h$ (i.e., eq. (3.32) is satisfied) then it is also conserved for $D_q^h(z,Q^2)$ in (3.39). Equations (3.41a) and (3.41b) also guarantee that (3.33) is satisfied.

At this point, for the benefit of my typist, let me define the convolution

$$C(z) = A \otimes B \equiv \int_z^1 \frac{dy}{y} A(z/y)B(y), \tag{3.42}$$

whereupon (3.39) becomes

$$D_q^h(z,Q^2) = D_{o,q}^h \otimes (1 + \frac{\alpha_s}{2\pi} \log(Q^2/m_g^2)P_{q\leftarrow q}+\alpha_s f_q)$$

$$+ D_{o,g}^h \otimes (\frac{\alpha_s}{2\pi} \log(Q^2/m_g^2)P_{g\leftarrow q}+\alpha_s f_g) \tag{3.43}$$

and

$$D_{NS}^h(z,Q^2) = D_{o,NS}^h \otimes (1 + \frac{\alpha_s}{2\pi} \log(Q^2/m_g^2)P_{q\leftarrow q}+\alpha_s f_q), \tag{3.44}$$

where the non-singlet fragmentation function is defined by

$$D_{NS}^h(z,Q^2) \equiv D_{q_i}^h(z,Q^2)-D_{\bar{q}_i}^h(z,Q^2). \tag{3.45}$$

As can be seen in (3.44), the experimental observable $D_{NS}^h(z,Q^2)$ appears to become infinite as $m_g \to 0$ like $\log(m_g^2)$. I will explain this in more detail later, but since we believe that all observable quantities should be well behaved as the mass of the gluon (and quarks) goes to zero, we must have made an "artificial" mass singularity in the way we have done the perturbative calculation. That is, we have divided $D_{NS}^h(z,Q^2)$ into two terms, $D_{o,NS}^h(z)$ and $\log(Q^2/m_g^2)P_{q\leftarrow q}(y)$, both of which diverge as $m_g \to 0$ but whose product is finite. The quantity $D_{o,NS}^h(z)$ must, therefore, have a perturbation series of the form

$$D_{o,NS}^h(z) = \tilde{D}_{o,NS}^h \otimes (1 + \frac{\alpha_s}{2\pi} \log(m_g^2/\Lambda^2)P_{q\leftarrow q}+\ldots), \tag{3.46}$$

where Λ is a mass scale that is related to the size of the hadrons in the outgoing jets and where $\tilde{D}_{o,NS}^h$ is now well behaved in the limit $m_g \to 0$. The mass singularities are now "factorized" off in

(3.44) in the following manner (this is a trivial property of logarithms at this order)

$$D^h_{NS}(z,Q^2) = D^h_{o,NS} \otimes (1 + \frac{\alpha_s}{2\pi}[\log(Q^2/\Lambda^2)+\log(\Lambda^2/m_g^2)]P_{q\leftarrow q})$$

$$= D^h_{o,NS} \otimes (1 + \frac{\alpha_s}{2\pi}\log(\Lambda^2/m_g^2)P_{q\leftarrow q})$$

$$\otimes (1 + \frac{\alpha_s}{2\pi}\log(Q^2/\Lambda^2)P_{q\leftarrow q}) + \text{order}(\alpha_s^2)$$

$$= \tilde{D}_{o,NS} \otimes (1 + \frac{\alpha_s}{2\pi}\log(Q^2/\Lambda^2)P_{q\leftarrow q}) + \text{order}(\alpha_s^2). \quad (3.47)$$

The mass singularity $\log(m_g^2)$ has been absorbed into the unknown function $\tilde{D}^h_{o,NS}(z)$. At this order the factorization of mass singularities is trivial. However, it has been shown that one can always (to all orders in perturbation theory) factorize out and absorb the mass singularities into "universal" fragmentation functions $\tilde{D}^h_{o,q}$ and $\tilde{D}^h_{o,g}$ [9-12].

Although $\tilde{D}^h_{o,q}$ and $\tilde{D}^h_{o,g}$ are unknown, we might hope to calculate the change of $D^h_q(z,Q^2)$ in (3.43) as we change Q^2 using perturbation theory. However, since $\alpha_s(Q^2)\log Q^2$ is of order 1 (i.e., $\alpha_s(Q^2) \propto 1/\log Q^2$), we cannot do this unless we can sum all terms of the form $[\alpha_s \log Q^2]^N$. To leading order, we see from (3.44) that

$$dD^h_{NS}(z,Q^2)/d\tau = \frac{\alpha_s}{2\pi}\tilde{D}^h_{o,NS} \otimes P_{q\leftarrow q} \quad (3.48a)$$

and similarly

$$dD^h_q(z,Q^2)/d\tau = \frac{\alpha_s}{2\pi}[\tilde{D}^h_{o,q} \otimes P_{q\leftarrow q} + \tilde{D}^h_{o,g} \otimes P_{g\leftarrow q}], \quad (3.48b)$$

where $\tau = \log(Q^2/\Lambda^2)$. As will become evident as we proceed[14,15], to sum over all terms of the form $[\alpha_s \log Q^2]^N$, we replace the "bare" fragmentation functions $\tilde{D}^h_{o,q}$ and $\tilde{D}^h_{o,g}$ in (3.48a) and (3.48b) by the observed functions at Q^2. As illustrated in Fig. 3.3, to leading order,

$$dD^h_{NS}(z,Q^2)/d\tau = \frac{\alpha_s(Q^2)}{2\pi} D^h_{NS}(Q^2) \otimes P_{q\leftarrow q} \tag{3.49a}$$

and

$$dD^h_{q_i}(z,Q^2)/d\tau = \frac{\alpha_s(Q^2)}{2\pi} [D^h_{q_i}(Q^2) \otimes P_{q\leftarrow q} + D^h_g(Q^2) \otimes P_{g\leftarrow q}], \tag{3.49b}$$

where I have used the definition of a convolution given by (3.42).

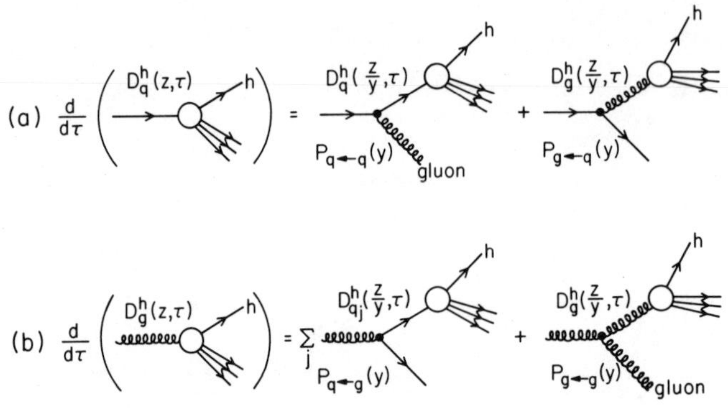

Fig. 3.3 - (a) Illustrates that the change of the quark fragmentation function, $D^h_q(z,\tau)$, w.r.t. $\tau = \log(Q^2/Q_o^2)$ is due to the Bremsstrahlung radiation of a gluon, $P_{q\leftarrow q}(y)$, and to the production of a gluon, $P_{g\leftarrow q}(y)$, which fragments into the observed hadrons.
(b) Illustrates that the change of the gluon fragmentation function, $D^h_g(z,\tau)$, w.r.t. τ is due to the production of a quark-antiquark pair, $P_{q\leftarrow g}(y)$, and to the Bremsstrahlung radiation of a gluon, $P_{g\leftarrow g}(y)$. This figure can be compared to Fig. 4.5 which illustrates the causes for the Q^2 change of the quark and gluon distributions. To leading order, the functions $P_{a\leftarrow b}(z)$ are the same.

Similarly, one can show that the Q^2 evolution of the gluon fragmentation functions is given by

$$dD^h_g(z,Q^2)/d\tau = \frac{\alpha_s(Q^2)}{2\pi} [\sum_{j=1}^{2n_f} D^h_{q_j}(Q^2) \otimes P_{q_j\leftarrow g} + D^h_g(Q^2) \otimes P_{g\leftarrow g}], \tag{3.49c}$$

which is also illustrated in Fig. 3.3. The functions $P_{q\leftarrow g}(y)$ and

$P_{g\leftarrow g}(y)$ will be written down in Section IV where we will work out the general solution of eqs. (3.49b) and (3.49c). Here we notice that if we define

$$\kappa = \frac{2}{\beta_o} \log\{\alpha_s(Q_o^2)/\alpha_s(Q^2)\} \tag{3.50}$$

so that to leading order $d\kappa/d\tau = \alpha_s/2\pi$, then eq. (3.49a) becomes

$$dD_{NS}^h/d\kappa = P \otimes D_{NS}^h, \tag{3.51}$$

where $P(y) = P_{q\leftarrow q}(y)$ and has the formal solution

$$D_{NS}^h(z,Q^2) = \exp(\kappa P \otimes) D_{NS}^h(Q_o^2). \tag{3.52}$$

Equation (3.52) relates $D_{NS}^h(z,Q^2)$ at two different values of Q^2. Given the fragmentation function at some reference point, say Q_o^2, one can calculate it at any larger Q^2 provided both Q^2 and Q_o^2 are large enough to justify leading order perturbation theory in $\alpha_s(Q^2)$. Equation (3.52) can be easily solved in terms of moments. In particular

$$M_{NS}^h(n,Q^2) = \exp(\kappa A_n^{NS}) M_{NS}^h(n,Q_o^2)$$

$$= [\log(Q^2/\Lambda^2)/\log(Q_o^2/\Lambda^2)]^{2A_n^{NS}/\beta_o} M_{NS}^h(n,Q_o^2) \tag{3.53}$$

where

$$M_{NS}^h(n,Q^2) = \int_o^1 z^{n-1} D_{NS}^h(z,Q^2)dz \tag{3.54a}$$

and the "anomalous dimensions", A_n^{NS}, and are given by

$$A_n^{NS} = \int_o^1 y^{n-1} P_{q\leftarrow q}(y)dy. \tag{3.54b}$$

The quantity β_o is given by (2.13).

The leading order Q^2 evolution formula (3.51) does not depend on the "constant" terms $f_q(y)$ and $f_g(y)$. However, these terms do play a role in next order. Rewriting (3.39) for the non-singlet fragmentation function gives

$$D^h_{NS}(z,Q^2) = \tilde{D}^h_{o,NS} + \frac{\alpha_s}{2\pi}\log(Q^2/\Lambda^2)\tilde{D}^h_{o,NS}\otimes P+\alpha_s\tilde{D}^h_{o,NS}\otimes f_q . \quad (3.55)$$

The change of $D^h_{NS}(z,Q^2)$ w.r.t. α_s is thus

$$dD^h_{NS}(z,Q^2)/d\alpha_s = \frac{1}{2\pi}\log(Q^2/\Lambda^2)D^h_{NS}(Q^2)\otimes P+D^h_{NS}(Q^2)\otimes f_q \quad (3.56a)$$

and in higher orders of perturbation theory, one has

$$dD^h_{NS}(z,Q^2)/d\alpha_s = [\frac{1}{2\pi}\log(Q^2/\Lambda^2)(P+\alpha_s R+\ldots)+(f+\alpha_s g+\ldots)]\otimes D^h_{NS}(Q^2), \quad (3.56b)$$

where $f(y) = f_q(y)$. The terms in (3.56b) containing a $\log Q^2$ are process independent but may depend on the regularization scheme. They arise from the parallel mass singularities. The "constant terms" $f(y)$, $g(y)$, .., etc. are, in general, process (and regularization) scheme dependent. To compute the rate of change of D^h_{NS} w.r.t. $\log Q^2$, we use perturbation theory (in α_s) to deduce how D^h_{NS} changes with α_s. Then we must include the added complication that α_s itself depends on $\log Q^2$ according to

$$\frac{d\alpha_s}{d\tau} = -\frac{\beta_o}{4\pi}\alpha_s^2(1+\frac{\beta_1}{4\pi\beta_o}\alpha_s+\ldots), \quad (3.57)$$

where I have used eq. (2.16a) and (2.17). Combining $dD/d\alpha_s$ and $d\alpha_s/d\tau$, we arrive at

$$dD^h_{NS}(z,Q^2)/d\tau = \frac{\alpha_s}{2\pi}P\otimes D^h_{NS}(Q^2) + \frac{\beta_o}{4\pi}\alpha_s^2[(\tau\alpha_s)R+f+\frac{\beta_1}{4\pi\beta_o}(\tau\alpha_s)P]\otimes D^h_{NS}(Q^2), \quad (3.58)$$

where I have kept terms of order α_s and α_s^2 (remember $\tau\alpha_s$ is of order 1). Since the Q^2 dependence of the fragmentation function is an observable, it must be the case that the right hand side of (3.58) is regularization scheme independent. This results because the higher order corrections to the anomalous dimensions (i.e., R) are scheme dependent as is f(y), but the combination of R and f that appears in (3.58) is not[6,13]. I will discuss this more in Section VI. In addition, since the constant terms (f and g) are process dependent, they are important (as we shall see) when comparing one process to another.

The fragmentation functions $D_i^h(z,Q^2)$ can be calculated in terms of those at Q_0^2 by the use of (3.49). However, care must be taken in choosing $D_i^h(z,Q_0^2)$ since one presumably must use the distribution of "primary" mesons before decay. In addition, one must guess at the distribution of hadrons in a gluon jet, $D_g^h(z,Q_0^2)$. Figure 3.4 shows the resulting Q^2 dependence of $D_u^{\pi^o}(z,Q^2)$ and $D_g^{\pi^o}(z,Q^2)$ for the particular reference momentum choices discussed in Refs. 16 and 17. The distribution of charged hadrons at Q_0^2 was adjusted to fit the data shown in Fig. 3.5. The gluon fragmentation function at Q_0^2 has been chosen to be steeper than the quark fragmentation function. That is, it is assumed that gluons fragment into fewer high z hadrons and are assumed to have a higher rapidity plateau than do quarks[18,19].

IV. DEEP INELASTIC SCATTERING

The Naive Parton Model

In the naive parton model, one defines parton distributions, $G_{h\to q}(x)$, as the number of quarks q with fraction of momentum between x and x + dx within a hadron of type h of high momentum. In particular, there are six functions necessary to describe the quark distributions in a proton:

$$u(x) \equiv G_{p\to u}(x), \qquad \bar{u}(x) \equiv G_{p\to\bar{u}}(x),$$

$$d(x) \equiv G_{p\to d}(x), \qquad \bar{d}(x) \equiv G_{p\to\bar{d}}(x),$$

$$s(x) \equiv G_{p\to s}(x), \qquad \bar{s}(x) \equiv G_{p\to\bar{s}}(x), \qquad (4.1a)$$

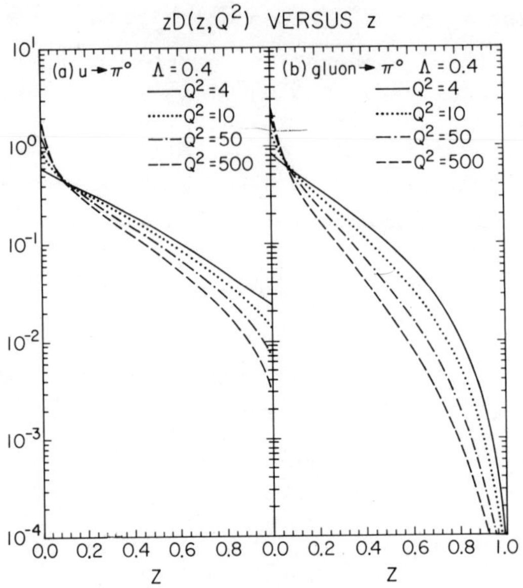

Fig. 3.4 – (a) The Q^2 dependence of the fragmentation function for a u-quark to a π^o, $D_u^{\pi^o}(z,Q^2)$, expected from QCD. The distributions at high Q^2 are calculated from the distribution at the reference momentum Q_o^2 = 4 GeV^2 using Λ = 0.4 GeV/c, where $D_q^h(z,Q_o^2)$ is taken from the analysis in Ref. 17.
(b) Same as (a) but for the gluon fragmentation function $D_g^{\pi^o}(z,Q^2)$.

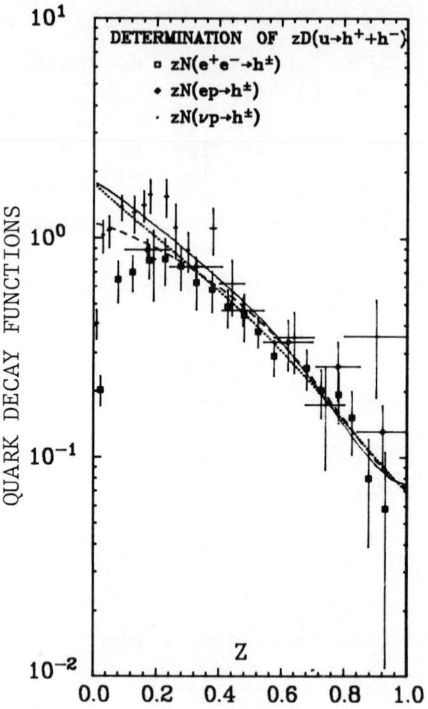

Fig. 3.5. Comparison of the charged particle distributions $zD_u^{h^+}(z,Q_o^2) + zD_u^{h^-}(z,Q_o^2)$ at Q_o^2 = 4 GeV^2 with data from $e^+e^- \to h^\pm + X$, $ep \to h^\pm + X$ and $\nu p \to h^\pm + X$. (The data used are described in Fig. 3 of Ref. 17).

where u, d and s refer to up, down and strange quarks, respectively, and $\bar{u}$, $\bar{d}$ and $\bar{s}$ to their antiquarks. The distribution of gluons within a proton is defined by

$$g(x) \equiv G_{p\to g}(x), \qquad (4.1b)$$

where g stands for gluon. These distributions satisfy the following sum rules:

$$\int_0^1 [u(x) - \bar{u}(x)]dx = 2 \tag{4.2a}$$

$$\int_0^1 [d(x) - \bar{d}(x)]dx = 1 \tag{4.2b}$$

$$\int_0^1 [s(x) - \bar{s}(x)]dx = 0. \tag{4.2c}$$

That is the net number of each kind of quark is just the number one arrives at in the simple non-relativistic quark model. In addition, momentum conservation implies

$$\int_0^1 \left\{ \sum_{i=1}^{n_f} x(G_{p\to q_i}(x) + G_{p\to\bar{q}_i}(x)) + xg(x)\right\} dx = 1, \tag{4.3}$$

where n_f is the number of quark flavors (i.e., u, d, s, ..., etc.). The distributions in a neutron are gotten from isospin symmetry, which implies that $G_{n\to u}(x) = G_{p\to d}(x) = d(x)$, $G_{n\to d}(x) = u(x)$, $G_{n\to s}(x) = s(x)$, etc.

In the naive parton model, complete knowledge of the deep inelastic structure functions for electron, neutrino and antineutrino scattering off protons and neutrons is sufficient to obtain u(x), d(x), $\bar{u}(x)$, $\bar{d}(x)$, g(x) and $s(x) + \bar{s}(x)$. For example, with the standard notation,

$$\nu W_2^{ep}(x) \equiv F_2^{ep}(x) = \frac{4}{9} x[u(x)+\bar{u}(x)] + \frac{1}{9} x[d(x)+\bar{d}(x)] + \frac{1}{9} x[s(x)+\bar{s}(x)], \tag{4.4a}$$

and

$$\nu W_2^{en}(x) \equiv F_2^{en}(x) = \frac{4}{9} x[d(x)+\bar{d}(x)] + \frac{1}{9} x[u(x)+\bar{u}(x)] + \frac{1}{9} x[s(x)+\bar{s}(x)], \tag{4.4b}$$

which are only functions of x and, in the naive parton model, do not depend separately on the energy loss of the leptons $\nu = E - E'$ or on the four-momentum transfer $q^2 = -Q^2$ [20]. This is because the basic interaction is assumed to be a photon of momentum q^2 interacting with a parton of momentum p ($p = \xi P$) with limited transverse momentum producing a parton of momentum $p' = p + q$. The condition that $(p')^2 = m^2$ implies

$$2\xi(P\cdot q) + q^2 + \xi^2 M^2 = m^2, \tag{4.5a}$$

which as $-q^2 \to \infty$ and $P\cdot q = M\nu \to \infty$ yields

$$\xi = \frac{-q^2}{2P\cdot q} = \frac{Q^2}{2M\nu} = x. \tag{4.5b}$$

QCD Corrections to Deep Inelastic Scattering

Gluon Contributions to Deep Inelastic Scattering. In QCD one must correct the naive parton model by including the possibility that a gluon in the initial proton can produce a quark-antiquark pair which the virtual photon then couples to. Gluons contribute to

$$\tilde{F}_1(x,Q^2) = 2F_1(x,Q^2) \tag{4.6a}$$

$$\tilde{F}_2(x,Q^2) = F_2(x,Q^2)/x \tag{4.6b}$$

(to order α_s) due to the subprocess $\gamma^* + g \to q + \bar{q}$ shown in Fig. 4.1.

The differential cross section for this process is given by

$$\frac{d\hat{\sigma}^{\Sigma}_{DIS}}{d\hat{t}} = \frac{\pi\alpha\alpha_s e_q^2}{(\hat{s}+Q^2)^2}\left(\frac{1}{8}\right)|A^{\Sigma}_{DIS}(\hat{s},\hat{t})|^2$$

$$= \frac{\pi\alpha\alpha_s e_q^2}{(\hat{s}+Q^2)^2}\left\{\frac{\hat{u}}{\hat{t}} + \frac{\hat{t}}{\hat{u}} + \frac{2Q^2}{\hat{t}\hat{u}}(\hat{t}+\hat{u}+Q^2)\right\}, \tag{4.7}$$

where the invariants $\hat{s}$, $\hat{t}$ and $\hat{u}$ are given by

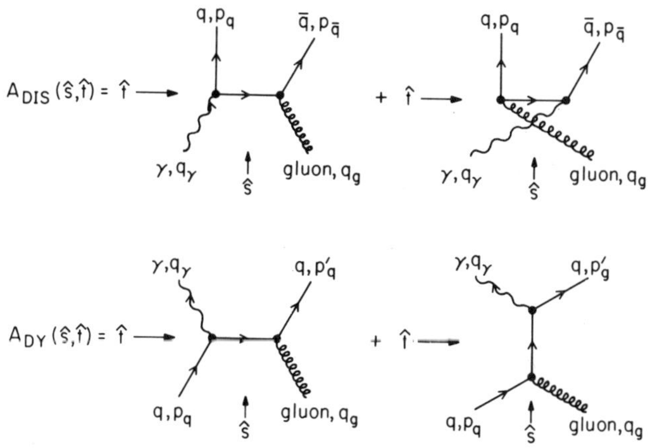

$$\hat{s} = (q_\gamma + q_g)^2$$

$$\hat{t} = (p_q - q_\gamma)^2$$

$$\hat{u} = (p_{\bar{q}} - q_\gamma)^2 \qquad (4.8)$$

and e_q^2 is the elastic charge of the quark q. In addition, the virtual photon momentum is given by $q_\gamma^2 = -Q^2$ and I have taken the gluon and quarks to have zero mass. The superscript Σ is to signify that in calculating (4.7), I have used $\Sigma\varepsilon_\mu\varepsilon_\nu^* = \delta_{\mu\nu}$ where ε_μ is the photon polarization. This means that σ_{DIS}^{Σ} is related to a particular combination of $\tilde{F}_1$ and $\tilde{F}_2$ which I will write down shortly in eqs. (4.16) and (4.20).

Fig. 4.1. Diagrams of order $\alpha_s(Q^2)$ that produce corrections to deep inelastic scattering (DIS) and to the "Drell-Yan" production of large mass muon pairs (DY). These diagrams produce corrections that are proportional to the probability of finding gluons with the initial hadron.

A problem now arises when we try to compute the total $\gamma^* + g$ cross section by integrating (4.7). The integral

$$\hat{\sigma}_{DIS}^{\Sigma}(\hat{s}) = \int_{\hat{t}_{max}}^{\hat{t}_{min}} d\hat{t}\, \frac{d\hat{\sigma}_{DIS}^{\Sigma}}{d\hat{t}}(\hat{s},\hat{t}) \qquad (4.9)$$

diverges logarithmically like $\log(\hat{t}_{min}/\hat{t}_{max})$ since for massless quarks and gluons $\hat{t}_{min} = 0$. As for the e^+e^- case in Section III, one must choose some procedure for removing (or regularizing) these infrared ($\hat{t} \to 0$, $\hat{u} \to 0$) divergences which occur. We will let the incoming gluon be slightly-off-shell and spacelike (i.e., $q_g^2 = -m_g^2$ in Fig. 4.1). (We cannot do as in e^+e^- and take $q_g^2 = m_g^2$ because then the incoming gluon then could actually decay into a quark-antiquark pair.) Taking $q_g^2 = -m_g^2$, we have

$$|A^{\Sigma}_{DIS}(\hat{s},\hat{t})|^2 = 8\{\frac{\hat{u}}{\hat{t}} + \frac{\hat{t}}{\hat{u}} + \frac{2Q^2}{\hat{t}\hat{u}}(\hat{t}+\hat{u}+Q^2) + \frac{2m_g^2}{\hat{t}\hat{u}}(\hat{t}+\hat{u}+m_g^2)$$

$$- Q^2 m_g^2 (\frac{1}{\hat{u}^2} + \frac{1}{\hat{t}^2} - \frac{4}{\hat{t}\hat{u}}) \qquad (4.10)$$

and $\hat{t}_{min}$ and $\hat{t}_{max}$ in (4.9) become (approximately)

$$\hat{t}_{min} = \hat{u}_{min} = -m_g^2 Q^2/(\hat{s}+Q^2)$$

$$\hat{t}_{max} = \hat{u}_{max} = -(\hat{s}+Q^2). \qquad (4.11)$$

The differential cross section can now be integrated to give

$$\hat{\sigma}^{\Sigma}_{DIS}(\hat{s}) = \frac{2\pi\alpha\alpha_s e_q^2}{(\hat{s}+Q^2)^2}\{-(\frac{2Q^2\hat{s}-(\hat{s}+Q^2)^2}{(\hat{s}+Q^2)})\log(\frac{(\hat{s}+Q^2)^2}{Q^2 m_g^2}) - 2(\hat{s}+Q^2)\}, \qquad (4.12)$$

where terms that vanish in the limit $m_g^2 \to 0$ have been dropped.

In particular

$$\int_{\hat{t}_{max}}^{\hat{t}_{min}} d\hat{t} = (\hat{s}+Q^2) \qquad (4.13a)$$

$$\int_{\hat{t}_{max}}^{\hat{t}_{min}} \frac{d\hat{t}}{\hat{t}} = \log(\frac{\hat{t}_{min}}{\hat{t}_{max}}) = -\log[\frac{(\hat{s}+Q^2)^2}{Q^2 m_g^2}] \qquad (4.13b)$$

$$m_g^2 \int_{\hat{t}_{max}}^{\hat{t}_{min}} \frac{d\hat{t}}{\hat{t}^2} = -\frac{(\hat{s}+Q^2)}{Q^2}. \qquad (4.13c)$$

We have thus succeeded in calculating $\hat{\sigma}^{\Sigma}_{DIS}(\hat{s})$. It contains two terms; one that diverges like $\log(m_g^2)$ as $m_g^2 \to 0$ and one that is finite in this limit (called the "constant" piece).

This subprocess must now be "embedded" in the desired observed process $\gamma^* + p \to X$. To do this, it is convenient to define

$$z \equiv \frac{Q^2}{2q_g \cdot q_\gamma} = \frac{x}{y} = Q^2/(\hat{s}+Q^2) \tag{4.14a}$$

where

$$x = \frac{Q^2}{2P \cdot q_\gamma} \tag{4.14b}$$

and y is the fraction of the proton momentum, P, carried by the gluon (i.e., $q_g = yP$) as illustrated in Fig. 4.2. In terms of z and Q^2, the total $\gamma^* + g$ cross section becomes

$$\hat{\sigma}^{\Sigma}_{DIS}(Q^2) = \frac{2\pi\alpha\alpha_s e_q^2}{Q^2} z\{(z^2+(1-z)^2)\log(\frac{Q^2}{m_g^2 z^2}) - 2\} \tag{4.15}$$

and if we now use the fact that the total γ^*p cross section can be related in general to the structure function by[20]

$$\frac{F_\Sigma(x,Q^2)}{x} = \int_x^{1.0} \frac{dy}{y} G_{p\to g}(y)(\frac{Q^2}{4\pi^2\alpha z})\hat{\sigma}_\Sigma(Q^2) \tag{4.16}$$

we arrive at a gluon contribution to F_Σ given by

$$\frac{F_\Sigma(x,Q^2)}{x} = 2e_q^2 \int_x^1 \frac{dy}{y} G^{(o)}_{p\to g}(y)\{\frac{\alpha_s}{2\pi} P_{q\leftarrow g}(z)\log \frac{Q^2}{m_g^2} + \alpha_s f_{\Sigma,g}(z)\}, \tag{4.17}$$

where I have included just one quark flavor and

$$P_{q\leftarrow g}(z) = \frac{1}{2}(z^2+(1-z)^2) \tag{4.18}$$

$$\alpha_s f_{\Sigma,g}(z) = -\frac{\alpha_s}{2\pi}\frac{1}{2}\{(z^2+(1-z)^2)(2\log z)+2\}. \tag{4.19}$$

As mentioned earlier, the structure function $F_\Sigma(x,Q^2)$ is related to $F_1(x,Q^2)$ and $F_2(x,Q^2)$. In particular

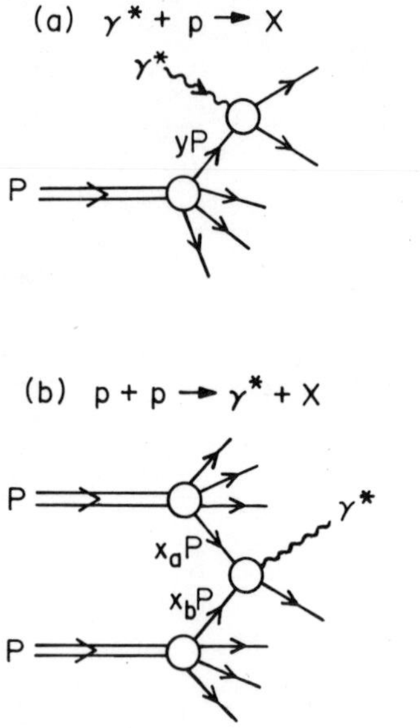

Fig. 4.2 - (a) Illustrates how the subprocess $\gamma^* + a \rightarrow b + c$ where a, b and c are constituents (quarks or gluons) is "embedded" within the experimentally measured process $\gamma^* + p \rightarrow X$, where P is a proton and y is the fraction of momentum of the proton carried by the constituent a.

(b) Illustrates how the constituent subprocess $a + b \rightarrow \gamma^* + c$ is "embedded" with the observed process $pp \rightarrow \gamma^* + X$. The virtual proton γ^* then fragments into a muon pair. The quantities x_a and x_b are the fractional momenta carried by constituents (quarks or gluons) a and b, respectively.

$$F_\Sigma(x,Q^2) = F_2(x,Q^2) - \frac{3}{2} F_L(x,Q^2), \tag{4.20a}$$

where the longitudinal structure function is given by

$$F_L(x,Q^2) = F_2(x,Q^2) - 2xF_1(x,Q^2). \tag{4.20b}$$

The perturbative calculation for F_L proceeds as before except now

$$\frac{d\hat{\sigma}_L^{DIS}}{d\hat{t}} = \frac{\pi\alpha\alpha_s e_q^2}{(\hat{s}+Q^2)^2} \hat{s} \tag{4.21}$$

and there are no divergent pieces. Integrating yields

$$\hat{\sigma}_L^{DIS} = \frac{\pi\alpha\alpha_s e_q^2}{(\hat{s}+Q^2)} \hat{s} = \pi\alpha_s\alpha(1-z). \tag{4.22}$$

By the use of

$$\frac{F_L(x,Q^2)}{x} = \int_x^{1.0} \frac{dy}{y} G_{p\to g}(y) \left(\frac{2z}{\pi^2\alpha}\right) \hat{\sigma}_L \tag{4.23}$$

we arrive at (again for one quark flavor)

$$\frac{F_L(x,Q^2)}{x} = 2e_q^2 \int_x^{1.0} \frac{dy}{y}$$

$$G_{p\to g}(y)\alpha_s f_{L,g}(z) \tag{4.24}$$

with

$$\alpha_s f_{L,g}(z) = \frac{\alpha_s}{2\pi} 2z(1-z). \tag{4.25}$$

Combining (4.17), (4.20) and (4.24) yields

$$\tilde{F}_1(x,Q^2) = 2F_1(x,Q^2) = 2e_q^2 \int_x^1 \frac{dy}{y} G_{p\to g}(y)$$

$$\left\{\frac{\alpha_s}{2\pi} P_{q\leftarrow g}(z)\log\frac{Q^2}{m_g^2} + \alpha_s f_{1,g}(z)\right\} \qquad (4.26a)$$

$$\tilde{F}_2(x,Q^2) = \frac{F_2(x,Q^2)}{x} = 2e_q^2 \int_x^1 \frac{dy}{y} G_{p\to g}(y)$$

$$\left\{\frac{\alpha_s}{2\pi} P_{q\leftarrow g}(z)\log\frac{Q^2}{m_g^2} + \alpha_s f_{2,g}(z)\right\} \qquad (4.26b)$$

with $z = \frac{x}{y}$ and $P_{q\leftarrow g}(z)$ given by eq. (4.18) and where

$$\alpha_s f_{2,g}(z) = \frac{\alpha_s}{2\pi}\left(\frac{1}{2}\right)[(z^2+(1-z)^2)(-2\log z)-2+6z-6z^2] \qquad (4.27a)$$

$$\alpha_s(f_{2,g}(z)-f_{1,g}(z)) = \frac{\alpha_s}{2\pi} 2z(1-z). \qquad (4.27b)$$

Contributions to Deep Inelastic Scattering from $\gamma^* + q \to q + g$. We must now correct the naive parton model by including the possibility that the quark can radiate a gluon before or after the interaction with the virtual photon γ^*. In this case we can regulate just as in the e^+e^- case. We regulate by giving the gluon a fictitious mass, $q_g^2 = m_g^2$. The differential cross section for the subprocess $\gamma^* + q \to q + g$ shown in Fig. 4.3 is given by

$$\frac{d\sigma^{\Sigma}_{DIS}}{d\hat{t}}(\hat{s},\hat{t}) = \frac{\pi\alpha\alpha_s e_q^2}{(\hat{s}+Q^2)^2}\left(\frac{8}{3}\right)\left\{-\frac{\hat{t}}{\hat{s}} - \frac{\hat{s}}{\hat{t}} + \frac{2Q^2(\hat{s}+\hat{t}+Q^2)}{\hat{s}\hat{t}} - \frac{m_g^2 Q^2}{\hat{t}^2} - \frac{m_g^2 Q^2}{\hat{s}^2}\right\}, \qquad (4.28)$$

where terms that contribute nothing to the total cross section, σ^{Σ}_{DIS}, in the limit $m_g \to 0$ have been dropped. Integrating over $\hat{t}$ as in eq. (4.9) with

$$t_{min} = -\beta z Q^2/(1-z)$$

$$t_{max} = -Q^2/z + \beta Q^2 \qquad (4.29)$$

yields

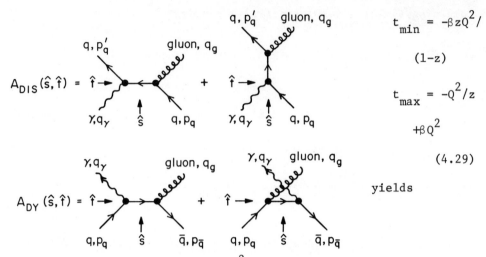

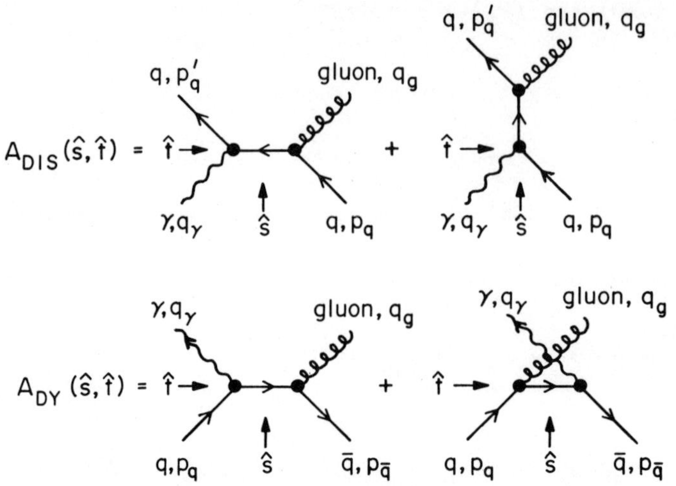

Fig. 4.3. Diagrams of order $\alpha_s(Q^2)$ that produce corrections to deep inelastic scattering (DIS) and to the "Drell-Yan" production of large mass muon pairs (DY) which are due to real gluon emission.

$$\hat{\sigma}^{\Sigma}_{DIS}(Q^2) = \frac{\pi\alpha\alpha_s e_q^2}{Q^2}\left(\frac{8}{3}\right)z\left\{\frac{(1+z^2)}{1-z}\log\left[\frac{Q^2(1-z)}{m_g^2 z^2}\right]\right.$$

$$\left. -\frac{3}{2}\frac{1}{1-z} + z + 1 + \beta\,\frac{2z^3-z^2}{(1-z)^2} + \frac{1}{2}\beta^2\,\frac{2z^3-z^4}{(1-z)^3}\right\}, \qquad (4.30)$$

where $\beta = m_g^2/Q^2$ and z is defined in (4.14a) but with q_g replaced by p_q. As we did in going from (3.19) to (3.22), the β and β^2 terms can be replaced by δ-function contributions leaving

$$\hat{\sigma}^{\Sigma}_{DIS}(Q^2) = \frac{\pi\alpha\alpha_s e_q^2}{Q^2}\left(\frac{8}{3}\right)z\left\{\frac{(1+z^2)}{1-z}\log(Q^2/m_g^2) + \frac{(1+z^2)}{1-z}\log(1-z)\right.$$

$$\left. -2\,\frac{(1+z^2)}{1-z}\log z - \frac{3}{2}\frac{1}{1-z} + z + 1 - \frac{5}{4}\,\delta(z-1)\right\}. \qquad (4.31)$$

The results are not complete, however, since we must add to them the contributions from the virtual gluon loop diagram in Fig. 4.4. These diagrams interfere with the Born term $\gamma^* + q \to q$ to produce $\delta(z-1)$

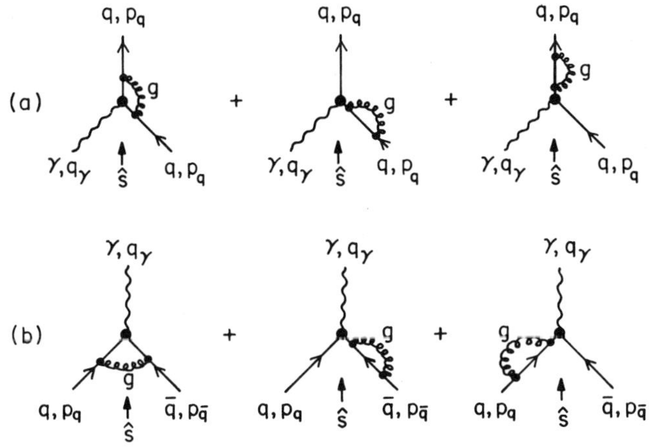

Fig. 4.4. Virtual gluon corrections to the deep inelastic scattering subprocess $\gamma^* + q \to q$ (a) and to the Drell-Yan subprocess $q + \bar{q} \to \gamma^*$ (b). These diagrams interfere with the order zero "Born diagrams" producing effects of order $\alpha_s(Q^2)$.

contributions of order α_s. From (3.25) we see that

$$\sigma^{DIS}_{virtual}(Q^2) = \frac{\pi\alpha\alpha_s e_q^2}{Q^2}\left(\frac{8}{3}\right)z \left\{-\log^2\beta - 3\log\beta - \frac{7}{2} - \frac{2\pi^2}{3}\right\} \delta(z-1), \quad (4.32)$$

where the extra factor of $-\pi^2$ comes from the change from timelike Q^2 in (3.25) to spacelike Q^2 in deep inelastic scattering (i.e., $\log^2(-Q^2) \to \log^2 Q^2 - \pi^2$). If we now combine (4.31) and (4.32) then insert into (4.16) and use the definitions of the "+ functions" in (3.36), we arrive at

$$\frac{F_\Sigma(x,Q^2)}{x} = e_q^2 \int_x^1 \frac{dy}{y} G^{(o)}_{p\to q}(y) \left\{\delta(z-1) + \frac{\alpha_s}{2\pi} P_{q\leftarrow q}(z)\log\left(\frac{Q^2}{m_g^2}\right) + \alpha_s f_{\Sigma,q}(z)\right\}, \quad (4.33)$$

where $z = x/y$ and

$$P_{q\leftarrow q}(z) = \left(\frac{4}{3}\right)\left\{\frac{1+z^2}{(1-z)_+} + \frac{3}{2}\,\delta(z-1)\right\} \quad (4.34a)$$

and

$$\alpha_s f_{\Sigma,q}(z) = \frac{\alpha_s}{2\pi}\left(\frac{4}{3}\right)\{(1+z^2)\left(\frac{\log(1-z)}{1-z}\right)_+ + \frac{1+z^2}{1-z}(-2\log z)$$

$$-\frac{3}{2}\frac{1}{(1-z)_+} + z + 1 - \left(\frac{2\pi^2}{3} + \frac{9}{4}\right)\delta(z-1)\}. \tag{4.34b}$$

The longitudinal differential cross section is

$$\frac{d\hat{\sigma}_L}{d\hat{t}} = \frac{\pi\alpha\alpha_s}{(\hat{s}+Q^2)^2} e_q^2\left(\frac{4}{3}\right)\{Q^2+\hat{t}+\hat{s}\} \tag{4.35}$$

which implies

$$\hat{\sigma}_L = \pi\alpha\alpha_s\left(\frac{4}{3}\right)e_q^2 \tag{4.36}$$

whereupon (4.23) yields (for one quark flavor)

$$\frac{F_L(x,Q^2)}{x} = e_q^2\int_x^1 \frac{dy}{y} G^{(o)}_{p\to q}(y)\alpha_s(Q^2)f_{L,q}(z), \tag{4.37}$$

with

$$\alpha_s f_{L,q}(z) = \frac{\alpha_s}{2\pi}\left(\frac{4}{3}\right)2z. \tag{4.38}$$

Finally, using (4.20), we arrive at for one quark flavor

$$\tilde{F}_1(x,Q^2) = 2F_1(x,Q^2) = e_q^2\int_x^1 \frac{dy}{y} G_{p\to q}(y)$$

$$\{\delta(z-1) + \frac{\alpha_s}{2\pi}P_{q\leftarrow q}(z)\log\frac{Q^2}{m_g^2} + \alpha_s f_{1,q}(z)\} \tag{4.39a}$$

$$\tilde{F}_2(x,Q^2) = \frac{F_2(x,Q^2)}{x} = e_q^2\int_x^1 \frac{dy}{y} G_{p\to q}(y)$$

$$\{\delta(z-1) + \frac{\alpha_s}{2\pi}P_{q\leftarrow q}(z)\log\frac{Q^2}{m_g^2} + \alpha_s f_{2,q}(z)\}, \tag{4.39b}$$

where $z = x/y$ and $P_{q\leftarrow q}(z)$ is given by (4.34a) and

$$\alpha_s f_{2,q}(z) = \frac{\alpha_s}{2\pi}\left(\frac{4}{3}\right)\left[-\frac{2(1+z^2)}{1-z}\log z - \frac{3}{2}\frac{1}{(1-z)_+} + (1+z^2)\left(\frac{\log(1-z)}{1-z}\right)_+ + 1 + 4z - \left(\frac{2\pi^2}{3} + \frac{9}{4}\right)\delta(z-1)\right] \tag{4.40a}$$

$$\alpha_s[f_{2,q}(z) - f_{1,q}(z)] = \frac{\alpha_s}{2\pi}\left(\frac{4}{3}\right)2z. \tag{4.40b}$$

The quantity $f_{2,q}(z)$ is regularization scheme dependent. However, $f_{2,q} - f_{1,q}$ is unique and the integral of $f_{2,q}$ which is

$$\int_0^1 f_{2,q}(z)dz = 0 \tag{4.41}$$

is also unique. We will see that this guarantees that the net number of quarks (i.e., $u-\bar{u}$, $d-\bar{d}$, etc.) in a proton is unchanged by order α_s corrections.

<u>Q^2 Evolution of Quark and Gluon Distributions</u>. If we now define quark distributions $G^{(1)}(x,Q^2)$ and $G^{(2)}(x,Q^2)$ by

$$\tilde{F}_1(x,Q^2) = 2F_1(x,Q^2) \equiv \sum_{i=1}^{n_f} e_{q_i}^2 \left(G^{(1)}_{p\to q_i}(x,Q^2) + G^{(1)}_{p\to\bar{q}_i}(x,Q^2)\right), \tag{4.42a}$$

and

$$\tilde{F}_2(x,Q^2) = \frac{F_2(x,Q^2)}{x} \equiv \sum_{i=1}^{n_f} e_{q_i}^2 \left(G^{(2)}_{p\to q_i}(x,Q^2) + G^{(2)}_{p\to\bar{q}_i}(x,Q^2)\right), \tag{4.42b}$$

then by combining eqs. (4.26) and (4.39), we arrive at

$$G^{(1)}_{p\to q}(x,Q^2) = \int_x^1 \frac{dy}{y} \{G^{(o)}_{p\to q}(y)[\delta(z-1) + \frac{\alpha_s}{2\pi} P_{q\leftarrow q}(z)\log \frac{Q^2}{m_g^2}$$

$$+ \alpha_s f_{1,q}(z)] + G^{(o)}_{p\to g}(y)[\frac{\alpha_s}{2\pi} P_{q\leftarrow g}(z)\log \frac{Q^2}{m_g^2}$$

$$+ \alpha_s f_{1,g}(z)]\}, \tag{4.43a}$$

and

$$G^{(2)}_{p\to q}(x,Q^2) = \int_x^1 \frac{dy}{y} \{G^{(o)}_{p\to q}(y)[\delta(z-1) + \frac{\alpha_s}{2\pi} P_{q\leftarrow q}(z)\log \frac{Q^2}{m_g^2}$$

$$+ \alpha_s f_{2,q}(z)] + G^{(o)}_{p\to g}(y)[\frac{\alpha_s}{2\pi} P_{q\leftarrow g}(z)\log \frac{Q^2}{m_g^2}$$

$$+ \alpha_s f_{2,g}(z)]\}, \tag{4.43b}$$

where $z = x/y$ and $P_{q\leftarrow q}(z)$ and $P_{q\leftarrow g}(z)$ are given by (4.34a) and (4.18), respectively, and the f-functions are given by (4.27) and (4.40). As explained in Section III, we can absorb the $\log(m^2)$ divergences into the unknown distributions $G^{(o)}(y)$. Then to order α_s, both $G^{(1)}(x,Q^2)$ and $G^{(2)}(x,Q^2)$ satisfy the evolution equations

$$dG_{p\to q_i}(x,Q^2)/d\tau = \frac{\alpha_s}{2\pi} [G_{p\to q_i}(Q^2) \otimes P_{q\leftarrow q} + G_{p\to g}(Q^2) \otimes P_{q\leftarrow g}] \tag{4.44a}$$

and

$$dG_{p\to g}(x,Q^2)/d\tau = \frac{\alpha_s}{2\pi} [\sum_{j=1}^{2n_f} G_{p\to q_j}(Q^2) \otimes P_{g\leftarrow q} + G_{p\to g}(Q^2) \otimes P_{g\leftarrow g}], \tag{4.44b}$$

as illustrated in Fig. 4.5 where $\tau = \log(Q^2/\Lambda^2)$ and where the convolution symbol $\otimes$ is defined in (3.42). The summing of all terms of the form τ^N has been accomplished by replacing the distributions

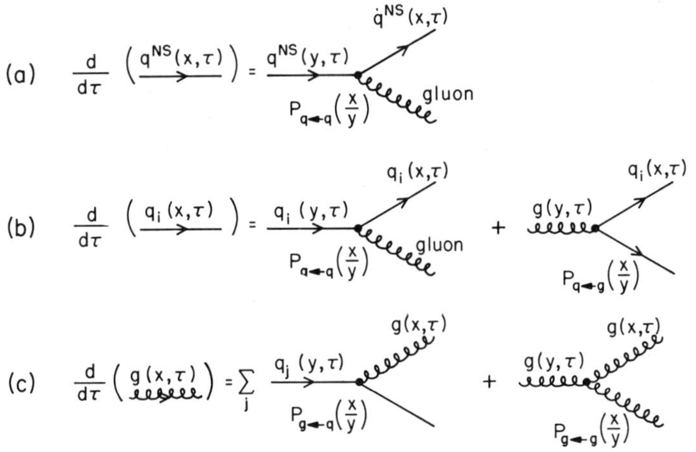

Fig. 4.5 - (a) Illustrates that the changes of the non-singlet quark distribution, $q^{NS}(x,\tau) = q(x,\tau)-\bar{q}(x,\tau)$, w.r.t. $\tau = \log(Q^2/Q_0^2)$ is due to the radiation of a gluon with $P_{q\leftarrow q}(z)$ being the probability of finding a quark with momentum fraction z "within" a quark.

(b) Illustrates that the change of a quark distribution, $q_i(x,\tau)$, w.r.t. τ is due to the Bremsstrahlung radiation of a gluon, $P_{q\leftarrow q}(z)$, and the production of quark-antiquark pairs from a gluon, $P_{q\leftarrow g}(z)$.

(c) Illustrates that the change of the gluon distribution, $g(x,\tau)$, w.r.t. τ is due to the Bremsstrahlung radiation of gluons from incident quarks, $P_{g\leftarrow q}(z)$, and from incident gluons, $P_{g\leftarrow g}(z)$.

$G^{(0)}(y)$ by $G(y,Q^2)$ on the right hand side of eqs. (4.44)[14,15]. The f-functions do not contribute to the Q^2 evolution at order α_s. They begin to contribute at order α_s^2 as illustrated in (3.57). The $P_{q\leftarrow q}$, $P_{q\leftarrow g}$ and $P_{g\leftarrow q}$ functions are given by (4.34a), (4.18) and (3.35b), respectively. Namely,

$$P_{q\to q}(z) = (\tfrac{4}{3})\left[\frac{1+z^2}{(1-z)_+} + \frac{3}{2}\delta(z-1)\right] = (\tfrac{4}{3})\left(\frac{1+z^2}{1-z}\right)_+ \tag{4.45a}$$

$$P_{q\leftarrow g}(z) = \frac{1}{2}\,(z^2 + (1-z)^2) \tag{4.45b}$$

$$P_{g\leftarrow q}(z) = (\tfrac{4}{3})\,\frac{1 + (1-z)^2}{z} \tag{4.45c}$$

The $P_{g\leftarrow g}$ function can be calculated in a similar manner and is given by[21]

$$P_{g\leftarrow g}(z) = 6\left[\frac{z}{(1-z)_+} + \frac{1-z}{z} + z(1-z) + \frac{(11 - \frac{2}{3}n_f)}{12}\delta(z-1)\right]. \tag{4.45d}$$

As Fig. 4.5 illustrates the quark distribution $q_i(x,\tau)$ changes with τ (or $\log Q^2$) due to the Bremsstrahlung of gluons $P_{q\to q}(z)$ and due to gluons that pair produce quarks of type q_i, $P_{g\to q}(z)$. The gluon distribution $g(x,\tau)$ changes with τ because of the Bremsstrahlung of gluons $P_{g\to g}(\tau)$ and from gluons that are produced from quark Bremsstrahlung, $P_{q\to g}(z)$. The equations are non-diagonal, both the quark and gluon distributions at $\tau + d\tau$ depend on the quark <u>and</u> gluon distributions at τ. It is easy to verify that

$$\int_0^1 P_{q\leftarrow q}(z)dz = 0 \tag{4.46a}$$

and that

$$\int_0^1 dz\ z[P_{q\leftarrow q}(z) + P_{g\leftarrow q}(z)] = 0 \tag{4.46b}$$

$$\int_0^1 dz\ z[2n_f\ P_{q\leftarrow g}(z) + P_{g\leftarrow g}(z)] = 0. \tag{4.46c}$$

If we define a "singlet" distribution by

$$G^s(x,Q^2) = \sum_{i=1}^{n_f}[G_{p\to q_i}(x,Q^2)+G_{p\to \bar{q}_i}(x,Q^2)] \tag{4.47a}$$

and a "non-singlet" distribution by

$$G_i^{NS}(x,Q^2) = G_{p\to q_i}(x,Q^2) - G_{p\to \bar{q}_i}(x,Q^2). \tag{4.47b}$$

Then we can write the matrix equation

$$d\underset{\sim}{G}(x,Q^2)/d\tau = \frac{\alpha_s(Q^2)}{2\pi} \underset{\sim}{P} \otimes \underset{\sim}{G}(Q^2) \qquad (4.48a)$$

and the "non-singlet" equation

$$dG_i^{NS}(x,Q^2)/d\tau = \frac{\alpha_s(Q^2)}{2\pi} P_{q\leftarrow q} \otimes G_i^{NS}(Q^2), \qquad (4.48b)$$

where

$$\underset{\sim}{G}(x,Q^2) = \begin{pmatrix} G^S(x,Q^2) \\ G_{p\to g}(x,Q^2) \end{pmatrix} \qquad (4.48c)$$

and

$$\underset{\sim}{P}(z) = \begin{pmatrix} P_{q\leftarrow q}(z) & 2n_f P_{q\leftarrow g}(z) \\ P_{g\leftarrow q}(z) & P_{g\leftarrow g}(z) \end{pmatrix} . \qquad (4.48d)$$

The total number of quarks of flavor i is

$$N_i(Q^2) = \int_0^1 G_i^{NS}(x,Q^2)dx, \qquad (4.49)$$

and from (4.46), we see that

$$\frac{dN_i(Q^2)}{d\tau} = 0. \qquad (4.50)$$

This means that if $N_u(Q_o^2) = 2$ for a proton as in (4.2a), it will remain 2 at any Q^2. In addition, (4.46a) and (4.46b) guarantee that

$$\frac{d}{d\tau}\int_0^1 dx\, x[\sum_{i=1}^{n_f}(G_{p\to q_i}(x,Q^2)+G_{p\to\bar{q}_i}(x,Q^2))+G_{p\to g}(x,Q^2)] = 0, \qquad (4.51)$$

so that the total momentum of the proton (i.e., of all partons) is unchanged as Q^2 changes. Defining

$$\kappa = \frac{2}{\beta_o} \log\{\alpha_s(Q_o^2)/\alpha_s(Q^2)\} \tag{4.52a}$$

and using $\alpha_s(Q^2) = 4\pi/(\beta_o \log Q^2/\Lambda^2)$, we have

$$\frac{d\kappa}{d\tau} = \frac{\alpha_s(Q^2)}{2\pi} \tag{4.52b}$$

so that eq. (4.48b) becomes

$$dG_i^{NS}(x,Q^2)/d\kappa = P_{q\leftarrow q} \otimes G_i^{NS}(Q^2), \tag{4.52c}$$

and the solution of this equation is

$$G_i^{NS}(x,Q^2) = \exp(\kappa P_{q\leftarrow q} \otimes) G_i^{NS}(Q_o^2). \tag{4.53a}$$

Similarly the solution of (4.48a) can be written

$$\underset{\sim}{G}(x,Q^2) = \exp(\kappa \underset{\sim}{P} \otimes) \underset{\sim}{G}(Q_o^2). \tag{4.53b}$$

These equations relate the distributions at Q^2 to those at some reference momentum Q_o^2.

The Moment Method. Equation (4.48) is usually written in terms of the moments of the parton distributions

$$M_n(Q^2) = \int_0^1 dx\, x^{n-1} G(x,Q^2), \tag{4.54}$$

which gives

$$\frac{dM_n^{NS}(Q^2)}{d\kappa} = A_n^{NS} M_n^{NS}(Q^2) \tag{4.55a}$$

and

$$\frac{d\underset{\sim}{M}_n(Q^2)}{d\kappa} = \underset{\sim}{A}_n \underset{\sim}{M}_n(Q^2), \tag{4.55b}$$

where

$$\underset{\sim}{M}_n(Q^2) = \begin{pmatrix} M_n^s(Q^2) \\ \\ M_n^g(Q^2) \end{pmatrix} \tag{4.55c}$$

and

$$\underset{\sim}{A} = \begin{pmatrix} A_n^{NS} & 2n_f A_n^{qg} \\ \\ A_n^{gq} & A_n^{gg} \end{pmatrix}, \tag{4.55d}$$

where M_n^{NS}, M_n^s and M_n^g are the moments of the non-singlet, singlet and gluon distributions, respectively. The "anomalous dimensions" A_n are given by

$$\underset{\sim}{A} = \int_0^1 dz\ z^{n-1} \underset{\sim}{P}(z). \tag{4.56}$$

We are left with three independent Q^2 evolution equations for the moments. Namely,

$$M_n^{NS}(Q^2) = \exp(\kappa A_n^{NS}) M_n^{NS}(Q_o^2) = (\alpha(Q_o^2)/\alpha(Q^2))^{2A_n^{NS}/\beta_o} M_n^{NS}(Q_o^2) \tag{4.57a}$$

and

$$M_n^+(Q^2) = \exp(\kappa A_n^+) M_n^+(Q_o^2)$$

$$M_n^-(Q^2) = \exp(\kappa A_n^-) M_n^-(Q_o^2), \tag{4.57b}$$

where $M_n^{\pm}(\tau)$ and $A_n^{\pm}$ are the eigenvectors and eigenvalues obtained upon diagonalizing eq. (4.54b). From (4.45) and (4.56) one obtains

$$A_n^{NS} = (\tfrac{4}{3})[-\tfrac{1}{2} + \frac{1}{n(n+1)} - 2\sum_{j=2}^{n} \tfrac{1}{j}] \tag{4.58a}$$

$$A_n^{gq} = (\tfrac{4}{3}) \frac{2 + n + n^2}{n(n^2-1)} \tag{4.58b}$$

$$A_n^{qg} = (\tfrac{1}{2}) \frac{2 + n + n^2}{n(n+1)(n+2)} \tag{4.58c}$$

$$A_n^{gg} = 3[-\tfrac{1}{6} + \frac{2}{n(n-1)} + \frac{2}{(n+1)(n+2)} - 2\sum_{j=2}^{n} \frac{1}{j} - \tfrac{1}{9} n_f]. \tag{4.58d}$$

The Convolution Method. R. P. Feynman, D. A. Ross and I have developed an alternative method for calculating the parton distributions at Q^2 in terms of those at Q_0^2 that does not involve taking moments[22]. The solution of eq. (4.52c) is given by (4.53a). Namely,

$$G^{NS}(x,Q^2) = \exp(\kappa P \otimes) G^{NS}(Q_0^2)$$

$$= G^{NS}(Q_0^2) + \kappa P \otimes G^{NS}(Q_0^2) + \tfrac{1}{2}\kappa^2 P \otimes P \otimes G^{NS}(Q_0^2) + \ldots,$$

where $P \equiv P_{q \leftarrow q}(z)$. (4.59)

As Table 4.1 below shows, κ is a small number so one might think that summing the full series in (4.59) is unnecessary and that the first two terms would be sufficient.

Table 4.1. Values of $\kappa = (6/25)\log[\log(Q^2/\Lambda^2)/\log(Q_0^2/\Lambda^2)]$ with Λ = 0.4 GeV and Q_0^2 = 4 GeV^2.

Q^2	κ
4	0
16	0.097
36	0.14
100	0.185
10^4	0.322
10^6	0.408

Unfortunately, the nature of $P_{q \leftarrow q}(z)$ forces one to sum the complete series. Suppose we keep only the first two terms, then from (3.37a) we see that as $x \to 1$

$$\frac{G(x,Q^2)}{G(x,Q_0^2)} = 1 + \kappa\log(1-x). \tag{4.60}$$

No matter how small κ is, log(1-x) becomes large so that one must include higher and higher terms in κ. This behavior arises because of the $1/(1-z)_+$ term in $P_{q\leftarrow q}(z)$. Perhaps we can find another function $P_A(z)$ that has the same singularity structure as $z \to 1$ but which we can explicitly do the sum so that

$$\exp(\kappa P_A \otimes)G \equiv R \otimes G. \tag{4.61}$$

It is not too hard to find such a function. Consider

$$P_A(z) = \frac{4}{3}\left[\frac{2}{(-\log z)_+}\right] \tag{4.62}$$

where the "+" is defined by (3.36a). This function has the property that its moments are

$$A_n = \int_0^1 dz\, z^{n-1} P_A(z) = -\frac{8}{3}\log(n) \tag{4.63}$$

so that

$$R_n = \int_0^1 dy\, y^{n-1} R(y) = \exp[-\frac{8}{3}\kappa\log(n)] = n^{-\frac{8}{3}\kappa}. \tag{4.64}$$

All we now need to do is to solve (4.64) for R(y). Using

$$\int_0^1 dy\, y^{n-1}(-\log y)^P = \frac{\Gamma(p+1)}{n^{p+1}} \tag{4.65}$$

we arrive at

$$R(y) = (-\log y)^{\frac{8}{3}\kappa-1}/\Gamma(\frac{8}{3}\kappa), \tag{4.66}$$

where R is defined in (4.61). Equation (4.59) can now be written

$$G(x,Q^2) = \exp(\kappa P_A \otimes)\exp(\kappa P_\Delta \otimes)G(Q_0^2) = R \otimes \exp(\kappa P_\Delta \otimes)G(Q_0^2) \tag{4.67}$$

where

$$P_\Delta(y) = P(y) - P_A(y) = \frac{4}{3}\left[\left(\frac{1+y^2}{1-y}\right)_+ - \frac{2}{(-\log y)_+}\right]. \tag{4.68a}$$

Since $P_A(y)$ and $P(y)$ have the same singularity structure as $y \to 1$ (i.e., $-\log(y) \sim 1-y$), the + functions on P_Δ can be removed leaving

$$P_\Delta(y) = \frac{4}{3}\left[\frac{1+y^2}{1-y} + \frac{2}{\log y} + \left(\frac{3}{2} - 2\gamma\right)\delta(y-1)\right], \tag{4.68b}$$

where γ is Eulers constant which arises from an integration by parts (after a change of variables to $x = -\log y$) and the use of

$$\int_0^\infty e^{-x}\log x dx = -\gamma. \tag{4.69}$$

The function $P_\Delta(y)$ is smooth and well behaved so we can expand $\exp(\kappa P_\Delta \otimes)$ and keep only the first two terms. This results in

$$G(x,Q^2) = R \otimes G(Q_0^2) + \kappa R \otimes G(Q_0^2) \otimes P_\Delta + \text{order } (\kappa^2). \tag{4.70}$$

Defining

$$\tilde{G}(x,Q^2) = R \otimes G(Q_0^2) = \int_x^1 \frac{dy}{y} G(\frac{x}{y},Q_0^2)\frac{(-\log y)^{\frac{8}{3}\kappa-1}}{\Gamma(\frac{8}{3}\kappa)} \tag{4.71a}$$

eq. (4.70) becomes

$$G(x,Q^2) = \tilde{G}(x,Q^2) + \kappa \int_x^1 \frac{dy}{y} \tilde{G}(\frac{x}{y},Q^2)P_\Delta(y) + \text{order } (\kappa^2), \tag{4.71b}$$

where $P_\Delta(y)$ is given by (4.68b) and κ is defined in (4.52a). Using this equation we can calculate $G(x,Q^2)$ in terms of a main term, $\tilde{G}(x,Q^2)$, and a correction term that is proportional to κ and should be small. If the correction term is not small, we could then calculate the order κ^2 term. We see that in order to calculate $G(x,Q^2)$

at some value of x and Q^2, we need to know $G(z,Q_0^2)$ only for values of z greater than x. In addition, we need to know κ which can be calculated from the knowledge of $\alpha_s(Q^2)$ (i.e., Λ). Thus given $G(x,Q_0^2)$ the one parameter Λ governs the Q^2 evolution. Table 4.2 shows a comparison of results obtained by the convolution method (to order κ) and the results obtained by an (exact) moment method. As can be seen, the convolution method is quite accurate even for κ as big as 0.35.

Table 4.2. Results at $Q^2 = 168{,}000\ \mathrm{GeV}^2$ Using the Non-singlet Q^2 Evolution Eq. (4.71) with the Reference Momentum $Q_0^2 = 4\ \mathrm{GeV}^2$ and $\Lambda = 0.4$ GeV ($\kappa = 0.35$). The Main Term $x\tilde{G}(x,Q^2)$ and the Order κ Correction Term are Shown Separately. The Final Result is Compared to the Exact Results from the Moment Method and the Reference Distribution $xG(x,Q_0^2)$ Taken from Ref. 23.

x	$x\tilde{G}(x,Q^2)$	Order (K) Correction Term	$xG(x,Q^2)$ Convolution Method	$xG(x,Q^2)$ Moment Method	$xG(x,Q_0^2$
0.1	0.431	-0.004	0.426	0.424	0.430
0.2	0.367	0.010	0.377	0.375	0.570
0.3	0.254	0.015	0.269	0.268	0.563
0.4	0.155	0.013	0.168	0.167	0.469
0.5	0.0832	0.0083	0.0915	0.0915	0.342
0.6	0.0377	0.0044	0.0421	0.0422	0.216
0.7	0.0132	0.0017	0.0149	0.0150	0.111
0.8	0.00289	0.00042	0.00331	0.00333	0.0393
0.9	0.00020	0.00003	0.00023	0.00023	0.0058

The singlet Q^2 evolution is a bit more complicated. Equation (4.67) is not valid in this case because the matrices $\underset{\sim}{P}_A$ and $\underset{\sim}{P}_\Delta$ do not commute. Instead we write (4.48a) as

$$d\underset{\sim}{G}(x,Q^2)/d\kappa = (\underset{\sim}{P}_A+\underset{\sim}{P}_\Delta) \otimes \underset{\sim}{G}(Q_0^2), \tag{4.72a}$$

where

$$\underset{\sim}{P}_A = \begin{pmatrix} a_q P_o & 0 \\ 0 & a_g P_o \end{pmatrix} \tag{4.72b}$$

with

$$P_o(z) = 1/(-\log z)_+ \tag{4.72c}$$

and

$$a_q = \frac{8}{3},\ a_g = 6. \tag{4.72d}$$

In this case

$$\underset{\sim}{P}_\Delta = \underset{\sim}{P} - \underset{\sim}{P}_A = \begin{pmatrix} P_{\Delta q}(z) & 2n_f P_{q\leftarrow g}(z) \\ P_{g\leftarrow q}(z) & P_{\Delta g}(z) \end{pmatrix}, \tag{4.73a}$$

where

$$P_{\Delta q}(z) = P_{q\leftarrow q}(z) - a_q P_o(z) = P_\Delta(z) \tag{4.73b}$$

with $P_\Delta(z)$ given in (4.68b) and

$$P_{\Delta g}(z) = P_{g\leftarrow g}(z) - a_g P_o(z) =$$

$$6\left[\frac{z}{1-z} + \frac{1}{\log z} + \frac{1-z}{z} + z(1-z) + \left(\frac{11}{12} - \frac{2}{36} n_f - \gamma\right)\delta(z-1)\right]. \tag{4.73c}$$

To order κ the solution of (4.72a) is given by

$$\underset{\sim}{G}(x,Q^2) = \exp(\kappa \underset{\sim}{P}_A \otimes) G(Q_o^2)$$

$$+ \kappa \int_0^1 d\rho \exp[\rho\kappa \underset{\sim}{P}_A \otimes] \underset{\sim}{P}_\Delta \exp[(1-\rho)\kappa \underset{\sim}{P}_A \otimes] \underset{\sim}{G}(Q_o^2) \tag{4.74}$$

which can easily be seen to yield

$$G_{p\to q_i}(x,Q^2) = \tilde{G}_{p\to q_i}(a_q,x,Q^2) + \kappa \int_x^1 \frac{dy}{y} \tilde{G}_{p\to q_i}(a_q,y,Q^2) P_{\Delta q}(\frac{x}{y})$$

$$+ \kappa \int_x^1 \frac{dy}{y} \tilde{\tilde{G}}_{p\to g}(y,Q^2) P_{q\leftarrow g}(\frac{x}{y}) \tag{4.75a}$$

$$G_{p\to g}(x,Q^2) = \tilde{G}_{p\to g}(a_g,x,Q^2) + \kappa \int_x^1 \frac{dy}{y} \tilde{G}_{p\to g}(a_g,y,Q^2) P_{\Delta g}(\frac{x}{y})$$

$$+ \kappa \sum_{j=1}^{2n_f} \int_x^1 \frac{dy}{y} \tilde{\tilde{G}}_{p\to q_j}(y,Q^2) P_{g\leftarrow q_j}(\frac{x}{y}), \tag{4.75b}$$

where

$$\tilde{G}_{p\to i}(a,x,Q^2) = \int_x^1 \frac{dy}{y} G_{p\to i}(\frac{x}{y},Q_0^2) \frac{(-\log y)^{a\kappa-1}}{\Gamma(aK)} . \tag{4.75c}$$

The quantity $\tilde{\tilde{G}}(x,Q^2)$ is new and is gotten by averaging $\tilde{G}(a,x,Q^2)$ over a. Namely,

$$\tilde{\tilde{G}}_{p\to i}(x,Q^2) = \frac{1}{(a_g-a_q)} \int_{a_q}^{a_g} \tilde{G}_{p\to i}(a,x,Q^2). \tag{4.75d}$$

Equation (4.75) expresses the distributions at Q^2 directly in terms of those at Q_0^2. These equations are not exact since I have neglected terms of order κ^2. However, they are very accurate in the range $0.05 \le x \le 1$ and $4.0 \le Q^2 \le 10^6$. One loses accuracy at very small x due to the 1/z terms in $P_{g\leftarrow g}(z)$ and $P_{g\leftarrow q}(z)$.

From (4.75) it is easy to deduce the large x behavior of $G(x,Q^2)$ since as $x \to 1$ $G(x,Q^2)$ approaches $\tilde{G}(a,x,Q^2)$. Suppose at Q_0^2 we have

$$G(x,Q_0^2) = A(1-x)^P \underset{x\to 1}{\approx} A(-\log x)^P. \tag{4.76}$$

In this case $\tilde{G}(a,x,Q^2)$ can be calculated analytically yielding[24]

$$\widetilde{G}(a,x,Q^2) = A\,\frac{\Gamma(p+1)}{\Gamma(p+1+a\kappa)}\,(-\log x)^{P+a\kappa} \underset{x\to 1}{\sim} C(1-x)^{P+\xi_a(Q^2)} \tag{4.77a}$$

where

$$\xi_a(Q^2) = a\kappa. \tag{4.77b}$$

For quarks we have (with $n_f = 4$)

$$\xi_q(Q^2) = \frac{16}{25}\log\{\log(Q^2/\Lambda^2)/\log(Q_o^2/\Lambda^2)\} \tag{4.77c}$$

and for gluons

$$\xi_g(Q^2) = 6\,\log\{\log(Q^2/\Lambda^2)/\log(Q_o^2/\Lambda^2)\}. \tag{4.77d}$$

The distributions become steeper as Q^2 increases with the gluon changing more rapidly.

Analysis of Deep Inelastic Electron and Muon Scattering. Following the analysis of G. C. Fox[23], the moments of the quark and gluon distributions

$$M_i(n,Q^2) = \int_0^1 dx\, x^{n-1} G_{p\to i}(x,Q^2) \tag{4.78a}$$

are given in terms of the moments at some reference momentum, Q_o^2, by

$$M_j(n,Q^2) = \sum_{i=1}^{9} M_i(n,Q_o^2) R_{ij}(n,Q^2,Q_o^2,\Lambda), \tag{4.78b}$$

where $R_{ij}(n,Q^2,Q_o^2,\Lambda^2)$ is the matrix constructed from eq. (4.55) and i corresponds to the constituent types ($u,d,s,c,\bar{u},\bar{d},\bar{s},\bar{c}$,glue). The matrix R_{ij} depends on $\alpha(Q^2)$ (i.e., on Λ) and on the calculable anomalous dimensions A_n^{NS}, A_n^{qg}, A_n^{gq} and A_n^{gg} given by (4.58). The resulting distributions at Q^2 are calculated in terms of those at Q_o^2 by diagonalizing (4.78b) and inverting (4.78a) by an inverse Mellin transform (eq. (13) of Ref. 23).

Figure 4.6 shows the expected dependence of $\nu W_2(x,Q^2)$ resulting from an analysis of ep and μp data. The x dependence of the parton distributions at the reference momenta, $Q_o^2 = 4\ \mathrm{GeV}^2$, was chosen to agree with experiment and Λ was varied to produce the observed amount of "scale breaking". The analysis of ep and μp data is sensitive to the gluon distribution only through diagrams like that of Fig. 4.1. We have taken

$$xG_{p\to g}(x,Q_o^2) = (1+9x)(1-x)^4, \tag{4.79}$$

however, the analysis of ep and μp is not sensitive to this precise choice. The resulting Q^2 dependence of $G_{p\to g}(x,Q^2) = g(x,Q^2)$ is shown in Fig. 4.6b. Figure 4.7 shows the behavior of the quark and antiquark distributions. Both $\nu W_2(x,Q^2)$ and $xg(x,Q^2)$ exhibit the characteristic rise at small x and decrease at large x as Q^2 increases.

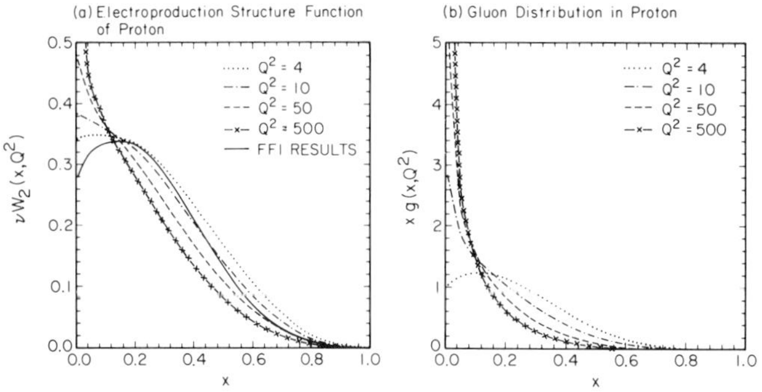

Fig. 4.6 - (a) Shows the predicted Q^2 dependence (scale breaking) of the electroproduction structure function for the proton, $\nu W_2(x,Q^2)$, arising from the constituent (quarks, antiquarks and gluons) distributions $G_i(x,Q^2)$ used in this analysis. The distributions at high Q^2 are calculated from the distributions at the reference momentum $Q_o^2 = 4\ \mathrm{GeV}^2$ using a QCD moment analysis with $\Lambda = 0.4$ GeV/c. In asymptotically free theories, one expects a decrease in the number of high x constituents and an increase in the number of low x constituents as Q^2 increases. Also shown is the value of $\nu W_2(x)$ (independent of Q^2) used in the quark-quark "black-box" model of FF1[28].
(b) Shows the predicted Q^2 dependence of the distribution of gluons within the proton $xG_{p\to g}(x,Q^2)$ used in this analysis. The distribution at high Q^2 is calculated in terms of a distribution at the reference momentum $Q_o^2 = 4\ \mathrm{GeV}^2$ chosen to be $xg(x,Q_o^2) = (1+9x)(1-x)^4$.

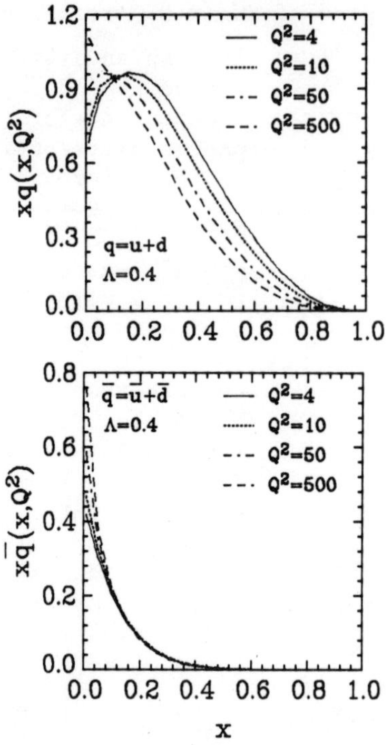

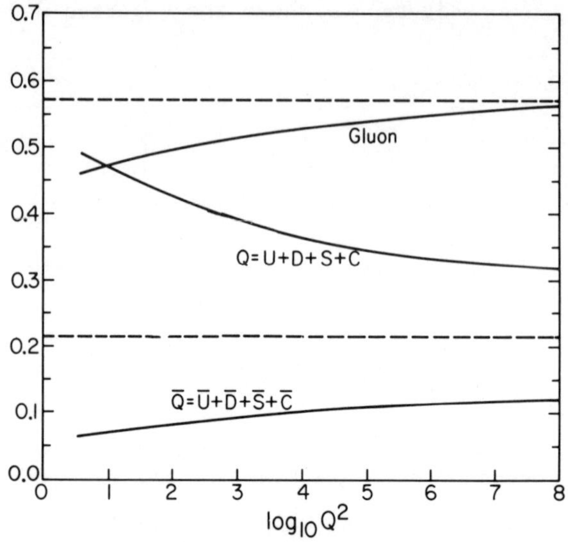

Fig. 4.8. Shows the predicted Q^2 dependence of the total momentum carried by quarks ($Q = U + D + S + C$), antiquarks ($\bar{Q} = \bar{U} + \bar{D} + \bar{S} + \bar{C}$) and gluons from a QCD analysis with $\Lambda = 0.4$ GeV/c. As Q^2 becomes large (very large!) $\bar{Q} \sim Q \sim 0.22$ and gluon ~ 0.57 for the number of quark flavors, n_f, equal to 4.

Fig. 4.7. Shows the predicted Q^2 dependence (scale breaking) of the quark ($q = u + d$) and antiquark ($\bar{q} = \bar{u} + \bar{d}$) distributions from the QCD moment analysis with $\Lambda = 0.4$ GeV/c.

Figure 4.8 shows how the total momentum carried by quarks, antiquarks and gluons within a proton is predicted to change with increasing Q^2. For $n_f = 4$, one expects the total momentum carried by quarks to approach that carried by antiquarks and to become 22%. The gluons carried (asymptotically) the remaining 57% of the proton momentum. However, as Fig. 4.8 indicates, the approach to asymptopia is quite gentle.

Fits to some of the existing ep and μp data are shown in Fig. 4.9 and Fig. 4.10. The "scale" breaking is as expected from QCD.

Some Order $\alpha_s(Q^2)$ Results for Deep Inelastic Electro-production. From (4.43) we see that

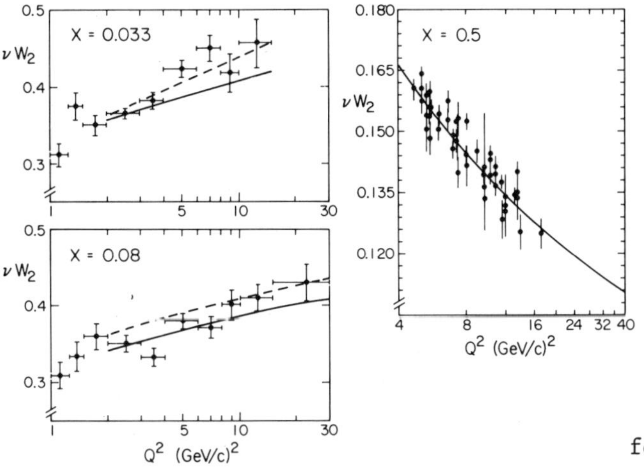

Fig. 4.9. Comparison of the scale breaking effects (Q^2 dependence) expected from an asymptotically free theory with data on ep and μp inelastic scattering at x = 0.033, 0.08 and 0.5. The theory comes from the analysis of Ref. 23 using Λ = 0.4 GeV/c (solid curve) and Λ = 0.5 GeV/c (dashed curve).

$$G^{(i)}_{p\to q}(x,Q^2) = \int_x^1 \frac{dy}{y} \{G_{p\to q}(y,Q^2)[\delta(z-1) + \alpha_s(Q^2) f_{i,q}(z)] + G_{p\to g}(y,Q^2)\alpha_s(Q^2) f_{i,g}(z)\} \quad (4.80)$$

for i = 1,2 with z = x/y and where the Q^2 dependences of $G^{(i)}(x,Q^2)$ and $G^{(i)}_{p\to g}(x,Q^2)$ are governed to order $\alpha_s(Q^2)$ by eq. (4.44) and where $G^{(i)}_{p\to q}(x,Q^2)$ are defined by (4.42). We are not quite finished since as mentioned earlier, the functions $f_{i,q}(z)$ and $f_{i,g}(z)$ in (4.80) are not unique. They depend on the particular "regularization" scheme we have employed. Only the $\Delta f(z)$ are unique. We will thus define the quark distributions determined by a measurement of $F_2(x,Q^2)$ in deep inelastic electron scattering to be the "reference" distributions[6]. That is, we will take

$$\frac{F_2(x,Q^2)}{x} \equiv \sum_{i=1}^{n_f} e^2_{q_i} (G_{p\to q_i}(x,Q^2) + G_{p\to \bar{q}_i}(x,Q^2)) \quad (4.81)$$

to be the definition of $G_{p\to q_i}(x,Q^2)$ to full order $\alpha_s(Q^2)$. With this definition, we have to order $\alpha_s(Q^2)$

$$G^{(2)}_{p\to q_i}(x,Q^2) = G_{p\to q_i}(x,Q^2) \quad (4.82a)$$

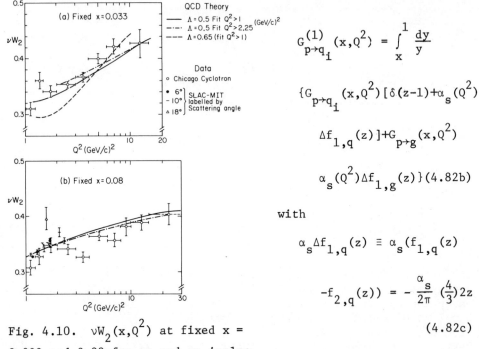

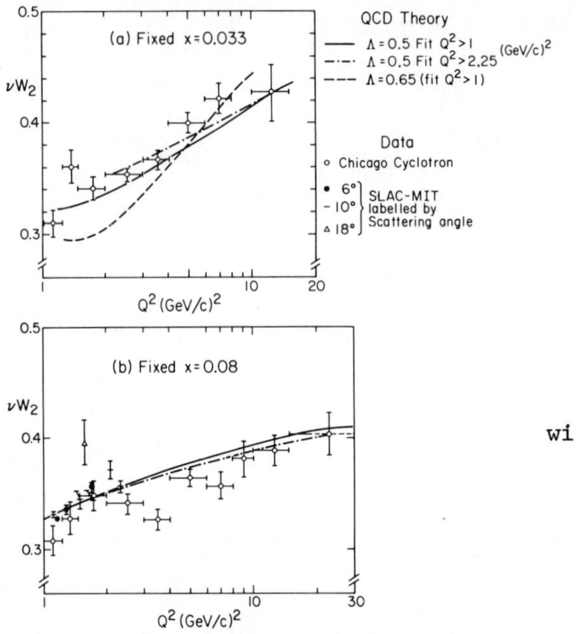

Fig. 4.10. $\nu W_2(x,Q^2)$ at fixed x = 0.033 and 0.08 for ep and μp inelastic scattering compared with the QCD predictions from Ref. 23.

$$G^{(1)}_{p\to q_i}(x,Q^2) = \int_x^1 \frac{dy}{y}$$

$$\{G_{p\to q_i}(x,Q^2)[\delta(z-1)+\alpha_s(Q^2)$$

$$\Delta f_{1,q}(z)]+G_{p\to g}(x,Q^2)$$

$$\alpha_s(Q^2)\Delta f_{1,g}(z)\} \quad (4.82b)$$

with

$$\alpha_s \Delta f_{1,q}(z) \equiv \alpha_s(f_{1,q}(z)$$

$$-f_{2,q}(z)) = -\frac{\alpha_s}{2\pi}\left(\frac{4}{3}\right)2z \quad (4.82c)$$

$$\alpha_s \Delta f_{1,g}(z) = \alpha_s(f_{1,g}(z)$$

$$-f_{2,g}(z)) = -\frac{\alpha_s}{2\pi}\,2z(1-z) \quad (4.82d)$$

and where $G^{(i)}_{p\to q_i}(x,Q^2)$ satisfy eq. (4.44). The $\Delta f(z)$ functions in (4.82) should now be unique.

Having developed all the machinery of eqs. (4.82), it is an easy matter to write down the expression for

$$R = \sigma_L/\sigma_T = F_L/2xF_1. \quad (4.83)$$

We have

$$\frac{F_L(x,Q^2)}{x} = -\alpha_s(Q^2)\int_x^1 \frac{dy}{y}\left\{\sum_{i=1}^{n_f} e_{q_i}^2 (G_{p\to q_i}(x,Q^2)+G_{p\to \bar{q}_i}(x,Q^2))\right.$$

$$\left.\Delta f_{1,q}(z)+\left(\sum_{i=1}^{2n_f} e_{q_i}^2\right) G_{p\to g}(x,Q^2)\Delta f_{1,g}(z)\right\} \tag{4.84}$$

or after substituting in for $\Delta f(z)$ and using (4.81) becomes[25]

$$F_L(x,Q^2) = \frac{\alpha_s(Q^2)}{2\pi} x^2 \int_x^1 \frac{dy}{y^3}\{\tfrac{8}{3} F_2(y,Q^2)+2a_e yG_{p\to g}(y,Q^2)(1-\tfrac{x}{y})\}, \tag{4.85a}$$

where

$$a_e = \left(\sum_{i=1}^{2n_f} e_i^2\right) = \frac{20}{9} \qquad (\text{for } n_f = 4). \tag{4.85b}$$

The quantity R is now given by

$$R = F_L/(F_L+F_2) = R_2/(1+R_2) \tag{4.86a}$$

where

$$R_2 = F_L/F_2 \tag{4.86b}$$

and F_2 is given by (4.81).

A particularly easy quantity to calculate is

$$\bar{R}_2(Q^2) = \bar{F}_L(Q^2)/\bar{F}_2(Q^2) \tag{4.87a}$$

with

$$\bar{F}_i(Q^2) = \int_0^1 F_i(x,Q^2)dx, \tag{4.87b}$$

for $i = 2,L$. Using (4.82c) and (4.82d), we see that

$$\alpha_s \int_0^1 z\Delta f_{1,q}(z)dz = -\frac{\alpha_s}{2\pi}\left(\frac{8}{9}\right) = -0.141\ \alpha_s \tag{4.88a}$$

$$\alpha_s \int_0^1 z\Delta f_{1,g}(z)dz = -\frac{\alpha_s}{2\pi}\left(\frac{1}{6}\right) = -0.027\ \alpha_s \tag{4.88b}$$

then from (4.85), we have

$$\bar{F}_L(Q^2) = \frac{\alpha_s}{2\pi}\left\{\left(\frac{8}{9}\right)\bar{F}_2(Q^2) + \left(\frac{1}{6}\right) a_e G(Q^2)\right\}, \tag{4.89a}$$

where $G(Q^2)$ is the total momentum carried by gluons:

$$G(Q^2) = \int_0^1 xG_{p\to g}(x,Q^2)dx. \tag{4.89b}$$

At $Q^2 = 16\ \text{GeV}^2$ our distributions give

$$\bar{F}_2(Q^2{=}16\ \text{GeV}) = 0.164 \tag{4.90a}$$

$$G(Q^2{=}16\ \text{GeV}) = 0.514 \tag{4.90b}$$

so

$$\bar{R}_2(Q^2{=}16\ \text{GeV}) = 0.107 \tag{4.90c}$$

$$\bar{R}(Q^2{=}16\ \text{GeV}) = 0.12. \tag{4.90d}$$

At this Q^2, about half of $\bar{F}_L(Q^2)$ is due to the gluon term $G(Q^2)$ and about half due to the quark term $\bar{F}_2(Q^2)$.

In all of this discussion, I have neglected corrections of order M^2/Q^2. Such contributions cannot be calculated by perturbation theory. An estimate of the $1/Q^2$ contribution to R is[20]

$$R(\text{primordial}) = 4\langle k_\perp^2\rangle_{\text{primordial}}/Q^2, \tag{4.91}$$

where $k_\perp$ is the non-perturbative "primordial" component to the transverse momentum of quarks within hadrons. The perturbative contribution to R behaves roughly as $\alpha_s(Q^2) \sim 1/\log(Q^2/\Lambda^2)$ so that at sufficiently large Q^2, one can deduce R. However, even at $Q^2 = 16$ GeV^2, R(primordial) $\simeq 0.06$ (using R(primordial) $\simeq 1/Q^2$) which is certainly not negligible compared to R(perturbation).

"Mini-analysis" of $xF_3(x,Q^2)$ from Neutrino and Antineutrino Nucleon Collisions. Equation (4.71) gives the non-singlet distribution at Q^2 in terms of the distribution at Q_o^2 provided we know Λ. If we fit the data at some Q_o^2, say $Q_o^2 = 20$ GeV^2, to determine $xF_3(x,Q_o^2)$ (xF_3 is a non-singlet distribution), then there is just one parameter, Λ, that determines the Q^2 evolution. Figure 4.11 shows a fit to the combined CDHS[26] and BEBC[27] neutrino data on xF_3 at $Q_o^2 = 20$ GeV^2. Using this fit, xF_3 is then predicted at $Q^2 = 60$ and 100 GeV^2 from eq. (4.71) and $\Lambda = 0.5$ GeV/c[28]. As seen in Fig. 4.11, the predictions are in quite good agreement with the data. In addition, since this analysis is performed for $Q^2 > 20$ GeV^2, it is not sensitive to corrections to the theory that are of order M^2/Q^2.

The analysis in Fig. 4.11 certainly supports QCD as the correct description of nature. However, if one asks how well can we determine the one parameter of the theory, Λ, from the neutrino data then the answer is disappointing. Figure 4.12 shows various fits to combined CDMS and BEBC data. In performing these fits, I have restricted myself to the range $Q^2 > 4$ GeV^2 and included some target mass corrections[28,29] (i.e., M^2/Q^2 effects) by using

$$xF_3(x,Q^2) = \frac{x^2}{\xi^2}\,\frac{1}{(1+4M^2x^2/Q^2)}\,F(\xi,Q^2)$$

$$+\frac{4M^2}{Q^2}\,\frac{x^3}{(1+4M^2x^2/Q^2)^{3/2}}\int_\xi^1 d\xi' F(\xi',Q^2)/\xi'^2, \qquad (4.92a)$$

where the scaling variable ξ is given by

$$\xi = \frac{Q^2}{2M^2x}\left[(1+4M^2x^2/Q^2)^{1/2}-1\right] \qquad (4.92b)$$

with M being the proton mass. The non-singlet function F in (4.92a) is given by $F(x,Q^2) = xG(x,Q^2)$ with $G(x,Q^2)$ satisfying eq. (4.71).

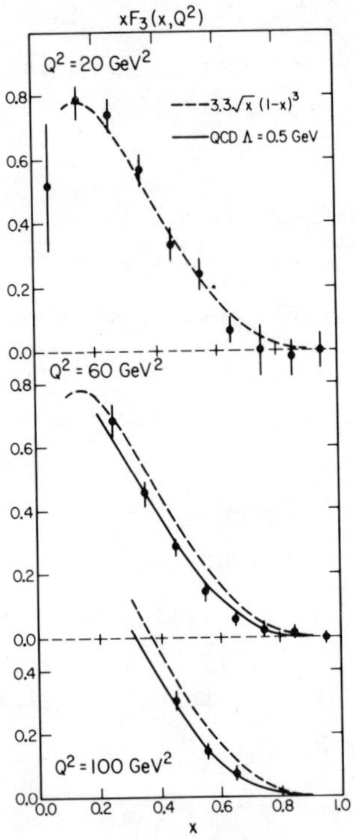

Fig. 4.11. Shows a fit of the form $xF_3(x,Q_o^2) = 3.3\sqrt{x}(1-x)^3$ at $Q_o^2 = 20\ GeV^2$ (dashed curves) together with the predicted values of $xF_3(x,Q^2)$ at $Q^2 = 60$ and $100\ GeV^2$ (solid curves) using eq. (4.71) with $\Lambda = 0.5\ GeV$[28]. The data are a combination of CDHS[26] and BEBC[27] measurements of xF_3 from neutrino and antineutrino nucleon interactions.

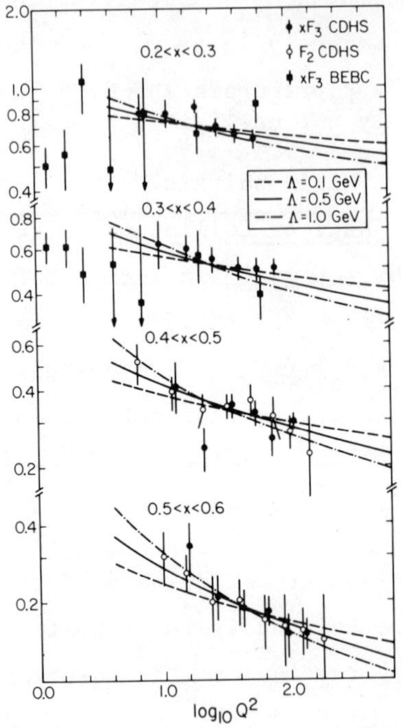

Fig. 4.12. Predictions of the Q^2 evolution of xF_3 from eq. (4.71) with $\Lambda = 0.1$, 0.5 and 1.0 GeV. Target mass corrections have been included by the use of eq. (4.92). The data are from BEBC[27] and CDHS[26].

Figure 4.12 indicates that one would need considerably more accurate data over a wide range of Q^2 before one could determine Λ with any precision from an analysis of scaling violations of $xF_3(x,Q^2)$. This is also evident in Fig. 4.13 where I plot the χ^2 per data point versus Λ for various fits to the data in Fig. 4.12. Values of Λ in the range 0.01 to about 0.8 result in acceptable fits.

<u>Some Order $\alpha_s(Q^2)$ Results for Deep Inelastic Neutrino and Antineutrino Nucleon Scattering</u>. The results for $G^{(1)}_{p\to q}(x,Q^2)$

and $G^{(2)}_{p\to q}(x,Q^2)$ given by (4.88) are the same for neutrino scattering.

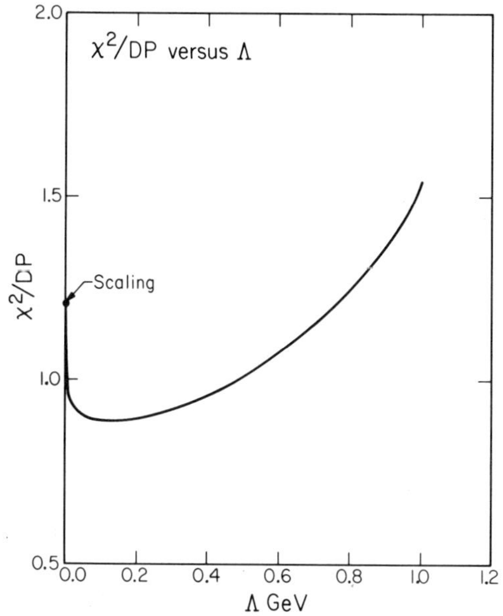

Fig. 4.13. Plot of the χ^2 per data point versus Λ arising from fits to the data in Fig. 4.12 (only $Q^2 > 4$ GeV^2 is used). Any fit with χ^2/DP $\leqslant 1.4$ is acceptable. The point $\Lambda = 0$ corresponds to scaling (ξ-scaling).

Here, however, we have one further structure function $F_3(x,Q^2)$. The quark probabilities for $F_3(x,Q^2)$ are

$$G^{(3)}_{p\to q_i}(x,Q^2)-G^{(3)}_{p\to\bar{q}_i}(x,Q^2) = \int_x^{1.0} \frac{dy}{y} (G_{p\to q_i}(y,Q^2) - G_{p\to\bar{q}_i}(y,Q^2))[\delta(z-1) + \alpha_s(Q^2)\Delta f_{3,q}(z)], \tag{4.93a}$$

where $z = x/y$ and where again the reference distribution is $G^{(2)}_{p\to q}(x,Q^2)$ in (4.82a). The function $\Delta f_{3,q}(z)$ is calculated as in the previous section with the result[6]

$$\alpha_s\Delta f_{3,q}(z) = \alpha_s(f_{3,q}(z)-f_{2,q}(z)) = \frac{\alpha_s}{2\pi}\left(-\frac{4}{3}\right)(1+z). \tag{4.93b}$$

The subprocess $\gamma^* + g \to q + \bar{q}$ does not contribute to (4.93a). It is now an easy matter to calculate neutrino and antineutrino observables to order $\alpha_s(Q^2)$. To leading order (and in the naive parton model), the structure functions

$$F_Q(x,Q^2) = \frac{1}{2}(F_2(x,Q^2)+xF_3(x,Q^2)) \tag{4.94a}$$

$$F_A(x,Q^2) = \frac{1}{2}(F_2(x,Q^2)-xF_3(x,Q^2)) \tag{4.94b}$$

measure the quark and antiquark distributions, respectively, within hadrons. Equation (4.93) shows that to order α_s, both F_Q and F_A receive contributions from quarks and antiquarks. In particular, for νN scattering

$$\frac{F_Q^{\nu N}(x,Q^2)}{x} = \int_x^{1.0} \frac{dy}{y} \{(G_{N\to d}(y,Q^2)+G_{N\to s}(y,Q^2))(2\delta(z-1) + \alpha_s(Q^2)\Delta f_{3,q}(z))-(G_{N\to\bar{u}}(y,Q^2) + G_{N\to\bar{c}}(y,Q^2))\alpha_s(Q^2)\Delta f_{3,q}(z)\}, \tag{4.95a}$$

$$\frac{F_A^{\nu N}(x,Q^2)}{x} = \int_x^{1.0} \frac{dy}{y} \{(G_{N\to\bar{u}}(y,Q^2)+G_{N\to\bar{c}}(y,Q^2))(2\delta(z-1) + \alpha_s(Q^2)\Delta f_{3,q}(z))-(G_{N\to d}(y,Q^2) + G_{N\to s}(y,Q^2))\alpha_s(Q^2)\Delta f_{3,q}(z), \tag{4.95b}$$

and for $\bar{\nu}N$ scattering, we have

$$\frac{F_Q^{\bar{\nu}N}(x,Q^2)}{x} = \int_x^{1.0} \frac{dy}{y} \{(G_{N\to u}(y,Q^2)+G_{N\to c}(y,Q^2))(2\delta(z-1) + \alpha_s(Q^2)\Delta f_{3,q}(z))-(G_{N\to\bar{d}}(y,Q^2) + G_{N\to\bar{s}}(y,Q^2))\alpha_s(Q^2)\Delta f_{3,q}(z)\} \tag{4.96a}$$

$$\frac{F_A^{\bar{\nu}N}(x,Q^2)}{x} = \int_x^{1.0} \frac{dy}{y} \{(G_{N\to\bar{u}}(y,Q^2)+G_{N\to\bar{s}}(y,Q^2))(2\delta(z-1) + \alpha_s(Q^2)\Delta f_{3,q}(z))-(G_{N\to u}(y,Q^2) + G_{N\to c}(y,Q^2))\alpha_s(Q^2)\Delta f_{3,q}(z)\}, \tag{4.96b}$$

where $z = x/y$. For the longitudinal structure function, we have for νN scattering

$$\frac{F_L^{\nu N}(x,Q^2)}{x} = 2\int_x^{1.0} \frac{dy}{y} \{(G_{N\to d}(y,Q^2)+G_{N\to s}(y,Q^2)+G_{N\to\bar{u}}(y,Q^2)$$

$$+ G_{N\to\bar{s}}(y,Q^2))(2\delta(z-1)+\alpha_s(Q^2)\Delta f_{1,q}(z))$$

$$+ 4G_{p\to g}(y,Q^2)\alpha_s(Q^2)\Delta f_{1,g}(z)\} \qquad (4.97)$$

and similarly $\bar{\nu}N$ scattering. If we use

$$\frac{F_2^{\nu N}(x,Q^2)}{x} = 2(G_{N\to d}(x,Q^2)+G_{N\to s}(x,Q^2)+G_{N\to\bar{u}}(x,Q^2)+G_{N\to s}(x,Q^2))$$

$$(4.98)$$

and substitute in for $\Delta f_{1,q}(z)$ and $\Delta f_{1,g}(z)$ from (4.82c,d), we arrive at

$$F_L^{(\nu,\bar{\nu})N}(x,Q^2) = \frac{\alpha_s(Q^2)}{2\pi} x^2 \int_x^1 \frac{dy}{y^3}$$

$$\{\frac{8}{3} F_2^{(\nu,\bar{\nu})N}(y,Q^2)+2a_\nu(1-\frac{x}{y})yG_{N\to g}(y,Q^2)\},$$

$$(4.99a)$$

which is the same as (4.85) except now

$$a_\nu = 8. \qquad (4.99b)$$

Figure 4.14 shows the predictions for $F_L(x,Q^2)$ at $Q^2 = 16\ \mathrm{GeV}^2$ resulting from (4.99).

Integrating F_A, F_Q and F_2 over x yields

$$\bar{F}_Q^{\nu N}(Q^2) = (U+D+2S)(1-0.0885\ \alpha_s(Q^2))$$

$$+ (\bar{U}+\bar{D}+2\bar{C})(0.0885\ \alpha_s(Q^2)) \qquad (4.100a)$$

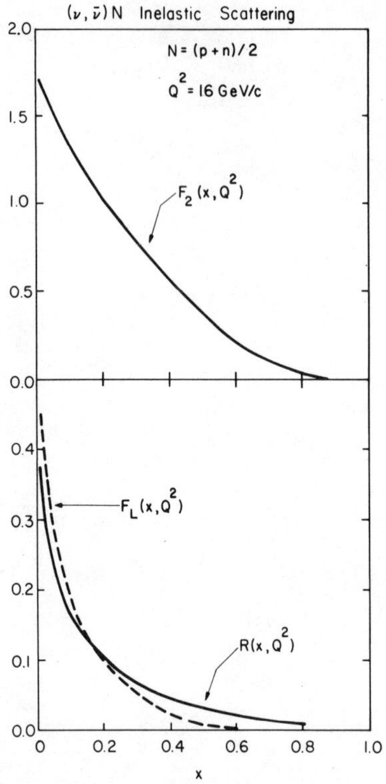

Fig. 4.14. Predictions for $F_2(x,Q^2)$, $F_L(x,Q^2)$ and $R(x,Q^2) = (F_2-2xF_1)/2xF_1$ for neutrino (or antineutrino) nucleon collisions at $Q^2 = 16$ GeV/c and where no effects of order M^2/Q^2 have been included.

$$\bar{F}_Q^{\bar{\nu}N}(Q^2) = (U+D+2C)(1-0.0885\ \alpha_s(Q^2))$$
$$+ (\bar{U}+\bar{D}+2\bar{S})(0.0885\ \alpha_s(Q^2)) \tag{4.100b}$$

$$\bar{F}_A^{\nu N}(Q^2) = (\bar{U}+\bar{D}+2\bar{C})(1-0.0885\ \alpha_s(Q^2))$$
$$+ (U+D+2S)(0.0885\ \alpha_s(Q^2)) \tag{4.100c}$$

$$\bar{F}_A^{\bar{\nu}N}(Q^2) = (\bar{U}+\bar{D}+2\bar{S})(1-0.0885\ \alpha_s(Q^2))$$
$$+ (U+D+2C)(0.0885\ \alpha_s(Q^2)) \tag{4.100d}$$

$$\bar{F}_L^{\nu N}(Q^2) = 0.141\ \alpha_s(Q^2)$$
$$(U+D+2S+\bar{U}+\bar{D}+2\bar{C})$$
$$+ 0.212\ \alpha_s(Q^2)G \tag{4.100e}$$

$$\bar{F}_L^{\bar{\nu}N}(Q^2) = 0.141\ \alpha_s(Q^2)$$
$$(U+D+2C+\bar{U}+\bar{D}+2\bar{S})$$
$$+ 0.212\ \alpha_s(Q^2)G, \tag{4.100f}$$

where $N = \frac{1}{2}(n+p)$ and where I have used (4.88) together with

$$\alpha_s \int_0^1 z\Delta f_{3,q}(z)dz = -\frac{10}{9}\frac{\alpha_s}{2\pi} = -0.177\ \alpha_s \tag{4.101}$$

and where U, D, S, etc., are the total fraction of momentum carried by u, d, s, quarks, respectively,

$$U(Q^2) = \int_0^1 xG_{p\to u}(x,Q^2)dx \tag{4.102}$$

and G is the fraction of momentum carried by gluons as in (4.90). At $Q^2 = 16\ \mathrm{GeV}^2$, our distributions give

$$U = 0.267 \quad D = 0.149 \quad S = \bar{S} = 0.016 \quad \bar{U} = 0.027$$

$$\bar{D} = 0.030 \quad G = 0.477 \quad C = \bar{C} = 0.006 \tag{4.103a}$$

which gives for an isoscalar target ($N = \frac{1}{2}(p-n)$)

$$\bar{F}_2^{\nu N} = \bar{F}_2^{\bar{\nu}N} = 0.527$$

$$\bar{F}_L^{\nu N} = \bar{F}_L^{\bar{\nu}N} = 0.057$$

$$\bar{F}_A^{\nu N} = 0.081$$

$$\bar{F}_A^{\bar{\nu}N} = 0.101$$

$$\bar{F}_Q^{\nu N} = 0.447$$

$$\bar{F}_Q^{\bar{\nu}N} = 0.426, \tag{4.103b}$$

where $\alpha_s(Q^2=16\ \mathrm{GeV}) = 0.327$ (i.e., $\Lambda = 0.4$ GeV/c). Thus at this Q^2

$$\bar{R} = (\bar{F}_2 - 2x\bar{F}_1)/2x\bar{F}_1 = 0.123 \tag{4.103c}$$

for νN and $\bar{\nu}N$ scattering. It is also interesting to note that

$$\bar{F}_A^{\nu N}/(\bar{U}+\bar{D}+2\bar{C}) = 1.16 \tag{4.103d}$$

at this Q^2. In the naive (leading order) parton model $\bar{F}_A^{\nu N} = \bar{U} + \bar{D} + 2\bar{C}$.

V. LARGE-MASS MUON PAIR PRODUCTION

QCD Perturbative Calculations

As a next example of QCD perturbation theory let us calculate the order $\alpha_s(Q^2)$ corrections to the production of large mass muon pairs in proton-proton collisions (the "Drell-Yan" process)[6,31-33]. Starting with the "Compton" term $g + q \to \gamma^* + q$ shown in Fig. 4.1 and taking the gluon off-mass shell $q_g^2 = -m_g^2$, we have

$$\frac{d\hat{\sigma}_{DY}^{c}}{d\hat{t}}(\hat{s},\hat{t}) = \frac{\pi\alpha\alpha_s e_q^2}{\hat{s}^2}\left(\frac{1}{3}\right)\left\{-\frac{\hat{t}}{\hat{s}} - \frac{\hat{s}}{\hat{t}} + \frac{2M^2}{\hat{s}\hat{t}}(\hat{s}+\hat{t}-M^2) - \frac{M^2 m_g^2}{\hat{t}^2}\right\}, \quad (5.1)$$

where $\hat{s} = (p_q+q_g)^2$, $\hat{t} = (q_\gamma-p_q)^2$, $\hat{u} = (q_\gamma-q_g)^2$, and where $q_\gamma^2 = M^2$ is the mass of the virtual photon and terms that give no contribution to the total rate

$$\hat{\sigma}_{DY}(\hat{s}) = \int_{\hat{t}_{max}}^{\hat{t}_{min}} d\hat{t}\, \frac{d\hat{\sigma}_{DY}}{d\hat{t}}(\hat{s},\hat{t}) \quad (5.2)$$

in the limit $m_g \to 0$ have been dropped. In this case

$$\hat{t}_{min} = -\frac{M^2 m_g^2}{\hat{s}} \quad (5.3a)$$

$$\hat{t}_{max} = M^2 - \hat{s}. \quad (5.3b)$$

Integration yields

$$\hat{\sigma}_{DY}^{c}(\hat{s}) = \frac{\pi\alpha\alpha_s e_q^2}{\hat{s}^2}\left(\frac{1}{3}\right)\left\{\left(\hat{s} + \frac{2M^4}{\hat{s}} - 2M^2\right)\log\left(\frac{\hat{s}(\hat{s}-m^2)}{M^2 m_g^2}\right) + \frac{1}{2}\frac{(\hat{s}-m^2)^2}{\hat{s}} + \frac{2M^2}{\hat{s}}(\hat{s}-m^2) - 1\right\}, \quad (5.4)$$

where I have used

$$\int_{t_{max}}^{t_{min}} d\hat{t} = \hat{s} - M^2, \quad \int_{t_{max}}^{t_{min}} tdt = -\frac{1}{2}(\hat{s}-m^2)^2$$

$$\int_{t_{max}}^{t_{min}} \frac{d\hat{t}}{\hat{t}} = \log\frac{t_{min}}{t_{max}} = -\log\left(\frac{(\hat{s}-M^2)\hat{s}}{M^2 m_g^2}\right)$$

$$m_g^2 \int_{t_{max}}^{t_{min}} \frac{dt}{t^2} = -\frac{\hat{s}}{M^2}. \tag{5.5}$$

To embed this subprocess inside the desired $p + p \to \mu^+\mu^- + X$ reaction, it is convenient to define

$$\hat{\tau} = M^2/\hat{s}, \tag{5.6a}$$

which is related to the observed $\tau = M^2/s$ by

$$\hat{\tau} = \tau/x_a x_b, \tag{5.6b}$$

since

$$\hat{s} = x_a x_b s, \tag{5.6c}$$

where, as illustrated in Fig. 4.2, x_a and x_b are the fraction of momentum of the initial protons carried by the gluon and quark, respectively. In terms of $\hat{\tau}$, (5.4) becomes

$$\hat{\sigma}^c_{DY}(\hat{s}) = \frac{\pi\alpha\alpha_s e_q^2}{\hat{s}}\left(\frac{1}{3}\right)\left\{(\hat{\tau}+(1-\hat{\tau})^2)\log\left(\frac{M^2(1-\hat{\tau})}{m_g^2\hat{\tau}^2}\right) - \frac{1}{2} + \hat{\tau} - \frac{3}{2}\hat{\tau}^2\right\}. \tag{5.7}$$

If we now use the relationship

$$\frac{d\sigma_{DY}}{dM^2}(s, M^2; pp\to\mu^+\mu^-+X) = \frac{\alpha}{3\pi M^2}\int_{\tau}^{1.0} dx_a \int_{\tau/x_a}^{1.0} dx_b$$

$$G_{p\to a}(x_a)G_{p\to b}(x_b)\hat{\sigma}_{DY}(\hat{s}; a+b\to\gamma^*+c), \tag{5.8}$$

where the $(\alpha/3\pi M^2)$ factor comes from integrating over the muon pair angular distribution, we arrive at (for one quark flavor)

$$\frac{d\sigma^{c}_{DY}}{dM^2}(s,M^2) = \left(\frac{4\pi}{9}\right)\frac{\alpha^2 e_q^2}{sM^2}\int_{\tau}^{1.0}\frac{dx_a}{x_a}\int_{\tau/x_b}^{1.0}\frac{dx_b}{x_b}\,(G^{(o)}_{p\to q}(x_a)G^{(o)}_{p\to g}(x_b)$$

$$+ (x_a \leftrightarrow x_b))\left(\frac{\alpha_s}{2\pi}P_{q\leftarrow g}(\hat{\tau})\log\frac{M^2}{m_g^2} + \alpha_s f_g^{DY}(\hat{\tau})\right), \qquad (5.9)$$

with

$$P_{q\leftarrow g}(\hat{\tau}) = \frac{1}{2}(\hat{\tau}+(1-\hat{\tau})^2) \qquad (5.10a)$$

which is identical to (4.18) and

$$\alpha_s f_g^{DY}(\hat{\tau}) = \frac{\alpha_s}{2\pi}\left(\frac{1}{2}\right)\{(\hat{\tau}^2+(1-\hat{\tau})^2)[-2\log\hat{\tau}+\log(1-\hat{\tau})] - \frac{1}{2} + \hat{\tau} - \frac{3}{2}\hat{\tau}^2\}. \qquad (5.10b)$$

The annihilation term, $q + \bar{q} \to \gamma^* + g$, shown in Fig. 4.3 yields the differential cross section

$$\frac{d\hat{\sigma}^{A}_{DY}}{d\hat{t}}(\hat{s},\hat{t}) = \frac{\pi\alpha\alpha_s e_q^2}{\hat{s}^2}\left(\frac{8}{9}\right)\{\frac{\hat{u}}{\hat{t}} + \frac{\hat{t}}{\hat{u}} + \frac{2M^4}{\hat{t}\hat{u}} - \frac{2M^2}{\hat{t}} - \frac{2M^2}{\hat{u}} - \frac{m_g^2M^2}{\hat{u}^2} - \frac{m_g^2M^2}{\hat{t}^2}\}, \qquad (5.11)$$

where the outgoing gluon has been given a fictitious mass $q_g^2 = m_g^2$. In addition, terms in (5.11) that do not contribute to the total integrated rate, $\hat{\sigma}^{A}_{DY}$, in the limit $m_g \to 0$ have been dropped.

The integration of (5.11) over $\hat{t}$ is a bit more difficult than the other cases we have considered. One must be careful to keep the exact form for t_{min} and t_{max}. Namely,

$$\hat{t}_{min,max} = -M^2\{(1-\hat{\tau}-\beta\hat{\tau}) \mp [(1-\hat{\tau})^2+\beta\hat{\tau}(\beta\hat{\tau}-2\hat{\tau}-2)]^{1/2}\}/2\hat{\tau} \qquad (5.12a)$$

with

$$\beta = m_g^2/M^2. \qquad (5.12b)$$

Integrating over $\hat{t}$ gives

$$\hat{\sigma}^A_{DY}(\hat{\tau}) = \frac{\pi\alpha\alpha_s}{\hat{s}}\left(\frac{8}{9}\right)\{2\left(\frac{1+\hat{\tau}^2}{1-\hat{\tau}}\right)\log(\hat{t}_{max}/\hat{t}_{min}) - 4(1-\hat{\tau})\} \tag{5.13}$$

and eq. (5.8) now results in (for one quark flavor)

$$\frac{d\sigma^A}{dM^2}(s,M^2) = \left(\frac{4\pi}{9}\right)\frac{\alpha^2 e_q^2}{sM^2}\int_\tau^{1.0}\frac{dx_a}{x_a}\int_{\tau/x_b}^{1.0}\frac{dx_b}{x_b}$$

$$(G^{(o)}_{p\to q}(x_a)G^{(o)}_{p\to\bar{q}}(x_b)+G^{(o)}_{p\to q}(x_b)G^{o}_{p\to q}(x_a))$$

$$\{\delta(\hat{\tau}-1) + \frac{\alpha_s}{\pi}\left(\frac{4}{3}\right)[\left(\frac{1+\hat{\tau}^2}{1-\hat{\tau}}\right)\log(\hat{t}_{max}/\hat{t}_{min})-4(1-\hat{\tau})]\}, \tag{5.14}$$

where the $\delta(\hat{\tau}-1)$ term comes from the leading $q + \bar{q} \to \gamma^*$ subprocess. We must now add the virtual gluon corrections shown in Fig. 4.4. Namely,

$$\sigma^A_{virtual} = \sigma_o \frac{\alpha_s}{2\pi}\left(\frac{4}{3}\right)\{-\log^2\beta - 3\log\beta - \frac{7}{2} + \frac{\pi^2}{3}\}\delta(\hat{\tau}-1), \tag{5.15}$$

where we use (3.25) rather than (4.32) since $q_\gamma^2 = M^2$ is timelike. Adding (5.15) into (5.4) and using the "+ functions" defined in (3.36) yields

$$\frac{d\sigma^A}{dM^2}(s,M^2) = \left(\frac{4\pi}{9}\right)\frac{\alpha^2 e_q^2}{sM^2}\int_\tau^{1.0}\frac{dx_a}{x_a}\int_{\tau/x_b}^{1.0}\frac{dx_b}{x_b}$$

$$(G^{(o)}_{p\to q}(x_a)G^{(o)}_{p\to\bar{q}}(x_b)+G^{(o)}_{p\to q}(x_b)G^{(o)}_{p\to\bar{q}}(x_a))$$

$$\{\delta(\hat{\tau}-1)+\frac{\alpha_s}{2\pi}P_{q\leftarrow q}(\hat{\tau})$$

$$(\log\frac{Q^2}{m_g^2} + \log\frac{Q^2}{m_g^2})+2\alpha_s f_q^{DY}(\hat{\tau})\}, \tag{5.16}$$

where

$$P_{q\leftarrow q}(\hat{\tau}) = \frac{4}{3}\{\frac{1+\hat{\tau}^2}{(1-\hat{\tau})_+} + \frac{3}{2}\,\delta(\hat{\tau}-1)\} \tag{5.17a}$$

and

$$\alpha_s f_q^{DY}(\hat{\tau}) = \frac{\alpha_s}{2\pi}\,(\frac{4}{3})\{2(1+\hat{\tau}^2)\left(\frac{\log(1-\hat{\tau})}{1-\hat{\tau}}\right)_+ -2(\frac{1+\hat{\tau}^2}{1-\hat{\tau}})\log\hat{\tau}$$

$$- 2(1-\hat{\tau}) - \frac{7}{4}\,\delta(\hat{\tau}-1)\}. \tag{5.17b}$$

The $\delta(\hat{\tau}-1)$ coefficient is deduced by noticing that

$$\alpha_s \int_0^1 f_q^{DY}(\hat{\tau})d\hat{\tau} = \frac{\alpha_s}{2\pi}\,(\frac{4}{3})(\frac{2\pi^2}{3} - \frac{7}{4}). \tag{5.18}$$

To arrive at (5.18) I have used the fact that

$$\int_{2\sqrt{\beta}}^1 \frac{\log(\hat{t}_{min}/\hat{t}_{max})}{1-\hat{\tau}}\,d\hat{\tau} = \frac{\pi^2}{12} - \log^2 2 + \int_{2\sqrt{\beta}}^1 \frac{\log(\beta\hat{\tau}^2/(1-\hat{\tau})^2}{1-\hat{\tau}}\,d\hat{\tau}. \tag{5.19}$$

As has been the case previously, the form of $f_q^{DY}(\hat{\tau})$ in (5.17b) is regularization scheme dependent; however, the integral of f_q^{DY} in (5.18) is unique. Also, $P_{q\leftarrow q}$ is unique and the same as in (3.35a) and (4.34a). Combining the annihilation piece (5.16) and the Compton piece (5.9) yields

$$\frac{d\sigma_{DY}}{dM^2}(s,M^2) = (\frac{4\pi}{9})\frac{\alpha^2 e_q^2}{sM^2}\int_\tau^{1.0}\frac{dx_a}{x_a}\int_{\tau/x_b}^{1.0}\frac{dx_b}{x_b}\{(G_{p\to q}^{(o)}(x_a)G_{p\to\bar{q}}^{(o)}(x_a)$$

$$+(x_a\leftrightarrow x_b))(\delta(\hat{\tau}-1)+\frac{\alpha_s}{2\pi}\,P_{q\leftarrow q}(\hat{\tau})\log(\frac{M^4}{m_g^4})+2\alpha_s f_q^{DY}(\hat{\tau}))$$

$$+ (G_{p\to q}^{(o)}(x_a)G_{p\to g}^{(o)}(x_b)+(x_a\leftrightarrow x_b))$$

$$(\frac{\alpha_s}{2\pi}\,P_{q\leftarrow g}(\hat{\tau})\log(\frac{M^2}{m_g^2})+\alpha_s f_g^{DY}(\hat{\tau}))\}. \tag{5.20}$$

If we now <u>define</u> "Drell-Yan" quark distributions by

$$\frac{d\sigma_{DY}}{dM^2}(s,M^2) = \frac{4\pi}{9}\frac{\alpha^2}{sM^2}\int_\tau^{1.0}\frac{dx_a}{x_a}\sum_{i=1}^{n_f} e_q^2[G^{DY}_{p\to q_i}(x_a,M^2)G^{DY}_{p\to\bar{q}_i}(x_b,M^2)$$

$$+ G^{DY}_{p\to q_i}(x_b,M^2)G^{DY}_{p\to\bar{q}_i}(x_a,M^2)], \qquad (5.21)$$

which is the usual formula where $x_a x_b = \tau$, then

$$G^{DY}_{p\to q_i}(x,M^2) = \int_x^{1.0}\frac{dy}{y}\{G^{(o)}_{p\to q_i}(y)[\delta(z-1)+\frac{\alpha_s}{2\pi}P_{q\leftarrow q}(z)\log(\frac{M^2}{m_q^2})$$

$$+ \alpha_s f_q^{DY}(z)]+G^{(o)}_{p\to g}(y)[\frac{\alpha_s}{2\pi}P_{q\leftarrow g}(z)\log(\frac{M^2}{m_g^2})+\alpha_s f_g^{DY}(z))\}$$

$$(5.22a)$$

and

$$G^{DY}_{p\to\bar{q}_i}(x,M^2) = \int_x^{1.0}\frac{dy}{y}\{G^{(o)}_{p\to\bar{q}_i}(y)[\delta(z-1)+\frac{\alpha_s}{2\pi}P_{q\leftarrow q}(z)\log(\frac{M^2}{m_g^2})$$

$$+ \alpha_s f_q^{DY}(z)]+G^{(o)}_{p\to g}(y)[\frac{\alpha_s}{2\pi}P_{q\leftarrow g}(z)\log(\frac{M^2}{m_g^2})+\alpha_s f_g^{DY}(z)]\},$$

$$(5.22b)$$

where $z = x/y$. One can easily verify this by substituting into (5.21) and keeping terms only to order α_s. We now play the same game of absorbing the $\log(m^2)$ divergences into the distributions $G^{(o)}(y)$ and use (4.44) to sum all the $\alpha_s(M^2)\log M^2$ terms. We arrive at

$$G^{DY}_{p\to q}(x,M^2) = \int_x^{1.0}\frac{dy}{y}\{G_{p\to q}(y,M^2)[\delta(z-1)+\alpha_s(M^2)\Delta f_q^{DY}(z)]$$

$$+ G_{p\to g}(y,M^2)\alpha_s(M^2)\Delta f_g^{DY}(z)\} \qquad (5.23a)$$

and similarly for $G^{DY}_{p\to\bar{q}}(x,M^2)$, and where

$$\alpha_s \Delta f_q^{DY}(z) = \alpha_s (f_q^{DY}(z) - f_{2,q}(z)) = \frac{\alpha_s}{2\pi}\left(\frac{2}{3}\right)[2(1+z^2)\left(\frac{\log(1-z)}{1-z}\right)_+$$

$$+ \frac{3}{(1-z)_+} - 6 - 4z + \left(\frac{4\pi^2}{3} + 1\right)\delta(z-1)], \qquad (5.23b)$$

$$\alpha_s \Delta f_g^{DY}(z) = \alpha_s (f_g^{DY}(z) - f_{2,g}(z)) = \frac{\alpha_s}{2\pi}\left(\frac{1}{2}\right)$$

$$[(z^2+(1-z)^2)\log(1-z) + \frac{1}{2}(9z^2 - 10z + 3)]. \qquad (5.23c)$$

In (5.23a), $G_{p\to q}(x,M^2)$ and $G_{p\to g}(x,M^2)$ are the quark distributions defined by F_2 in ep scattering in (4.81) with Q^2 replaced by M^2. The evolution with M^2 is governed by the "renormalization" eqs. (4.44). Substituting (5.23a) back into (5.21) gives

$$\frac{d\sigma_{DY}}{dM^2}(s,M^2) = \frac{4\pi}{9}\frac{\alpha^2}{sM^2}\int_\tau^{1.0}\frac{dx_a}{x_a}\sum_{i=1}^{n_f} e_{q_i}^2 [G_{p\to q_i}(x_a,M^2) G_{p\to \bar{q}_i}(x_b,M^2) +$$

$$(x_a \leftrightarrow x_b)] + \frac{4\pi\alpha^2}{9sM^2}\int_\tau^{1.0}\frac{dx_a}{x_a}\int_{\tau/x_a}^{1.0}\frac{dx_b}{x_b}\sum_{i=1}^{n_f} e_q^2 [G_{p\to q_i}(x_a,M^2)$$

$$G_{p\to \bar{q}_i}(x_b,M^2) + (x_a \leftrightarrow x_b)] 2\alpha_s \Delta f_q^{DY}(\hat{\tau}) + \frac{4\pi\alpha^2}{9sM^2}\int_\tau^{1.0}\frac{dx_a}{x_a}$$

$$\int_{\tau/x_a}^{1.0}\frac{dx_b}{x_b}\sum_{i=1}^{n_f} e_q^2 [G_{p\to q_i}(x_a,M^2) G_{p\to g}(x_b,M^2) + (x_a \leftrightarrow x_b)]$$

$$\alpha_s \Delta f_g^{DY}(\hat{\tau}) + \frac{4\pi\alpha^2}{9sM^2}\int_\tau^{1.0}\frac{dx_a}{x_a}\int_{\tau/x_a}^{1.0}\frac{dx_b}{x_b}\sum_{i=1}^{n_f} e_q^2$$

$$[G_{p\to \bar{q}_i}(x_a,M^2) G_{p\to g}(x_b,M^2) + (x_a \leftrightarrow x_b)] \alpha_s \Delta f_g^{DY}(\hat{\tau}).$$

$$(5.24)$$

The first term is Politzer's[34] leading order result that the Drell-Yan rate is given by the usual parton model formula but with the "renormalization" group improved distributions $G(x,M^2)$ with Q^2 replaced by M^2 [35]. The second, third and fourth terms are the order α_s corrections to this result. Let us now examine some of the phenomenology of the production of large mass muon pairs in pp collisions.

Leading Order Analysis of Muon Pair Production

QCD Factorization and the Total Muon Pair Rate. We begin our phenomenological investigation of the production of large mass muon pairs in hadron-hadron collisions by examining first the leading order QCD predictions. Then we will look at some of the order α_s corrections. As we saw in the preceding section, to leading order, the quark and gluon distributions, $G_i(x,Q^2)$, determined in Section IV from an analysis of electroproduction are processes independent. Parton distributions determined from one process (like $ep \to e + X$) can, to leading order, be used to make predictions elsewhere (like for $pp \to \mu^+\mu^- + X$ or $pp \to \pi^0 + X$).

The total muon pair cross section (integrated over all muon pair $p_\perp$) is given to leading order (i.e., order $\alpha_s(Q^2)\log Q^2$) by[34,35]

$$\sigma_{tot}(s,M^2,y) \equiv \frac{d\sigma}{dM^2 dy}(s,M^2,y)$$

$$= \frac{4\pi\alpha^2}{3sM^2}\left(\frac{1}{3}\right)\sum_i e_i^2\{G_{p\to q_i}(x_a,Q^2)G_{p\to\bar{q}_i}(x_b,Q^2)+(q_i\leftrightarrow\bar{q}_i)\}, \tag{5.25a}$$

where y is the rapidity of the muon pair of mass M and

$$x_a = \sqrt{\tau}\,\exp(y), \qquad x_b = \sqrt{\tau}\,\exp(-y), \tag{5.25b}$$

and where the parton distributions $G(x,Q^2 = M^2)$ are the "renormalization group improved" functions given in Section IV. One does not see the "Compton" term, $g + q \to q + \gamma^*$, explicitly in (5.25) since to leading order it is included in the probability of finding an antiquark in the proton. As illustrated in Fig. 5.1, the divergent pieces (behaving like $\alpha_s(Q^2)\log Q^2$) of gluon Bremsstrahlung from the incoming antiquark and incoming gluon quark-antiquark pair production

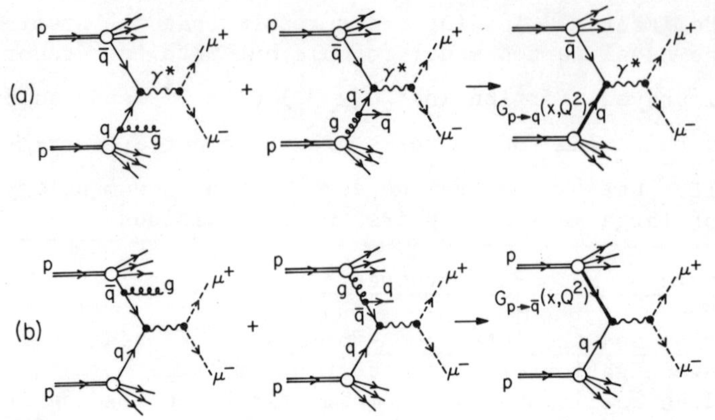

Fig. 5.1 - (a) Illustration of two diagrams for $pp \to \mu^+\mu^- + X$ which are included (divergent parts) in the "renormalization group" improved quark distributions, $G_{p\to q}(x,Q^2)$, when one calculates the contribution from the subprocess $q\bar{q} \to \gamma^* \to \mu^+\mu^-$.

(b) Illustration of two diagrams which are included (divergent parts) in the "renormalization group" improved antiquark distributions, $G_{p\to\bar{q}}(x,Q^2)$.

are absorbed into and generate the Q^2 dependence of $G_{p\to\bar{q}}(x,Q^2)$[34-36]. Figure 5.2 shows the data on $d\sigma/dMdy$ for $pN \to \mu^+\mu^- + X$ at $W = 27.4$ GeV and $y = 0$ together with the predictions from eq. (5.25) using the parton distributions $G_i(x,Q^2)$ determined in Section IV. The agreement of the leading order prediction with the magnitude of the muon-pair cross section is reasonable, although it does depend sensitively on the antiquark distributions which are not determined too well.

<u>Muon Pairs Produced at Large Mass and Large $p_\perp$.</u> Effects due to the transverse momentum, $k_\perp$, of quarks and gluons within hadrons can sometimes be very important. In QCD, transverse momentum of partons can arise in two ways. Firstly, in, for example, a proton beam, quarks are confined in the transverse direction to within the proton radius. Therefore, from the uncertainty principle, they must have some transverse momentum. This momentum, called primordial, is intrinsic to the basic parton wave function inside the proton. It involves small Q^2 values and thus cannot be calculated using perturbation theory. At present, it must be viewed as unknown but bounded (falling off like an exponential or Gaussian in $k_\perp$). Secondly, in QCD, one expects to receive an "effective" $k_\perp$ of quarks in protons due to the hard Bremsstrahlung of gluons which can be calculated perturbatively if the momentum transfers are large.

A particularly nice place to study the interplay between these two components of transverse momentum is in the production large mass muon pairs in pp collisions, $pp \to \mu^+\mu^- + X$. Many people have

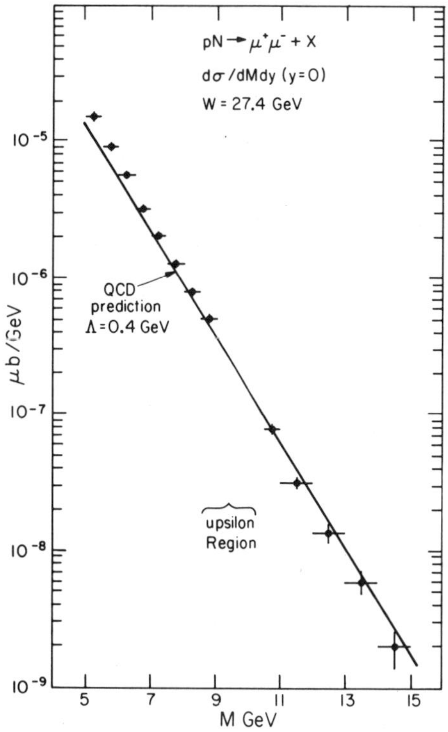

Fig. 5.2. Leading order QCD prediction from eq. (5.25) of the dimuon mass spectrum at $y = 0$ and $W = \sqrt{s} = 27.4$ GeV using the "renormalization group" improved quark and antiquark distributions $G_i(x,Q^2)$ from eqs. (4.75a) and (4.75b). The data are from Ref. 44.

analyzed this process in terms of QCD[36-41]. The analysis presented here follows closely the work of Altarelli, Parisi and Petronzio[36].

The perturbative component of the transverse momentum of muon pairs is generated by the two-to-two constituent subprocess $q\bar{q} \to \gamma^* +$ gluon ("annihilation") and gluon + $q \to \gamma^* + q$ ("Compton"), where the virtual photon, γ^*, then decays into a $\mu^+\mu^-$ pair. Other graphs are higher order in α_s and have been neglected[42]. The cross sections for these processes are given by

$$\frac{d\sigma_A}{dM^2 d\hat{t}} = \left(\frac{8}{27}\right)\left(\frac{\alpha^2 \alpha_s e_q^2}{M^2 \hat{s}^2}\right)\left(\frac{2M^2\hat{s} + \hat{u}^2 + \hat{t}^2}{\hat{t}\hat{u}}\right) \tag{5.26a}$$

$$\frac{d\sigma_C}{dM^2 d\hat{t}} = \left(\frac{1}{9}\right)\left(\frac{\alpha^2 \alpha_s e_q^2}{M^2 \hat{s}^2}\right)\left(\frac{2M^2\hat{u} + \hat{s}^2 + \hat{t}^2}{-\hat{t}\hat{s}}\right), \tag{5.26b}$$

where $\hat{s}$, $\hat{t}$ and $\hat{u}$ are the usual Mandelstam invariants and M^2 is the invariant mass squared of the muon pair and where A and C refer to annihilation and Compton, respectively. The cross section for producing muon pairs of mass M^2, rapidity y, and transverse momentum $p_\perp$ at a center of mass energy squared s is then given by

$$\sigma_P(s,M^2,y,p_\perp) \equiv \frac{d\sigma_P}{dM^2 dy d^2p_\perp}(s,M^2,y,p_\perp) = \sum_{q=u,d,s} \int_{x_a^{min}}^{1.0} dx_a$$

$$\{G_{p\to q}(x_a,M^2) G_{p\to\bar{q}}(x_b,M^2) + (\bar{q}\leftrightarrow q)\} \frac{x_a x_b}{(x_a - x_1)}$$

$$\left(\frac{1}{\pi}\frac{d\sigma_A}{dM^2 d\hat{t}}\right) + \sum_{\substack{q=u,d,s,\\ \bar{u},\bar{d},\bar{s}}} \int_{x_a^{min}}^{1.0} dx_a \{G_{p\to q}(x_a,M^2)$$

$$G_{p\to g}(x_b,M^2) + (g\leftrightarrow q)\} \frac{x_a x_b}{(x_a - x_1)} \left(\frac{1}{\pi}\frac{d\sigma_C}{dM^2 d\hat{t}}\right), \tag{5.27}$$

where $x_a^{min} = (x_1-\tau)/(1-x_2)$ and $x_b = (x_a x_2 - \tau)/(x_a - x_1)$. The quantities x_1 and x_2 are given by

$$x_1 = \frac{1}{2}(x_\perp^2 + 4\tau)^{1/2} e^{y} \tag{5.28a}$$

$$x_2 = \frac{1}{2}(x_\perp^2 + 4\tau)^{1/2} e^{-y}, \tag{5.28b}$$

where $x_\perp = 2p_\perp/\sqrt{s}$ and $\tau = M^2/s$[43]. The subprocess invariants are given by

$$\hat{s} = x_a x_b s \tag{5.29a}$$

$$\hat{t} = -x_a s x_2 + M^2 \tag{5.29b}$$

$$\hat{u} = -x_b s x_1 + M^2. \tag{5.29c}$$

In (5.27), the label P refers to "perturbative contribution" and where the "renormalization improved" parton distributions $G(x,Q^2 = M^2)$ from (4.44) and the running coupling constant $\alpha_s(M^2)$ from (2.12) are used. It is clear from the work of Politzer[34] and Sachrajda[35] that here one should use the "renormalization improved" $G(x,Q^2)$ and $\alpha_s(Q^2)$ functions. However, here there are two large invariants M^2

and $p_\perp^2$ and it is not clear whether the Q^2 dependence evolves according to $G(x,Q^2=M^2)$ or $G(x,Q^2=p_\perp^2)$ or some other combination. It, of course, does not matter in leading order and at very large values of M^2 and $p_\perp^2$ but it certainly matters for the phenomenology at existing M^2 and $p_\perp^2$ values. This question can be answered only by calculating higher order corrections to the large $p_\perp$ Drell-Yan rate. Various choices for Q^2 will produce different higher order corrections. Presumably one would pick that choice of Q^2 that minimized higher order terms.

The perturbative contributions from (5.27) are shown in Fig. 5.3. They are absolutely normalized and agree roughly with the data at large $p_\perp$. They, however, have the wrong shape at small $p_\perp$ and diverge at $p_\perp = 0$. This infrared difficulty besets other QCD perturbative calculations. We will see in a later section that the $p_\perp = 0$ divergence will disappear when the higher orders of perturbation theory are summed. Also one expects that non-perturbative phenomena at small $p_\perp$ will regularize this singularity leaving a smooth $p_\perp$ distribution.

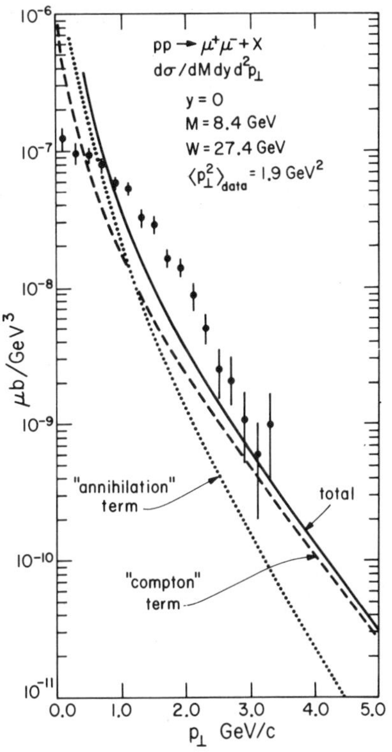

Fig. 5.3. The distribution in transverse momentum, $p_\perp$, of the $\mu^+\mu^-$ pair produced in pp collisions at W = 27.4 GeV together with the QCD perturbative predictions. The "Compton" and "annihilation" terms are given by the dashed and dotted curves, respectively.

The soft, non-perturbative, component of $k_\perp$ (the primordial $k_\perp$) can be used to regularize $\sigma_P(s,M^2,y,p_\perp)$ and produce a finite distribution in $p_\perp$ at all $p_\perp$. This can be done crudely by convoluting $\sigma_P(p_\perp)$ with a quark primordial motion given, for instance, by

$$f(k_\perp^2) = \exp[-k_\perp^2/(4\sigma_q^2)]/(4\pi\sigma_q^2), \tag{5.30}$$

where for a single constituent in a proton, one has

$$\langle k_\perp^2 \rangle_{\text{primordial}} = 2\sigma_q^2. \tag{5.31}$$

The result is

$$\sigma(s,M^2,y,p_\perp^2) = \int d^2k_\perp \sigma_p(M^2,s,y,(\vec{p}_\perp-\vec{k}_\perp)^2)f(k_\perp^2)$$

$$= \int d^2k_\perp \sigma_p(M^2,s,y,(\vec{p}_\perp-\vec{k}_\perp)^2)[f(k_\perp^2) - f(p_\perp^2)]$$

$$+ f(p_\perp^2) \int d^2k_\perp \sigma_p(M^2,s,y,(\vec{p}_\perp-\vec{k}_\perp)^2), \qquad (5.32)$$

where I have added and subtracted the second term in (5.23). Actually, in doing this convolution, I should have added to σ_p a contribution of order α_s arising from the vertex correction to the subprocess $q\bar{q} \to \gamma^*$. This contribution also diverges but only contributes at $p_\perp = 0$ (i.e., has a $\delta(p_\perp)$). It does not contribute to the first term in (5.32) since $[f(k_\perp^2) - f(p_\perp^2)]\delta(k_\perp^2)$ is zero, but when added to the second term yields

$$\sigma(s,M^2,y,p_\perp^2) = \int d^2q_\perp \sigma_p(M^2,s,y,q_\perp^2)[f((\vec{p}_\perp-\vec{q}_\perp)^2) - f(p_\perp^2)]$$

$$+ f(p_\perp^2)\sigma_{tot}(s,M^2,y), \qquad (5.33)$$

where I have defined $\vec{q}_\perp = \vec{p}_\perp - \vec{k}_\perp$ (one must keep track of the vector direction) and where $\sigma_{tot}(s,M^2,y)$ is given by definition by (5.25).

Both terms in (5.33) are now finite at all $p_\perp$ and one is left with one additional parameter, the primordial $\langle k_\perp \rangle_{p\to q}$ in (5.31), to be adjusted to fit the data. The fit to the data is shown in Fig. 5.4 and yields $\sigma_q = 0.48$ GeV or

$$\langle k_\perp \rangle_{primordial} = \sqrt{\pi/2}\, \sigma_q \approx 600 \text{ MeV}. \qquad (5.34)$$

This means that the mean $p_\perp^2$ of the $\mu^+\mu^-$ data, which at this energy and $y = 0$ is about 1.9 GeV^2, results from about 0.9 GeV^2 due to primordial motion and about 1.0 GeV^2 due to the hard QCD subprocesses. Figure 5.4 shows also the second term in (5.33) which would be the prediction if only primordial motion (with $\sigma_q = 0.48$ GeV) were present.

Clearly, this smearing procedure which is used to regulate the divergences is a bit ad hoc and the fit to the shape of the $p_\perp$ spectrum shown in Fig. 5.4 cannot be viewed as a success of QCD. One could have fit the same data with a Gaussian with $\sigma_q = 0.677$ GeV.

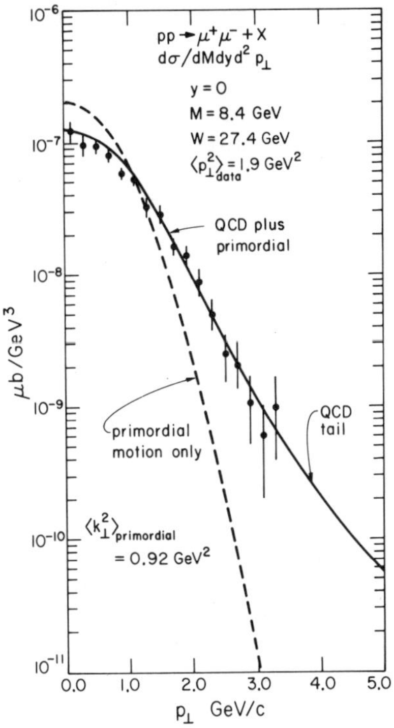

Fig. 5.4. The same data as Fig. 5.3 together with the perturbative QCD contributions folded with a Gaussian momentum spectrum with $\langle k_\perp \rangle_{h\to q} = \langle k_\perp \rangle_{h\to \bar{q}} = 600$ MeV (solid curve). The dashed curve results from the primordial motion only with no perturbative QCD terms.

The real test of the presence of the QCD component to the effective $k_\perp$ of partons comes from examining the energy dependence of the muon pair $p_\perp$ spectrum. Predictions for this are shown in Fig. 5.5. One expects to see a flatter $p_\perp$ spectra as the energy increases (and M^2 is fixed). Recent data on $pN \to \mu^+\mu^- + X$ at 200, 300 and 400 GeV[44] do show a mean $p_\perp$ that increases with increasing energy in agreement with the QCD expectations.

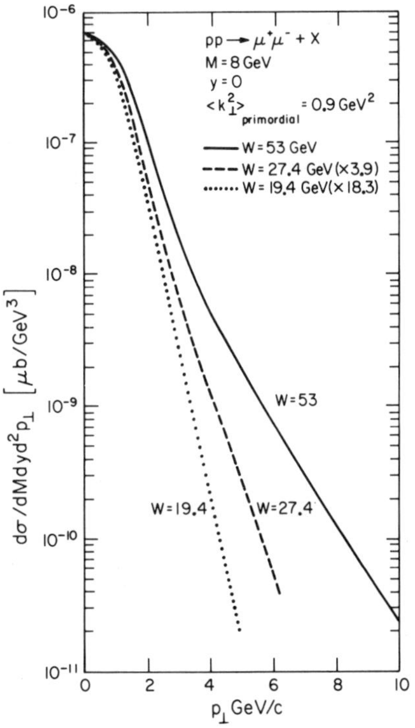

Fig. 5.5. Energy dependence of the large $p_\perp$ tail expected for $pp \to \mu^+\mu^- + X$ from the QCD perturbative contributions folded with a primordial transverse momentum with $\langle k_\perp \rangle_{h\to q} = 600$ MeV for each parton.

<u>"Scale" Breaking in $pp \to \mu^+\mu^- + X$.</u> It is interesting to look at the "scale-breaking" expected in QCD for observables in $pp \to \mu^+\mu^- + X$. In the naive parton model, one expected

$$M^4 \frac{d\sigma}{dM^2 dy}(x,M^2,y) = f(\tau,y), \tag{5.35}$$

to be only a function of $\tau = M^2/s$ and y and not to depend separately on $W = \sqrt{s}$. As shown in Fig. 5.6, one now expects small scale breaking effects. At small τ one sees a slight rise with increasing W and at large τ a decrease with increasing W. The effects are small. They are comparable to the breaking of $\nu W_2(x,Q^2)$ in Fig. 4.6 and Fig. 4.9 and will probably not be seen experimentally for quite some time.

Other muon pair observables show larger scale breaking effects. For example, for the annihilation and Compton subprocesses dimensional counting yields

$$M^4 p_\perp^2 \frac{d\sigma}{dM^2 dy d^2 p_\perp} = F(\tau, y, x_\perp), \tag{5.36a}$$

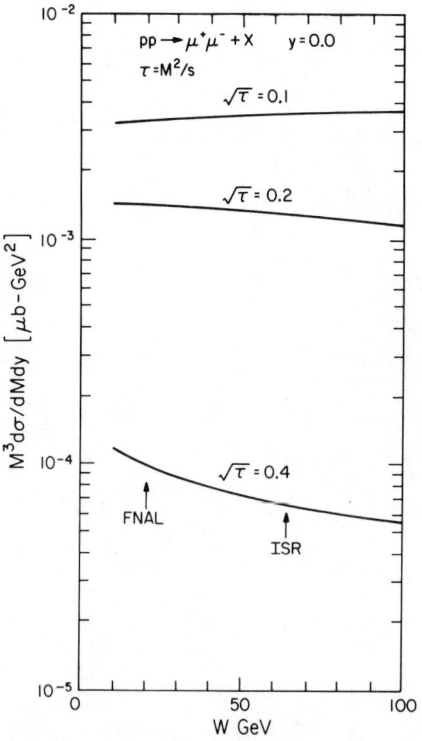

Fig. 5.6. Expected "scale breaking" of the quantity $M^3 d\sigma/dMdy$ for $pp \to \mu^+\mu^- + X$ at $y = 0$ from the QCD non-scaling structure functions with $\Lambda = 0.4$ GeV/c. Scaling would predict this quantity to be independent of W at fixed τ.

or where $x_\perp = 2p_\perp/W$ or

$$W^5 \frac{d\sigma}{dMdyd^2 p_\perp} = \tilde{F}(\tau, y, x_\perp). \tag{5.36b}$$

For asymptotically free theories, (5.36) does not scale. Figure 5.7 shows that for fixed $x_\perp$ and τ, $W^5 d\sigma/dMdyd^2p_\perp$ decreases as W increases and approaches a constant ("scaling" result) asymptotically[36,39]. As can be seen in this figure, the primordial motion produces an additional "scale breaking" term. (If there were only primordial motion, then $W^5 d\sigma/dMdyd^2p_\perp$ would go to zero at increasing W (fixed τ and $x_\perp$) like $1/W^2$.) If one could observe experimentally this decrease and approach to a constant for $\tilde{F}$ in (5.36b), it would certainly be support for QCD. However, such measurements are a long way off. They require, for example, at $x_\perp = 0.2$ and $\sqrt{\tau} = 0.2$, comparing a point at FNAL $W = 19.4$, $M = 3.9$ GeV and $p_\perp = 1.94$ GeV/c with a point at ISR $W = 53$ GeV, $M = 10.6$ GeV, and $p_\perp = 5.3$ GeV/c!

Order α_s Corrections to the Total Muon Pair Rate[6,31-33]

We have seen that the leading order QCD predictions agree

qualitatively with the data on large-mass muon pair production. However, perturbative predictions to a given order are only meaningful if the next order (and all higher order) corrections are small. Since we are dealing with an expansion in $\alpha_s(Q^2)$ (or $1/\log Q^2$) and since $\log Q^2/\Lambda^2$ is not very large, we might worry that higher order corrections may be quite important. We shall now examine the order α_s corrections to the total muon pair cross section, $d\sigma_{DY}/dM^2$. The results are somewhat disturbing. The order α_s corrections are, in certain kinematic regions, equal to or greater than the leading order term. This probably means that one is not justified in dropping order α_s^2 terms (etc.) and that, at present, the normalization of the muon pair cross section cannot be predicted from the parton distributions determined from the electroproduction data.

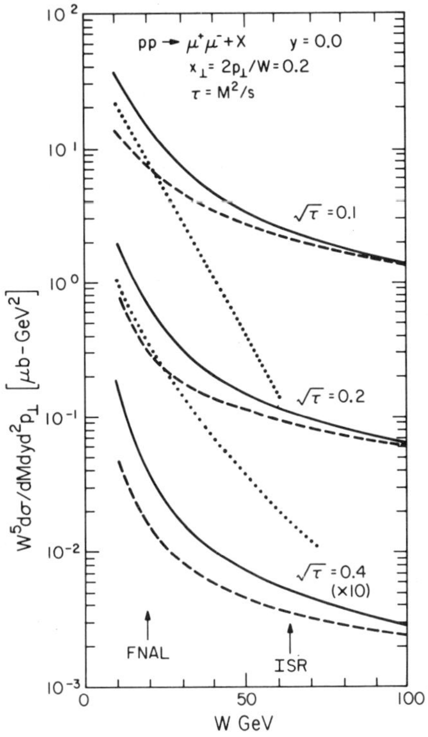

Fig. 5.7. Expected "scale breaking" of the quantity $W^5 d\sigma/dMdyd^2p_\perp$ for $pp \to \mu^+\mu^- + X$ at $y = 0$ and $x_\perp = 0.2$ with $\Lambda = 0.4$ GeV/c. The solid (dashed) curves are the results after (before) smearing with a primordial parton transverse momentum with $\langle k_\perp\rangle_{h\to q} = 600$ MeV for each parton. The dotted curves would be expected if primordial motion alone were responsible for the muon pair $p_\perp$. Scaling would predict this quantity to be independent of W at fixed τ and $x_\perp$.

To leading order the total muon pair rate is

$$\frac{d\sigma_{DY}}{dM^2}(s,M^2) = \frac{4\pi}{9}\frac{\alpha^2}{sM^2}\int_\tau^1 \frac{dx_a}{x_a}\sum_{i=1}^{n_f} e_{q_i}^2 \, [G_{p\to q_i}(x_a,M^2)\, G_{p\to\bar{q}_i}(x_b,M^2) + (x_a \leftrightarrow x_b)], \quad (5.37)$$

where $x_b = \sqrt{\tau}/x_a$ and where $G(x,M^2)$ are the "renormalized improved" parton distributions $G(x,Q^2=M^2)$ from (4.44). To order α_s, $d\sigma_{DY}/dM^2$ receives corrections from the annihilation and virtual diagrams in Fig. 4.3 and Fig. 4.4 (i.e., $\Delta f_q^{DY}(z)$) and from the Compton diagram in Fig. 4.1 (i.e., $\Delta f_g^{DY}(z)$) and is given by (5.24). Figure 5.8 compares the leading order predictions with the full order α_s results for $d\sigma_{DY}/dM$. For $0.1 \leqslant \sqrt{\tau} \leqslant 0.6$, the order α_s results are roughly a factor of 2 larger than the leading order predictions. This can be seen more clearly in Fig. 5.9 where I have plotted the ratio of the full order α_s to the leading order predictions. As can be seen, the factor of 2 comes from the Δf_g^{DY} term in (5.24). The gluon corrections Δf_g^{DY} actually reduce the leading order predictions slightly. For $0.1 \leqslant \sqrt{\tau} \leqslant 0.6$, the large correction from the Δf_q^{DY} term comes primarily from the δ-function piece in (5.23b). Namely,

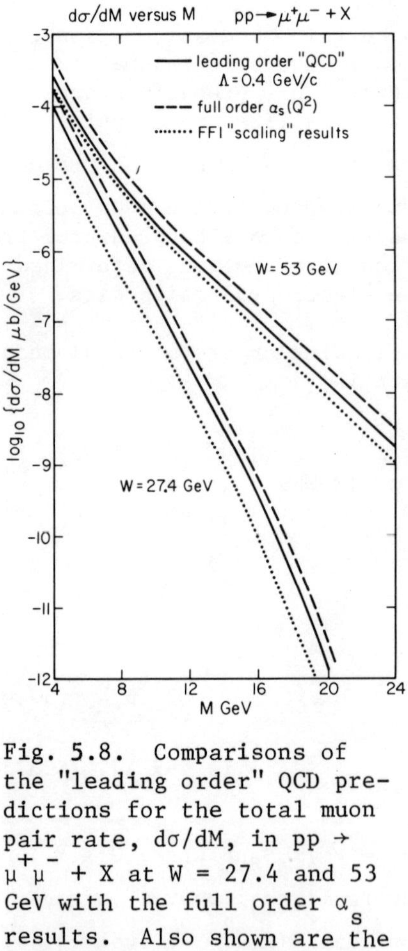

Fig. 5.8. Comparisons of the "leading order" QCD predictions for the total muon pair rate, dσ/dM, in pp → $\mu^+\mu^-$ + X at W = 27.4 and 53 GeV with the full order α_s results. Also shown are the "naive" parton model predictions using the scaling parton distributions of FF1[30].

$$[\alpha_s \Delta f_q^{DY}(z)]_{\delta\text{-function}} = \frac{\alpha_s}{2\pi}\left(\frac{2}{3}\right)\left(\frac{4\pi^2}{3} + 1\right)\delta(z-1) \approx 1.5\ \alpha_s . \tag{5.38}$$

As shown in Fig. 5.10, this term alone results in

$$\frac{G^{DY}(x,Q^2)}{G(x,Q^2)} \approx 1 + 1.5\ \alpha_s(Q^2), \tag{5.39a}$$

or

$$\frac{d\sigma_{DY}/dM(\text{order } \alpha_s)}{d\sigma_{DY}/dM(\text{leading})} \approx 1 + 3\ \alpha_s(M^2) \tag{5.39b}$$

and since $\alpha_s(Q^2{=}M^2) \approx 1/3$, this is quite a large effect. It probably means that unless one can sum these contributions to all orders in α_s then the total muon pair cross section,

$d\sigma_{DY}/dM^2$, cannot be determined from the electroproduction distributions until $\alpha_s(Q^2=M^2) \approx 1/10$ or $M^2 \approx 10^5$ GeV^2!

In addition, the $(\log(1-z)/(1-z))_+$ term in (5.23b) causes the order α_s results to deviate more and more from the leading order as $\sqrt{\tau} \to 1$ (see Fig. 5.9). This term results in (see (3.37b))

$$\frac{G^{DY}(x,Q^2)}{G(x,Q^2)} \underset{x\to 1}{\approx} 1 + \frac{\alpha_s(Q^2)}{2\pi} \left(\frac{4}{3}\right) \log^2(1-x) \tag{5.40}$$

and hence the order $\alpha_s(Q^2)$ corrections become arbitrarily large as $x \to 1$ as seen in Fig. 5.10. No matter how small $\alpha_s(Q^2)$ is, the order α_s term can be made to dominate over the leading term by taking x close to one. This is not quite as disturbing as at first it appears. It simply reflects the fact that $G^{DY}(x,Q^2)$ has a different shape as $x \to 1$ than does $G(x,Q^2)$. The former does not approach zero as fast as the latter. If one views the order α_s corrections as a slight change in the shape as $x \to 1$, then maybe the perturbation result is meaningful in spite of the large ratio in (5.40).

In fact, one can view the order α_s corrections in a different, and

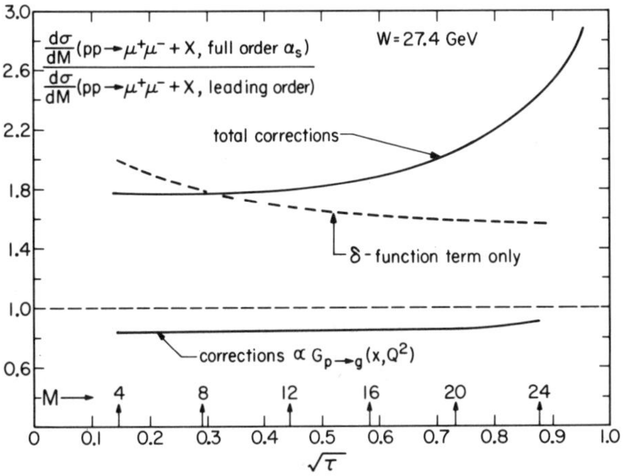

Fig. 5.9. Ratio of the complete order α_s results to the leading order QCD predictions for the total muon pair rate, $d\sigma/dM$, in $pp \to \mu^+\mu^- + X$ at W = 27.4 GeV. The contributions from the δ-function term in Δf_q^{DY} and the contributions from the gluon term Δf_g^{DY} are shown separately. The results are plotted versus both the muon pair mass, M, and $\sqrt{\tau}$, where $\tau = M^2/s$.

somewhat less disturbing, manner. Instead of plotting the ratio of the corrected to the uncorrected predictions as I have done in Fig. 5.9 and Fig. 5.10, one might ask how much one must change M in Fig. 5.8 to get from the leading order to the order α_s results. Since $d\sigma_{DY}/dM$ is a steeply falling function, the change in M is not great, only about 7% at M = 12 GeV. Thus we can predict the rate $d\sigma_{DY}/dM$ but with about a 7 - 10% uncertainty as to the precise value of M at which the rate is actually attained. Viewed in this way, the results of this section don't seem as disturbing. However, until we really understand the physics behind these large order α_s corrections and perhaps learn to sum them to all orders or find that the order α_s^2 corrections are small, I feel we cannot predict $d\sigma_{DY}/dM$ from the parton distributions determined in electroproduction!

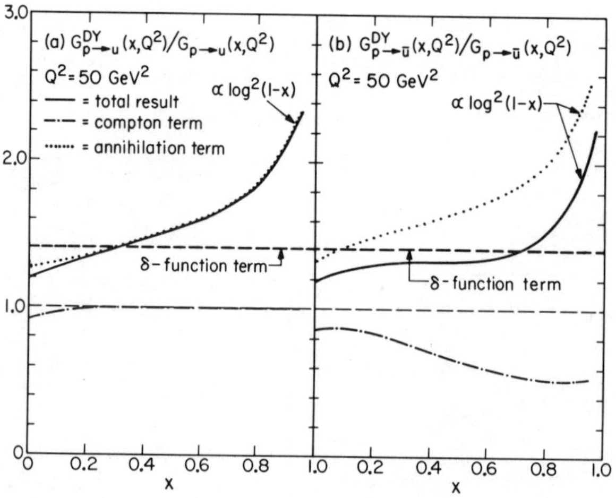

Fig. 5.10. Ratio of $G^{DY}_{p\to u}(x,Q^2)/G_{p\to u}(x,Q^2)$ and $G^{DY}_{p\to\bar{u}}(x,Q^2)/G_{p\to\bar{u}}(x,Q^2)$ versus x at $Q^2 = 50\ \text{GeV}^2$. The functions $G^{DY}(x,Q^2)$ are the full order α_s distributions defined by (5.23a) for the total muon pair rate and $G(x,Q^2)$ are the leading order QCD parton distributions defined for electroproduction. The contributions from the annihilation term, Δf_q^{DY}, and the Compton term, Δf_g^{DY}, are shown separately. Also shown is the contribution from the δ-function term alone in Δf_q^{DY}. The functions $G^{DY}_{p\to u}$ and $G^{DY}_{p\to\bar{u}}$ are not separate observables and only have real significance when convoluted together to produce the total rate, $d\sigma_{DY}/dM^2$, in (5.21).

I should say, however, that the $G^{DY}(x,Q^2)$ functions in (5.23) refer only to the total muon pair cross section, $d\sigma_{DY}/dM^2$, in (5.24). I have not yet looked at the order α_s corrections to other muon pair observables like

$d\sigma_{DY}/dM^2dy$ in (5.25). One may be able to find observables where the order α_s corrections are not large and hence make an accurate prediction. Furthermore, if one is not interested in the normalization but rather some other property like the $p_\perp$ spectrum of the muon pair[45] or the M^2 "scaling" behavior, then one might still have some predictive power[46].

Summing Leading Logarithms[7,8,47-51]

In the past year progress has been made at understanding better the divergence that occurs at $p_\perp = 0$ in the order α_s perturbative diagrams (see Fig. 5.3). Integrating (5.27) over y and M^2, we arrive at (for one quark flavor)[39]

$$\frac{d\sigma}{dp_\perp^2} \approx \frac{1}{3} e_q^2 \frac{4\alpha_s}{3\pi} \int dx_a \int dx_b\, G_{p\to q}(x_a) G_{p\to\bar{q}}(x_b)$$

$$\frac{1}{(\hat{s}-M^2)} \left[\log\left(\frac{2p_\perp^2\sqrt{\tau}}{s(1-\tau)}\right) + \text{order}(\hat{s}-M^2)\right]$$

$$\approx -\frac{4\alpha_s}{3\pi} \frac{\sigma_o(\tau)}{p_\perp^2} \log\left(\frac{2p_\perp^2\sqrt{\tau}}{s(1-\tau)}\right) \tag{5.41a}$$

for the $p_\perp$ spectrum of the virtual γ^* (or the $\mu^+\mu^-$ pair) arising from the annihilation term and where

$$\sigma = \sigma_o(\tau) = \frac{1}{3} e_q \int dx_a \int dx_b\, G_{p\to q}(x_a) G_{p\to\bar{q}}(x_b)\,\delta(1-M^2/\hat{s}), \tag{5.41b}$$

with $\tau = M^2/s$. The expression in (5.41) is valid provided $p_\perp^2 = \hat{t}\hat{u}/\hat{s}$ is small and $\hat{s} \approx M^2$ and arises from the region of phase space in which the outgoing gluon (in Fig. 4.3) is both soft and nearly collinear to the incoming q and $\bar{q}$.

The general form of the perturbation series for the γ^* $p_\perp$ spectrum, $p_\perp^2/\sigma\; d\sigma/dp_\perp^2$, is

$$\alpha_s[a_1\omega+b_1]+\alpha_s^2[a_2\omega^3+b_1\omega^2+c_1\omega+d_1]+\alpha_s^3[a_3\omega^5+b_3\omega^4+\ldots]+\ldots. \tag{5.42}$$

with $\omega = \log(s/p_\perp^2)$. Given that we are in a region where $\omega >> 1$, then the first terms in each bracket are the "leading log terms". The first term in the first bracket (with $a_1 = 4/3\pi$) corresponds to eq. (5.41a). The appearance of $\log(s/p_\perp^2)$ in (5.42) rather than $s(1-\tau)/(2p_\perp^2\sqrt{\tau})$ as in (5.41a) is a beyond leading log result (i.e., $b_1 = \log[(1-\tau)/2\sqrt{\tau}]$).

It is relatively easy to show that the leading log terms in (5.42) exponentiate leaving[48-51]

$$\frac{1}{\sigma_o(\tau)}\frac{d\sigma}{dp_\perp^2} \approx \frac{4\alpha_s}{3\pi}\frac{1}{p_\perp^2}\log(s/p_\perp^2)\exp[-\frac{2\alpha_s}{3\pi}\log^2(s/p_\perp^2)] \tag{5.43}$$

as the leading log result. Instead of diverging as $p_\perp \to 0$ as did (5.41), the leading log sum vanishes as $p_\perp \to 0$. This is physically sensible since if you allow for an infinite number of soft gluon emissions, it becomes impossible that the initial quark and antiquark remain precisely back-to-back.

Two conditions must be satisfied for the leading log result to be valid. First, one must be in the region

$$\log(s/p_\perp^2) >> 1 \tag{5.44a}$$

to insure that in the first bracket in (5.42) one can neglect the order α_s term relative to the order $\alpha_s\omega$ term. Then the condition that subleading log terms like $\alpha_s^2\omega^2$ should be unimportant (compared with the leading sum $\omega\exp(-2\alpha_s\omega^2/3\pi)$) becomes

$$\alpha_s(s')\log(s/p_\perp^2) << 1. \tag{5.44b}$$

Here it is not clear what one should use for s' (i.e., $s' = s$, $s' = p_\perp^2$, etc.). This choice has an effect beyond leading log order. However, it is clear that the two constraints on the validity of the leading log approximation overlap only at very high energy[49].

At each order in perturbation theory, the γ^* transverse momentum spectrum diverges as $p_\perp \to 0$. For modest $p_\perp$, the leading log sum (5.43) exhibits the usual $1/p_\perp^2$ behavior found at order α_s, but for $p_\perp \leq (p_\perp)_{peak}$, where

$$(p_\perp^2)_{peak} \approx \frac{s}{2} \exp[-3\pi/4\alpha_s]$$

$$\approx \frac{s}{2} \left(\frac{\Lambda^2}{s}\right)^{25/16}, \qquad (5.45)$$

radiation damping occurs and the spectrum goes rapidly to zero. This is all well and good, however, the leading log perturbative methods used to derive (5.43) become invalid at such small $p_\perp$ values. At these low $p_\perp$ non-perturbative hadronic effects undoubtedly dominate. That is, eq. (5.43) must still be smeared over the unknown non-perturbative primordial $k_\perp$ of the quark and antiquark within the initial protons, as we did in the previous section for the order α_s term. Now, however, it takes somewhat smaller values of the mean primordial $k_\perp$ in order to fit the data shown in Fig. 5.4[48,52].

VI. QCD PERTURBATION THEORY

General Formalism

In the previous sections I have worked out several examples using QCD perturbation theory. I would like now to write down the general QCD perturbative formalism. It is instructive to compare the formalism to the examples we have already worked out. I will not derive the general formalism but refer the reader to the original papers by Georgi and Politzer[1,53,54] and Gross and Wilczek[2]. In addition, there are recent review articles by A. Peterman[55] and A. J. Buras[56] that are quite useful. Also, see the lectures by C. H. Llewellyn Smith[15].

We will be interested in the three structure functions

$$\tilde{F}_1(x,Q^2) \equiv 2F_1(x,Q^2) \qquad (6.1a)$$

$$\tilde{F}_2(x,Q^2) \equiv F_2(x,Q^2)/x \qquad (6.1b)$$

$$\tilde{F}_3(x,Q^2) \equiv F_3(x,Q^2). \qquad (6.1c)$$

The moments of these distributions are defined by

$$M_i(n,Q^2) = \int_0^1 dx x^{n-1} \tilde{F}_i(x,Q^2). \qquad (6.2)$$

For simplicity let me consider only the non-singlet distribution (4.47b) and write the moments as

$$M^{NS}(n,Q^2) = \int_0^1 dx x^{n-1}\tilde{F}_{NS}(x,Q^2)$$

$$\equiv C_n^{NS}(Q^2/\mu^2,g^2)q_o^{(n)}, \tag{6.3}$$

where g^2 is the strong interaction coupling and μ^2 is the (arbitrary) renormalization point (as in Section II) and where the presently uncalculable non-perturbative part of the non-singlet moments has been absorbed into the (unknown) moments $q_o^{(n)}$. Perturbation theory will be applied to the moments $C_n^{NS}(Q^2/\mu^2,g^2)$ which is a function of Q^2/μ^2 and the coupling. Since the theory is renormalizable, C_n^{NS} cannot depend on the choice of μ^2. The equation which expresses this fact is the renormalization group equation

$$(\mu\frac{\partial}{\partial\mu} + \beta(g)\frac{\partial}{\partial g} - \gamma_n(g))C_n^{NS}(Q^2/\mu^2,g^2) = 0, \tag{6.4}$$

which is just the total derivative of C_n^{NS} w.r.t. μ. It has the general solution

$$C_n^{NS}(Q^2/\mu^2,g^2) = C_n^{NS}(1,\bar{g}^2)\mathrm{EXP}\{-\int_{\bar{g}(\mu^2)}^{\bar{g}(Q^2)} \frac{\gamma_n(t)}{\beta(t)}dt\}, \tag{6.5}$$

where $\beta(g)$ governs how the effective strong interaction coupling $\bar{g}^2(Q^2)$ depends on $\tau = \log(Q^2/\mu^2)$. Namely,

$$\frac{d\bar{g}^2}{d\tau} = \bar{g}\beta(\bar{g}) \tag{6.6}$$

with $\bar{g}(\tau=0) = g$.

From (6.3) and (6.5), it is easy to see that the change of M_n^{NS} w.r.t. $\alpha_s(Q^2) = \bar{g}^2(Q^2)/4\pi$ is given by

$$\frac{dM^{NS}(n,Q^2)}{d\alpha_s(Q^2)} = (-\frac{2\pi\gamma_n(\bar{g})}{\bar{g}\beta(\bar{g})} + \frac{1}{C_n^{NS}(1,\bar{g}^2)}\frac{d\bar{C}_n^{NS}(1,\bar{g}^2)}{d\alpha_s(Q^2)})M^{NS}(n,Q^2). \tag{6.7}$$

The functions $\gamma_n(\bar{g}^2)$, $\bar{g}\beta(\bar{g})$ and $C_n^{NS}(1,\bar{g}^2)$ can all be expanded in powers of $\alpha_s(Q^2)$ yielding

$$\gamma_n(\bar{g}) = \gamma_0^{(n)} \frac{\alpha_s(Q^2)}{4\pi} + \gamma_1^{(n)} \left(\frac{\alpha_s(Q^2)}{4\pi}\right)^2 + \ldots \tag{6.8a}$$

$$\bar{g}\beta(\bar{g}) = -\beta_0(\alpha_s(Q^2))^2 - \left(\frac{\beta_1}{4\pi}\right)(\alpha_s(Q^2))^3 + \ldots. \tag{6.8b}$$

$$C_n^{NS}(1,\bar{g}^2) = 1 + B_n^{NS} \frac{\alpha_s(Q^2)}{4\pi} + \ldots \ . \tag{6.8c}$$

In Section II, we discussed the effective strong interaction coupling $\bar{g}^2(Q^2)$. The strong interaction coupling calculated to leading order from (6.6) and (6.8b) is as in (2.12)

$$\alpha_o(Q^2) = \frac{4\pi}{\beta_0 \log Q^2/\Lambda^2} \tag{6.9}$$

with

$$\beta_0 = (33-2n_f)/3. \tag{6.10}$$

In the next order (i.e., α_o^2), we have

$$\alpha(Q^2) = \alpha_o(Q^2)\left(1 - \frac{\beta_1}{\beta_0}\frac{\alpha_o(Q^2)}{4\pi} \log\log(Q^2/\Lambda^2)\right) \tag{6.11}$$

with

$$\beta_1 = (306-38n_f)/3. \tag{6.12}$$

Now going back to (6.7) and inserting (6.8) we have, to lowest order

$$\alpha_s(Q^2) \frac{dM^{NS}(n,Q^2)}{d\alpha_s(Q^2)} = \left(\frac{\gamma_0^{(n)}}{2\beta_0}\right) M^{NS}(n,Q^2) \tag{6.13}$$

which implies

$$M^{NS}(n,Q^2) = q_n^{NS}(\alpha_s(Q^2))^{d_n}, \tag{6.14a}$$

where $q_n^{NS} = q_o^{(n)}$ are unknown. Comparing two different Q^2 values yields

$$M^{NS}(n,Q^2) = M^{NS}(n,Q_o^2)[\alpha(Q^2)/\alpha(Q_o^2)]^{d_n}$$

$$= M^{NS}(n,Q_o^2)[\log(Q^2/\Lambda^2)/\log(Q_o^2/\Lambda^2)]^{-d_n}, \tag{6.14b}$$

with

$$d_n = \frac{\gamma_o^{(n)}}{2\beta_o} = -\frac{2A_n^{NS}}{\beta_o}, \tag{6.14c}$$

where A_n^{NS} has been defined earlier in eq. (4.57a). As discussed in Section IV, eq. (6.14) can be written in terms of a convolution as

$$\frac{dq^{NS}(x,\tau)}{d\tau} = \frac{\alpha_s(Q^2)}{2\pi}\int_x^1 \frac{dy}{y}\, q^{NS}(y,\tau)P_{q\leftarrow q}(\tfrac{x}{y}) \tag{6.15a}$$

provided

$$A_n^{NS} = \int_0^1 z^{n-1}P_{q\leftarrow q}(z)dz \tag{6.15b}$$

and where the quark distribution, $q^{NS}(x,\tau)$, is defined by (4.47b).

Notice that

$$[\alpha_s(Q^2)/\alpha_s(Q_o^2)]^{d_n} = [1 - \frac{\beta_o}{4\pi}\alpha_s(Q^2)\log(Q^2/Q_o^2)]^{d_n}$$

$$\approx 1 + \frac{\alpha_s(Q^2)}{2\pi}A_n^{NS}\log(Q^2/Q_o^2) + \ldots. \tag{6.16}$$

is a power series in $\alpha_s(Q^2)\log Q^2 \sim 0(1)$. Thus by the use of the renormalization group equation, we have succeeded in summing all terms of the form $(\alpha_s(Q^2)\log Q^2)^N$. If we instead had simply calculated using ordinary perturbation theory only to order $\alpha_s(Q^2)\log Q^2$, we would have arrived at

$$q^{NS}(x,\tau) = \int_x^1 \frac{dy}{y} q_o^{NS}(y)\{\delta(z-1) + \frac{\alpha_s(Q^2)}{2\pi} P_{q\leftarrow q}(z)\tau\} \tag{6.17a}$$

where $\tau = \log(Q^2/Q_o^2)$ or in terms of moments

$$M^{NS}(n,Q^2) = q_o^{(n)} \, (1 + \frac{\alpha_s(Q^2)}{2\pi} A_n^{NS} \log(Q^2/Q_o^2)). \tag{6.17b}$$

Actually, this is not correct. As we saw in Section IV, ordinary perturbative calculations for massless quarks and gluons diverge logarithmically. To calculate with perturbation theory in QCD, one must adopt some procedure for handling these infrared divergences. Politzer has suggested a scheme for removing (or "regularizing") these divergences that is similar to renormalization. One takes the incoming quark and gluons for any basic subprocess to be slightly off mass shell and space-like, for example, $p_q^2 = -m_q^2$. The perturbative results then are finite and to order $\alpha_s(Q^2)\log Q^2$ have a form similar to (6.17b). Namely,

$$M^{NS}(n,Q^2) = \tilde{q}_o^{(n)} (1 + \frac{\alpha(Q^2)}{2\pi} A_n^{NS} \log(Q^2/m_q^2)). \tag{6.18}$$

Since we expect that the physical observable $M^{NS}(n,Q^2)$ in (6.18) to be finite in the limit $m_q \to 0$, it must be that we have made an "artificial" divergence in the way we have done the perturbation calculation. That is, we have divided $M^{NS}(n,Q^2)$ into two pieces $\tilde{q}_o^{(n)}$ and $A_n^{NS}\log(Q^2/m_q^2)$ both of which diverge as $m_q \to 0$ but whose product is finite. After all, the $\tilde{q}_o^{(n)}$ are unknown and arbitrary. According to Politzer, we should now "renormalize" $\tilde{q}_o^{(n)}$ in (6.18). We imagine that it has an expansion of the form

$$\tilde{q}_o^{(n)} = q_o^{(n)} (1 + \frac{\alpha_s}{2\pi} A_n^{NS} \log(m_q^2/Q_o^2) + \ldots) \tag{6.19}$$

where $q_o^{(n)}$ are finite in the $m_q^2 \to 0$ limit. Equation (6.18) now becomes (to order α_s) equivalent to eq. (6.17b). The unwanted $\log(m_q^2)$ divergence has been absorbed into the arbitrary and unknown $\tilde{q}_o^{(n)}$.

This seems a reasonable thing to do particularly since we will only be interested in the change of the distribution $q^{NS}(x,Q^2)$ relative to what it is at Q_o^2. The functions $P_{q\leftarrow q}(z)$ in (6.15a) are calculated from ordinary perturbation theory by picking out the coefficient of the $\log(Q^2/m_q^2)$ term in eq. (6.17a) in a calculation to order $\alpha(Q^2)\log Q^2$. Equation (6.15) is then used to sum all the orders of $(\alpha_s(Q^2)\log Q^2)^N$. This is precisely what we did in Sections III, IV and V except in some cases we regularized by allowing the gluon to have a small mass. In this case, m_q is replaced by m_g in eqs. (6.18) and (6.19).

Including the next order in (6.7) and (6.8) results in the differential equation

$$\alpha_s(Q^2)\,\frac{dM^{NS}(n,Q^2)}{d\alpha_s(Q^2)} = \left\{\frac{\gamma_o^{(n)}}{2\beta_o} + (h_n+B_n^{NS})\alpha_s(Q^2)/4\pi\right\}M^{NS}(n,Q^2), \quad (6.20)$$

where

$$h_n = \frac{\gamma_1^{(n)}}{2\beta_o} - \frac{\beta_1\gamma_o^{(n)}}{2\beta_o^2}. \quad (6.21)$$

The solution of which is

$$M^{NS}(n,Q^2) = q_n^{NS}[\alpha_s(Q^2)/\alpha_s(Q_o^2]^{d_n}\exp[(h_n+B_n)\alpha_s(Q^2)/4\pi] \quad (6.22a)$$

or

$$M^{NS}(n,Q^2) = M^{NS}(n,Q_o^2)[\alpha_s(Q^2)/\alpha_s(Q_o^2)]^{d_n}\exp[(h_n+B_n)(\alpha_s(Q^2) - \alpha_s(Q_o^2))/4\pi]$$

$$\approx M^{NS}(n,Q_o^2)[\alpha_s(Q^2)/\alpha_s(Q_o^2)]^{d_n} [1+(h_n+B_n)(\alpha_s(Q^2)-\alpha_s(Q_o^2))/4\pi]. \quad (6.22b)$$

Expanding (6.22b) in powers of $\alpha_o(Q^2)$ yields (to order $\alpha_o^2(Q^2)$)

$$M^{NS}(n,Q^2) = M^{NS}(n,Q_o^2)\,[\alpha_o(Q^2)/\alpha_o(Q_o^2)]^{d_n}$$

$$[1+(h_n+B_n+\ell_n)(\alpha_o(Q^2)-\alpha_o(Q_o^2)/4\pi] \tag{6.22c}$$

with

$$\ell_n = -\frac{\beta_1\gamma_o^{(n)}}{2\beta_o^2}\log\log Q^2/\Lambda^2. \tag{6.23}$$

I should point out that the individual terms h_n and B_n in (6.22) are, in general, not unique[6,13]. They will depend on the particular "regularization" scheme one has used. The sum, $h_n + B_n$, however, will be unique.

Since $\alpha_s(Q^2) - \alpha_s(Q_o^2)$ is of order $\alpha_o^2(Q^2)$, the corrections of (6.22) due to h_n and B_n are of order $\alpha_o^2(Q^2)$. Suppose, however, we are interested in comparing distributions determined in process A, $M_A^{NS}(n,Q^2)$, with those determined in process B, $M_B^{NS}(n,Q^2)$, then from (6.22b) we have (to order $\alpha_s(Q^2)$)

$$M_B^{NS}(n,Q^2) - M_A^{NS}(n,Q^2) = M_A^{NS}(n,Q^2)\Delta B_n^{NS}\alpha_s(Q^2)/4\pi, \tag{6.24a}$$

where

$$\Delta B_n^{NS} \equiv B_n^{NS,B} - B_n^{NS,A}. \tag{6.24b}$$

Thus the corrections due to B_n are of order $\alpha_s(Q^2)$ when comparing one process with another. Notice that since $\gamma_o^{(n)}$, $\gamma_1^{(n)}$, β_o and β_1 in (6.8a) are process independent, h_n cancels out when comparing one process with another. As before, eq. (6.24) can be written in terms of a convolution yielding

$$q_B^{NS}(x,\tau) = \int_x^1 \frac{dy}{y}\, q_A^{NS}(y,\tau)\{\delta(z-1)+\alpha_s(Q^2)\Delta f_q(z)\}, \tag{6.25a}$$

where $z = x/y$ and

$$\Delta B_n^{NS} = 4\pi\int_o^1 z^{n-1}\Delta f_q(z)dz. \tag{6.25b}$$

To order $\alpha_o(Q^2)$, both $q_B^{NS}(x,\tau)$ and $q_A^{NS}(x,z)$ satisfy (6.15).

Thus if we are interested in calculating to full order $\alpha_o(Q^2)$, we must pick a "reference reaction" A (we follow Ref. 6 and choose F_2 in ep scattering) to define the quark distributions. They will evolve with Q^2 according to (6.15). The quark distributions in other processes (like F_1 in ep scattering or F_3 in νp scattering, etc.) can be calculated from (6.25) if we know the functions $\Delta f_q(z)$ (or equivalently ΔB_n).

To calculate $\Delta f(z)$, we notice that if we calculate using ordinary perturbation theory and keep terms of order $\alpha_s(Q^2)\log Q^2$ <u>and</u> terms of order $\alpha_s(Q^2)$ then we arrive at

$$q^{NS}(x,Q^2) = \int_x^1 \frac{dy}{y}\, \tilde{q}_o^{NS}(y)$$

$$[\delta(z-1) + \frac{\alpha_s}{2\pi} P_{q\leftarrow q}(z)\log(Q^2/m_q^2) + \alpha_s \Delta f(z)] \tag{6.26a}$$

or

$$M^{NS}(n,Q^2) = \tilde{q}_o^{(n)}[1 + \frac{\alpha_s}{2\pi} A_n^{NS}\log(Q^2/m_q^2) + \frac{\alpha_s(Q^2)}{4\pi}\Delta B_n], \tag{6.26b}$$

where again we have regularized by taking the incoming partons off-shell. After absorbing the $\log(m_q^2)$ divergence into the $\tilde{q}_o^{(n)}$ by the use of (6.19), we arrive at

$$q^{NS}(x,Q^2) = \int_x^{1.0} \frac{dy}{y}\, q_o^{NS}(y)[\delta(z-1) + \frac{\alpha_s(Q^2)}{2\pi} P_{q\leftarrow q}(z)\tau$$

$$+ \alpha_s(Q^2)\Delta f(z)]. \tag{6.26c}$$

In the more general case (like the singlet distributions) where there is mixing between initial quark, $G_{p\to q}^{(o)}(y)$, and gluon, $G_{p\to g}^{(o)}(y)$, distributions, one has

$$G_{p\to q}(x,Q^2) = \int_x^1 \frac{dy}{y} \{G^{(o)}_{p\to q}(y)[\delta(z-1) + \frac{\alpha_s}{2\pi} P_{q\leftarrow q}(z)\tau + \alpha_s(Q^2)\Delta f_q(z)]$$

$$+ G^{(o)}_{p\to g}(y)[\frac{\alpha_s}{2\pi} P_{g\leftarrow q}(z)\tau + \alpha_s(Q^2)\Delta f_g(z)]\}, \qquad (6.26d)$$

with $\tau = \log Q^2/Q_0^2$. From here we can read off not only the $P_{q\leftarrow q}(z)$ and $P_{g\leftarrow q}(z)$ functions that generate the Q^2 evolution of $G_{p\to q}(x,Q^2)$ according to (4.29) but also the terms $\Delta f_q(z)$, $\Delta f_g(z)$ that are proportional to $\alpha_s(Q^2)$.

In general, in any problem there will be a sum of terms of the form $[\alpha_s(Q^2)\log Q^2]^N$ that are all of leading order 1 and are summed by the use of eq. (4.44). In addition, there are terms proportional to $[\alpha_s(Q^2)]^N$. These are not summed but since $\alpha_s(Q^2)$ is small, hopefully one only needs to calculate to order $\alpha_o(Q^2)$ (or maybe $\alpha_o^2(Q^2)$) to get an accurate result.

The KLN Theorem[57,58]

In the examples that we have examined in Sections III, IV and V, we found that in some cases the perturbative results were infrared finite. The e^+e^- total cross section was such an example. It contains only ultraviolet divergences that renormalize the effective coupling, $\alpha_s(Q^2)$. As Q^2 increases, $\alpha_s(Q^2)$ decreases and $R = \sigma(e^+e^- \to \text{hadrons})/\sigma(e^+e^- \to \mu^+\mu^-)$ approaches the naive parton model result. In contrast to this was the situation encountered in calculating QCD corrections to the quark distributions $G_{p\to q}(x)$. Here the perturbative results contained mass singularities that were "absorbed" into the unknown reference distributions. As Q^2 increases, $G_{p\to q}(x,Q^2)$ does not approach the naive parton model result but deviates more and more from it. In this section we will examine when one expects to encounter mass singularities. In addition, I will try and placate those readers who are uneasy about "absorbing" infinities into unknown distributions. We will examine a "regularization" scheme that is free of mass singularities.

Let me begin by defining precisely what is meant by infrared divergence and mass singularity. If infinities occur in the limit that the boson (gluon) mass, m_g, goes to zero while the fermion mass is not zero, then we say the calculation is "infrared" divergent. If the calculation diverges as the fermion (quark) mass (and the gluon mass) goes to zero, then there are mass singularities. Both cases are actually mass singularities; one when the boson mass goes to zero, one when the fermion (and the boson) mass goes to zero. In QED one only encounters infrared divergences since electrons have mass and the photons do not couple to each other. In QCD we are concerned with the mass singularities that occur as the quark and gluon masses go to zero.

The Kinoshita-Lee-Nauenberg (KLN) theorem states[57,58] that if one considers the scattering of an incoming state A to an outgoing state B by an S-matrix, S_{BA}, then the quantity

$$\sum_{D(E_A)} \sum_{D(E_B)} |S_{BA}|^2 \tag{6.27}$$

contains no mass singularities provided we sum over both subsets $D(E_A)$ and $D(E_B)$ which consist of all states that are degenerate (in the limit $m_g \to 0$, $m_q \to 0$) with the initial state A and the final state B. For example, a state of a quark with a three-momentum $\vec{p}$ is degenerate with the state which consists of a quark with momentum $\vec{p} - \vec{k}$ and a gluon with momentum $\vec{k}$ provided $m_q = 0$ ($m_g = 0$) and the angle between $\vec{p}$ and $\vec{k}$ is zero.

In QED to arrive at results that are infrared finite, one needs only sum over initial or final degenerate states. Both

$$\sum_{D(E_a)} |S_{BA}|^2 ; \quad \sum_{D(E_b)} |S_{BA}|^2 \tag{6.28}$$

are infrared finite.

A corollary to the KLN theorem is that observables that do not discriminate between initial and final states differing by the inclusion of a very low energy gluon or by the replacement of one quark by two collinear (quark + gluon) partons with the same total energy are infrared and mass finite.

Let us now go back and examine the calculation of the e^+e^- total rate. Since we asked for the total rate, we automatically summed over all final state degenerate states (Fig. 3.1). In addition, since the incoming state contains only leptons and no colored objects (quarks and gluons), there are no initial degenerate states to sum over. Therefore, according to KLN, the answer is mass finite which is what we found.

On the other hand, when we computed the corrections to $G_{p\to q}(x)$ in Section IV, we included only the diagrams in Fig. 6.1a and the virtual gluon diagrams in Fig. 4.4. Hence we did sum over final degenerate states, but we did not sum over initial degenerate states. Thus according to (6.27), the results are not mass finite. At the order we worked at, we had no triple gluon couplings, so the theory was similar to QED. Thus from (6.28) the results were infrared finite. The fact that we satisfied (6.28) is important since this insured that we did not get a log squared in the final answer - only a single log. Had we gotten a log squared, we could not have "factored" out the mass singularity as we did in writing, for example, (3.47)[59].

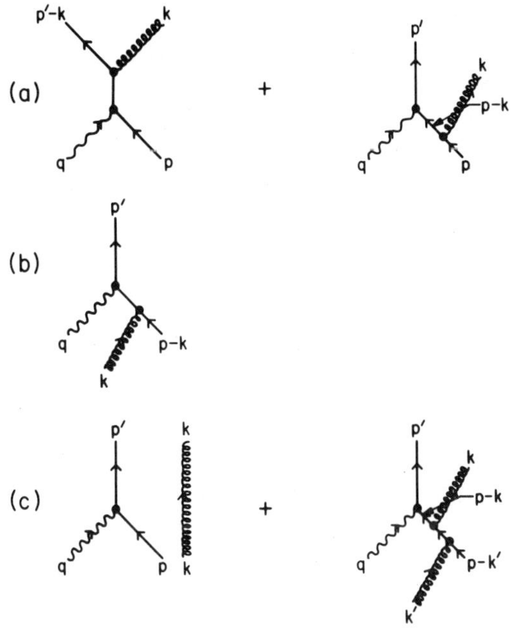

Fig. 6.1 - (a) Diagrams for γ^* + quark $\to$ quark + glue that are included when calculating the order α_s corrections to the deep inelastic structure functions, $G_{p\to q}(x,Q^2)$. The incoming virtual photon has momentum $Q^2 = -q^2$ and the incoming quark (within the proton) has momentum p.

(b) Order α_s diagram that is degenerate (when $\vec{p}$ and $\vec{k}$ are parallel) with the first diagram in (a). Here the gluon of momentum k and quark of momentum p-k are both within the initial proton.

(c) Diagrams whose order α_s interference term is degenerate (when $\vec{p}$ and $\vec{k}'$ are parallel) with the second diagram in (a).

To arrive at a result that is mass finite, we must include the diagram in Fig. 6.1(b) which is degenerate (when $\vec{p}$ and $\vec{k}$ are parallel) with the first diagram in Fig. 6.1(a). In addition, we must include the interference between the two diagrams in Fig. 6.1(c) which is degenerate (when $\vec{p}$ and $\vec{k}'$ are parallel) with the second diagram in Fig. 6.1(a). This means that to get a mass finite result, we must consider the case where the incoming proton contains both a quark and a gluon and we must weigh all states with a quark and N parallel gluons of a given energy the same. Now whether or not we are left with any terms of the form $\alpha_s(Q^2)\log Q^2$ or not depends on how we construct the initial state proton. Suppose "protons" are objects of fixed angular size δ. Then presumably we arrive at something like

$$\alpha_s(Q^2)\log(Q^2/m_g^2) + \alpha_s(Q^2)\log(m_g^2/\delta^2E^2) = \alpha_s(Q^2)\log(Q^2/\delta^2E^2), \tag{6.29}$$

where the first term is our result (4.31) for diagram 6.1a and the second term comes from the remaining terms in Fig. 6.1 and E is, say, the energy of the proton. Adding the two terms given a mass finite result, but in this case we would arrive at a scaling parton distribution since $\log(Q^2/E^2) \approx 1$. Now suppose "protons" are objects of fixed transverse size, which we characterize by some $\langle k_\perp^2\rangle$ of partons within the incident proton. We then have

$$\alpha_s(Q^2)\log(Q^2/m_g^2) + \alpha_s(Q^2)\log(m_g^2/\langle k_\perp^2\rangle) = \alpha_s(Q^2)\log(Q^2/\langle k_\perp^2\rangle). \tag{6.30}$$

The result is mass finite and we have merely replaced the fictitious gluon mass, m_g^2, by something that represents the size of hadrons, $\langle k_\perp^2\rangle$. According to this logic, the reason one gets scale breaking in deep inelastic scattering is because protons have fixed transverse size (i.e., have some fixed radius). We also see that if you do not like factorizing off and absorbing mass singularities into the unknown reference distributions, you can keep all the diagrams necessary to get a mass finite result. However, you will arrive at the same results and it is much more work[60]!

VII. LARGE $p_\perp$ MESON AND "JET" PRODUCTION IN HADRON-HADRON COLLISIONS

The QCD Approach

In the naive parton model, the large $p_\perp$ production of hadrons in the process $A + B \to h_1 + h_2 + X$ is described by the diagram in Fig. 7.1b. The large-transverse-momentum reaction is assumed to occur as the result of a single large-angle scattering $a + b \to c + d$ of constituents a and b, followed by the decay or fragmentation of constituent c into the trigger hadron h_1 and constituent d into the "away-side" hadrons, h_2, This results in the four jet structure in Fig. 7.1a. The invariant cross section for the process $A + B \to h + X$ is given by

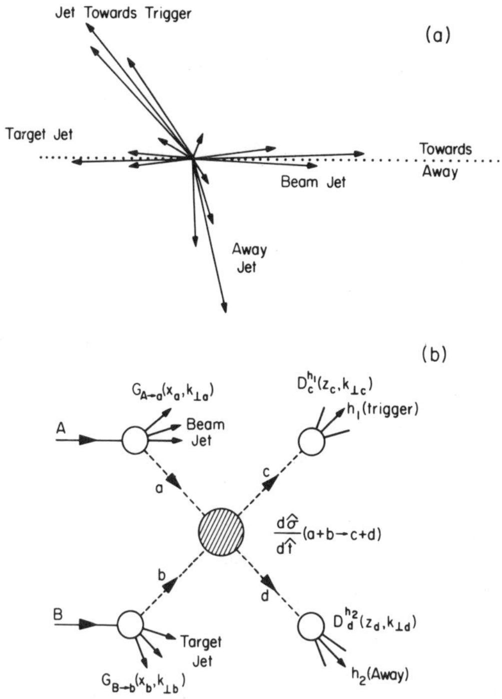

Fig. 7.1 - (a) Illustration of the four jet structure resulting from a beam hadron (entering at left along dotted line) colliding with a target hadron (entering at right along dotted line) in the CM frame: two jets with large $p_\perp$ (collection of particles moving roughly in the same direction), one called the "toward" (trigger) side and one on the "away" side; and two jets with small $p_\perp$ that result from the break up of the beam and target hadrons (usually referred to as the "soft hadronic" background).

(b) Illustration of the underlying structure of the large $p_\perp$ process $A + B \to h_1 + h_2 + X$. The large $p_\perp$ trigger hadron h_1 occurs as the result of a large angle scattering of constituents ($q_a + q_b \to q_c + q_d$), followed by the decay of fragmentation of constituent c into a towards side jet of hadrons (one being the trigger h_1) and constituent d into an away side jet of hadrons (one being h_2). The quantities x_a, x_b, $k_{\perp a}$, $k_{\perp b}$ are the longitudinal fraction of the incoming hadrons A, B momentum and perpendicular momentum of constituents a, b and z_c, z_d, $k_{\perp c}$, $k_{\perp d}$ are the fraction of the outgoing constituents longitudinal momentum and perpendicular momentum carried by the detected hadrons h_1 and h_2.

$$E \frac{d\sigma}{d^3p}(s,p_\perp,\theta_{cm}) =$$

$$\int dx_a \int dx_b G_{A\to a}(x_a) G_{B\to b}(x_b) D_c^h(z_c) \frac{1}{z_c} \frac{1}{\pi} \frac{d\hat{\sigma}}{d\hat{t}}(\hat{s},\hat{t}), \qquad (7.1)$$

where $\hat{s}$, $\hat{t}$ are the usual invariants but for the constituent two-to-two subprocess $d\hat{\sigma}/d\hat{t}$ $(a + b \to c + d)$. The quantities x_a and x_b are the fractional momentum carried by the constituents a and b, respectively, and z_c is the fraction of the outgoing constituent momentum that appears in the hadron, h.

In the theory of QCD, the constituent subprocess, $a + b \to c + d$, must be corrected for the emission of gluons. That is one must include the higher order subprocesses $a + b \to c + d + g$, $a + b \to c + d + g + g$, etc., where g is a gluon. As discussed by C. Sachrajda[61], to leading order these processes modify (7.1) in the manner illustrated in Fig. 7.2[62]. For example, summing all the (divergent) gluon radiation from the outgoing quark in $q + q \to q + q$ generates the "renormalization group improved" scale breaking fragmentation function $D_q^h(z,Q^2)$ (Fig. 7.2a). Summing over all the (divergent) gluon radiation from the incoming quarks generates the "renormalization group improved" scale breaking quark distributions $G_{p\to q}(x,Q^2)$ (Fig. 7.2b). The divergent piece of the diagram

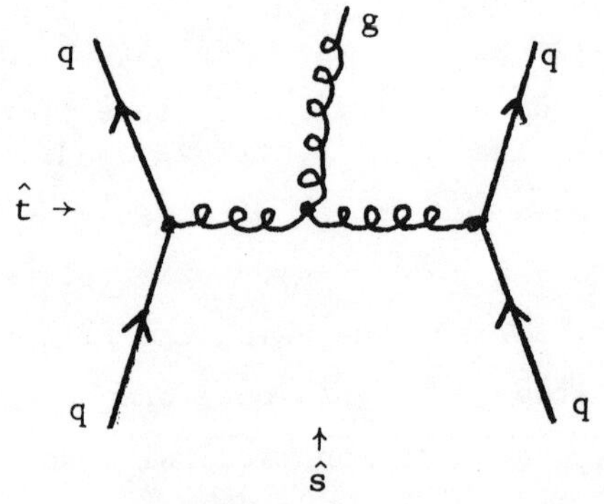

(7.2)

is contained in the "renormalization improved" gluon distribution $G_{p\to g}(x,Q^2)$ (Fig. 7.2c), where the constituent subprocess is now $g + q \to g + q$. Other higher order corrections to the gluon propagators and to the vertices generate the "renormalized" coupling $\alpha_s(Q^2)$ as

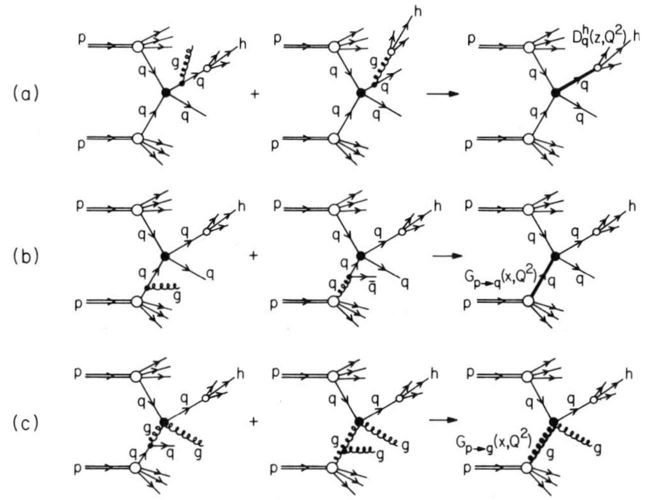

Fig. 7.2 - (a) Illustration of two diagrams for pp → h + X which are included (divergent parts) in the "renormalization group" improved quark fragmentation function, $D_q^h(z,Q^2)$, when one calculates the contribution from the subprocess qq → qq.

(b) Illustration of two diagrams for pp → h + X which are included (divergent parts) in the "renormalization group" improved quark distribution, $G_{p\to q}(x,Q^2)$, when one calculates the contribution from the subprocess qq → qq.

(c) Illustration of two diagrams for pp → h + X which are included (divergent parts) in the "renormalization group" improved gluon distribution, $G_{p\to g}(x,Q^2)$, when one calculates the contribution from the subprocess qg → qg.

discussed in Section II. Thus to leading order in QCD, eq. (7.1) becomes

$$E\,\frac{d\sigma}{d^3p}\,(s,p_\perp,\theta_{cm}) = \sum_{a,b} \int dx_a \int dx_b\, G_{A\to a}(x_a,Q^2)\, G_{B\to b}(x_b,Q^2)\, D_c^h(z_c,Q^2)\, \frac{1}{z_c}\,\frac{1}{\pi}\,\frac{d\hat{\sigma}}{d\hat{t}}\,(\hat{s},\hat{t}), \qquad (7.3)$$

where $G(x,Q^2)$ and $D(z,Q^2)$ are the "renormalization group improved" parton distributions and fragmentation functions, respectively, and where one includes all seven subprocesses: qq → qq, $\bar{q}q \to \bar{q}q$, $\bar{q}\bar{q} \to \bar{q}\bar{q}$, gq → gq, $g\bar{q} \to g\bar{q}$, $gg \to \bar{q}q$, $\bar{q}q \to gg$, and gg → gg. Each 2 → 2 differential cross section, $d\hat{\sigma}/d\hat{t}$, is calculated to lowest order in perturbation theory with an effective coupling constant $\alpha_s(Q^2)$ as in (2.12). These cross sections have been calculated previously by Cutler and Sivers[63] and by Combridge, Kripfganz and Ranft[64] and all behave as $\hat{s}^{-2}$ at fixed $\hat{t}/\hat{s}$ (and for constant α_s).

For ep collisions, Q is the 4-momentum transfer from the electron to the quark and for $pp \to \mu^+\mu^- + X$, it is the mass of the muon pair. On the other hand, the correct kinematic quantity to use for Q^2 in the constituent subprocesses contributing to large $p_\perp$ meson production in pp collisions is not at present known. (As I pointed out previously, changing the definition of Q^2 is a higher than leading order effect.)

For our analyses of high $p_\perp$[16,65,66], we have taken for definiteness

$$Q^2 = 2\hat{s}\hat{t}\hat{u}/(\hat{s}^2 + \hat{t}^2 + \hat{u}^2), \tag{7.4}$$

where $\hat{s}$, $\hat{t}$ and $\hat{u}$ are the usual Mandelstam invariants but for the constituent subprocesses. This arbitrariness in the form for Q^2 makes predictions at low Q^2 (i.e., low $p_\perp$) in hadron-hadron collisions uncertain.

Smearing

There is considerable experimental evidence that the constituents inside the proton have a large effective internal transverse momentum[67-69]. Effects due to the transverse momentum of quarks within hadrons, $(k_\perp)_{h\to q}$, and of hadrons within the outgoing jets, $(k_\perp)_{q\to h}$, called "smearing" effects are particularly important for large $p_\perp$ calculations. This is due to the "trigger bias" which selects the configuration in which the initial quarks (or gluons) are already moving toward the trigger. As for the muon pairs, in QCD, this transverse momentum of the partons can arise from two sources (illustrated in Fig. 7.3).

Firstly, in a proton beam, quarks are confined in the transverse direction to within the proton radius. Therefore, from the uncertainty principle, they must have some transverse momentum. This "primordial" momentum is intrinsic to the basic parton "wave function" inside the proton. As illustrated in Fig. 7.3a, one might expect the wave function to have a term where the trigger parton $k_\perp$ is balanced by another constituent (or constituents) which has the opposite $k_\perp$ and most of the remaining longitudinal momentum. Consider now the plane formed by the beam, target and a 90° trigger hadron (called the x-z plane in Fig. 7.3).

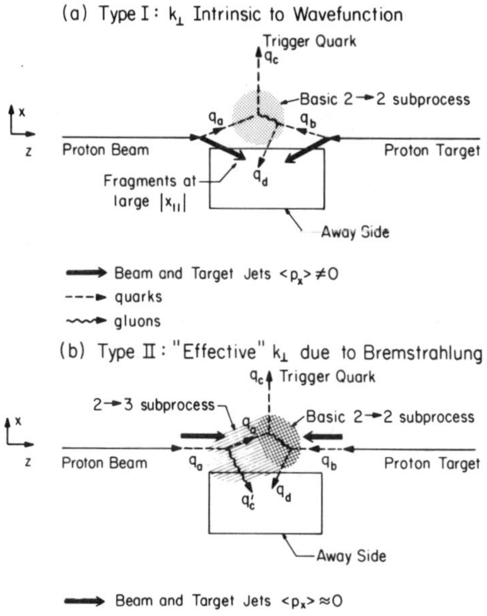

Fig. 7.3 - (a) Illustration of the non-perturbative ("primordial") component of the transverse momentum of quarks within proton that is intrinsic to the wave function of the proton. One expects this transverse momentum to be balanced by the remaining constituents in the proton which can, in turn, fragment into particles at high $x_{\parallel}$. The away-side consists of the recoiling quark, q_d, and two slightly shifted jets, one from the beam and one from the target.

(b) Illustration of a perturbative component to the transverse momentum of a quark with a hadron which is due to the Bremsstrahlung of a gluon before the basic 2 → 2 scattering occurs. In this case, the trigger quark is balanced by two away-side jets, one from the quark q_d and one from the radiated gluon q_c'.

Typically, the trigger arises from the fragmentation of a constituent with $k_{\perp x} > 0$ which is balanced by the remaining constituents having $k_{\perp x} < 0$. One expects to see this negative $k_{\perp x}$ as a shift in the beam and target jets at large $|x_{\parallel}|$. This shift (i.e., nonzero $\langle k_{\perp x}\rangle$) of the beam jet as one increases the $p_\perp$ of a 90° trigger has recently been observed by the BFS group at ISR[70] (see Fig. 7.5 of Ref. 66).

Secondly, in QCD, one expects to receive an "effective" $k_\perp$ of quarks in protons due to the Bremsstrahlung of gluons. This perturbative term, which arises from diagrams like those in Fig. 7.2, is illustrated in Fig. 7.3b. It corresponds to including two particle to three or more particle processes (2 → 3) rather than just the two particle to two particle 2 → 2 scatterings. For such subprocesses, the $k_\perp$ of the quark q_a is balanced by a gluon jet on the away-side which subsequently fragments into many low momentum hadrons. In addition, the mean value of the effective $k_\perp$ is expected to depend on the x value of quark q_a and on the Q^2 for the processes. Separating the origin of the transverse momenta into Type I and Type II as seen in Fig. 7.3 is, of course, a bit

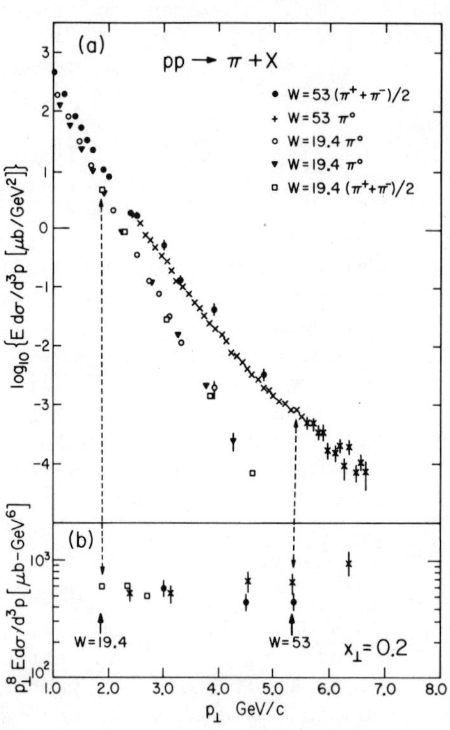

Fig. 7.4 - (a) The invariant cross section for $pp \to \pi + X$ at $W = \sqrt{s} = 19.4$ and 53 GeV/c with $\theta_{cm} = 90°$ (open squares = Ref. 95, solid dots = Ref. 96, crosses = Ref. 97, solid triangles = Ref. 98, open circles = Ref. 99).

(b) Plot of the invariant cross section for $pp \to \pi + X$ at $x_\perp = 0.2$ and $\theta_{cm} = 90°$ times $p_\perp^8$ versus $p_\perp$.

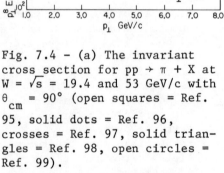

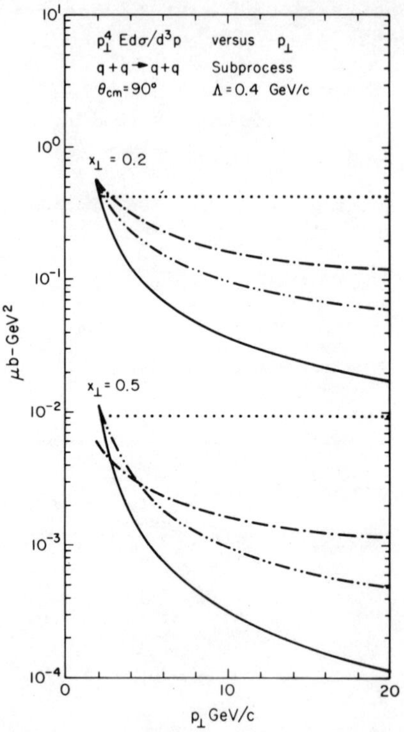

Fig. 7.5. Behavior of the quantity $p_\perp^4\ Ed\sigma/d^3p$ for $pp \to \pi^o + X$ at $\theta_{cm} = 90°$ and $x_\perp = 0.2$ and 0.5 arising from the QCD subprocess $q + q \to q + q$. The dot-dashed and dot-dot-dashed curves result when one includes a running coupling constant $\alpha_s(Q^2)$ and when one includes both $\alpha_s(Q^2)$ plus the non-scaling structure functions $G(x, Q^2)$, respectively. The solid curves arise from including a running coupling constant $\alpha_s(Q^2)$ plus $G(x,Q^2)$ and in addition include the scale breaking fragmentation functions $D(z,Q^2)$. The dotted curves indicate perfect "scaling."

artificial since both mechanisms occur simultaneously.

The analysis of large $p_\perp$ meson production is not yet as complete as the discussion of $pp \to \mu^+\mu^- + X$ presented in Section V. We have not separated the "effective" $k_\perp$ of the quarks and gluons in the initial hadrons in the "primordial" and "perturbative" components. For the present, we parameterize the transverse momentum of the partons (quarks and gluons) by a Gaussian with $\langle k_\perp \rangle_{h\to q} = 848$ MeV which produces for $pp \to \mu^+\mu^- + X$ a mean $p_\perp^2$ of 1.9 GeV^2 in agreement with the data in Fig. 5.4. We take this distribution to be independent of x and Q^2 and to be the same for quarks, antiquarks and gluons in the proton[71]. In so doing, we are <u>not</u> handling properly the x and Q^2 dependence of the high $k_\perp$ tails expected from QCD Bremsstrahlung. The next step would, of course, be to calculate and include explicitly the $2 \to 3$ subprocess expected by QCD (like $qq \to qqg$, etc.) and smear the results with the "primordial" $k_\perp$ only (which is presumably smaller than the effective 848 MeV we now use). Care would have to be taken since one would include only the non-divergent parts of the $2 \to 3$ subprocesses. The divergent parts have already been included by the use of the "renormalization improved" distributions $G(x,Q^2)$ and fragmentation functions $D(z,Q^2)$. For the present, however, we merely use the data in Fig. 5.4 to give an "effective" $k_\perp$ distribution and include explicitly only $2 \to 2$ subprocesses.

It is clear from examining data that smearing effects are present in nature. However, no one yet really knows how to handle the primordial $\langle k_\perp \rangle$ properly theoretically[73]. Different models for smearing give quite differing results[74-77]. Surely our crude way of handling it is not precisely correct. Present day calculations are uncertain in the low $p_\perp$ ($p_\perp \leqslant 4$ GeV/c) region where these effects are important.

The emission of gluons after the hard scattering ($2 \to 2$) subprocesses induces an "effective" $k_\perp$ of the hadrons that fragment from the outgoing quarks because one is sometimes really seeing two jets rather than one. As for the quark distributions in the proton, we do not include these effects (we also neglect the interferences that arise between the amplitude for emitting gluon before and after the hard $2 \to 2$ process) and for the present take the transverse momentum distribution of hadrons from outgoing quarks (and gluons) to be a Gaussian with $\langle k_\perp \rangle_{q\to h} = 439$ MeV independent of z or Q^2. Again, this is not precisely correct and should be improved upon in later work.

<u>Single Particle Production</u>

<u>The Cross Section</u>. In the naive parton model, the quark distribution and fragmentation function in (7.1) scale. In this case, the invariant cross section for producing large $p_\perp$ mesons directly

reflects the energy dependence of the constituent cross section, $d\hat{\sigma}/d\hat{t}$. If this latter cross section behaves as

$$d\hat{\sigma}/d\hat{t} = h(\hat{t}/\hat{s})/\hat{s}^n, \tag{7.5a}$$

then the former behaves as

$$Ed\sigma/d^3p = f(x_\perp,\theta_{cm})/p_\perp^N, \tag{7.5b}$$

where $N = 2n$, $x_\perp = 2p_\perp/W$ and $W = \sqrt{s}$.

The experimental determination of the effective $p_\perp$ power index N in (7.5b) involves the comparison of data at different energies W and different $p_\perp$ <u>but</u> at the same ratio $x_\perp = 2p_\perp/W$ and the same angle θ_{cm}. For example, Fig. 7.4 illustrates how one determines the $p_\perp$ dependence at $x_\perp = 0.2$. Data at the lowest Fermilab energy W = 19.4 GeV and $p_\perp$ = 1.94 GeV/c are compared with the ISR energy W = 53 GeV at $p_\perp$ = 5.3 GeV/c. As can be seen in the lower part of this figure, where I have multiplied the cross section by $p_\perp^8$, the W = 53 GeV, $p_\perp$ = 5.3 GeV/c point is about a factor of $(1.94/5.3)^8$ times the W = 19.4 GeV, $p_\perp$ = 1.94 GeV/c point. By combining these two points with points from other energies, one finds that the data at $x_\perp = 0.2$ behave roughly like $p_\perp^{-8}$ over the range $2 \leqslant p_\perp \leqslant 6$ GeV/c and $\theta_{cm} = 90°$. Thus the naive expectation from field theory that $n \approx 2$ is not seen in the data.

However, one cannot use dimensional counting in QCD. There is an intrinsic mass scale Λ in (2.12) that is generated by the interaction of quarks and gluons. Figure 7.5 shows the behavior of the single particle invariant cross section, $pp \to \pi^o + X$, arising from the subprocess $qq \to qq$. The cross section, $Ed\sigma/d^3p$, is plotted at $\theta_{cm} = 90°$ and $x_\perp$ = 0.2 and 0.4 versus $p_\perp$ and is multiplied by $p_\perp^4$ which would be independent of $p_\perp$ if "scaling" were valid (dotted curves). The dot-dashed and dot-dot-dashed curves illustrate the effects of including the running coupling constant $\alpha_s(Q^2)$ and $\alpha_s(Q^2)$ plus the scale breaking quark distributions $G(x,Q^2)$, respectively. The solid curves are the results after one includes a running coupling constant $\alpha_s(Q^2)$ plus scale breaking quark distributions $G(x,Q^2)$ plus scale breaking quark fragmentation functions $D(z,Q^2)$. The curves decrease with increasing $p_\perp$ eventually approaching a $p_\perp^{-4}$ behavior asymptotically at very large $p_\perp$.

Figure 7.6 shows the final results for $p_\perp^4\, E d\sigma/d^3p$ for $pp \to \pi^0 + X$ at $\theta_{cm} = 90°$ and $x_\perp = 0.2$ where now all seven subprocesses discussed in the previous section are included. The dash-dot and solid curves are the results before and after the addition scale breaking due to the transverse momentum of the quarks and gluons within the initial protons (smearing) have been included with $\langle k_\perp \rangle_{h\to q} = 848$ MeV and $\Lambda = 0.4$ GeV. The dashed curve is the final result (after smearing) with $\Lambda = 0.6$ GeV and the dotted curve shows a $p_\perp^{-8}$ behavior. The final result exhibits a rough $p_\perp^{-8}$ behavior over the range $2 \le p_\perp \le 6$ with an approach to $p_\perp^{-4}$ at very high $p_\perp$ ($p_\perp \gtrsim 14$ GeV/c).

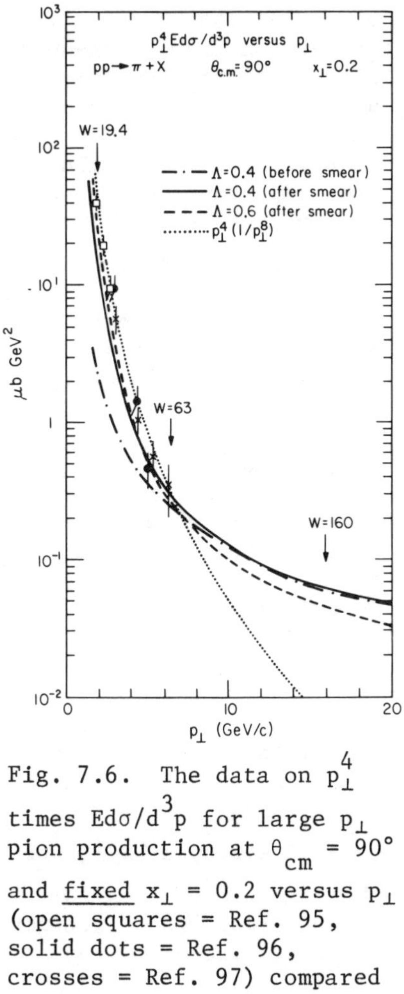

Fig. 7.6. The data on $p_\perp^4$ times $E d\sigma/d^3p$ for large $p_\perp$ pion production at $\theta_{cm} = 90°$ and <u>fixed</u> $x_\perp = 0.2$ versus $p_\perp$ (open squares = Ref. 95, solid dots = Ref. 96, crosses = Ref. 97) compared with the predictions (with absolute normalization) of a model that incorporates all the features expected from QCD. The dot-dashed and solid curves are the results before and after smearing with $\langle k_\perp \rangle_{h\to q} = 848$ MeV, respectively, using $\Lambda = 0.4$ GeV/c and the dashed curves are the results using $\Lambda = 0.6$ GeV/c (after smearing). The dotted curve is $p_\perp^4(1/p_\perp^8)$.

The QCD results for $p_\perp^4\, E d\sigma/d^3p$ for $pp \to \pi^0 + X$ in Fig. 7.6 can be compared to the prediction for $W^5\, d\sigma/dMdyd^2p_\perp$ for $pp \to \mu^+\mu^- + X$ in Fig. 5.7. The nature of the expected scale breaking is similar except the "breaking" is somewhat larger in the $pp \to \pi^0 + X$ case. This is due to the additional "scale breaking" from the fragmentation functions and from a trigger bias effect which makes $pp \to \pi^0 + X$ more sensitive to transverse momentum effects. When one performs a high $p_\perp$ experiment, one sits at large $p_\perp$ and waits for an event. This biases one in favor of the configuration where the initial partons are already moving toward the trigger and so smearing makes a large effect in these experiments.

The data on $E d\sigma/d^3p$ at <u>fixed</u> W = 19.4 and 53 GeV versus $p_\perp$ are compared with the theory in Fig. 7.7. The agreement is quite good. The results before smearing are shown by the dot-dashed curves. Smearing has little effect for $p_\perp \geq 4.0$ GeV/c at W = 53 GeV but has a sizable effect (even at $p_\perp$ = 6.0 GeV/c) at W = 19.4 GeV due to the steepness of the cross section at this low energy. At $p_\perp$ = 2.0 and W = 19.4 GeV, smearing increases the cross section by about a factor of 10. The contributions to the total invariant cross section from quark-quark elastic scattering (plus $q\bar{q} \to q\bar{q}$ and $\bar{q}\bar{q} \to \bar{q}\bar{q}$) are shown by the dotted curves. As noted by several authors, gluons make important contributions to the cross section at small $x_\perp$ ($x_\perp \leq 0.4$)[63,64].

The disagreement in the normalization of the theory seen in Figs. 7.6 and 7.7 at low $p_\perp$ (the Λ = 0.4 GeV/c solution is about a factor of 2 low at $p_\perp$ = 2 GeV/c and W = 53 GeV) is not significant. The Λ = 0.6 GeV solution agrees better at low $p_\perp$ but the theory at present cannot be calculated precisely at these low $p_\perp$ values. At low $p_\perp$, the results depend to sensitivity on things like the unknown gluon distributions, the precise shape of the transverse momentum distributions, the low $\hat{s}$ and $\hat{t}$ cut-off employed[74-77], the choice for Q^2, and also higher order QCD corrections may be important. (For $\alpha_s(Q^2) \gtrsim 0.3$ non-perturbative effects may begin to play a role.) It may well be that the scattering of quarks and gluons as described by the leading QCD subprocesses is responsible for all the cross section down to $p_\perp$'s as low as 1.0 or 2.0 GeV/c; one simply cannot say at present.

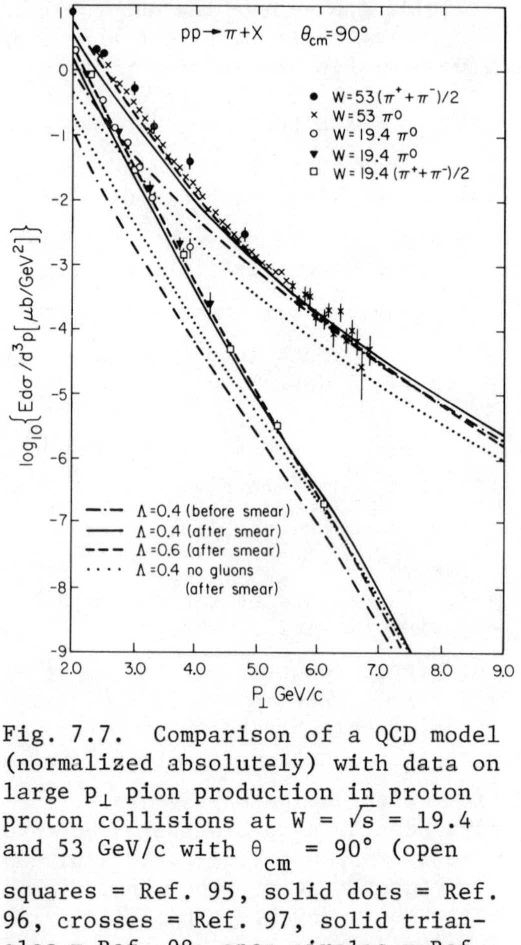

Fig. 7.7. Comparison of a QCD model (normalized absolutely) with data on large $p_\perp$ pion production in proton proton collisions at $W = \sqrt{s}$ = 19.4 and 53 GeV/c with θ_{cm} = 90° (open squares = Ref. 95, solid dots = Ref. 96, crosses = Ref. 97, solid triangles = Ref. 98, open circles = Ref. 99). The dot-dashed and solid curves are the results before and after smearing, respectively, using Λ = 0.4 GeV/c and $\langle k_\perp \rangle_{h\to q}$ = 848 MeV and the dashed curves for Λ = 0.6 GeV/c (after smearing). The contribution arising from quark-quark, quark-antiquark and antiquark-antiquark scattering (i.e., no gluons) is shown by the dotted curves (after smearing).

On the other hand, other non-leading subprocesses may play a role at low $p_\perp$ for single particle triggers particularly if these subprocesses produce single particles without being suppressed by a fragmentation function. One process that behaves like $1/p_\perp^6$ (before scale breaking) and must occur at some level is $q + g \to \pi + q$. Here one only needs to assume that quarks and gluons within the incident protons are important but estimates give a contribution from this subprocess that is only about 1/100 the size of the π cross section at W = 53 GeV, θ_{cm} = 90° and $p_\perp$ = 2.0 GeV/c. Other "CIM" subprocesses like, $\pi + q \to \pi + q$ (see Fig. 7.11), behave roughly like $1/p_\perp^8$ but require knowledge of the probability of finding a π within a proton, $G_{p\to\pi}(x)$, and are thus difficult to estimate. However, recent calculations by Blankenbecler, Brodsky and Gunion[78] and by D. Jones and J. F. Gunion[79] indicate that CIM terms may be important at low $p_\perp$ with the leading QCD processes dominating for $p_\perp \geq 5$ GeV/c. Personally, I don't believe that including CIM terms is a consistent way of handling the higher order corrections to QCD.

Figure 7.8 shows a comparison of the predicted and experimental behavior of $p_\perp^8$ times $Ed\sigma/d^3p$ for $pp \to \pi + X$ at θ_{cm} = 90° and $x_\perp$ = 0.2, 0.35 and 0.5 versus $p_\perp$. The dot-dashed and solid curves are the final results before and after smearing, respectively, with $\Lambda = 0.4$. The dashed curves are the results (after smearing) using $\Lambda = 0.6$. For the range $2.0 \leq p_\perp \leq 6.0$ GeV/c at $x_\perp = 0.2$, and $4.0 \leq p_\perp \leq 10.0$ GeV/c at $x_\perp = 0.5$, the results are roughly independent of $p_\perp$ (when multiplied by $p_\perp^8$). However, this $p_\perp^{-8}$ behavior is only a "local" effect. It holds only over a small range of $p_\perp$ (at low $p_\perp$); the region depending somewhat on $x_\perp$. As $p_\perp$ increases, the predictions approach the expected $p_\perp^{-4}$ behavior. The new data from ISR (triangles[80] and open circles[81]) shown in Fig. 7.8 do show an increase from the flat ($1/p_\perp^8$) behavior at large $p_\perp$ in agreement with the QCD expectations.

Certainly it is important to measure the cross section at large $p_\perp$ to establish if it is approaching the QCD prediction of $1/p_\perp^4$ (actually any field theory predicts this). Figure 7.9 shows the very recent data[82] from ISR on $pp \to \pi^o + X$ at large $p_\perp$ together with predictions from the QCD approach. The agreement is quite good. In addition, these data scale roughly like $1/p_\perp^6$ as indicated by the open circles which are the W = 53 GeV data scaled to W = 62.4 GeV using $1/p_\perp^6$. The theoretical predictions also behave roughly like $1/p_\perp^6$ in this kinematical region. I am a bit reluctant to claim that QCD for large $p_\perp$ meson production is now experimentally verified, but the

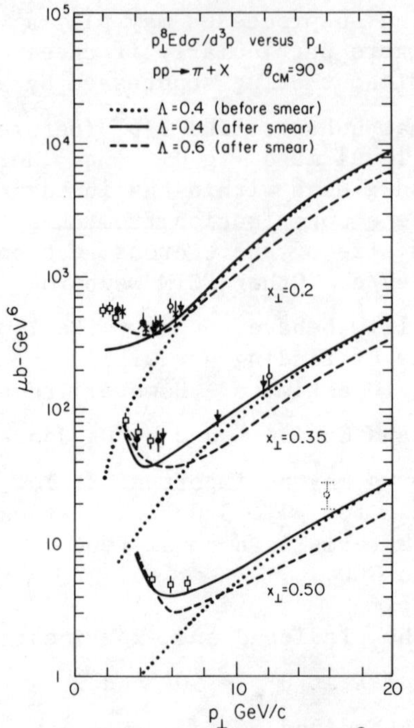

Fig. 7.8. The data on $p_\perp^8$ times $E d\sigma/d^3p$ for large $p_\perp$ pion production at $\theta_{cm} = 90°$ and <u>fixed</u> $x_\perp = 0.2$, 0.35 and 0.5 versus $p_\perp$ (open squares = Ref. 95, solid dots = Ref. 96, crosses = Ref. 97) compared with the predictions (with absolute normalization) of a model that incorporates all the features expected from QCD. The dot-dashed and solid curves are the results before and after smearing, respectively, using $\Lambda = 0.4$ GeV/c and the dashed curves are the results using $\Lambda = 0.6$ GeV/c (after smearing). Recent data (triangles = Ref. 81, open circles = Ref. 80) from the ISR do show a deviation from a straight line ($1/p_\perp^8$) behavior as expected from QCD.

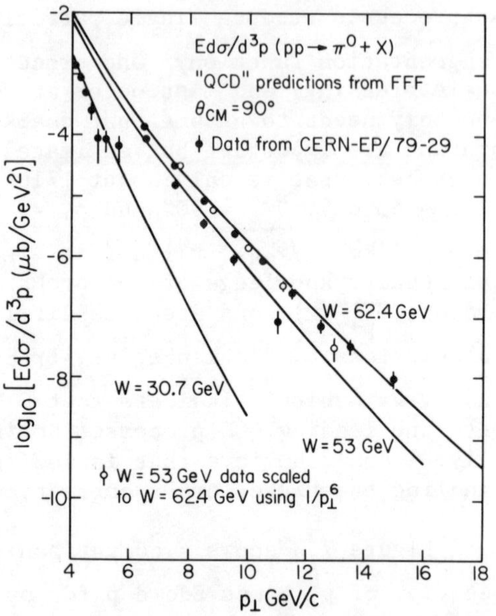

Fig. 7.9. Comparison of recent ISR data on $pp \to \pi^0 + X$ at $\theta_{cm} = 90°$ and W = 30.7, 53 and 62.4 GeV (solid dots; Ref. 82) with the "QCD" predictions from FFF (Ref. 16). The open circles are the data at W = 53 GeV scaled to W = 62.4 GeV using $1/p_\perp^6$, which is roughly the $p_\perp$ dependence of the theory in this region.

recent data from ISR are certainly encouraging for QCD.

<u>Particle Ratios</u>. As can be seen from Fig. 7.7, gluons make important contributions to the single particle rate at low $x_\perp$. However, since the gluon fragmentation function has been chosen to be considerably smaller at large z than the quark fragmentation function (see Fig. 3.4), an experiment demanding a large $p_\perp$ meson trigger is "biased" in favor of the toward-side constituent being a quark rather than

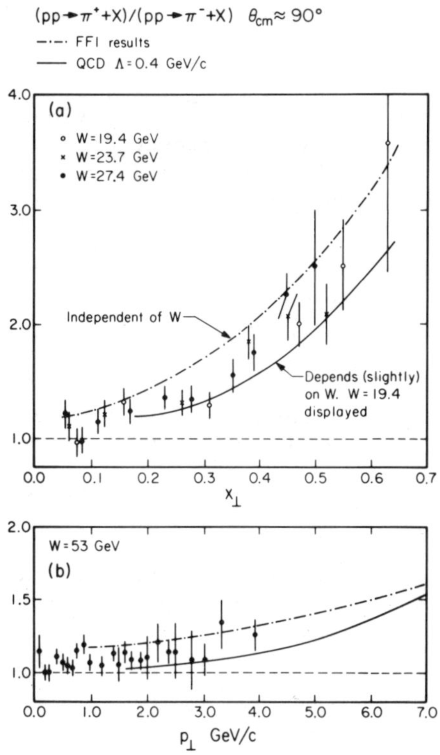

Fig. 7.10 - (a) Comparison of the data on the π^+/π^- ratio in pp collisions at $\theta_{cm} = 90°$ versus $x_\perp$ with the quark-quark scattering model of FF1[30] (which is independent of W at fixed $x_\perp$) and the QCD results using $\Lambda = 0.4$ GeV/c. The QCD results are plotted for W = 19.4 GeV and are not precisely independent of W. The π^+/π^- ratio increases at fixed $x_\perp$ and θ_{cm} as W increases (by about 20% in going from 19.4 to 53 GeV).

(b) Comparison of the data at W = 53 GeV on $pp \to (\pi^+/\pi^-) + X$ at $\theta_{cm} = 90°$ versus $p_\perp$ with the FF1 model and with the QCD results.

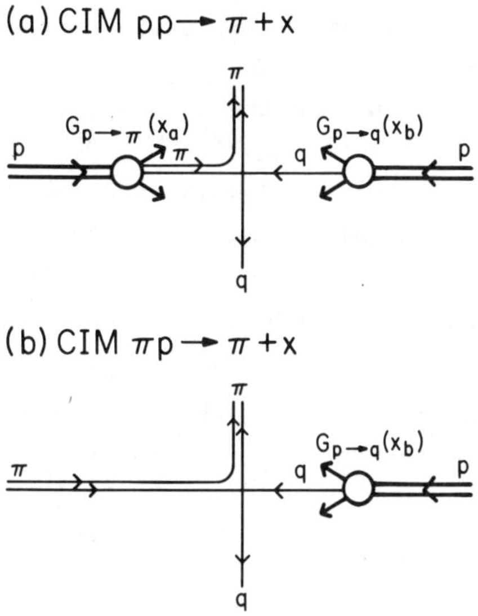

Fig. 7.11 - (a) Constituent interchange model (CIM) contribution[78,79] to large $p_\perp$ π production in proton-proton collisions from the subprocess $\pi + q \to \pi + q$.

(b) Same as (a) but for pion-proton collisions.

a gluon. At W = 53 GeV, a π^0 trigger with $p_\perp = 4.0$ GeV/c, the toward constituent is a quark (or antiquark) 72% of the time while the away constituent is a gluon 62% of the time.

This "bias" for quarks rather than gluons in the single-particle triggers means that particle ratio predictions are not very different from a model with only quarks (and antiquarks). As shown in Fig. 7.10, $pp \to (\pi^+/\pi^-) + X$ ratio predictions from QCD are only slightly smaller than the quark-quark "black-box" approach[30]. Both are in acceptable agreement with the data.

The π^-/π^+ ratio for large $p_\perp$ pions produced in π^-p collisions provides an interesting test as to the relative importance of CIM terms compared with the leading order QCD subprocesses. If the subprocess $\pi + q \to \pi + q$ in Fig. 7.11a is normalized so as to give a significant contribution to the single particle rate in $pp \to \pi + X$, then it will make a large contribution also to $\pi^-p \to \pi^- + X$[79]. This is because in the pp case, the subprocess is weighted by the small probability $G_{p\to\pi}(x)$, whereas in the π^-p case the beam supplies the π^-. Because of this, the CIM model predicts a very large π^-/π^+ ratio in π^-p collisions since the double charge exchange process $\pi^- + q \to \pi^+ + q$ is forbidden. On the other hand, the leading order QCD diagrams yield about equal amounts of π^- and π^+ for $\pi^-p \to \pi^\mp + X$. Recent data from FNAL[83] on $\pi^-p \to (\pi^-/\pi^+) + X$ at 200 GeV/c and $\theta_{cm} \approx 90°$ are shown in Fig. 7.12 together with the CIM and QCD estimates. The data indicate that the CIM subprocess $\pi + q \to \pi + q$ does not play a significant role in producing pions at 90° in π^-p (and therefore also pp) collisions.

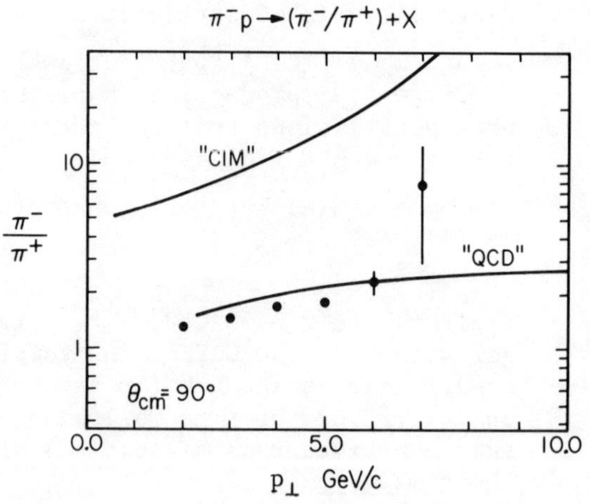

Fig. 7.12. Comparison of the "CIM" and "QCD" predictions for $\pi^-p \to (\pi^-/\pi^+) + X$ at $\theta_{cm} = 90°$ and $p_{beam} = 400$ GeV/c with the recent FNAL data of Ref. 83.

P-out Distributions

The use of an effective transverse momentum $\langle k_\perp \rangle_{h\to q} = 848$ MeV as determined from the muon pair data results in mean P-out values are quite large. (P-out is defined in Fig. 7.13.) As can be seen in Fig. 7.14, the new results are in reasonable agreement with the hadron data, although the predicted <P-out> is still a bit too small. Some of the discrepancy may be due to contributions from the beam and target jets which have not been included. Also, it may be that one should be using a slightly larger $\langle k_\perp \rangle_{h\to q}$ in the ISR range W = 53 GeV since the

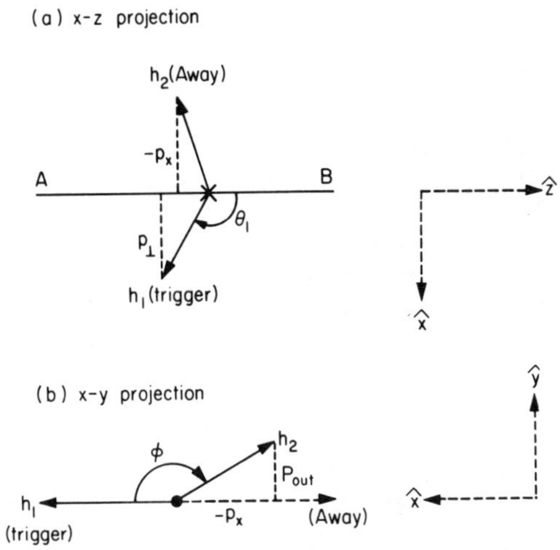

Fig. 7.13. Definition of kinematic variables used in describing the process $A + B \to h_1 + h_2 + X$: (a) x - z projection, where the beam, target and trigger hadron h_1 form this plane: (b) x - y projection.

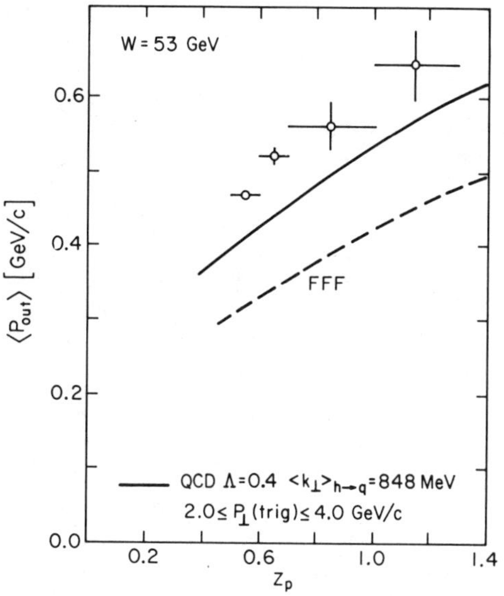

Fig. 7.14. The dependence on z_p of the mean value of the $|P_{out}|$ of away-side charged hadrons at W = 53 GeV and $2.0 \leq p_\perp(\text{trig}) \leq 4.0$ GeV/c with θ_1 averaged over 45° and 20° from the CCHK collaboration[68] on $pp \to h_1^\pm + h_2^\pm + X$, where $z_p = -p_x(\text{away})/p_\perp(\text{trig})$ (see Fig. 7.13). The predictions from the QCD approach at $\theta_1 = 45°$ with $\Lambda = 0.4$ GeV/c, and $\langle k_\perp\rangle_{h\to q} = 848$ MeV, and $\langle k_\perp\rangle_{q\to h} = 439$ MeV (solid curve) and the results of FFF1[67] (dashed curve) curve are shown.

effective $\langle k_\perp\rangle$ distribution should increase slightly with increasing energy as seen in Fig. 5.5.

We have not really done a proper analysis of the P-out distribution. What one should do (and many are undoubtedly working on this) is to include 2 → 3 process explicitly in the large $p_\perp$ analysis similar to what was done for $pp \to \mu^+\mu^- + X$. The primordial motion could then be set at the value determined from $pp \to \mu^+\mu^- + X$ (i.e., $\langle k_\perp\rangle_{\text{primordial}} = 600$ MeV or perhaps less). One would then predict a large momentum tail to the P-out distribution. It would not be expected to be bounded (like a Gaussian) and the tail would increase at increasing energies. There is, at present, no experimental evidence for a large momentum tail to the P-out distribution, although <P-out> is large. As Fig. 7.15 shows, it looks

Gaussian, but so does the $\mu^+\mu^-$ spectrum in Fig. 5.5.

The "Jet" Cross Section

A dramatic prediction of the QCD parton approach is the size of the cross section for producing a jet (parton = quark, antiquark or gluon) of momentum $p_\perp$ compared to that for producing a single particle at the same $p_\perp$. In this approach, the single particle trigger always comes from a parton carrying more $p_\perp$ (typically about 15% more for quarks and greater for gluons) than the trigger particle. Furthermore, the chance of a parton fragmenting into hadrons in such a way that one particle carries almost all the momentum is small (only a few percent) as can be seen in Fig. 3.5. These two effects combine to give the large $\sigma(pp \to \text{jet} + X)/\sigma(pp \to \pi^o + X)$ ratio shown in Fig. 7.16[84]. In the QCD approach, this ratio does not scale (i.e., it is a function of $x_\perp$, θ_{cm} and W).

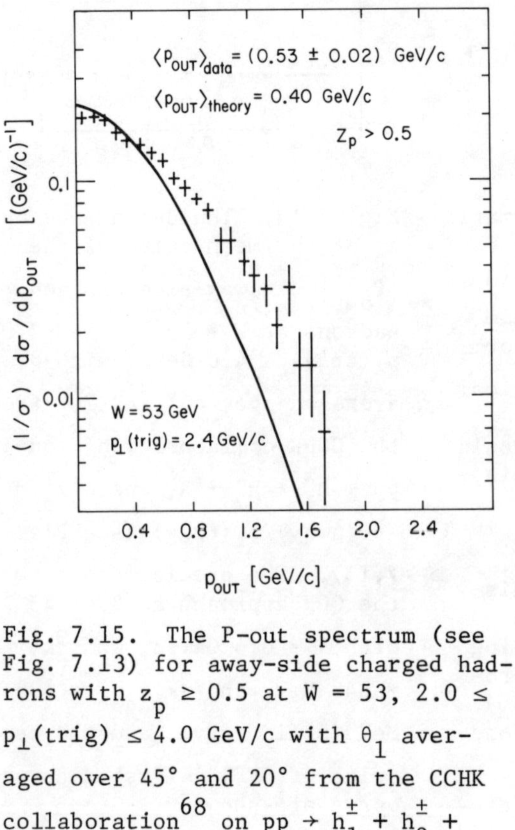

Fig. 7.15. The P-out spectrum (see Fig. 7.13) for away-side charged hadrons with $z_p \geq 0.5$ at W = 53, $2.0 \leq p_\perp(\text{trig}) \leq 4.0$ GeV/c with θ_1 averaged over 45° and 20° from the CCHK collaboration[68] on $pp \to h_1^\pm + h_2^\pm + X$. The prediction from the QCD approach at $\theta_1 = 45°$ with $\Lambda = 0.4$ GeV/c, $\langle k_\perp\rangle_{h\to q} = 848$ MeV and $\langle k_\perp\rangle_{q\to h} = 439$ MeV is shown by the solid curve.

The cross section for producing a "jet" of particles whose transverse momentum sum to give $p_\perp$ has been measured now by two groups[69,85] and is shown in Fig. 7.17. The measured jet rate is several orders of magnitude greater than the π^o rate and is in qualitative agreement with the QCD predictions.

It is extremely difficult to make precise quantitative comparisons with the jet data in Fig. 7.17. Theoretically what is shown in Fig. 7.16 and Fig. 7.17 is the cross section for producing a quark (or gluon) with a given momentum (in Fig. 7.17 it is divided by the π^o cross section at the same momentum). However, as discussed in Ref. 17, quarks of a given momentum (equal to their

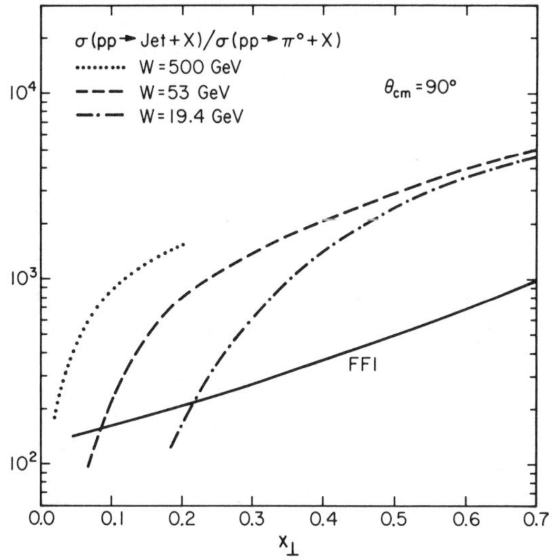

Fig. 7.16. Prediction of the jet to single π^0 ratio at $\theta_{cm} = 90°$ versus $x_\perp$ for W = 500, 53 and 19.4 GeV from the QCD approach using $\Lambda = 0.4$ GeV/c. The jet cross section is defined as the cross section for producing a parton (quark, antiquark and glue) with the given $x_\perp = 2\, p_\perp/W$. Also shown is the prediction from the quark scattering model of FF1[30] which is independent of W at fixed $x_\perp$ and θ_{cm}.

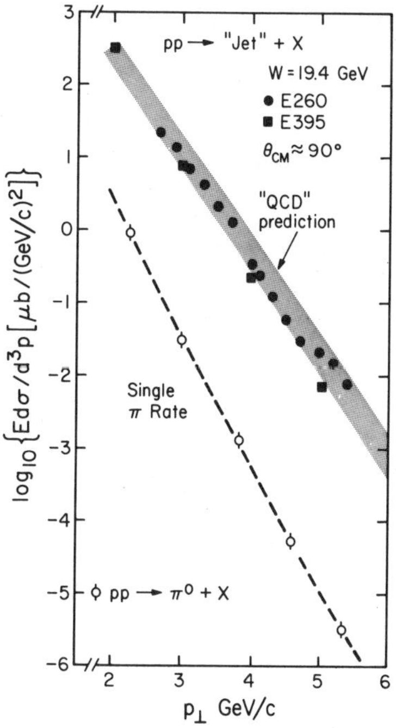

Fig. 7.17. Comparison of the jet and single π^0 cross sections measured at 200 GeV/c (W = 19.4 GeV) and $\theta_{cm} \approx 90°$. The jet data are from two FNAL experiments, E260[85] and E395[69], where a jet is defined as the sum of all particles into their respective detectors. Also shown is the QCD prediction for the cross section of producing a parton (quark, antiquark or gluon) at $\theta_{cm} = 90°$ and W = 19.4 GeV with a given energy equal $p_\perp$.

energy) cannot produce jets with the momentum of all particles equal to the energy of all particles. Our jet model[17] gives $E_{tot} - p_{z_{tot}} \approx$ 1.2 GeV for quark jets. Since the cross section for producing jets falls so steeply, the cross section for producing a jet with a given $p_{z_{tot}}$ is considerably smaller than that for producing one with a given E_{tot}. As explained in Ref. 86, it is the former that is more closely connected to what is measured experimentally. At W = 19.4 GeV/c, the cross section to produce a jet where $p_{z_{tot}}$ = 5 GeV/c at 90° is about 10 times smaller than the cross section to produce a jet whose E_{tot} = 5 GeV/c (see Fig. 7.18). The difference between E_{tot} and $p_{z_{tot}}$ of a jet arises, of course, from low momentum particles that have energy due to their mass (or $k_\perp$) but have little momentum p_z. This is tangled with the experimental uncertainty in all hadron jet experiments concerning low $p_\perp$ particles. One cannot be sure that one is not losing the low $p_\perp$ jet particles that are not well collimated or gaining low $p_\perp$ background from the beam and target jets in Fig. 7.1. Only by doing a very careful analysis, including the precise acceptances of a given experiment, can one make any quantitative statements.

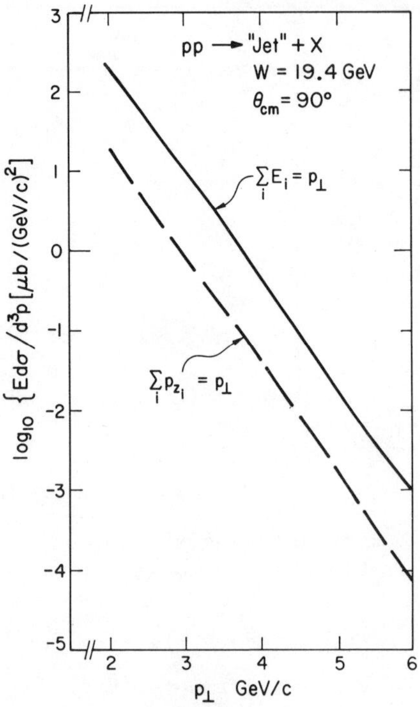

Fig. 7.18. Comparison of two different definitions of the cross section for producing a "jet" at θ = 90° and W = 19.4 GeV. The first is the cross section for producing a jet of particles whose total energy is $p_\perp$; the second is the cross section for producing a jet of particles whose $\hat{z}$-component (component along the quark direction of momentum sums to give $p_\perp$. The quark jet model in Ref. 17 was used to calculate the difference between the total energy and total p_z of all hadrons in a jet.

The "bias" in favor of toward-side quarks, discussed previously does not occur when one triggers on jets rather than on single particles and thus gluons make up a sizable fraction of the jet cross section. With our guesses for the gluon distributions, gluons are responsible for 73% of the jet triggers at $p_\perp$ = 4 GeV/c, W = 53 GeV and θ_{cm} = 90°. Even at higher $x_\perp$ values like $p_\perp$ = 6.0 GeV/c, W = 19.4 GeV, θ_{cm} = 90°, gluons still make up 45% of the jets. One might hope someday to distinguish experimentally between gluon and quark jets. The gluon jets are assumed to have a higher multiplicity of particles each with lower momentum on the average. In addition, unlike the quark jets discussed in Ref. 17, gluon jets will carry on the average no net charge (or strangeness, etc.).

Very High Energy Expectations

Figure 7.8 shows that the QCD predictions begin to deviate from a $1/p_\perp^8$ behavior (at fixed $x_\perp$) as $p_\perp$ increases yielding a much larger cross section than expected from a $p_\perp^{-8}$ model. This is also seen in Fig. 7.19 where the QCD predictions for $p_\perp^8$ times $Ed\sigma/d^3p$ versus $p_\perp$ at $x_\perp = 0.05$ and $\theta_{cm} = 90°$ are plotted. At W = 500 GeV, the QCD results are a factor of 100 greater than a straight $(1/p_\perp^8)$ extrapolation and show a factor of 1000 increase at W = 1000 GeV. Figure 7.20 shows the predictions for 90° π^o and jet production at W = 53, 500 and 1000 GeV versus $p_\perp$. The $p_\perp$ = 30 GeV/c 90° π^o cross section at W = 500 GeV is predicted in the QCD approach to be about the same magnitude as that measured at $p_\perp$ = 6.0 GeV/c at Fermilab (W = 19.4 GeV)!

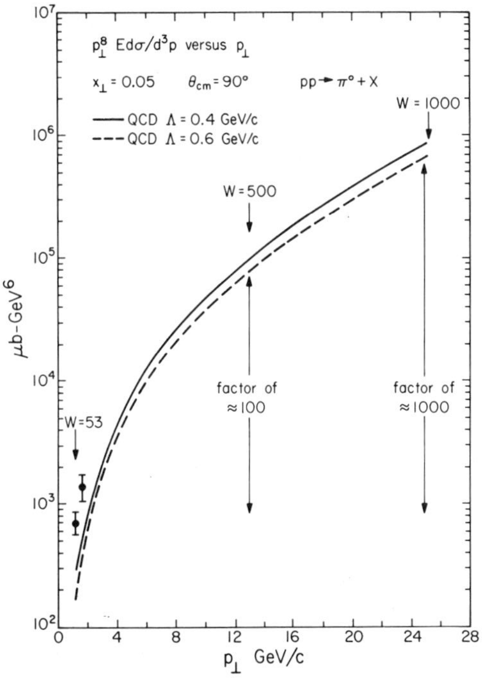

Fig. 7.19. The behavior of $p_\perp^8$ times the 90° single π^o cross section, $Ed\sigma/d^3p$, at $x_\perp = 0.05$ versus $p_\perp$ calculated from the QCD approach with Λ = 0.4 GeV/c (solid curve) and Λ = 0.6 GeV/c (dashed curve). The two low $p_\perp$ data points are at W = 53 and 63 GeV[96]. The predictions are a factor of 100 (1000) times larger than the flat $(p_\perp^{-8})$ extrapolation to W = 500 GeV (1000 GeV).

It is not clear yet precisely what the quark and gluon jets will look like at very high $p_\perp$ (like $p_\perp$ = 30 GeV/c). If QCD is correct, they will certainly not look like the well collimated $\langle k_\perp \rangle_{q\to h}$ = 430 MeV objects used in this analysis. At $p_\perp$ = 30 GeV/c, they should "appear" to be fatter. This is because as the $p_\perp$ of the outgoing quark increases, it becomes increasingly likely that it radiate a hard gluon and become two jets (one quark and one gluon). Then, this quark or gluon might radiate producing still more subjets. This is the same mechanism that is responsible for the scale breaking of the fragmentation functions $D(z,Q^2)$ discussed in Section III. The net result is that most of the time it will look as if there is one somewhat fatter jet (with the fatness increasing with increasing momentum); however,

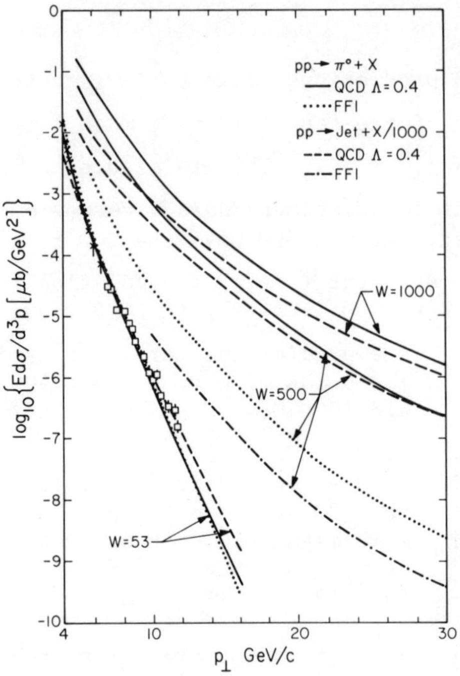

Fig. 7.20. Comparison of the results on the 90° π^o cross section, $Ed\sigma/d^3p$, from the QCD approach with $\Lambda = 0.4$ GeV/c (solid curve) and the quark-quark "black-box" model of FFF1 (dotted curves)[67]. Both models agree with the data at W = 53 GeV (crosses = Ref. 97) where the open squares are the "preliminary" data from the CCOR collaboration[81] normalized to agree with the lower $p_\perp$ experiments. The QCD approach results in much larger cross sections than the FFF1 model at W = 500 and 1000 GeV. The FFF1 results at 1000 GeV (not shown) are only slightly larger than the results at 500 GeV. Also shown are the cross sections for producing a jet at 90° (divided by 1000) as predicted by the QCD approach (dashed curves) and the FFF1 model (dot-dashed curve).

occasionally when the radiation is hard enough, one will see two or three subjets.

Figure 7.21 shows the leading order QCD predictions for $pp \to \pi^o + X$ and $pp \to$ Jet + X at W = 500, 1000, 2000 and 40,000 GeV. One should not take these predictions too seriously since I have extrapolated the leading order formulas quite a long way. Higher order corrections could be important for such long extrapolations (i.e., $\alpha_s(Q^2) - \alpha_s(Q_o^2)$ is large - see (6.22)). Nevertheless, it is clear that QCD predicts quite large cross sections at these energies. This is also seen in Fig. 7.22 and Fig. 7.23 where I have plotted the integrated spectrum for $pp \to$ Jet + X.

<u>Direct Photons at Large $p_\perp$</u>[87-91]

As shown in Fig. 7.7, gluons are responsible for a sizable portion of the large $p_\perp$ π^o cross section. In fact, the dominant subprocess at W = 53 GeV, $p_\perp = 4$ GeV/c, $\theta_{cm} = 90°$ is quark-gluon scattering, $g + q \to g + q$. If gluons participate in this subprocess, then necessarily they must produce direct large $p_\perp$ photons by the process, $g + q \to \gamma + q$, as illustrated in Fig. 7.24. Even though the process $g + q \to \gamma + q$ is down by α_{QED}/α_s relative to $g + q \to g + q$, when comparing the rate for large $p_\perp$ photons to that for producing, say π^o's, it is

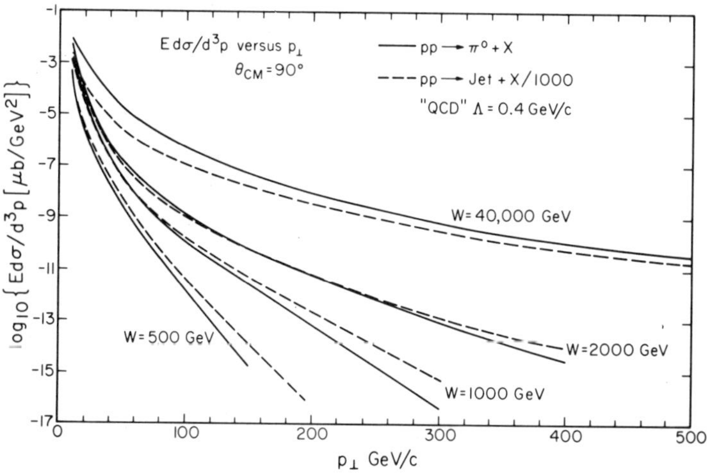

Fig. 7.21. Leading order QCD predictions for pp → π^0 + X (solid curves) and pp → Jet + X (divided by 1000, dashed curves) at 90° using Λ = 0.4 GeV/c. The predictions are plotted versus $p_\perp$ and given at W = $\sqrt{s}$ = 500, 1000, 2000 and 40,000 GeV.

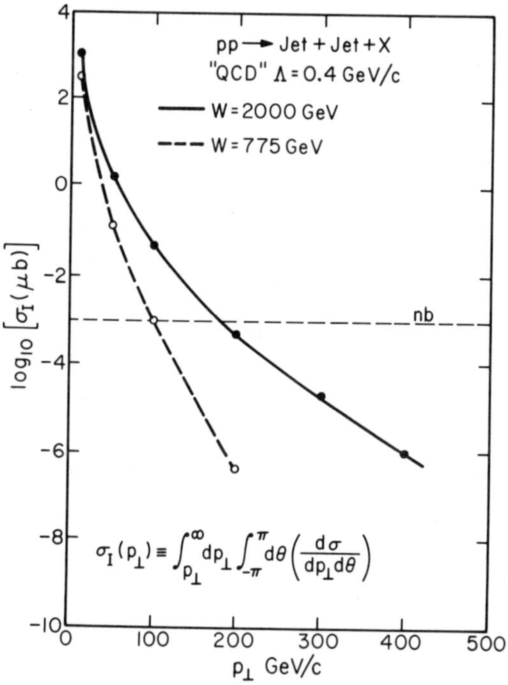

Fig. 7.22. Leading order QCD predictions for the integrated spectrum in pp → Jet + X at W = $\sqrt{s}$ = 775 and 2000 GeV.

enhanced since this latter must proceed via a quark or gluon fragmentation function.

Figure 7.25 shows the predicted invariant cross sections, $Ed\sigma/d^3p$, for pp → γ + X at W = 53 and 19.4 GeV and θ_{cm} = 90° (before and after smearing with $\langle k_\perp \rangle_{h\to q}$ = 848 MeV) compared to the observed π^0 rate. At $p_\perp$ = 14 GeV/c, θ_{cm} = 90° and W = 53 GeV, one expects about as many direct photons as π^0's! These photon events are quite distinctive. They occur with a photon at large $p_\perp$

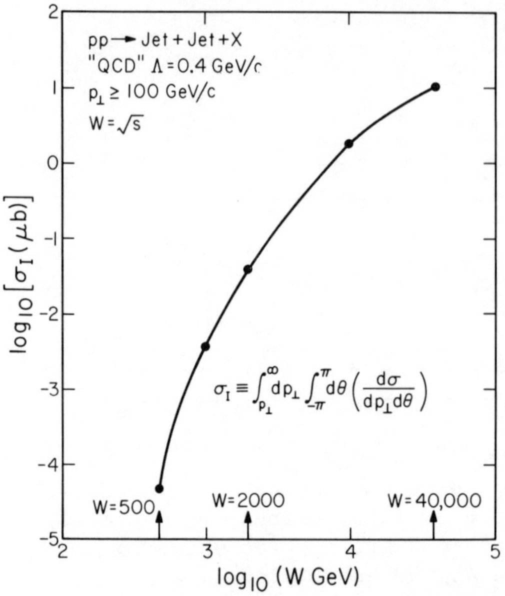

Fig. 7.23. Leading order QCD predictions for the total rate for producing jets with $p_\perp$ greater than 100 GeV/c versus $W = \sqrt{s}$.

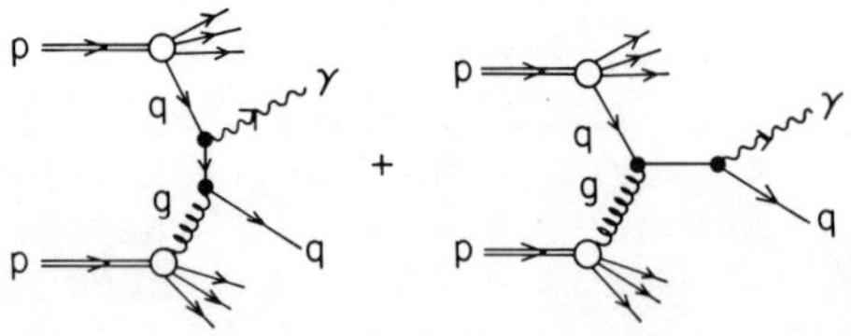

Fig. 7.24. Diagrams for the production of direct photons, γ, at large $p_\perp$ in pp collisions from the "Compton" subprocess $gq \to \gamma q$.

on the trigger side with no accompanying toward-side hadrons. The away-side hadrons come from the fragmentation of a quark. On the other hand, the production of photons due to Bremsstrahlung from, say, $q + q \to q + q$ results in events where the photon is produced in association with other trigger-side hadrons. In addition, the γ/π^0 rate for photons produced via Bremsstrahlung is only about 2 - 5%.

It is interesting that in ep collisions one probes the quark distributions with an incoming virtual photon, while studying large $p_\perp$ real photons in $pp \to \gamma + X$, one can probe the distribution of gluons within the proton through the subprocess $g + q \to \gamma + q$. If one does not find a reasonable rate for producing γ's at large $p_\perp$, the QCD approach as I have outlined it will be in trouble.

Three Large $p_\perp$ Jets in pp Collisions

Measurements of P-out. In the QCD approach, one expects a broad P-out distribution for mesons produced out of the production plane in pp collisions like that observed in Fig. 7.15. This lack of coplanarity is due to the presence of two-to-three subprocess like $qq \to qqg$ as illustrated in Fig. 7.2 and Fig. 7.3b. One expects a large momentum tail of the P-out distribution due, for example, to the Bremsstrahlung radiation of a hard gluon from an outgoing quark in $qq \to qq$. The tail should increase with increasing trigger $p_\perp$.[92]

Measurements of the Three Particle Cross Section. One way to test for three jet events is to simply measure the rate for simultaneously producing three large $p_\perp$ particles all at large angles to each other[92-94]. For example, one could measure the rate for producing three particles all with $p_\perp > 3$ GeV/c and all at 90° to the beam and all 120° degrees apart. The rate will, of course, be small, but it will be orders of magnitude larger if three jet events exist than if they don't. Efforts to estimate these rates are in progress.

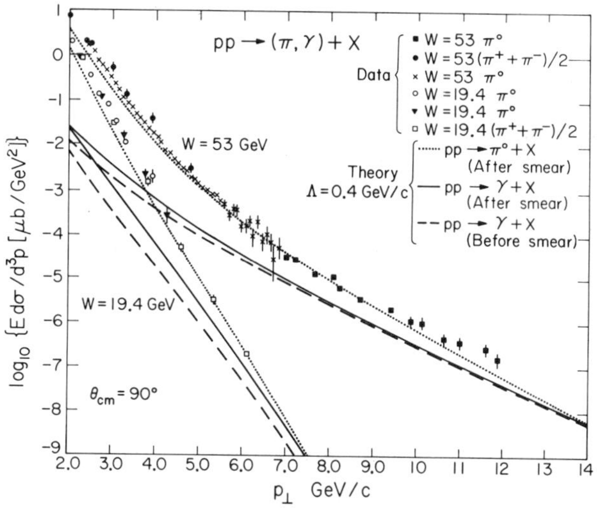

Fig. 7.25. Predicted cross section, $Ed\sigma/d^3p$, for producing direct photons, $pp \to \gamma + X$, at $\theta_{cm} = 90°$ and $W = 19.4$ and 53 GeV, where the photons are produced by the "Compton" subprocess $gq \to \gamma q$ illustrated in Fig. 7.24. The results are shown before and after smearing with $\langle k_\perp \rangle_{h\to q} = \langle k_\perp \rangle_{h\to g} = 848$ MeV and are compared with the data on $pp \to \pi^0 + X$ from Figs. 7.7 and 7.20.

VIII. SUMMARY AND CONCLUSIONS

Many of the calculations discussed here should be considered as "crude" phenomenological attempts to examine experimental consequences of QCD. The theory of QCD is, however, more than a phenomenological model. It is a precise and complete theory purporting to be an ultimate explanation of all hadronic experiments at all energies, high and low. There are many reasons to hope and expect it to be right. The question is, is it indeed right? Mathematical complexity has, so far, prevented quantitative testing its correctness. The theory itself is remarkably simple and beautiful; however, what it predicts is not yet clearly known. Nevertheless, its property of asymptotic freedom leads one to expect that phenomena of high momentum transfer should be analyzable (by perturbation) and some applications of the theory have been examined here. Unfortunately, most processes involve both low and high energy aspects and ways of separating the low energy (or Q^2) pieces that cannot be calculated by perturbation from the high Q^2 perturbative corrections are just now becoming understood.

One reason to view many of the present day applications of QCD as preliminary is because calculations to leading order are only meaningful if one shows that the next order corrections are indeed small. For example, although to leading order the antiquark distributions as measured in neutrino and antineutrino interactions are the same as the antiquark distribution one should use in the Drell-Yan calculation, there are higher order corrections that vitiate this direct connection. In fact, calculations indicate that for the total muon pair rate, $d\sigma_{DY}/dM^2$, the order α_s corrections are quite large. So large that one must go to $M^2 \approx 10{,}000\ \mathrm{GeV}^2$ before one could believe the leading order result. This may be true for other observables and other processes, and, if so, it represents an important new problem for perturbative QCD. Since we are making an expansion in $1/\log Q^2$, one must go to large $\log Q^2$ (not just large Q^2) before one can neglect higher terms and trust the leading order results (in leading order, we have already summed all terms of the form $(\alpha(Q^2) \log Q^2)^N$). I believe that for those cases where the coefficient of the order α_s term is large, one must develop means of summing the perturbation series if one is going to make precise predictions at present or foreseeable Q^2 values.

One place where the next order corrections are not too important (for $0.2 \leq x \leq 0.8$) is in the QCD equations for the Q^2 dependence for the parton distributions in lepton scattering experiments[100]. Recent data from ep, μp and νp experiments do show clearly the "scale" breaking effects expected from QCD. However, the bulk of the data is at relatively low Q^2 and results are sensitive to $1/Q^2$ contributions that are difficult, if not impossible, to calculate (they involve knowledge of, for example, the primordial transverse momentum). In addition, the high Q^2 data are not accurate enough to allow a precise determination of the parameter Λ. Any Λ in the range $0.01 \leqslant \Lambda \leqslant 0.8$ is acceptable.

The transverse momentum of muon pairs is certainly larger than one would have expected from the naive parton model; however, the experiments have not really seen the high $p_\perp$ tail predicted by QCD. There is some evidence to suggest that $\langle p_\perp \rangle_{\mu^+\mu^-}$ does increase with increasing energy (fixed M^2) as expected. But, there are no data to check the approach to a constant of $W^5\, d\sigma/dMdyd^2k_\perp$ as W increases, shown in Fig. 5.7. Finally, there is the question of why the primordial transverse momentum still comes out as large as 600 MeV even after one includes the first order QCD perturbative corrections.

At one time it was thought that the experimentally observed $p_\perp^{-8}$ behavior of large $p_\perp$ meson production in hadron-hadron collisions might pose a problem for QCD. However, we now see that there is no problem[101]. The energy ($p_\perp$) of existing experiments is too low and there are too many non-asymptotic effects acting. All the scale breaking effects act in the same direction to produce an effective apparent $p_\perp$ power that is roughly eight at low $p_\perp$. In addition, the predicted size of the invariant cross section is just about right. Results closer to a $p_\perp^{-4}$ fall off should appear only at much higher $p_\perp$. Furthermore, one has indirect evidence from large $p_\perp$ correlations that gluons as well as quarks must be included in a description of the data.

The uncertainty in the perturbative parameter Λ and the necessity of separating off uncalculable low Q^2 phenomena from the calculable perturbative high Q^2 parts make precise perturbative predictions difficult. Most perturbative predictions are qualitative and agree with data qualitatively. This is unsatisfactory, but I see no improvement of this situation in the near future. The theory will be tested by an increasing number of qualitative successes over a wide range of phenomena rather than one or two precise tests. Precise tests will come when we understand the low Q^2 non-perturbative regime.

We conclude that there is no evidence from eN, μN, νN, $\bar{\nu}$N interactions or $pp \to \mu^+\mu^- + X$, or large $p_\perp$ production in hadron-hadron experiments against QCD. On the contrary, the overall picture favors a QCD approach. However, most of the conclusive and exciting predictions of the theory have not yet been seen. The next generation proton-proton machines should see hundreds or even thousands of times more mesons at large $p_\perp$ than expected from extrapolations of existing data. One should occasionally see three distinct jets in e^+e^- collisions and at large $p_\perp$ in pp collisions. One should see gluon jets as well as quark jets.

QCD is not just "another theory." If it is not the correct description of nature then it will be quite some time before another candidate theory emerges.

Acknowledgments

It is a pleasure to thank my collaborators at Caltech, R. P. Feynman, G. C. Fox, D. A. Ross and S. Wolfram without whom these lectures could not have been given. In addition, I would like to acknowledge useful discussions with A. J. Buras, R. K. Ellis, D. Gross and H. D. Politzer. I have benefited greatly from the parallel lectures by S. D. Ellis and C. H. Llewllyn Smith. We all said the same thing (I think) but from three different points of view. I thank Rajan Gupta for a careful reading of sections of this write-up. Finally, let me congratulate K. T. Mahanthappa on a most enjoyable and stimulating summer workshop.

Footnotes and References

1. H. D. Politzer, Physics Reports 14C (1974).
2. D. Gross and F. Wilczek, Phys. Rev. D8, 3633 (1973); Phys. Rev. D9, 980 (1974).
3. M. Gell-Mann, Phys. Rev. 125, 1067 (1962).
4. C. G. Callan, R. F. Dashen and D. J. Gross, Lectures presented at the La Jolla Institute, 1978; AIP Conference Proceedings, No. 55, Subseries No. 18.
5. R. D. Field, Lectures presented at the La Jolla Institute, 1978; AIP Conference Proceedings, No. 55, Subseries No. 18.
6. G. Altarelli, R. K. Ellis and G. Martinelli, "Large Perturbative Corrections to the Drell-Yan Process in QCD," MIT preprint CTP# 776 (1979); G. Altarelli, R. K. Ellis, G. Martinelli and So-Young Pi, "Processes Involving Fragmentation Functions Beyond the Leading Order in QCD," MIT preprint CTP#793. Also, G. Altarelli, R. K. Ellis and G. Martinelli, Nucl. Phys. B143, 521 (1978), Erratum, Nucl. Phys. B146, 544 (1978).
7. Geoffrey C. Fox and Stephen Wolfram, Nucl. Physics B149, 413 (1979).
8. C. L. Basham, L. S. Brown, S. D. Ellis and S. T. Love, Phys. Rev. D17, 2298 (1978); C. L. Basham, L. S. Brown, S. D. Ellis and S. T. Love, Phys. Rev. Letters 41, 1585 (1978).

9. R. K. Ellis, H. Georgi, M. Machacek, H. D. Politzer and G. G. Ross, Phys. Lett. 78B, 281 (1978); Nucl. Phys. B152, 285 (1979).
10. D. Amati, R. Petronzio and G. Veneziano, Nucl. Phys. B146, 29 (1978).
11. S. Libby and G. Sterman, Phys. Rev. D18, 3252 (1978).
12. A. H. Mueller, Phys. Rev. D18, 3705 (1978).
13. W. Bardeen, A. J. Buras, D. W. Duke and T. Muta, Phys. Rev. D18, 3998 (1978).
14. S. D. Ellis, These proceedings.
15. C. H. Llewellyn Smith, These proceedings. Also, A. B. Carter and C. H. Llewellyn Smith, "Perturbative QCD in a Covariant Gauge," University of Oxford preprint 34/79 (1979).
16. (FFF) R. P. Feynman, R. D. Field and G. C. Fox, Phys. Rev. D18, 3320 (1978).
17. R. D. Field and R. P. Feynman, Nucl. Phys. B136, 1 (1978).
18. See K. Konishi, A. Ukawa and G. Veneziano, "A Simple Algorithm for QCD Jets," CERN preprint TH2509 (1978); K. Konishi, A. Ukawa and G. Veneziano, "Jet Calculus: A Simple Algorithm for Resolving QCD Jets," Rutherford preprint RL-79-026 (1979).
19. S. Brodsky and J. Gunion, Phys. Rev. Letters 37, 402 (1976); S. Brodsky, invited talk at the XII Rencontre de Moriond (1977), (SLAC-PUB-1937).
20. R. P. Feynman, "Photon-Hadron Interactions," (Benjamin, Reading, Mass., 1972).
21. G. Altarelli and G. Parisi, Nucl. Phys. B126, 298 (1977).
22. R. D. Field and D. A. Ross (in preparation).
23. G. C. Fox, Nucl. Phys. B131, 107 (1977).
24. D. J. Gross, Phys. Rev. Letters 32, 1071 (1974).
25. G. Altarelli and G. Martinelli, Phys. Letters 76B, 89 (1978).
26. J. G. H. de Groot et al., Z. Physik C1, 143 (1979).
27. P. Bosetti et al., Nucl. Phys. B142, 1 (1978).
28. See also, L. F. Abbott, "Topics in QCD Phenomenology of Deep-Inelastic Scattering," invited talk presented at the Orbis Scientiae, Coral Gables, 1979; L. F. Abbott and R. M. Barnett, SLAC-PUB-2325 (submitted to Ann. of Phys.).
29. R. Barbieri, J. Ellis, M. K. Gaillard and G. G. Ross, Phys. Letters 64B, 171 (1976).
30. (FF1) R. D. Field and R. P. Feynman, Phys. Rev. D15, 2590 (1977).
31. J. Kubar-Andre and F. E. Paige, Phys. Rev. D19, 221 (1979).
32. J. Abad and B. Humpert, Phys. Letters 78B, 627 (1978).
33. K. Harada, T. Kaneko and N. Sakai, CERN preprint TH2619 (1979) and erratum.
34. H. D. Politzer, Nucl. Phys. B129, 301 (1977) and CALT-68-628 (1977).
35. C. T. Sachrajda, Phys. Letters 73B, 185 (1978).
36. G. Altarelli, G. Parisi and R. Petronzio, Phys. Lett. 76B, 351 (1978); Phys. Lett. 76B, 356 (1978); R. Petronzio, CERN preprint TH-2495 (1978).

37. H. Fritzsch and P. Minkowski, Phys. Lett. 73B, 80 (1978).
38. C. Michael and T. Weiler, contribution to the XIIIth Rencontre de Moriond, Les Arcs, France (1978).
39. K. Kajantie and R. Raitio, Nucl. Phys. B136, 72 (1978); K. Kajantie, J. Lindfors and R. Raitio, Helsinki report HU-TFT-78-18; K. Kajantie and J. Lindfors, Helsinki report HU-TFT-78-33.
40. F. Halzen and D. Scott, University of Wisconsin report COO-881-21 (1978).
41. E. L. Berger, "Massive Lepton Pair Production in Hadronic Collisions," invited talk at the International Conference at Vanderbilt University (1978) ANL-HEP-PR-78-12; E. L. Berger, "Tests of QCD in the Hadroproduction of Massive Lepton Pairs," ANL-HEP-PR-78-18.
42. It is a bit dangerous to neglect the subprocess $q + q \rightarrow q + q + \gamma^*$, for although this subprocess is down by α_s from $g + q \rightarrow \gamma^* + q$ and $q + \bar{q} \rightarrow \gamma^* + g$, there are many more quarks at high x in a proton than there are gluons or antiquarks. However, recently it has been shown that the subprocess $q + q \rightarrow q + q + \gamma^*$ makes a negligible contribution over most of the kinematic range. See, J. Kripfganz and A. P. Contogouris, McGill University preprint (1979).
43. The region of integration in eq. (5.27) is discussed in detail in Ref. 39.
44. D. C. Hom et al., Phys. Rev. Lett. 36, 1239 (1976) and 37, 1374 (1976); S. W. Herb et al., Phys. Rev. Lett. 39, 252 (1977); W. R. Innes et al., Phys. Rev. Lett. 39, 1240 (1977).
45. This is assuming that the order α_s^2 corrections to the $p_\perp$ spectrum are not so large as to vitiate the conclusions based on calculations of order α_s.
46. One must be careful for if the Δf_q^{DY} (or equivalently the ΔB_n^{DY}) are too large, then according to (6.22b) there will be a sizable correction to the Q^2 evolution formula.
47. Y. L. Dokshitser, D. I. D'Yakonov and S. I. Troyan, "Inelastic Processes in Quantum Chromodynamics," from the XIIIth Winter School of Leningrad B. P. Konstantinov Institute of Nuclear Physics (1978), SLAC translation #183.
48. G. Parisi and R. Petronzio, "Small Transverse Momentum Distributions in Hard Processes," CERN-TH-2627 (1979).
49. Geoffrey C. Fox and Stephen Wolfram, "A Gallimaufry of e^+e^- Annihilation Event Shapes," CALT-68-723 (1979).
50. C. Y. Lo and J. D. Sullivan, "Transverse Momentum Distributions in Drell-Yan Processes," University of Illinois preprint, ILL-(th)-79-22 (1979).
51. Stephen Wolfram, "Jet Development in Leading Log QCD," CALT-68-740 (1979).
52. For further leading log applications, see Ref. 51.

53. H. D. Politzer, Phys. Lett. 70B, 430 (1977); H. Georgi and H. D. Politzer, Phys. Rev. Lett. 40, 3 (1978).
54. H. Georgi and H. D. Politzer, Phys. Rev. D14, 1829 (1976); A. De Rújula, H. Georgi and H. D. Politzer, Ann. Phys. (N.Y.) 103, 351 (1977).
55. A. Peterman, "Renormalization Group and the Deep Structure of the Proton," CERN-TH-2581 (1978).
56. A. J. Buras, "Asymptotic Freedom in Deep Inelastic Processes in the Leading Order and Beyond," Lectures given at the VIth International Workshop on Weak Interactions, Iowa, 1978, Fermilab-Pub-79/17-THY.
57. T. Kinoshita, J. Math. Phys. 3, 650 (1962).
58. T. D. Lee and M. Nauenberg, Phys. Rev. 133B, 1547 (1964).
59. In higher orders in QCD presumably summing over initial or final degenerate states (as in (6.28)) is sufficient to remove leading double logs and allow a factorizing result. See, Refs. 9-12.
60. It is not at all obvious that the constant terms (i.e., the $\Delta f(z)$ in (6.26)) are the same in this scheme. In principle, they could depend on, for example, the joint probability of finding a quark at x_1 and a gluon at x_2 in the incident proton.

That this does not happen in deep inelastic scattering is guaranteed by the operator product expansion (see Ref. 55). That this does not happen in processes like Drell-Yan is not so clear, but since the diagram structure is similar (see Ref. 15), presumably everything is OK and the constant terms are calculable and unique.
61. C. T. Sachrajda, Phys. Letters 76B, 100 (1978).
62. See also, W. Furmanski, "Large $p_\perp$ Jet Cross-Section from QCD," Phys. Letters 77B, 312 (1978) and Jagellonian University preprints TPJU-10/78, TPJU-11/78 and TPJU-12/78.
63. R. Cutler and D. Sivers, Phys. Rev. D16, 679 (1977); Phys. Rev. D17, 196 (1978).
64. B. L. Combridge, J. Kripfganz and J. Ranft, Phys. Lett. 234 (1977).
65. R. D. Field, Phys. Rev. Letters 40, 997 (1978).
66. G. C. Fox, "Application of Quantum Chromodynamics to High Transverse Momentum Hadron Production," invited talk at the Orbis Scientiae 1978 (Coral Gables) CALT-68-643.
67. (FFF1) R. P. Feynman, R. D. Field and G. C. Fox, Nucl. Phys. B128, 1 (1977).
68. M. Della Negra et al., (CCHK Collaboration), Nucl. Phys. B127, 1 (1977).
69. Fermilab-Lehigh-Pennsylvania-Wisconsin Collaboration, talks given by W. Selove and A. Erwin at the symposium on Jets in High Energy Collisions, Niels Bohr Institute-Nordita, Copenhagen, July 1978; also, M. D. Corcoran et al., "A Study of Parton Transverse Momentum Using Jets from Hadron Interactions," University of Wisconsin preprint (1979).
70. M. G. Albrow et al., Nucl. Phys. B135, 461 (1978).

71. It is quite possible that quarks, antiquarks and gluons do not all have the same "effective" $k_\perp$ spectra. For example, the larger mean $k_\perp$ of the Upsilon compared with the non-resonant background might be interpreted by saying that gluons have a larger effective $\langle k_\perp \rangle$ than do quarks. This approach has been adapted by the Florida State group in Ref. 72.
72. J. F. Owens, E. Reya and M. Gluck, FSU preprint 77-09-07 (1978); J. F. Owens and J. D. Kimel, FSU HEP 78-03-30 (1978).
73. Recently some attempts have been made to handle the primordial $k_\perp$ more properly. See, for example, Howard Georgi and Jon Sheiman, "Transverse Momentum Distributions in Lepton-Hadron Scattering from QCD," HUTP-78/A034.
74. F. Halzen, G. A. Ringland and R. G. Roberts, Phys. Rev. Lett. 40, 991 (1978).
75. J. F. Gunion, "The Interrelationship of the Constituent Interchange Model and Quantum Chromodynamics," presented at the discussion meeting on Large Transverse Momentum Phenomena, SLAC, January 1978; R. Horgan, W. Caswell and S. J. Brodsky, SLAC-PUB (1978).
76. K. Kinoshita and Y. Kinoshita, "Effects of Parton Transverse Momenta on Hadronic Large $p_\perp$ Reactions," preprint submitted to the XIX International Conference on High Energy Physics, Tokyo, 1978.
77. R. R. Horgan and P. N. Scharbach, "Transverse Momentum Fluctuations and High $p_\perp$ Processes in Quantum Chromodynamics," SLAC-PUB-2188 (1978).
78. R. Blankenbecler, S. J. Brodsky and J. F. Gunion, "The Magnitudes of Large Transverse Momentum Cross Sections," SLAC-PUB-2057 (1977).
79. D. Jones and J. F. Gunion, "The Transition from Constituent Interchange to QCD $p_\perp^{-4}$ Dominance at High Transverse Momentum," SLAC-PUB-2157 (1978).
80. A. G. Clark et al., Phys. Letters 74B, 267 (1978); A. G. Clark, invited talk at the symposium on Jets in High Energy Collisions, Niels Bohr Institute, Nordita, Copenhagen, July 1978.
81. CERN-Columbia-Oxford-Rockefeller Experiment, reported by L. Di Lella in the Workshop on Future ISR Physics, Sept. 14-21, 1977, edited by M. Jacob; and the talk by L. Di Lella at the symposium on Jets in High Energy Collisions, Niels Bohr Institute-Nordita, Copenhagen, July 1978; also see the invited talk by M. J. Tannenbaum at the XIV Rencontre de Moriond Conference (1979), COO-2232A-79.
82. C. Kourkoumelis et al., "Inclusive π^0 Production at Very Large $p_\perp$ at the ISR," CERN-EP/79-29 (1979).
83. H. J. Frisch, "Precise Measurements of High $p_\perp$ Single Particle Spectra in $\pi^- p$ Collisions at Fermilab (E258)," invited talk presented at the XIV Rencontre de Moriond Conference (1979).
84. This was also predicted in a less model dependent way by D. S. Ellis, M. Jacob and P. V. Landshoff, Nucl. Phys. B108, 93

(1976); M. Jacob and P. V. Landshoff, Nucl. Phys. B113, 395 (1976).

85. C. Bromberg et al., Phys. Rev. Letters 38, 1447 (1977); C. Bromberg et al., CALT-68-613 (to be published in Nucl. Phys.).

86. G. C. Fox, "Recent Experimental Results on High Transverse Momentum Scattering from Fermilab," invited talk given at the Argonne APS Meeting (1977).

87. G. Farrar and S. Frautschi, Phys. Rev. Letters 36, 1017 (1976); G. R. Farrar, Phys. Letters 67B, 337 (1977).

88. F. Halzen and D. M. Scott, Phys. Rev. Letters 40, 1117 (1978); University of Wisconsin preprint COO-881-21 (1978).

89. C. O. Escobar, Nucl. Phys. B98, 173 (1975); Phys. Rev. D15, 355 (1977).

90. R. Ruckl, S. J. Brodsky and J. F. Gunion, "The Production of Real Photons at Large Transverse Momentum in pp Collisions," UCLA preprint (1978).

91. H. Fritzsch and P. Minkowski, CERN preprint TH2320 (1978).

92. Z. Kunszt and E. Pietarinen, "Production of Three Large $p_\perp$ Jets in Hadron-Hadron Collisions," DESY preprint 79/34 (1979); J. F. Gunion, Z. Kunszt and E. Pietarinen, "Comment on the ISR $\pi^0\pi^0$ Azimuthal Correlation Data," DESY preprint 79/35 (1979).

93. J. Kripfganz and A. Schiller, "QCD Three-Jet Production in Large $p_\perp$ Processes," Leipzig preprint KMU-HEP-78-10 (1978).

94. Thomas Gottschalk and Dennis Sivers, "Basic Processes and Formalism for the Hadronic Production of Three Large $p_\perp$ Jets," ANL-HEP-PR-79-07 (1979).

95. J. W. Cronin et al., (CP Collaboration), Phys. Rev. D11, 2105 (1975); D. Antreasyan et al., Phys. Rev. Lett. 38, 112 (1977); Phys. Rev. Lett. 38, 115 (1977).

96. B. Alper et al., (BS Collaboration), Nucl. Phys. B100, 237 (1975).

97. F. W. Busser et al., Nucl. Phys. B106, 1 (1976).

98. G. Donaldson et al., Phys. Rev. Lett. 36, 1110 (1976).

99. D. C. Carey et al., Fermilab Report No. FNAL-PUB-75120-EXP. (1975).

100. D. A. Ross, "The Effects of Higher Order QCD Corrections in Deep Inelastic Scattering," CALT-68-699 (1979); D. A. Ross and C. T. Sachrajda, CERN preprint TH2565 (1978).

101. See also, A. P. Contogouris, R. Gaskell and S. Papadopoulos, McGill University preprints; A. P. Contogouris, McGill University preprint (1978); J. Ranft and G. Ranft, Leipzig preprint (1978); R. Raitio and R. Sosnowski, University of Helsinki preprint HU-TFT-77-22, invited talk given at the Workshop on Large $p_\perp$ Phenomena, University of Bielefeld, Sept. 5-8, 1977.

TOPICS IN PERTURBATIVE QCD BEYOND THE LEADING ORDER

Andrzej J. Buras

Fermi National Accelerator Laboratory

Batavia, Illinois

1. GENERAL OVERVIEW

Quantum Chromodynamics (QCD) is the most promising candidate for a theory of strong interactions. In these lectures we shall discuss higher order QCD predictions* for

a) Inclusive deep-inelastic scattering

$$\begin{aligned} eh &\to e + \text{anything} , \\ \nu h &\to \mu^- + \text{anything} , \end{aligned} \tag{1.1}$$

etc.

We shall also present the basic structure of QCD formulae for semi-inclusive processes such as:

b) $$e^+e^- \to h + \text{anything} \tag{1.2}$$

c) $$eh_1 \to e + h_2 + \text{anything} \tag{1.3}$$

d) $$h_1h_2 \to \mu^+\mu^- + \text{anything} \tag{1.4}$$

and

e) $$e^+e^- \to h_1 + h_2 + \text{anything} \tag{1.5}$$

where h_i stand for hadrons.

*For recent reviews see refs. 1-3, 57.

Finally we shall make a list of some recent higher order QCD calculations.

In the simple parton model,[4] in which strong interactions and mass scales are neglected, the cross-sections for processes (1.1)-(1.5) are expressed in terms of parton distributions and parton fragmentation functions as follows*:

a) $$\sigma_h^{DIS}(x) = \sum_i^f e_i^2 [x q_i^h(x) + x \bar{q}_i^h(x)] \qquad (1.6)$$

for deep-inelastic scattering (1.1),

b) $$\sigma_h^{e^+e^-}(z) = \sum_i^f e_i^2 [z D_{q_i}^h(z) + z D_{\bar{q}_i}^h(z)] \qquad (1.7)$$

for semi-inclusive e^+e^- annihilation (1.2),

c) $$\sigma_{h_1h_2}^{DIS}(x,z) = \sum_{i=1}^f e_i^2 [x q_i^{h_1}(x) z D_{q_i}^{h_2}(z) + x \bar{q}_i^{h_1}(x) z D_{\bar{q}_i}^{h_2}(z)] \qquad (1.8)$$

for semi-inclusive deep-inelastic scattering (1.3),

d) $$\frac{d\sigma}{dQ^2} = \int \frac{dx_1}{x_1} \int \frac{dx_2}{x_2} \sigma_{h_1h_2}^{\mu^+\mu^-}(x_1,x_2) \qquad (1.9)$$

with ($\tau = Q^2/s$)

$$\sigma_{h_1h_2}^{\mu^+\mu^-}(x_1,x_2) = \sum_{i=1}^f e_i^2 \left\{ \left[x_1 q_i^{h_1}(x_1)\right] \times \left[x_2 \bar{q}_i^{h_2}(x_2)\right] \delta\left(1 - \frac{\tau}{x_1x_2}\right) + \text{"}1 \leftrightarrow 2\text{"} \right\} \qquad (1.10)$$

---

*In order to simplify the presentation we suppress obvious factors such as $(4\pi\alpha^2)/(3Q^2)$ in Eqs. (1.7), (1.9) and (1.11) and color factors: "3" in Eqs. (1.7) and (1.11) and "1/3" in (1.9). Furthermore to unify notation we denote the well-known structure functions $F_2(x)$ and $F_2(x,z)$ by $\sigma_h^{DIS}(x)$ and $\sigma_{h_1h_2}^{DIS}(x,z)$ respectively. Finally, unless otherwise specified, we restrict our discussion to the transverse parts of the cross-sections for processes (1.2) and (1.5).

for massive μ-pair production (1.4),[5] and

$$\sigma^{e^+e^-}_{h_1h_2}(z_1,z_2) = \sum_{i=1}^{f} e_i^2 \left\{ \left[z_1 D^{h_1}_{q_i}(z_1) \right] \left[z_2 D^{h_2}_{\bar{q}_i}(z_2) \right] \right\} + \text{"}1 \leftrightarrow 2\text{"} \tag{1.11}$$

for two-hadron semi-inclusive e^+e^- annihilation (1.5).

In the above equations $q_i^h(x)$ and $\bar{q}_i^h(x)$ are the parton distributions (quark, antiquark) which measure the probability for finding a parton of type i in a hadron h with the momentum fraction x. Similarly $D^h_{q_i}(z)$ and $D^h_{\bar{q}_i}(z)$ are the fragmentation functions which measure the probability for a quark q_i or antiquark $\bar{q}_i$ to decay into a hadron h carrying the fraction z of the quark or antiquark momentum respectively. Finally e_i stand for quark charges and f denotes number of flavors.

The following properties of Eqs. (1.6)-(1.11) deserve attention:

i) Bjorken scaling: parton distributions and parton fragmentation functions depend only on x and z respectively.

ii) Factorization between

x and z in $\sigma^{DIS}_{h_1h_2}(x,z)$

x_1 and x_2 in $\sigma^{\mu^+\mu^-}_{h_1h_2}(x_1,x_2)$

and

z_1 and z_2 in $\sigma^{e^+e^-}_{h_1h_2}(z_1,z_2)$.

iii) The building blocks of all parton model formulae are <u>universal</u> (process independent) parton distributions and fragmentation functions. Therefore taking into account ii) we observe that if we can extract all parton distributions from inclusive deep-inelastic processes (a)) and fragmentation functions from e^+e^- annihilation (b)), then the cross-sections for the remaining processes listed above can be predicted. This implies that in the simple parton model there are relations between various processes.

iv) The gluon distribution G(x), and gluon fragmentation function $D^h_G(z)$ do not enter any of the formulae above.

v) Finally all cross-sections above can be reproduced from diagrams of Fig. 1 by using the "Feynman rules" of Fig. 2.

It is well known that in QCD quark distributions and quark fragmentation functions acquire Q^2 dependence and it is of interest and importance to ask:

— whether QCD predictions for semi-inclusive processes amount to using these Q^2 dependent functions in the parton model formulae (1.6-1.11);

— whether factorization properties listed under ii) (and correspondingly relations between various processes) are still satisfied;

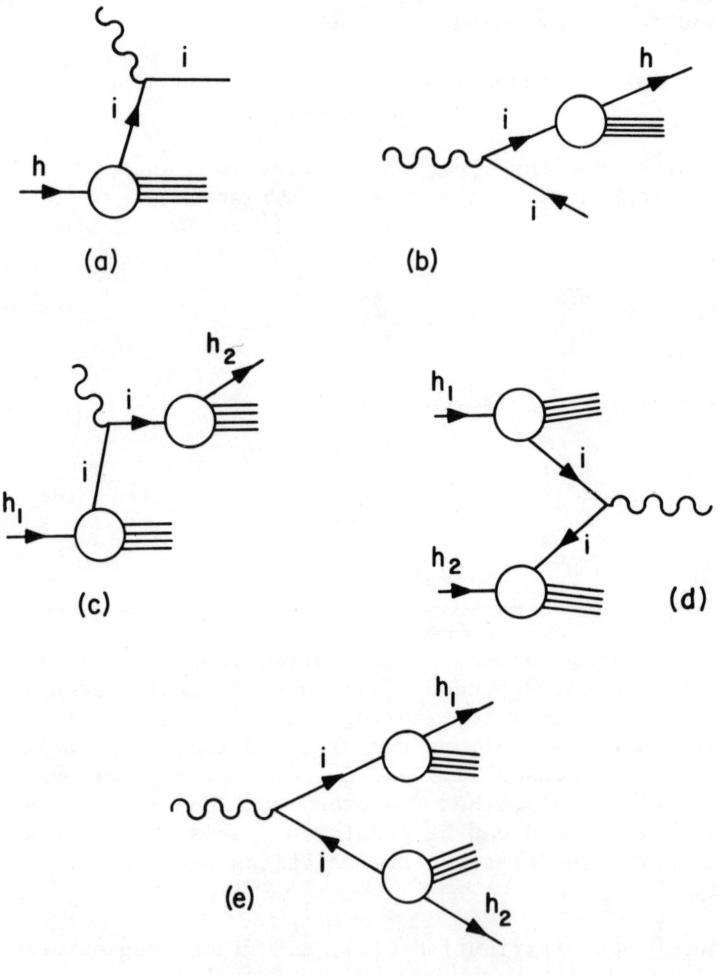

Fig. 1. Parton model diagrams for processes (1.1)-(1.5).

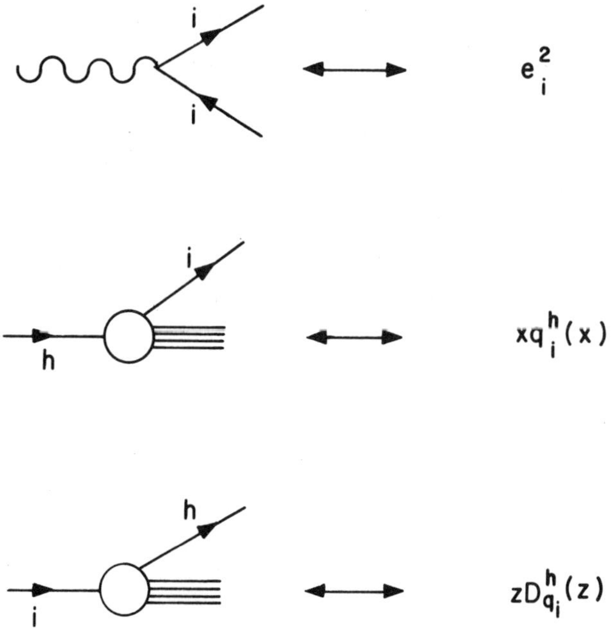

Fig. 2. Parton model rules.

— whether gluon distributions and gluon fragmentation functions explicitly enter QCD formulae, and

— whether one can find a simple extension of the rules of Fig. 2 which would allow us to construct in an easy way cross-sections for the processes a)-e) in QCD.

In these lectures we shall address these questions on two levels:

a) so-called leading order of asymptotic freedom and

b) next to leading order

with particular emphasis on the latter case.

To be more specific, if the QCD predictions for the moments of deep-inelastic structure functions are given as follows

$$\int_0^1 dx\, x^{n-2} F(x,Q^2) = A_n [\ln Q^2]^{-d_n} \left[1 + \frac{f_n}{\ln Q^2} + \ldots\right] \qquad (1.12)$$

then keeping only "1" on the r.h.s. of Eq. (1.12) corresponds to the leading order whereas the second term $f_n/\ln Q^2$ stands for the next to leading order corrections. The numbers A_n, d_n and f_n will be discussed in subsequent sections.

Before going into details it is perhaps useful to get a general overview and list how the parton model properties above are modified in QCD.

The cross-sections for the processes (1.1)-(1.3) are given in QCD as follows*

$$\sigma_h^{DIS}(x,Q^2) = \sum_j \int_x^1 \frac{d\xi}{\xi} \sigma_P^j\left(\frac{x}{\xi}, Q^2\right) \left[\xi f_j^h(\xi,Q^2)\right] \tag{1.13}$$

$$\sigma_h^{e^+e^-}(z,Q^2) = \sum_j \int_z^1 \frac{d\xi}{\xi} \tilde{\sigma}_P^j\left(\frac{z}{\xi}, Q^2\right) \left[\xi D_j^h(\xi,Q^2)\right] \tag{1.14}$$

and

$$\sigma_{h_1h_2}^{DIS}(x,z,Q^2) = \sum_{jk} \int_x^1 \frac{d\xi_1}{\xi_1} \int_z^1 \frac{d\xi_2}{\xi_2} \tilde{\sigma}_P^{jk}\left(\frac{x}{\xi_1}, \frac{z}{\xi_2}, Q^2\right) \left[\xi_1 f_j^{h_1}(\xi_1,Q^2)\right] \left[\xi_2 D_k^{h_2}(\xi_2,Q^2)\right] . \tag{1.15}$$

The processes are shown schematically in Fig. 3. Similar equations exist for massive μ-pair production and two hadron semi-inclusive e^+e^- annihilation. In Eqs. (1.13)-(1.15) the sums run over quarks, antiquarks and gluons and $f_i^h(\xi,Q^2)$ denote generally parton distributions. Furthermore

— σ_P^j is the photon-parton j cross-section.

— $\tilde{\sigma}_P^j$ is the cross-section for the production of the parton j in e^+e^- annihilation, and

— $\tilde{\sigma}_P^{jk}$ stands for the photon-parton j cross-section with the parton k in the final state.

---

*There are so many papers on this subject that we cannot list all of them here. Representative are Refs. 6-8, 37 where further references can be found.

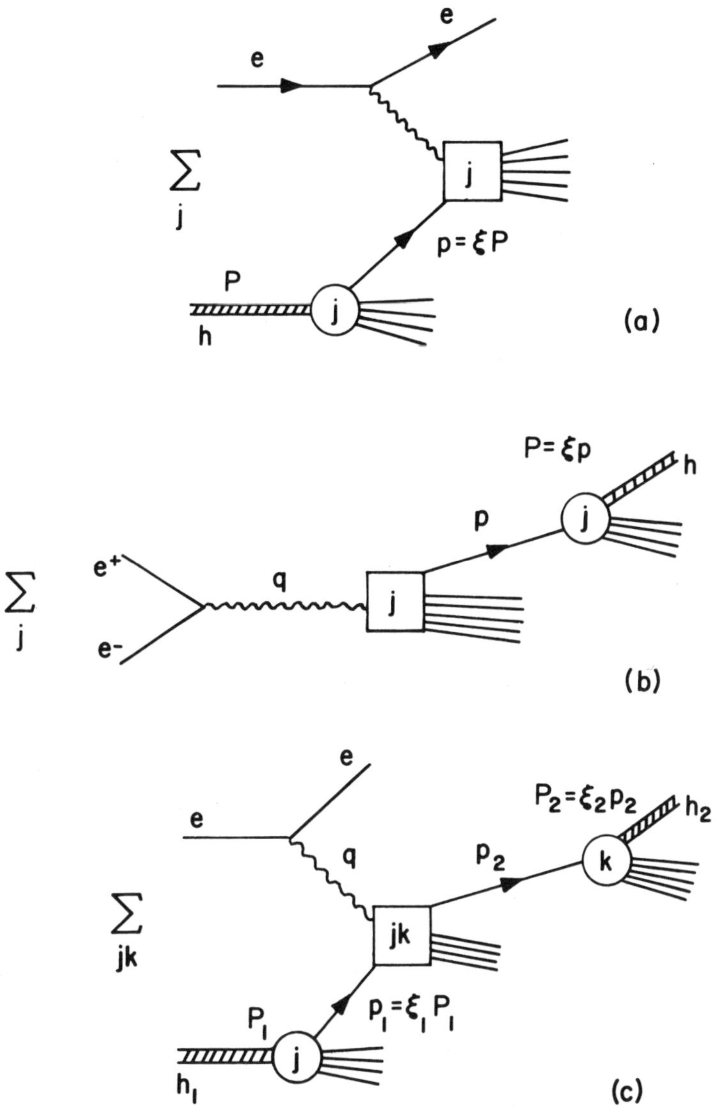

Fig. 3. Illustration of the r.h.s. of Eqs. (1.13), (1.14) and (1.15). The sums run over quarks and gluons. The circles stand for parton distributions or parton fragmentation functions. The squares denote the parton cross-sections.

Depending on the order considered (leading order, next-to-leading order) there are different rules for the parton cross-sections, parton distributions and parton fragmentation functions, which are the building blocks of the QCD formulae above.

For the first two orders the rules in question are as follows:

a) Leading Order Rules

Rule 1 (Parton cross-sections)

$$\sigma_P^j\left(\frac{x}{\xi}, Q^2\right) = \begin{cases} e_j^2\delta(1 - \frac{x}{\xi}) & j = q,\bar{q} \\ 0 & j = G \end{cases} \tag{1.16}$$

$$\tilde{\sigma}_P^j\left(\frac{z}{\xi}, Q^2\right) = \begin{cases} e_j^2\delta(1 - \frac{z}{\xi}) & j = q,\bar{q} \\ 0 & j = G \end{cases} \tag{1.17}$$

$$\tilde{\sigma}_P^{jk}\left(\frac{x}{\xi_1}, \frac{z}{\xi_2}, Q^2\right) = \begin{cases} e_j^2\delta_{jk}\delta\left(1-\frac{x}{\xi_1}\right)\delta\left(1-\frac{z}{\xi_2}\right) & j = q,\bar{q} \\ 0 & j = G \end{cases} \tag{1.18}$$

etc.

Rule 2 (Parton distributions and parton fragmentation functions)

The Q^2 evolution of the parton distributions is governed by certain equations. In the case of non-singlet quark distributions

$$\Delta_{ij}(x,Q^2) = q_i(x,Q^2) - q_j(x,Q^2) \tag{1.19}$$

these equations have a very simple form

$$<\Delta_{ij}(Q^2)>_n = <\Delta_{ij}(Q_o^2)>_n \left[\frac{\ln \frac{Q^2}{\Lambda^2}}{\ln \frac{Q_o^2}{\Lambda^2}}\right]^{-d_n} \tag{1.20}$$

where

$$<\Delta_{ij}(Q^2)>_n \equiv \int_0^1 dx\, x^{n-1}\Delta_{ij}(x,Q^2) , \tag{1.21}$$

Q_o^2 is a reference momentum at which $\Delta_{ij}(x,Q_o^2)$ is to be taken from the data, d_n are known numbers and Λ is a scale parameter to be discussed later on. More complicated equations exist for the sums of quark and antiquark distributions (so-called singlet distributions). In the form of integro-differential equations they are often called Altarelli-Parisi equations.[9] Similar equations exist for the Q^2 evolution of the parton fragmentation functions. For nonsinglet combinations of fragmentation functions the equations are the same as the Eq. (1.20). For the singlet combinations of fragmentation functions there are slight modifications of the Altarelli-Parisi equations which are discussed in ref. 10. Using the rules 1 and 2 in Eqs. (1.13) and (1.15) we observe that the leading order corresponds to the parton model formulae (1.6)-(1.8) with $q_i^h(x)$ and $D_{q_i}^h(z)$ replaced by $q_i^h(x,Q^2)$ and $D_{q_i}^h(z,Q^2)$ which have calculable Q^2 dependence given by the equations mentioned above. Consequently all the parton model properties (except for the breakdown of scaling) are still satisfied. The following comment is, however, necessary.

On the basis of these results one could expect that if we extract parton distributions from inclusive deep-inelastic scattering and fragmentation functions from e^+e^- annihilation then the cross-sections for the processes (1.3)-(1.5) and in particular their Q^2 dependence can be predicted. This is not exactly true. The reason is, as we shall discuss in next sections, that in the leading order the numerical values of the scale parameter Λ extracted from the data need not be the same for different processes. Therefore a meaningful comparison of scaling violations in different processes can only be made if at least next to leading order corrections are included.[11] This point will be discussed in detail later on. The phenomenological applications of asymptotic freedom in the leading order have already been discussed in other lectures at this Summer School[12] and we shall not present them here.

b) Next to Leading Order Rules

Rule 1' (Parton cross-sections)

$$\sigma_P^j\left(\frac{x}{\xi},Q^2\right) = \begin{cases} \delta\left(1-\frac{x}{\xi}\right) + \bar{g}^2(Q^2)\,b_q\left(\frac{x}{\xi}\right) & j = q,\bar{q} \\ \bar{g}^2(Q^2)\,b_G\left(\frac{x}{\xi}\right) & j = G \end{cases} \tag{1.22}$$

$$\tilde{\sigma}_P^j\left(\frac{z}{\xi},Q^2\right) = \begin{cases} \delta\left(1-\frac{z}{\xi}\right) + \bar{g}^2(Q^2)\,d_q\left(\frac{z}{\xi}\right) & j = q,\bar{q} \\ \bar{g}^2(Q^2)\,d_G\left(\frac{z}{\xi}\right) & j = G \end{cases} \tag{1.23}$$

$$\tilde{\sigma}_P^{jk}\left(\frac{x}{\xi_1},\frac{z}{\xi_2},Q^2\right)=\begin{cases}\delta\left(1-\frac{x}{\xi_1}\right)\delta\left(1-\frac{z}{\xi_2}\right)+\bar{g}^2(Q^2)f_{qq}\left(\frac{x}{\xi_1},\frac{z}{\xi_2}\right) & j=k=q,\bar{q}\\ \bar{g}^2(Q^2)f_{qG}\left(\frac{x}{\xi_1},\frac{z}{\xi_2}\right) & j=q,\,k=G\\ \bar{g}^2(Q^2)f_{Gq}\left(\frac{x}{\xi_1},\frac{z}{\xi_2}\right) & j=G,\,k=q\end{cases}\tag{1.24}$$

where b_q, d_q, f_{qq}, f_{qG} and f_{Gq} are calculable functions to be discussed below. Similar rules exist for processes (1.4) and (1.5).

Rule 2' (Parton distributions and parton fragmentation functions)

The Q^2 evolution of parton distributions and parton fragmentation functions is governed by new equations which differ from the leading order equations (Rule 2) by calculable corrections of order $\bar{g}^2(Q^2)$. These new equations will be discussed in the course of these lectures.

The new features of Rules 1' and 2' not encountered in the parton model and in the leading order are as follows.

i) Factorization between x and z is broken through the functions f_{ij}. Similarly the factorization between x_1 and x_2 in the process (1.4), and between z_1 and z_2 in the process (1.5) is broken by calculable corrections to the parton cross-sections.

ii) The explicit $\bar{g}^2(Q^2)$ corrections to the parton cross-sections depend on the definition of parton distributions beyond the leading order. In other words the rules 1' and 2' are not independent of each other and must be consistent with each other in order that a physical answer independent of any particular definition is obtained for the measurable quantities as $\sigma_h^{DIS}(x,Q^2)$, $\sigma_h^{e^+e^-}(z,Q^2)$, etc.

iii) The $\bar{g}^2(Q^2)$ corrections to various parton cross-sections depend on the process considered. There exist however certain relations between some of the parton cross-sections (see Section 6).

iv) Because the parton cross-sections involving gluons are non-zero the gluon distributions and gluon fragmentation function enter the QCD formulae explicitly.

This completes the general overview. In what follows we shall show systematically how to obtain the rules 1' and 2'. In Section 2 we recall the ingredients of the formal approach to deep-inelastic scattering based on operator product expansion and renormalization group equations. In Sections 3 and 4 the inclusive deep-inelastic scattering beyond the leading order is discussed in some detail. Subsequently in Section 5 we present the basic structure of next to leading order calculations for semi-inclusive processes. In Section 6 we will list other recent higher order QCD calculations. We end our lectures with a brief summary and outlook.

2. BASIC FORMALISM

In this section we shall recall the basic formal tools used to extract QCD predictions for deep-inelastic scattering. Generalization to other processes will be discussed in Section 5.

Let us imagine that we want to find QCD predictions for deep-inelastic structure functions. We can proceed as follows:

Step 1

We consider the spin-averaged amplitude $T_{\mu\nu}$ for the forward scattering of a weak or electromagnetic current J_μ. The amplitude $T_{\mu\nu}$ can be decomposed into invariant amplitudes as follows:

$$\begin{aligned} T_{\mu\nu} &= i \int d^4x\, e^{iq\cdot x} \langle p|T(J_\mu(x)J_\nu(0))|p\rangle_{\text{spin averaged}} \\ &= e_{\mu\nu}T_L(Q^2,\nu) + d_{\mu\nu}T_2(Q^2,\nu) - i\varepsilon_{\mu\nu\alpha\beta}\frac{p_\alpha q_\beta}{\nu}T_3(Q^2,\nu) \ . \end{aligned} \tag{2.1}$$

Here $\nu = p\cdot q$, $Q^2 = -q^2$ and $|p\rangle$ is, for instance, a proton state. The tensors $e_{\mu\nu}$, $d_{\mu\nu}$ and $\varepsilon_{\mu\nu\alpha\beta}$ are well known.

Step 2

We expand the product of currents, which enters Eq. (2.1) as a sum of products of local operators O_i^n of definite spin n times certain coefficient functions $\tilde{C}_n^i$ called Wilson coefficient functions.[13] We write symbolically this operator product expansion (OPE) as follows:

$$J(x)J(0) = \sum_{i,n} \tilde{C}_n^i(x^2)O_i^n \ . \tag{2.2}$$

The sum in Eq. (2.2) runs over spin n, twist 2 operators* such as the fermion nonsinglet operator O^n_{NS} and the singlet fermion and gluon operators O^n_ψ and O^n_G respectively.

For readers less familiar with the operator product expansion we only recall that the matrix elements of local operators between proton states can be interpreted as moments of parton distributions (see Section 3). Notice that any deep-inelastic structure functions can be expressed through three types of parton distributions: non-singlet, singlet quark and gluon distributions. Correspondingly there are three types of operators O^n_{NS}, O^n_ψ and O^n_G. A careful discussion of operator product expansion can be found for instance in ref. 1, 13 and 14.

Step 3

Inserting Eq. (2.2) into (2.1) and using dispersion relations between deep-inelastic structure functions and the invariant amplitudes of Eq. (2.1) we obtain[15]

$$M_k(n,Q^2) \equiv \int_0^1 dx\, x^{n-2} F_k(x,Q^2) = \sum_i A^i_n(\mu^2) C^i_{k,n}\left(\frac{Q^2}{\mu^2}, g^2\right) \tag{2.3}$$

where $C^i_{k,n}$ are fourier transforms of the coefficient functions in Eq. (2.2) and A^i_n are the matrix elements of operators O^n_i between the hadronic state $|p\rangle$. Furthermore g is the renormalized quark-gluon coupling constant and μ^2 is the subtraction scale at which the theory is renormalized. The important property of Eq. (2.3) is the factorization of non-perturbative pieces $A^i_n(\mu^2)$ from perturbatively calculable coefficient functions $C^i_{k,n}(Q^2/\mu^2,g^2)$.

Step 4

We decompose $F_k(x,Q^2)$ into a sum of singlet and non-singlet (under flavor symmetry) contributions as follows

$$F_k(x,Q^2) = F^{NS}_k(x,Q^2) + F^s_k(x,Q^2)\ . \tag{2.4}$$

We have in an obvious notation:

$$M^{NS}_k(n,Q^2) = A^{NS}_n(\mu^2) C^{NS}_{k,n}\left(\frac{Q^2}{\mu^2}, g^2\right) \tag{2.5}$$

---

*Neglecting operators of higher twist corresponds to neglecting contributions which, with increasing Q^2, decrease as inverse powers of Q^2.

and

$$M_k^s(n,Q^2) = A_n^\psi(\mu^2) C_{k,n}^\psi\left(\frac{Q^2}{\mu^2}, g^2\right) + A_n^G(\mu^2) C_{k,n}^G\left(\frac{Q^2}{\mu^2}, g^2\right) . \tag{2.6}$$

Step 5

We use renormalization group equations, which govern the Q^2 dependence of $C_{k,n}^i(Q^2/\mu^2, g^2)$. These equations are given as follows[16]:

$$\left[\mu\frac{\partial}{\partial\mu} + \beta(g)\frac{\partial}{\partial g} - \gamma_{NS}^n(g)\right] C_{k,n}^{NS}\left(\frac{Q^2}{\mu^2}, g^2\right) = 0 , \tag{2.7}$$

and

$$\left[\mu\frac{\partial}{\partial\mu} + \beta(g)\frac{\partial}{\partial g}\right] C_{k,n}^i\left(\frac{Q^2}{\mu^2}, g^2\right) = \sum_j \gamma_{ji}^n(g) C_{k,n}^j\left(\frac{Q^2}{\mu^2}, g^2\right) \quad i,j = \psi, G . \tag{2.8}$$

Notice that the Q^2 dependence of $C_{k,n}^\psi$ and $C_{k,n}^G$ is governed by two coupled renormalization group equations due to the mixing of the operators O_ψ^n and O_G^n under renormalization. $\gamma_{NS}^n(g)$ is the anomalous dimension of O_{NS}^n and $\gamma_{ij}^n(g)$ are the elements of the 2 x 2 anomalous dimension matrix $\hat{\gamma}^n(g)$. $\beta(g)$ is the well-known renormalization group function which governs the Q^2 evolution of the effective coupling constant $\bar{g}^2(Q^2)$:

$$\frac{d\bar{g}^2}{dt} = \bar{g}\beta(g) \; ; \; \bar{g}(t = 0) = g \equiv \bar{g}(\mu^2) \tag{2.9}$$

where $t = \ln Q^2/\mu^2$.

The solutions of Eqs. (2.7) and (2.8) can be written in terms of $\bar{g}^2(Q^2)$ as follows:

$$C_{k,n}^{NS}\left(\frac{Q^2}{\mu^2}, g^2\right) = C_{k,n}^{NS}(1,\bar{g}^2)\exp\left[-\int_{\bar{g}(\mu^2)}^{\bar{g}(Q^2)} dg' \frac{\gamma_{NS}^n(g')}{\beta(g')}\right] \tag{2.10}$$

and

$$\vec{C}_{k,n}\left(\frac{Q^2}{\mu^2}, g^2\right) = \left[T \exp\int_{\bar{g}(Q^2)}^{\bar{g}(\mu^2)} dg' \frac{\hat{\gamma}^n(g')}{\beta(g')}\right] \vec{C}_{k,n}(1,\bar{g}^2) \tag{2.11}$$

where $\vec{C}_{k,n}$ is the column vector whose components are $C^{\psi}_{k,n}$ and $C^{G}_{k,n}$. Equations (2.9)-(2.11) combined with Eqs. (2.4)-(2.6) give us general expressions for the Q^2 evolution of the moments of the deep-inelastic structure functions in terms of the renormalization group functions $\beta(g)$, $\gamma^n_{NS}(g)$ and $\hat{\gamma}^n(g)$ and the coefficient functions $C^i_{k,n}(1,\bar{g}^2)$.

Step 6

In order to find explicit Q^2 dependence of $M_k(n,Q^2)$ we have to calculate $\beta(g)$, $\gamma^n_{NS}(g)$, $\hat{\gamma}^n(g)$ and $C^i_{k,n}(1,\bar{g}^2)$. This is done in perturbation theory in g. We shall discuss explicit examples below.

Step 7

So far our discussion was very formal. In order to have relation with the parton picture of Section 1 we can cast the formal expressions for the Q^2 dependent structure functions into the parton model-like formulae with effective Q^2 dependent parton distributions. We shall discuss such expressions below.

3. NEXT TO LEADING ORDER ASYMPTOTIC FREEDOM CORRECTIONS TO DEEP-INELASTIC SCATTERING (NON-SINGLET CASE)

3.1 Derivation of Basic Formulae

In order to find explicit expressions for the leading and next to leading contributions to $C^{NS}_{k,n}(Q^2/\mu^2,g^2)$ as given by Eq. (2.10) we expand $\gamma^n_{NS}(\bar{g})$, $\beta(\bar{g})$ and $C^{NS}_{k,n}(1,\bar{g}^2)$ in powers of $\bar{g}^2$

$$\gamma^n_{NS}(\bar{g}) = \gamma^{(0),n}_{NS}\frac{\bar{g}^2}{16\pi^2} + \gamma^{(1),n}_{NS}\left(\frac{\bar{g}^2}{16\pi^2}\right)^2 + \ldots \tag{3.1}$$

$$\beta(\bar{g}) = -\beta_0\frac{\bar{g}^3}{16\pi^2} - \beta_1\frac{\bar{g}^5}{(16\pi^2)^2} + \ldots \tag{3.2}$$

and (through order $\bar{g}^2$)

$$C^{NS}_{k,n}(1,\bar{g}^2) = \delta^k_{NS}\left[1 + \frac{\bar{g}^2}{16\pi^2}B^{NS}_{k,n}\right] \qquad k = 2,3\ . \tag{3.3}$$

Here δ^k_{NS} are constants which depend on weak and electromagnetic charges.

For the exponential in Eq. (2.10) we obtain (through order $\bar{g}^2$), with $\mu^2 = Q_0^2$

$$\exp\left[\int_{\bar{g}^2(Q_0^2)}^{\bar{g}^2(Q^2)} dg' \frac{\gamma_{NS}^n(g')}{\beta(g')}\right] = \left[1 + \frac{[\bar{g}^2(Q^2) - \bar{g}^2(Q_0^2)]}{16\pi^2} Z_n^{NS}\right]\left[\frac{\bar{g}^2(Q^2)}{\bar{g}^2(Q_0^2)}\right]^{d_{NS}^n}, \tag{3.4}$$

where

$$Z_n^{NS} = \frac{\gamma_{NS}^{(1),n}}{2\beta_o} - \frac{\gamma_{NS}^{(0),n}}{2\beta_o^2}\beta_1 \quad ; \quad d_{NS}^n = \frac{\gamma_{NS}^{(0),n}}{2\beta_o} . \tag{3.5}$$

Combining Eqs. (3.3), (3.4), (2.10) and (2.3) we obtain

$$M_k^{NS}(n,Q^2) = \delta_{NS}^k \bar{A}_n^{NS}(Q_o^2)\left[1 + \frac{[\bar{g}^2(Q^2) - \bar{g}^2(Q_o^2)]}{16\pi^2} R_{k,n}^{NS}\right]\left[\frac{\bar{g}^2(Q^2)}{\bar{g}^2(Q_o^2)}\right]^{d_{NS}^n}, \tag{3.6}$$

where

$$R_{k,n}^{NS} = B_{k,n}^{NS} + Z_n^{NS} , \tag{3.7}$$

$$\bar{A}_n^{NS}(Q_o^2) = A_n^{NS}(Q_o^2)\left[1 + \frac{\bar{g}^2(Q_o^2)}{16\pi^2} B_{k,n}^{NS}\right] \tag{3.8}$$

and $\bar{g}^2(Q^2)$ is to be calculated by means of Eq. (2.9) with the β function given by Eq. (3.2). In phenomenological applications it is often convenient to insert into Eq. (3.6) the explicit expression for $\bar{g}^2(Q^2)$,

$$\frac{\bar{g}^2(Q^2)}{16\pi^2} = \frac{1}{\beta_o \ln\frac{Q^2}{\Lambda^2}} - \frac{\beta_1}{\beta_o^3}\frac{\ln\ln\frac{Q^2}{\Lambda^2}}{\ln^2\frac{Q^2}{\Lambda^2}} + O\left(\frac{1}{\ln^3\frac{Q^2}{\Lambda^2}}\right) \tag{3.9}$$

with the result

$$M_k^{NS}(n,Q^2) = \delta_{NS}^k \bar{A}_n^{NS}(Q_o^2)\left[1 + \frac{R_{k,n}^{NS}(Q^2)}{\beta_o \ln\frac{Q^2}{\Lambda^2}} - \frac{R_{k,n}^{NS}(Q_o^2)}{\beta_o \ln\frac{Q_o^2}{\Lambda^2}}\right]\left[\frac{\ln\frac{Q^2}{\Lambda^2}}{\ln\frac{Q_o^2}{\Lambda^2}}\right]^{-d_{NS}^n} \tag{3.10}$$

where

$$R_{k,n}^{NS}(Q^2) = R_{k,n}^{NS} - \frac{\beta_1}{2\beta_0^2}\gamma_{NS}^{0,n}\ln\ln\frac{Q^2}{\Lambda^2} . \tag{3.11}$$

Equation (3.10) is the basic formula of this section. The value of Q_0^2 in Eq. (3.10) is arbitrary as required by the renormalization group equations and the predictions for $M_k^{NS}(n,Q^2)$ should be independent of it. Therefore it is sometimes convenient to get rid of Q_0^2 by writing Eq. (3.10) as

$$M_k^{NS}(n,Q^2) = \delta_{NS}^k \bar{A}_n^{NS}\left[1+\frac{R_{k,n}^{NS}(Q^2)}{\beta_0 \ln\frac{Q^2}{\Lambda^2}}\right]\left[\ln\frac{Q^2}{\Lambda^2}\right]^{-d_{NS}^n} \qquad k = 2,3 . \tag{3.12}$$

Here $\bar{A}_n^{NS}$ are constants (independent of Q_0^2).

3.2 Discussion of Basic Properties and Subtle Points

a) We first notice that in order to find the next to leading order corrections to the non-singlet structure functions, one has to calculate the two-loop contributions to $\gamma_{NS}^n(g)$ and $\beta(g)$ and one-loop corrections to $C_{k,n}^{NS}(1,\bar{g}^2)$ i.e. the parameters $\gamma_{NS}^{(1),n}$, β_1 and $B_{k,n}^{NS}$ respectively. The parameters $\gamma_{NS}^{(0),n}$ and β_0 are known already from the leading order calculations.[17]

b) The two-loop contributions to the β function, i.e. the parameter β_1, has been calculated in ref. 18 and for an $SU(3)_c$ gauge theory with f flavors is given by

$$\beta_1 = 102 - \frac{38}{3} f . \tag{3.13}$$

It should be remarked that β_1 as well as $\gamma_{NS}^{0,n}$ and β_0 are renormalization prescription-and gauge-independent.

c) The parameters $B_{k,n}$ and $\gamma_{NS}^{(1),n}$ are separately renormalization prescription dependent (i.e. they depend on the way one renormalizes the quantities used to calculate them (see Appendix)) and in principle gauge dependent. However as shown by Floratos, Ross and Sachrajda[19] the quantity

$$B_{k,n}^{NS} + \frac{\gamma_{NS}^{(1),n}}{2\beta_0} \tag{3.14}$$

is renormalization prescription independent if (of course) $B^{NS}_{k,n}$ and $\gamma^{(1),n}_{NS}$ are calculated in the same renormalization scheme. Consequently the parameters $R^{NS}_{k,n}$ of Eq. (3.7) are renormalization prescription independent. From this we can draw two lessons:

1) care must be taken that $B^{NS}_{k,n}$ and $\gamma^{(1),n}_{NS}$ are calculated in the same renormalization scheme and

2) without doing explicit calculations one cannot a priori neglect either of the two quantities $B^{NS}_{k,n}$ and $\gamma^{(1),n}_{NS}/2\beta_0$ in any higher order formulae. The reason is that in some schemes the two-loop contribution is dominant in the sum (3.14) whereas in other schemes $B^{NS}_{n,k}$ is more important.

The method of calculation of the parameters $B^{NS}_{k,n}$ is described in the Appendix.

d) The full calculation of the sum in Eq. (3.14) has been performed in the literature only in the 't Hooft's minimal subtraction scheme (MS)*. The parameters $B^{NS}_{k,n}$ have been calculated in ref. 20 and recalculated in ref. 21. The authors of ref. 19 have calculated the two-loop anomalous dimensions $\gamma^{(1),n}_{NS}$. The latter calculation is particularly complicated. Calculations of the parameters $B^{NS}_{k,n}$ in different renormalization schemes have been done in refs. 22-25. However these results cannot be combined with the two-loop anomalous dimensions of ref. 19. In spite of this the results of refs. 22-25 will turn out to be useful in the study of QCD effects of other processes (see Section 5).

e) In the minimal subtraction scheme one obtains[19,20]

$$B^{NS}_{2,n} = \bar{B}^{NS}_{2,n} + \tfrac{1}{2}\gamma^{0,n}_{NS}(\ell n\, 4\pi - \gamma_E) \tag{3.15}$$

and

$$B^{NS}_{3,n} = B^{NS}_{2,n} - \frac{4}{3}\,\frac{4n+2}{n(n+1)} \tag{3.16}$$

where

---

*In this scheme the Feynman diagrams are evaluated, using dimensional regularization, in $d = 4 - \varepsilon$ dimensions and singularities are extracted as poles $1/\varepsilon$, $1/\varepsilon^2$ etc. The minimal subtraction then means that the amplitudes are renormalized by simply subtracting the pole parts $1/\varepsilon$, $1/\varepsilon^2$ etc.

$$\bar{B}^{NS}_{2,n} = \frac{4}{3}\left\{3\sum_{j=1}^{n}\frac{1}{j} - 4\sum_{j=1}^{n}\frac{1}{j^2} - \frac{2}{n(n+1)}\sum_{j=1}^{n}\frac{1}{j} + 4\sum_{s=1}^{n}\frac{1}{s}\sum_{j=1}^{n}\frac{1}{j}\right.$$

$$\left. + \frac{3}{n} + \frac{4}{(n+1)} + \frac{2}{n^2} - 9\right\} \qquad (3.17)$$

and $\gamma_E = 0.5777$ is the Euler-Mascheroni constant. The "strange" terms $(\ell n\, 4\pi - \gamma_E)$ arise from the expansion of $\Gamma(\varepsilon/2)\cdot(4\pi)^{\varepsilon/2}$ around $\varepsilon = 0$. The numerical values of $B^{NS}_{2,n}$, $\bar{B}^{NS}_{2,n}$ and Z^{NS}_n for the renormalization scheme in question are collected for some values of n in Table 1. We observe that the terms Z^{NS}_n are small compared to the parameters $B^{NS}_{2,n}$ and $\bar{B}^{NS}_{2,n}$. In order to understand the difference between $B^{NS}_{2,n}$ and $\bar{B}^{NS}_{2,n}$ we now turn to the last important point which is related to the parameter Λ and to the arbitrariness in the definition of the effective coupling constant.

3.3 Parameter Λ and Various Definitions of $\bar{g}^2(Q^2)$

The effect of the redefinition of the scale parameter Λ is equivalent through order $\bar{g}^2(Q^2)$ to the shift of $R^{NS}_{k,n}(Q^2)$ in Eq. (3.12) by a constant amount proportional to $\gamma^{0,n}_{NS}$.[20] In fact rescaling Λ in Eq. (3.12) to Λ' by

$$\Lambda = \Lambda' \exp(-\tfrac{1}{2}\kappa) \qquad (3.18)$$

Table 1. Numerical values of the parameters $B_{2,n}$, $\bar{B}_{2,n}$ and Z^{NS}_n (see Eqs. (3.15), (3.17) and (3.5) respectively) for various values of n and f = 4. The values of Z^{NS}_n are obtained on the basis of results of ref. 19 and correspond to minimal subtraction scheme.

n	$B_{2,n}$	$\bar{B}_{2,n}$	Z^{NS}_n
2	7.39	0.44	1.65
4	19.70	6.07	2.05
6	28.77	11.18	2.16
8	35.96	15.53	2.25

where κ is a constant, and dropping terms of order $\bar{g}^4(Q^2)$ generated by this rescaling one obtains

$$M_k^{NS}(n,Q^2) = \delta_{NS}^k \bar{A}_n^{NS} \left[1 + \frac{R'^{NS}_{k,n}(Q^2)}{\beta_0 \ \ln \frac{Q^2}{\Lambda'^2}} \right] \left[\ln \frac{Q^2}{\Lambda'^2} \right]^{-d^n_{NS}} \tag{3.19}$$

where

$$R'^{NS}_{k,n}(Q^2) = R^{NS}_{k,n}(Q^2) - \tfrac{1}{2}\gamma^{0,n}_{NS}\kappa \ . \tag{3.20}$$

The Λ' thus corresponds to the $\bar{g}^2$ corrections given by Eq. (3.20) and $\bar{g}^2(Q^2)$ having the form of Eq. (3.9) with Λ replaced by Λ'. To be more specific let us denote by Λ_{MS} the scale parameter which corresponds to $R^{NS}_{k,n}$ given by Eq. (3.7) with $B^{NS}_{k,n}$ given by Eqs. (3.15) and (3.16). This is so called minimal scheme (MS) for Λ. In the literature two other schemes have been discussed:

— $\overline{MS}$ scheme[20] for which the parameters $R^{NS}_{k,n}$ are replaced by

$$\bar{R}^{NS}_{k,n} = R^{NS}_{k,n} - \tfrac{1}{2}\gamma^{0,n}_{NS}(\ln 4\pi - \gamma_E) \tag{3.21}$$

and the corresponding Λ denoted by $\Lambda_{\overline{MS}}$, and

— MOM (momentum subtraction scheme)[26,27] for which $R^{NS}_{k,n}$ are replaced by*

$$R^{NS}_{k,n}\Big|_{MOM} = R^{NS}_{k,n} - \tfrac{1}{2}\gamma^{0,n}_{NS}(a) \ , \quad a = 3.5 \tag{3.22}$$

and the corresponding Λ denoted by Λ_{MOM}. Notice that the $\overline{MS}$ scheme corresponds to dropping terms in Eq. (3.15) involving factors $(\ln 4\pi - \gamma_E) \approx 1.95$.

For the three schemes considered MS, $\overline{MS}$ and MOM the effective coupling constant is given by Eq. (3.9) with Λ replaced by Λ_{MS}, $\Lambda_{\overline{MS}}$ and Λ_{MOM} respectively. It is obvious that since the functional forms of $R^{NS}_{k,n}$, $\bar{R}^{NS}_{k,n}$ and $R^{NS}_{k,n}|_{MOM}$ are different from

---

*The $\bar{g}^2(Q^2)$ defined by momentum subtraction is gauge dependent but the gauge dependence is weak. The value 3.5 corresponds to the Landau gauge.

each other so will be the free parameters Λ_{MS}, $\Lambda_{\overline{MS}}$ and Λ_{MOM} extracted from the data. Needless to say the three schemes considered are equivalent representations of next to leading corrections. On the other hand they correspond to different estimates of the higher order terms $O(\bar{g}^4(Q^2))$ not included in the analysis. In the next subsection we shall present numerical values of Λ_i, effective coupling constants and explicit higher order corrections for the schemes considered.

Since the explicit higher order corrections and the parameter Λ are related to each other we conclude that one cannot discuss numerical values of Λ in a theoretically meaningful way without calculating at least next to leading order corrections[11] and without specifying the definition of the effective coupling constant.[20]

Once a definition of $\bar{g}^2(Q^2)$ is made and is used in calculations of higher order corrections in various processes it is possible to make a meaningful comparison of higher order corrections to various processes.[11] We shall see that these corrections are generally different for different processes. This teaches us that it is unjustified in principle to use the same value of Λ in the leading order expressions for different processes. On the other hand, once higher order corrections are included in the analysis and $g^2(Q^2)$ is defined in a universal way, it is justified to use the same value of Λ in different processes.

Let us summarize two basic lessons of this section

i) the parameters $\gamma_{NS}^{(1),n}$ and $B_{k,n}^{NS}$ have to be calculated in the same renormalization scheme;

ii) there is a well-defined dependence of the functional form of the explicit higher order corrections on the definition of $\bar{g}^2(Q^2)$ or, equivalently, on Λ.

3.4 Phenomenology and Numerical Estimates

The following values for Λ_{MS}, $\Lambda_{\overline{MS}}$ and Λ_{MOM} have been obtained on the basis of BEBC data[28] for the moments of $F_3^{\nu,\bar{\nu}}$:[3,20]

$$\Lambda_{MS} = 0.40 \text{ GeV} \; ; \; \Lambda_{\overline{MS}} = 0.50 \text{ GeV} \; ; \; \Lambda_{MOM} = 0.85 \text{ GeV}^* \; . \qquad (3.23)$$

---

* These values are shown here as an example of a fit to a particular set of data. The CDHS data[29] lead for instance to smaller values of Λ. For a careful phenomenological study of both CDHS and BEBC data with higher order effects and mass corrections included we refer the reader to the recent paper by Abbott and Barnett.[30]

We recall that the leading order analysis leads to $\Lambda_{LO} = 0.7$ GeV.[28] The error bars for the values of Λ are 0(0.05 GeV). All three schemes agree with the BEBC data for $n \leq 5$. Higher moments are discussed below. The effective coupling constants in the three schemes considered for $\Lambda_{MS} = 0.40$, $\Lambda_{\overline{MS}} = 0.50$ and $\Lambda_{MOM} = 0.85$ are plotted in Fig. 4. We observe the following inequalities $\bar{g}^2(Q^2)|_{MOM} > \bar{g}^2(Q^2)|_{\overline{MS}} > \bar{g}^2(Q^2)|_{MS}$ which correspond to $R^{NS}_{2,n} > \bar{R}^{NS}_{2,n} > R^{NS}_{2,n}|_{MOM}$. Furthermore in all cases considered the effective coupling constant is smaller than that given by the leading order expression (the first term on the r.h.s. of Eq. (3.9)) with $\Lambda_{LO} = 0.7$ GeV.

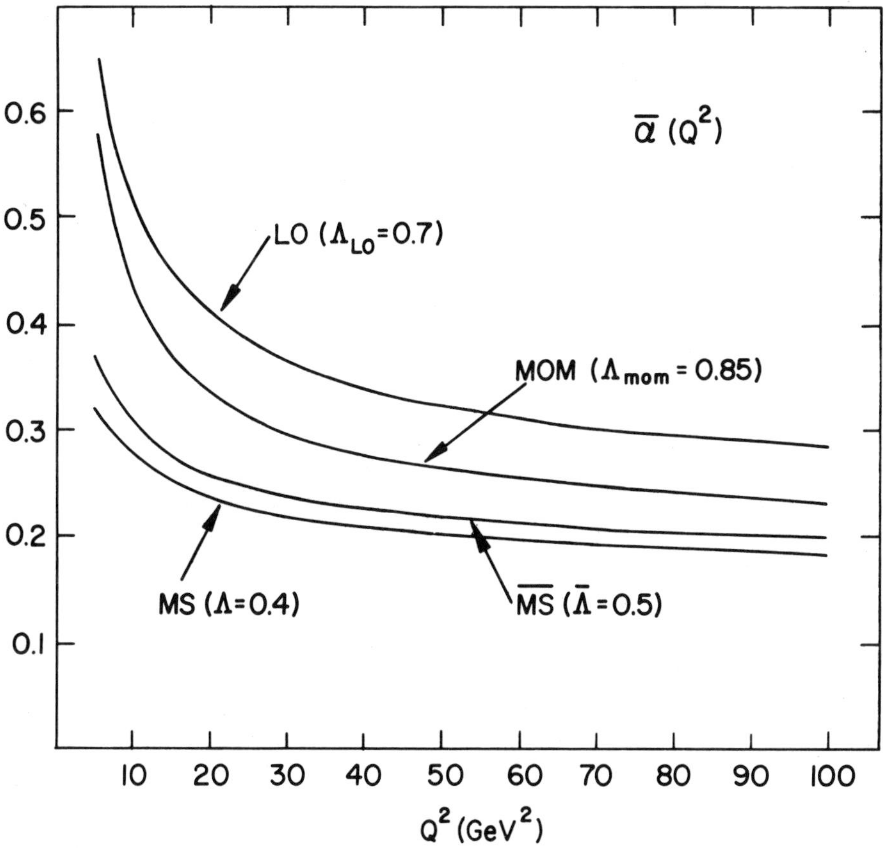

Fig. 4. The effective coupling constant $\bar{\alpha}(Q^2)$ as extracted from the BEBC data for the leading order (L.O.), MS scheme, $\overline{MS}$ scheme and momentum subtraction scheme MOM.

It is instructive to calculate the term

$$1 + \frac{R^{NS}_{k,n}(Q^2)}{\beta_o \ln \frac{Q^2}{\Lambda_i^2}} \qquad i = MS, \overline{MS}, MOM \tag{3.24}$$

in Eq. (3.12) which is equal unity in the leading order. The result is shown in Table 2.

Table 2. The values of the quantity $1 + \frac{R^{NS}_{2,n}(Q^2)}{\beta_o \ln \frac{Q^2}{\Lambda^2}}$ as a function of n and Q^2 in various schemes: $\overline{MS}$ ($\overline{\Lambda}$ = 0.5 GeV), MOM (Λ = 0.85 GeV) and MS (Λ = 0.4 GeV).

n	Scheme	Q^2 [GeV^2] 5	10	50	200
	$\overline{MS}$	0.97	0.96	0.95	0.95
2	MOM	0.68	0.73	0.80	0.83
	MS	1.20	1.15	1.09	1.07
	$\overline{MS}$	1.10	1.05	0.99	0.97
4	MOM	0.63	0.65	0.71	0.75
	MS	1.53	1.42	1.26	1.20
	$\overline{MS}$	1.24	1.15	1.05	1.01
6	MOM	0.69	0.68	0.71	0.74
	MS	1.79	1.62	1.40	1.30
	$\overline{MS}$	1.37	1.25	1.11	1.06
8	MOM	0.79	0.74	0.73	0.75
	MS	2.00	1.79	1.52	1.39

We conclude that in the expansion in the inverse powers of logarithms the next-to-leading order corrections to $M_2^{NS}(n,Q^2)$ calculated in the momentum subtraction scheme with a = 3.5 are larger than those in the $\overline{MS}$ scheme but smaller than in the MS scheme.

On the other hand in the expansion in $\bar{g}^2(Q^2)$ [see Eq. (3.6)] the momentum subtraction scheme leads to a better convergence of the perturbative series than $\overline{MS}$ and MS schemes.[26,27,58]

From the discussion in Subsec. 3.3 it is clear that if the coefficient of $1/(\beta_0 \ln Q^2/\Lambda^2)$ in Eq. (3.12) were independent of Q^2 and had exactly the same n dependence as $\gamma_{NS}^{0,n}$, then all $\bar{g}^2$ corrections could be absorbed in the parameter Λ, and the higher order formula would look like the leading order expression. Conversely, we could say that the leading order formula assumes that the next-to-leading order corrections have the same n dependence as the $\gamma_{NS}^{0,n}$. Therefore it is of interest to see whether the next-to-leading order corrections, which we have calculated in this section, exhibit a non-trivial n-dependence different from $\gamma_{NS}^{0,n}$.

This is most conveniently done by putting Eq. (3.12) into a form of a leading order expression[11,20,59]

$$M_k^{NS}(n,Q^2) = \delta_{NS}^k \bar{A}_n^{NS} \left[\ln \frac{Q^2}{\Lambda_n^{(k)}(Q^2)}\right]^{-d_{NS}^n} \tag{3.25}$$

with

$$\Lambda_n^{(k)}(Q^2) = \Lambda \exp\left[\frac{R_{k,n}^{NS}(Q^2)}{\gamma_{NS}^{0,n}}\right] \approx \Lambda \exp\left[\frac{B_{k,n}^{NS}}{\gamma_{NS}^{0,n}}\right] \equiv \Lambda_n^{(k)} . \tag{3.26}$$

The second relation in Eq. (3.26) is a good approximation if the quantities Z_n^{NS} in Eq. (3.7) are calculated in 't Hooft's minimal subtraction scheme. Needless to say the n-dependence of $\Lambda_n(Q^2)$ or Λ_n is independent of the definition of $\bar{g}^2(Q^2)$. In Fig. 5 the formula (3.26) is compared[31] with Λ_n extracted from the data of BEBC and CDHS for F_3^ν and Fermilab and SLAC ep, en and μp and μn data for F_2^{p-n}. We observe a remarkable agreement of Eq. (3.26) with the data for F_2^{p-n}. The BEBC and in particular CDHS data do not show very clear n-dependence of Λ_n, although at lower values of n they are consistent with Eq. (3.26). We may conclude in particular, on the basis of F_2^{p-n}, that there are indications in

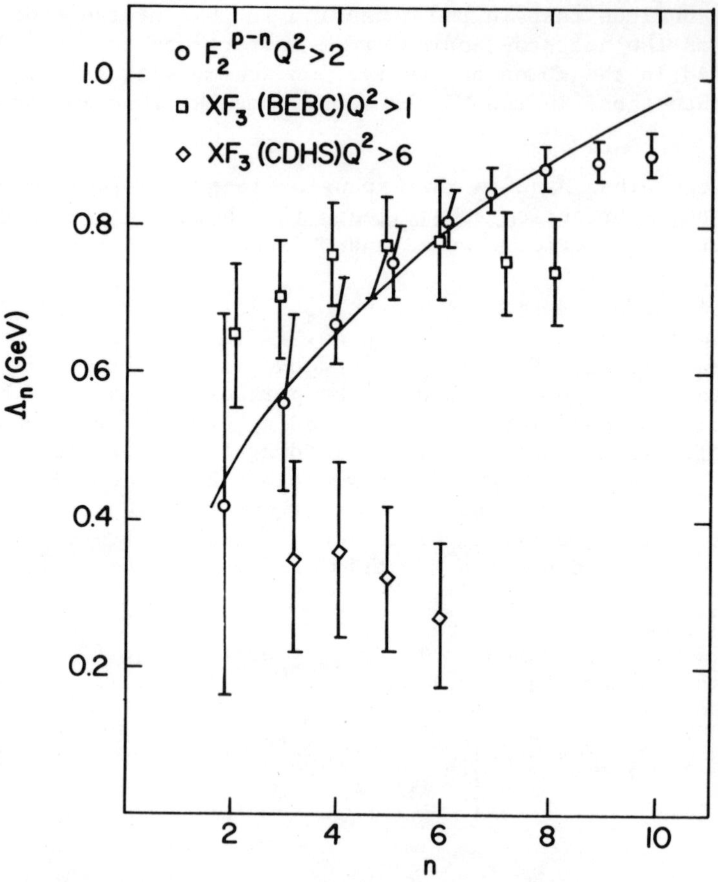

Fig. 5. Experimental Λ_n values obtained by Duke and Roberts using the data of BEBC (open box), CDHS (open diamond) and the entire SLAC data (Ref. 31). However, see Refs. 35 and 59.

the data for the n-dependence of Λ_n as predicted by QCD. It is of interest to see whether the new μ-experiments at CERN and Fermilab will confirm these results.

3.5 Parton Distributions Beyond Leading Order

So far our discussion of higher order corrections was very formal. We shall now express Eq. (3.6) in terms of parton distributions and parton cross-sections to obtain a more intuitive formula (1.13).* Let us first recall that the parametrization of

*In this section we discuss only non-singlet parts of Eq. (1.13).

the QCD predictions (3.6), (3.10) or (3.12) in terms of an effective $\bar{g}^2(Q^2)$ and explicit higher order corrections depends on the definition of $\bar{g}^2(Q^2)$. Similarly the parametrization of QCD predictions in terms of "effective" parton distributions and parton cross-sections depends on the definition of parton distributions.

In order to illustrate this point, we consider the moments of a non-singlet structure function which in the leading order is expressed through the moments of a non-singlet quark distribution $\Delta(x,Q^2)$ as follows

$$M^{NS}(n,Q^2) = \delta_{NS}^{(k)} A_n^{NS}(Q_o^2) \left[\frac{\ln \frac{Q^2}{\Lambda^2}}{\ln \frac{Q_o^2}{\Lambda^2}} \right]^{-d_{NS}^n} \equiv \delta_{NS}^{(k)} \langle\Delta(Q^2)\rangle_n \qquad (3.27)$$

with $\delta_{NS}^{(k)}$ being a charge factor; e.g. $\delta_{NS}^{(k)} = 1/6$ for F_2^{ep}. Notice that

$$\langle\Delta(Q^2)\rangle_n = A_n^{NS}(Q_o^2) \left[\frac{\ln \frac{Q^2}{2}}{\ln \frac{Q_o^2}{\Lambda^2}} \right]^{-d_{NS}^n} \equiv A_n^{NS}(Q^2) \qquad (3.28)$$

i.e. the moments of a non-singlet parton distribution are equal to matrix elements of a non-singlet operator normalized at Q^2.

Beyond the leading order there are various ways of defining parton distributions. We shall discuss here two examples:

a) Generalization of Eq. (3.28)[32]

$$\langle\Delta(Q^2)\rangle_n^{(a)} = A_n^{NS}(Q_o^2) \exp\left[- \int_{\bar{g}^2(Q_o^2)}^{\bar{g}^2(Q^2)} dg' \frac{\gamma_{NS}^n(g')}{\beta(g')} \right] = A_n^{NS}(Q^2) \qquad (3.29)$$

where through order $\bar{g}^2$, the exp ( ) is given by Eq. (3.4). Furthermore $A_n^{NS}(Q_o^2) = \langle\Delta(Q_o^2)\rangle_n^{(a)}$. In terms of $\langle\Delta(Q^2)\rangle_n^{(a)}$ we have, on the basis of Eq. (2.5) and (2.10),

$$M_k^{NS}(n,Q^2) = \langle\Delta(Q^2)\rangle_n^{(a)} C_{k,n}^{NS}(1,\bar{g}^2) \qquad (3.30)$$

which when inverted gives the form of Eq. (1.13)

$$F_k^{NS}(x,Q^2) = \int_x^1 \frac{d\xi}{\xi} \left[\xi\Delta^{(a)}(\xi,Q^2)\right] \sigma_{P,k}^{NS}(\frac{x}{\xi}, Q^2) \ . \tag{3.31}$$

Here

$$\int_0^1 dx\ x^{n-2}\, \sigma_{P,k}^{NS}(x,Q^2) = C_{k,n}^{NS}(1,\bar{g}^2) \tag{3.32}$$

with $C_{k,n}^{NS}(1,\bar{g}^2)$ given by Eqs. (3.3), (3.15) and (3.16).

b) In the second example[21,24] one absorbs all higher order corrections to $M_2^{NS}(n,Q^2)$ into the parton distributions i.e.

$$M_2^{NS}(n,Q^2) = \langle\Delta(Q^2)\rangle_n^{(b)}\, \delta_{NS}^{(2)} \ . \tag{3.33}$$

$\langle\Delta(Q^2)\rangle_n^{(b)}$ is obtained by comparing Eq. (3.33) with Eq. (3.10) for k = 2.

Notice that the Q^2 dependence of $\Delta^{(a)}(x,Q^2)$ and of $\Delta^{(b)}(x,Q^2)$ are different from each other and so are the corresponding parton cross-sections. In the second example the parton cross-section is trivial for k = 2. (b_q in Eq. (1.22) is zero.) For $k \neq 2$ there are some non-zero $\bar{g}^2$ corrections to parton cross-sections even in the second example due to the fact that $B_{n,k}^{NS}$ are different for different k. The parton distributions and parton cross-sections in the example b) are separately renormalization prescription and gauge independent. This is not the case in the example a) but the renormalization prescription and gauge dependences of $\Delta^{(a)}(\xi,Q^2)$ and of $\sigma_{P,k}^{NS}$ cancel in the final expression for $F_k^{NS}(x,Q^2)$ (Eq. 3.31). Since one can define parton distributions in many ways anyhow, one should not worry about this renormalization prescription dependence of parton distributions in example a).

We next notice that whereas the input distributions at $Q^2 = Q_0^2$ in the example b) will be the same as in the leading order (i.e. the data does not change) for k = 2, the input distributions in the example a) will differ considerably at low Q^2 and large x from those used in the leading order phenomenology. The reason is that even in the $\overline{MS}$ scheme $C_n^{NS}(1,\bar{g}^2)$ differs considerably from 1 for low Q^2 and large n. On the other hand the example a) turns out to be useful for inversion of moments if $Z_{n,k}^{NS}$ are calculated in the 't Hooft's scheme. In this scheme $Z_{n,k}^{NS}$ are small (see Table 1) and the equation for the Q^2 dependence of the parton distributions is essentially the same as the leading order equation (3.28) with $\bar{g}^2(Q^2)$ now given by Eq. (3.9). Therefore the standard techniques used to invert moments in the leading order can also be used successfully here. In particular one can find analytic expressions for $\Delta^{(a)}(x,Q^2)$ which to a good approximation represent the exact QCD predictions. The function $C^{NS}_{n,k}(1,\bar{g}^2)$ which contains all non-trivial n-dependence of higher order corrections, can be inverted exactly (analytically). As a result of this procedure one obtains approximate analytic expressions for $\Delta^{(a)}(x,Q^2)$ and exact expressions for $\sigma^{NS}_k(x,Q^2)$. Inserting these two functions into Eq. (3.31) leads to $F^{NS}_k(x,Q^2)$. Details of this inversion method can be found in Ref. 33. For different inversion methods see Refs. 35 and 60.

4. SINGLET SECTOR BEYOND THE LEADING ORDER

The study of next-to-leading order QCD corrections to the singlet structure functions is much more complicated than for the non-singlet structure functions due to the mixing between fermion singlet and gluon operators (see Eq. 2.8). The derivation of the formal and parton model-like expressions can be found in refs. 3, 21, 34 and 35. Here we shall only make a few remarks.

i) The analysis of singlet contributions requires the calculation of the two-loop anomalous dimension matrix and of the one-loop corrections to the fermion singlet and gluon Wilson coefficient functions $C^{\psi}_{k,n}(1,\bar{g}^2)$ and $C^{G}_{k,n}(1,\bar{g}^2)$. As in the non-singlet case, one has to take care that all these quantities are calculated in the same renormalization scheme. The full answer has been obtained in the literature again only in 't Hooft's scheme.[20,21*] The calculation of the two-loop anomalous dimension matrix performed in ref. 21 is particularly complicated and involves O(100) two-loop diagrams.

ii) It turns out that the formal expressions for the Q^2 evolution of the moments of singlet structure functions are very simple[34] e.g.

$$M^S_2(n,Q^2) = \delta^{(2)}_{\psi}\bar{A}^-_n \left[1 + \frac{R^-_{2,n}(Q^2)}{\beta_o \ln \frac{Q^2}{\Lambda^2}}\right]\left[\ln\frac{Q^2}{\Lambda^2}\right]^{-d^n_-}$$
$$+ \delta^{(2)}_{\psi}\bar{A}^+_n \left[1 + \frac{R^+_{2,n}(Q^2)}{\beta_o \ln \frac{Q^2}{\Lambda^2}}\right]\left[\ln\frac{Q^2}{\Lambda^2}\right]^{-d^n_+} \qquad (4.1)$$

---

*For calculations of $C^G_{k,n}(1,\bar{g}^2)$ in other renormalization schemes we refer the reader to refs. 22-25, and 36.

where $\bar{A}_n^{\mp}$ and Λ are the only free parameters. The next to leading order corrections turn out to be of the same order as in the non-singlet sector although their n dependence in particular at low values of n is different due to the mixing. Numerical estimates of these corrections are given in ref. 34.

iii) The equations for the Q^2 evolution of the singlet parton distributions turn out to be much more complicated than Eq. (4.1). As in the non-singlet sector they depend on the definition of parton distributions. Explicit expressions for the singlet analogs of examples a) and b) of the previous section are presented in refs. 3 and 21 respectively.

iv) Equation (3.31) is now generalized to

$$F_2^S(x,Q^2) = \int_x^1 \frac{d\xi}{\xi} [\xi\Sigma(\xi,Q^2)\sigma_{P,2}^{\psi}(\frac{x}{\xi},Q^2) + \xi G(\xi,Q^2)\sigma_{P,2}^{G}(\frac{x}{\xi},Q^2)] \tag{4.2}$$

where $\Sigma(\xi,Q^2)$ and $G(\xi,Q^2)$ are singlet fermion and gluon distributions respectively. The important point is that only the sum of the two terms in Eq. (4.2) does not depend on the definition of parton distributions. The separation of F_2^S into quark and gluon contributions depends on the other hand on the definition of parton distributions. Some phenomenological applications of the singlet formulae can be found in the paper by Anderson et al.[31] and in ref. 35.

5. SEMI-INCLUSIVE PROCESSES

5.1 Massive μ-pair Production

We begin our discussion of semi-inclusive processes with the massive μ-pair production in hadron-hadron collisions. In the leading order of asymptotic freedom one sums the QCD diagrams contributing to this process to all orders in g^2 and keeps only the leading logarithms. In the course of the calculation one encounters mass singularities which are factored out and absorbed in the wave functions of the incoming hadrons.* As a result[6,38]

---

*Since factorization of mass singularities has been discussed in other lectures at this Summer School in detail,[12] we do not demonstrate it here. Proofs of factorization of mass singularities to all orders can be found in refs. 7, 8 and 37.

the standard Drell-Yan formula (1.10) is reproduced with the scale independent quark distributions replaced by the Q^2 dependent quark distributions with Q^2 dependence governed by the leading order formula (1.20) and its generalization to the singlet sector.

In order to discuss the structure of the corresponding calculations in the next-to-leading order it is useful to take moments in τ

$$\sigma_n(Q^2) = \int d\tau\, \tau^n \frac{d\sigma}{dQ^2} . \tag{5.1}$$

Formally $\sigma_n(Q^2)$ can be written as follows[8]

$$\sigma_n(Q^2) = \sum_{i,j} A^{(1)}_{n,i}(\mu^2) A^{(2)}_{n,j}(\mu^2) C^{ij}_n\left(\frac{Q^2}{\mu^2}, g^2\right) \tag{5.2}$$

where the sums run over $\bar{q}$, q and G, and the indices (1) and (2) label the incoming hadrons.

The expansion in Eqs. (5.2) is analogous to the operator product expansion of Eq. (2.3). The A_n's which correspond to matrix elements of local operators in Eq. (2.3) are called cut vertices.[8] They are incalculable in perturbation theory. In more intuitive language they can be related to the moments of the parton distributions at $Q^2 = \mu^2$. The $C^{ij}_n(Q^2/\mu^2, g^2)$ are the coefficient functions in the cut vertex expansion in Eq. (5.2). They satisfy[8] renormalization group equations similar to the ones satisfied by $C^i_n(Q^2/\mu^2, g^2)$. $C^{ij}_n(Q^2/\mu^2, g^2)$ are calculable in perturbation theory. Expansion (5.2) expresses the factorization of the non-perturbative pieces (cut vertices) from the perturbatively calculable pieces (coefficient functions) just as the operator product expansion. In what follows we shall discuss the calculation of $C^{ij}_n(Q^2/\mu^2, g^2)$ in the framework of two different approaches.

Approach I[39]

We begin by expressing Eq. (5.2) in terms of parton distributions. It turns out that the cut vertices for the incoming hadrons are the same as the matrix elements of local operators which enter the discussion of deep inelastic scattering.

Consequently the anomalous dimensions which enter the calculation of $C^{ij}_n(Q^2/\mu^2, g^2)$ are the ones which we encountered already in deep-inelastic scattering. Therefore if we put $\mu^2 = Q^2$ in Eq. (5.2) we can write

$$\sigma_n(Q^2) = \sum_{i,j} A_{n,i}^{(1)}(Q^2)A_{n,j}^{(2)}(Q^2)C_n^{ij}(1,\bar{g}^2) \tag{5.3}$$

$$\equiv \sum_{i,j} \langle f_i(Q^2)\rangle_n^{(1)}\langle f_j(Q^2)\rangle_n^{(2)} C_n^{ij}(1,\bar{g}^2) \tag{5.4}$$

where $\langle f_i(Q^2)\rangle_n$ are the moments of parton distributions (quarks and gluons) defined as in example a) of Section 3 (Eqs. (3.29, 3.30)). $C_n^{ij}(1,\bar{g}^2)$ can be interpreted in analogy with Eq. (3.32) as the moments of cross-sections for parton j-parton i scattering or annihilation with a $\mu^+\mu^-$ pair in the final state. We know already the Q^2 dependence of $\langle f_i(Q^2)\rangle_n$ with next-to-leading order corrections included (example a) of Sections 3 and 4). What remains to be done is to calculate the coefficients $C_n^{ij}(1,\bar{g}^2)$. The procedure for calculation of $C_n^{ij}(1,\bar{g}^2)$ is a straightforward generalization of the procedure used in the calculation of $C_n^{i}(1,\bar{g}^2)$ for deep-inelastic scattering which we outlined in the Appendix. Let us illustrate this procedure with $i = q$ and $j = \bar{q}$, i.e. $C_n^{q\bar{q}}(1,\bar{g}^2)$ which through order $\bar{g}^2$ has the expansion

$$C_n^{q\bar{q}}(1,\bar{g}^2) = 1 + \frac{\bar{g}^2}{16\pi^2} B_n^{q\bar{q}} \ . \tag{5.5}$$

In order to find $B_n^{q\bar{q}}$ one considers $q\bar{q}$ annihilation in which case we have

$$\sigma_n^{q\bar{q}}(Q^2) = A_{n,q}^{(1)}(\mu^2)A_{n,\bar{q}}^{(2)}(\mu^2)C_n^{q\bar{q}}\left(\frac{Q^2}{\mu^2}, g^2\right) \ . \tag{5.6}$$

Taking incoming quarks off-shell ($p_1^2 < 0$, $p_2^2 < 0$) one obtains in analogy with (A.2)

$$\sigma_n^{q\bar{q}}(Q^2) = 1 + \frac{g^2}{16\pi^2}\left[-\frac{1}{2}\gamma_{qq}^{0,n}\ln\frac{Q^2}{-p_1^2} - \frac{1}{2}\gamma_{qq}^{0,n}\ln\frac{Q^2}{-p_2^2} + \bar{\sigma}_n^{q\bar{q}}\right] \tag{5.7}$$

where $\bar{\sigma}_n^{q\bar{q}}$ are independent of p_i^2 and Q^2, and $\gamma_{qq}^{0,n} = \gamma_{NS}^{0,n}$. Next using Eq. (A.4)* for $A_{n,q}^{(1)}$ and $A_{n,\bar{q}}^{(2)}$ we obtain from (5.6) and (5.7), after putting $Q^2 = \mu^2$

$$B_n^{q\bar{q}} = \bar{\sigma}_n^{q\bar{q}} - 2A_{nq}^q \ . \tag{5.8}$$

---

* g^2 corrections to $A_{n,q}^{(1)}$ and $A_{n,\bar{q}}^{(2)}$ are equal to each other in this order and equal to g^2 corrections to the matrix elements of non-singlet operator taken between quark states (see Eq. (A.4)).

$B_n^{q\bar{q}}$ depends on the renormalization scheme through A_{nq}^{q}. This renormalization prescription dependence is however cancelled in the final expression (5.4) by that of $\langle f_i(Q^2)\rangle_n$ if the two-loop anomalous dimensions which enter Eq. (3.29) are calculated in the same scheme as $A_{n,q}^{q}$. To complete the calculation of next to leading order corrections to massive μ-pair production one has to calculate

$$C_n^{qG}(1,\bar{g}^2) = \frac{\bar{g}^2}{16\pi^2} B_n^{qG} \quad . \tag{5.9}$$

As the reader may easily check in this case in an obvious notation

$$B_n^{qG} = B_n^{\bar{q}G} = \bar{\sigma}_n^{qG} - A_{nG}^{q} \tag{5.10}$$

with A_{nG}^{q} known already from the analysis of deep-inelastic scattering (Eq. A.6). In summary in this approach (dropping summation over flavors) we have

$$\sigma_n(Q^2) = \langle q(Q^2)\rangle_n \langle \bar{q}(Q^2)\rangle_n \left[1 + \frac{\bar{g}^2}{16\pi^2} B_n^{q\bar{q}}\right]$$
$$+ \langle G(Q^2)\rangle_n \left[\langle q(Q^2)\rangle_n + \langle \bar{q}(Q^2)\rangle_n\right] \left[\frac{\bar{g}^2}{16\pi^2} B_n^{qG}\right] + O(\bar{g}^4) \tag{5.11}$$

with $B_n^{q\bar{q}}$ and B_n^{qG} given by Eqs. (5.8) and (5.10) respectively, and the parton distributions defined as in example a) of Section 3.

<u>Approach II</u>[24,25]

If the parton distributions are defined as in example b) of Sections 3 and 4 (i.e. the next to leading order corrections to F_2 of deep-inelastic scattering are absorbed totally in the definition of <u>quark</u> distributions) then Eq. (5.11) is replaced by

$$\sigma_n(Q^2) = \langle \tilde{q}(Q^2)\rangle_n \langle \tilde{\bar{q}}(Q^2)\rangle_n \left[1 + \frac{\bar{g}^2}{16\pi^2} \tilde{B}_n^{q\bar{q}}\right]$$
$$+ \langle \tilde{G}(Q^2)\rangle_n \left[\langle \tilde{q}(Q^2)\rangle_n + \langle \tilde{\bar{q}}(Q^2)\rangle_n\right] \left[\frac{\bar{g}^2}{16\pi^2} \tilde{B}_n^{qG}\right] \tag{5.12}$$

where the parton distributions are modified relative to those in (5.11) and $\tilde{B}_n^{q\bar{q}}$ and $\tilde{B}_n^{qG}$ are

$$\tilde{B}_n^{q\bar{q}} = B_n^{q\bar{q}} - 2B_n^q = \bar{\sigma}_n^{q\bar{q}} - 2\bar{\sigma}_n^q \tag{5.13}$$

and

$$\tilde{B}_n^{qG} = B_n^{qG} - B_n^G = \bar{\sigma}_n^{qG} - \bar{\sigma}_n^G \quad . \tag{5.14}$$

B_n^q, B_n^G, $\bar{\sigma}_n^q$ and $\bar{\sigma}_n^G$ are defined in the Appendix. We observe that in this approach the A_n's do not need to be calculated to obtain $\tilde{B}_n^{q\bar{q}}$ and $\tilde{B}_n^{qG}$. Furthermore, contrary to the previous approach, the parton distributions and parton cross-sections ($\tilde{B}_n^{q\bar{q}}$, $\tilde{B}_n^{qG}$) are separately renormalization and regularization scheme* independent. It should be, however, stressed that in both (5.11) and (5.12) only $\sigma_n(Q^2)$ is independent of the definition of parton distributions. The separation of $\sigma_n(Q^2)$ into qq and qG term depends as we have seen on the definition of parton distributions used. In particular the qG term and the next to leading order corrections to the Q^2 evolution of quark distributions in the $q\bar{q}$ term depend on each other.

As yet nobody has presented the full numerical calculation of next to leading order corrections to $\sigma_n(Q^2)$.** Numerical estimates of the separate corrections to parton cross-sections are however known. In particular the parameters $B_n^{q\bar{q}}$ and $\tilde{B}_n^{q\bar{q}}$ obtained by various groups[24,25,39,41] are so large that the authors conclude that at present values of Q^2 the perturbative calculations cannot be trusted. Further details can be found in refs. 24, 25, 39.

5.2 Processes Involving Fragmentation Functions[8,42,43,44]

Here we shall comment briefly on the processes (1.2), (1.3) and (1.5). We begin with the semi-inclusive e^+e^- annihilation. The expression for the moments of $\sigma_h^{e^+e^-}(z, Q^2)$ which we denote by $\sigma_h^{e^+e^-}(n,Q^2)$ can be written formally as[8]

---

*If of course the same regularization schemes are used for $\bar{\sigma}_n^{q\bar{q}}$ and $\bar{\sigma}_n^q$, and $\bar{\sigma}_n^{qG}$ and $\bar{\sigma}_n^G$.

**In all papers on higher order corrections to Drell-Yan process the leading order formulae for the Q^2 evolution of parton distributions have been used.

$$\sigma_h^{e^+e^-}(n,Q^2) = \sum_i V_n^i(\mu^2)\tilde{C}_n^i\left(\frac{Q^2}{\mu^2}, g^2\right) \tag{5.15}$$

$$\equiv \sum V_n^i(Q^2)\tilde{C}_n^i(1,\bar{g}^2) \qquad i = q,\ \bar{q},\ G\ . \tag{5.16}$$

Equation (5.15) is the analogue of Eqs. (2.3) and (5.2) with $V_n^i(\mu^2)$ being called time-like cut vertices.[8] $\tilde{C}_n^i(Q^2/\mu^2,g^2)$ are the corresponding coefficient functions. In analogy with Eq. (5.4), $V_n^i(Q^2)$ can be interpreted as the moments of Q^2-dependent fragmentation functions with $\tilde{C}_n^i(1,\bar{g}^2)$ as the moments of the cross-section for the production of parton i in e^+e^- annihilation. Inverting the moments one obtains Eq. (1.14). The structure of next to leading order QCD corrections to the process in question is similar to that in deep-inelastic scattering. The questions of renormalization prescription dependence of separate quantities entering (5.16) and the freedom in defining fragmentation functions beyond the leading order also arise here. The full study of next to leading order corrections to the process above has not yet been completed. Missing is the calculation of two-loop anomalous dimensions of time-like cut vertices. This implies that we do not know at present the full next to leading order corrections to the Q^2 evolution of fragmentation functions, which enter the QCD formulae for processes (1.2), (1.3) and (1.5). What we know, however, are the relevant parton cross-sections which contribute to the processes in question. They have been most extensively studied by the authors of ref. 43. The calculations of next to leading order corrections to parton cross-sections relevant for the processes (1.2) and (1.3) can also be found in refs. 42 and 44. The method used in these calculations is the extension of the approach II discussed previously to processes involving fragmentation functions. One defines fragmentation functions by absorbing all next to leading order QCD corrections to $e^+e^- \to h$ + anything into quark fragmentation functions. The cross-sections for the processes (1.3) and (1.5) can then be expressed through so defined fragmentation functions and parton distributions which we discussed previously. The form of the resulting expression for $eh_1 \to eh_2$ + anything is shown in Eq. (1.15). A similar equation exists for $e^+e^- \to h_1 + h_2$ + anything with $f_j^{h_1}$ replaced by $D_j^{h_1}$ and $\tilde{\sigma}_p^{jk}$ by parton cross-sections for production of partons j and k in e^+e^- annihilation. The latter parton cross-section we shall denote by $\hat{\sigma}_p^{jk}$. As in the case of massive μ-pair production $\tilde{\sigma}_p^{jk}$ and $\hat{\sigma}_p^{jk}$ calculated in the approach in question are regularization and renormalization scheme independent.

Let us enumerate some properties of $\tilde{\sigma}_p^{jk}$ and $\hat{\sigma}_p^{jk}$:

i) The $\bar{g}^2$ corrections to these cross-sections are only large at kinematical boundaries.

ii) There is a breakdown of factorization in x and z in $\tilde{\sigma}_P^{jk}$ which is of order 10-20% for small and moderate x and z but larger when both x and z are large.

iii) The authors of ref. 43 have found the following relations between $\tilde{\sigma}_P^{jk}$ and $\hat{\sigma}_P^{jk}$

$$\tilde{\sigma}_P^{jk}(\frac{1}{x}, z, Q^2) = -\frac{1}{x}\hat{\sigma}_P^{jk}(x, z, Q^2) \qquad j,k = q, \bar{q}, G . \quad (5.17)$$

Equation (5.17) is analogous to the parton model and leading order relations connecting deep-inelastic structure functions and structure functions for $e^+e^- \to h$ + anything, which have been proposed by Gribov and Lipatov[45] and Drell, Levy and Yan.[46] It should be remarked that the relations like (5.17) are shown to hold only for properly defined parton cross-sections and are not expected to hold beyond the leading order for parton distributions and parton fragmentation functions themselves.

For further details related to the processes (1.2), (1.3) and (1.5) we refer the interested reader to refs. 42-44.

6. OTHER HIGHER ORDER CALCULATIONS

There are a few recent higher order calculations which we did not discuss in the previous section. We shall comment here on them very briefly.

a) Violations of Parton Model Sum Rules

Gross-Llewellyn-Smith and Bjorken Sum Rule relations are violated beyond the leading order as follows[20,24]

$$\int_0^1 dx\,[F_3^{\bar{\nu}p} + F_3^{\nu p}] = 6\left[1 - \frac{12}{(33-2f)\ln\frac{Q^2}{\Lambda^2}}\right] \qquad (6.1)$$

$$\int_0^1 dx\,[F_1^{\bar{\nu}p} - F_1^{\nu p}] = 1 - \frac{8}{(33-2f)\ln\frac{Q^2}{\Lambda^2}} \,. \qquad (6.2)$$

This leads to 20% and 10% corrections respectively at values of $Q^2 \sim O(10\ \mathrm{GeV}^2)$.

There are also corrections to Callan-Gross relation[47] which turn out to be much smaller than the violation of this relation seen in the data.

b) Higher Order Corrections to Photon-Photon Scattering

The process $\gamma + \gamma \to$ hadrons can be measured in $e^+e^- \to e^+e^-$ + hadrons.[48] When one photon has large Q^2 and the other is close to its mass-shell the photon-photon process can be viewed as deep-inelastic scattering on a photon target. It turns out that due to the point-like character of the photon the dominant contribution to photon-photon scattering at large Q^2 can be exactly calculated in QCD with the result

$$\int_0^1 dx\, x^{n-2} F_2^\gamma(x,Q^2) = \alpha^2 \left[a_n \ln \frac{Q^2}{\Lambda^2} + \tilde{a}_n \ln\ln \frac{Q^2}{\Lambda^2} + b_n + O\left(\frac{1}{\ln \frac{Q^2}{\Lambda^2}} \right) \right]. \tag{6.3}$$

Here F_2^γ is the photon structure function. a_n have been calculated in ref. 49. The $\tilde{a}_n$ and b_n have been obtained in ref. 50. The exact values of b_n depend on the definition of $\bar{g}^2(Q^2)$ or equivalently Λ. Taking the values of Λ extracted from deep-inelastic scattering in the same scheme for $\bar{g}^2(Q^2)$ in which b_n's are calculated,[50] one can make definite predictions about the moments of Eq. (6.3). The corrections turn out to be slightly bigger than corresponding corrections in deep-inelastic scattering and suppress F_2^γ at large values of x. In the $\overline{\text{MS}}$ scheme b_n's are negative but in the MOM scheme they are positive. However when the corresponding values of $\Lambda(\Lambda_{\overline{MS}} = 0.5,\ \Lambda_{MOM} = 0.85$ GeV) are inserted in Eq. (6.3) the same predictions[3] are obtained for $F_2^\gamma(x,Q^2)$. Bigger corrections are found in the MS scheme.

c) α_s^2 Corrections to $e^+e \to$ hadrons

Recently α_s^2 corrections to $R = \sigma(e^+e^- \to \text{hadrons})/\sigma(e^+e^- \to \mu^+\mu^-)$ have been calculated with the result:[51]

$$R = 3\,\Sigma\, Q_i^2 \left[1 + \frac{\alpha_s}{\pi} + A\left(\frac{\alpha_s}{\pi}\right)^2 \right] \tag{6.4}$$

where, for four flavors, A = 5.6, 1.5 and -1.7 for MS, $\overline{\text{MS}}$ and MOM schemes respectively. Taking the corresponding values for Λ from deep-inelastic scattering we observe that α_s^2 corrections to R are small.

d) Large QCD corrections to $Q\bar{Q} \to 2$ gluon decay have been reported by the authors of ref. 52.

e) Higher order corrections to the polarized electroproduction structure functions have been calculated in ref. 53.

f) qq contributions (order $\bar{g}^4(Q^2)$) to massive muon production have been calculated in ref. 54. They turn out to be small.

g) Finally the next to leading order corrections to large $p_\perp$ processes are being performed.[55]

7. SUMMARY

In these review lectures we have discussed higher order QCD predictions for inclusive deep-inelastic scattering. We have also presented the basic structure of QCD formulae and the methods of corresponding calculations for semi-inclusive processes. We have seen that the structure of QCD formulae with higher order corrections taken into account is fairly complicated and involves many features not encountered in the leading order. These new features include:

i) gauge and renormalization-prescription dependences of separate elements of the physical expressions;

ii) well-defined dependence of the functional form of the explicit higher order corrections on the definition of $\bar{g}^2(Q^2)$ or, equivalently, on Λ;

iii) freedom in the definition of parton distributions and parton fragmentation functions beyond the leading order approximation.

These features have to be kept in mind when carrying out calculations to make sure that various parts of the higher order calculations are compatible with each other. Only then can a physical result be obtained which is independent of gauge, renormalization scheme, particular definition of $\bar{g}^2(Q^2)$, and particular definition of the parton distributions.

We have seen that the higher order corrections are quite large and, moreover, that there are some indications for their presence in the deep-inelastic scattering data. This is most clearly seen in the n-dependence of the parameter Λ_n extracted from the data on the basis of the leading order formulae. This n-dependence agrees well with that obtained from higher order calculations.

In some processes such as massive μ-pair production the next to leading order corrections turn out to be too large at present values of Q^2 for the perturbative calculations to be trusted. This is also the case of η_c decay. We think these processes deserve further study.

The calculation of two-loop anomalous dimensions to the Q^2 evolution of fragmentation functions is very desirable. Also more phenomenology of next to leading order corrections to various processes should be done.

In our lectures we have not discussed the mass effects and higher twist operators effects, which at values of Q^2 of $O(5\ \text{GeV}^2)$ are of some importance. They have been recently discussed in refs. 30 and 56 which the interested reader may consult. At low values of Q^2 also nonperturbative effects should be taken into account.

In spite of the fact that there is still much to be done, both theoretically and phenomenologically, we believe that a lot of progress has been made in the past few years in understanding QCD effects in the inclusive and semi-inclusive processes and we are looking forward to the summer schools next year when surely more progress will be reported.

ACKNOWLEDGMENTS

It is a great pleasure to thank Dennis Creamer for a very careful reading of the manuscript, critical comments and numerous discussions. In preparing this manuscript I also benefitted from conversations with many other colleagues from the Theory Group at Fermilab and in particular with Andrzej Bialas and Bill Bardeen. I would also like to thank Professor K.T. Mahanthappa for inviting me to this wonderful, in many respects, summer school.

APPENDIX Procedure for the Calculation of $B_{k,n}^{NS}$

We first notice that in order to find $B_{k,n}^{NS}$ as defined in Eq. (3.3) it is sufficient to calculate $C_{2k,n}^{NS}(Q^2/\mu^2, g^2)$ in perturbation theory to order g^2 and put $Q^2 = \mu^2$. This is obvious from Eqs. (2.9) and (2.10). In order to calculate $C_{k,n}^{NS}(Q^2/\mu^2, g^2)$ in perturbation theory we calculate first the virtual Compton amplitude for quark photon scattering. Taking the incoming quarks slightly off-shell ($p^2 < 0$) in order to regulate mass-singularities we obtain for $T_k(Q^2, \nu)$

$$T_k(Q^2,\nu) \equiv \sum_n \frac{1}{x^n} \sigma^q_{k,n}\left(\frac{Q^2}{p^2}, g^2\right) \tag{A.1}$$

with*

$$\sigma^q_{k,n}\left(\frac{Q^2}{p^2}, g^2\right)_\nu \equiv \sigma^q_n\left(\frac{Q^2}{p^2}, g^2\right) = 1 + \frac{g^2}{16\pi^2}\left[-\frac{1}{2}\gamma^{0,n}_{NS}\ln\left(\frac{Q^2}{-p^2}\right) + \bar{\sigma}^q_n\right] \tag{A.2}$$

where $\bar{\sigma}^q_n$ are constant terms and $x = Q^2/(2p\cdot q)$. The diagrams contributing to σ^q_n in order g^2 are shown in Fig. 6. Now in accordance with the operator product expansion

$$\sigma^q_n\left(\frac{Q^2}{p^2}, g^2\right) = C^{NS}_n\left(\frac{Q^2}{\mu^2}, g^2\right) A^{NS}_n\left(\frac{p^2}{\mu^2}, g^2\right) \tag{A.3}$$

where $C^{NS}_n(Q^2/\mu^2, g^2)$ is the same coefficient function as in Eq. (2.5) but $A^{NS}_n(p^2/\mu^2, g^2)$ are the matrix elements of non-singlet operator sandwiched between quark states instead of proton states as in Eq. (2.5). The coefficient functions are the same as before because they do not depend on the state between which the operator product expansion is sandwiched. $A^{NS}_n(p^2/\mu^2, g^2)$ can be evaluated in perturbation theory (see diagrams in Fig. 7) with the result**

$$A^{NS}_n\left(\frac{p^2}{\mu^2}, g^2\right) = 1 + \frac{g^2}{16\pi^2}\left[\frac{1}{2}\gamma^{0,n}_{NS}\ln\frac{-p^2}{\mu^2} + A^q_{nq}\right] \tag{A.4}$$

where A^q_{nq} are independent of p^2. Combining Eqs. (A.2)-(A.4) and using (3.3) we obtain

$$B^{NS}_n = \bar{\sigma}^q_n - A^q_{nq} . \tag{A.5}$$

---

*We drop the index k to simplify notation and drop the charge factor δ^k_{NS}.

**Generally A^i_{nj} are the constant pieces in order g^2 of the matrix element of operator O_i sandwiched between j-state.

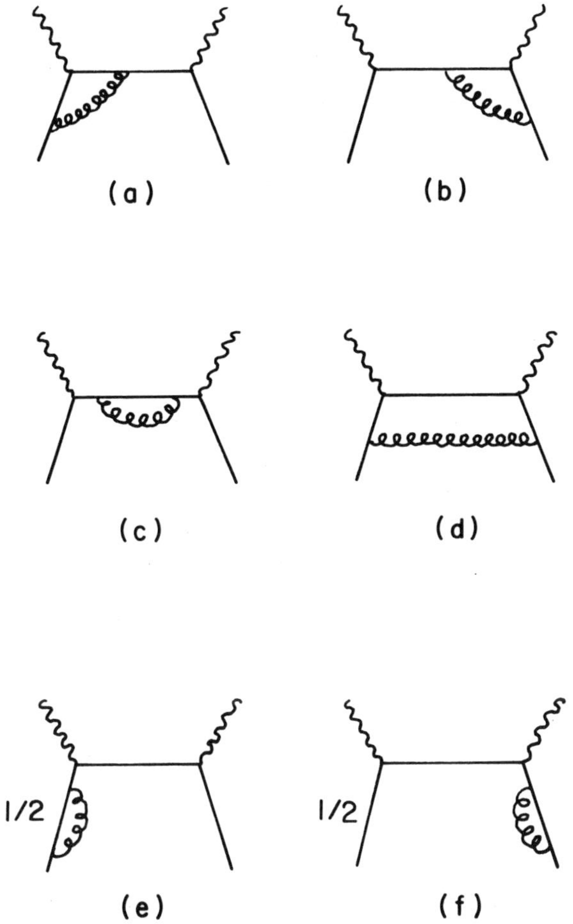

Fig. 6. Diagrams entering the calculation of $\sigma^q_{k,n}(Q^2/p^2,g^2)$ of Eq. (A.2).

The point is that A^q_{nq} depends on the renormalization scheme used to render finite the result of the calculation of diagrams in Fig. 7. In so-called $p^2 = -\mu^2$ subtraction schemes A^q_{nq} is put to zero. In the 't Hooft's scheme it is non-zero. The evaluation of the g^2 corrections to the gluon coefficient function proceeds in a similar way with the result

$$C^G_n(1,\bar{g}^2) = \frac{\bar{g}^2}{16\pi^2} B^G_n \; ; \; B^G_n = \bar{\sigma}^G_n - A^q_{nG} \qquad \text{(A.6)}$$

where $\bar{\sigma}^G_n$ is obtained by calculating photon-gluon scattering to order g^2 and A^q_{nG} is the constant piece in the matrix element of

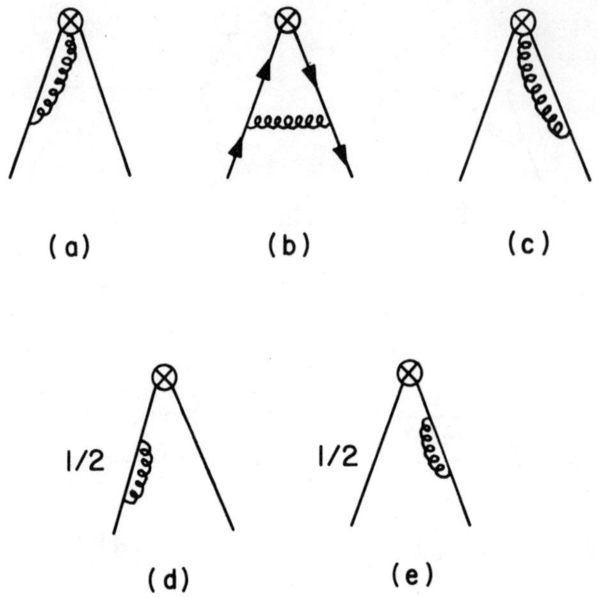

Fig. 7. Diagrams entering the calculation of $A_n^{NS}(p^2/\mu^2, g^2)$ of Eq. (A.4).

quark operator between gluon states. Eq. (A.6) is explicitly derived in refs. 3, 20. Details of the calculations using this method can be found for instance in refs. 3 and 20.

REFERENCES

1. A. Peterman, Phys. Rep. 53C, 157 (1979).

2. R.D. Field, CALT-68-696.

3. A.J. Buras, Fermilab-Pub-79/17-THY, to appear in Reviews of Modern Physics, Vol. 42, No. 1.

4. R.P. Feynman, "Photon-Hadron Interactions," (W.A. Benjamin, New York, 1972).

5. S.D. Drell and T. Yan, Ann. Phys. (N.Y.) 66, 578 (1971).

6. H.D. Politzer, Nucl. Phys. B129, 301 (1977); C.T. Sachrajda, Phys. Lett. 73B, 185 (1978); Phys. Lett. 76B, 100 (1978).

7. R.K. Ellis, H. Georgi, M. Machacek, H.D. Politzer, and G.G. Ross, Phys. Lett. 78B, 281 (1978); Nucl. Phys. B152, 285 (1979).

8. A.H. Mueller, Phys. Rev. D18, 3705 (1978); S. Gupta and A.H. Mueller, Phys. Rev. D20, 118 (1979).

9. G. Altarelli and G. Parisi, Nucl. Phys. B126, 298 (1977).

10. J.F. Owens, Phys. Lett 76B, 85 (1978); T. Uematsu, Phys. Lett. 79B, 97 (1978).

11. M. Bace, Phys. Lett. 78B, 132 (1978).

12. See the lectures of S. Ellis, R.D. Field and C.H. Llewellyn Smith at this Summer School.

13. K. Wilson, Phys. Rev. 179, 1499 (1969).

14. W. Zimmerman, in Lectures on Elementary Particles and Quantum Field Theory, (MIT Press, Cambridge, 1971); D.J. Gross, in Methods in Field Theory, Les Houches, 1975, ed. by R. Balian and J. Zinn-Justin (North-Holland, Amsterdam, 1979); J. Ellis, in Lectures on Weak and Electromagnetic Interactions at High Energies, Les Houches, 1976, ed. by R. Balian and C.H. Llewellyn Smith.

15. N. Christ, B. Hasslacher and A. Mueller, Phys. Rev. D6, 3543 (1972).

16. Renormalization group equations as applied to QCD are discussed in detail by H.D. Politzer, Physics Reports 14, 129 (1974) and D.J. Gross and F. Wilczek, Phys. Rev. D9, 980 (1974). See also refs. 1 and 3.

17. H.D. Politzer, Phys. Rev. Lett. 30, 1346 (1973); D.J. Gross and F. Wilczek, Phys. Rev. Lett. 30, 1323 (1973); H. Georgi and H.D. Politzer, Phys. Rev. D9, 416 (1974).

18. W. Caswell, Phys. Rev. Lett. 33, 244 (1974); D.R.T. Jones, Nucl. Phys. B75, 531 (1974).

19. E.G. Floratos, D.A. Ross and C.T. Sachrajda, Nucl. Phys. B129, 66 (1977) and Erratum, Nucl. Phys. B139, 545 (1978).

20. W.A. Bardeen, A.J. Buras, D.W. Duke and T. Muta, Phys. Rev. D18, 3998 (1978).

21. E.G. Floratos, D.A. Ross and C.T. Sachrajda, Nucl. Phys. B152, 483 (1979).

22. A. De Rujula, H. Georgi and H.D. Politzer, Ann. of Phys. 103, 315 (1977).

23. M. Calvo, Phys. Rev. D15, 730 (1977).

24. G. Altarelli, R.K. Ellis and G. Martinelli, Nucl. Phys. B143, 521 (1978) and Erratum, Nucl. Phys. B146, 544 (1978).

25. J. Kubar-Andre and F.E. Paige, Phys. Rev. D19, 221 (1979).

26. R. Barbieri, L. Caneschi, G. Curci and E. d'Emilio, Phys. Lett. 81B, 207 (1979).

27. W. Celemaster and R.J. Gonsalves, Phys. Rev. Lett. 42, 1435 (1979).

28. P.C. Bosetti et al., Phys. Lett 70B, 273 (1977).

29. J.G.H. DeGroot et al., Zeitschrift für Physik C, Volume I, 143 (1979).

30. L.F. Abbott and R.N. Barnett, SLAC-PUB-2325 (1979).

31. D.W. Duke and R.G. Roberts, Phys. Lett. 85B, 289 (1979). Similar analyses with the same conclusions have been done by H.L. Anderson et al., Fermilab-Pub-79/30-EXP.

32. L. Baulieu and C. Kounnas, Nucl. Phys. B141, 423 (1978); J. Kodaira and T. Uematsu, Nucl. Phys. B141, 497 (1978).

33. A. Bialas and A.J. Buras, Fermilab-Pub-79/73-THY.

34. W.A. Bardeen and A.J. Buras, Phys. Lett. 86B, 61 (1979).

35. D.W. Duke and R.G. Roberts, Rutherford Preprint, RL-79-073.

36. R.L. Kingsley, Nucl. Phys. B60, 45 (1973); E. Witten, Nucl. Phys. B104, 445 (1976).

37. D. Amati, R. Petronzio and G. Veneziano, Nucl. Phys. B140, 54 (1978); Nucl. Phys. B146, 29 (1978); S.B. Libby and G. Sterman, Phys. Rev. D18, 3252, 4737 (1978). A simple proof is presented by Llewellyn Smith in lectures in this Summer School.

38. Yu. L. Dokshitser, D.I. Dyakonov and S.I. Troyan, SLAC-TRANS-183.

39. K. Harada, T. Kaneko and N. Sakai, Nucl. Phys. B155, 169 (1978).

40. G. Altarelli, R.K. Ellis and G. Martinelli, MIT preprint, MIT-CTP-776 (1979).

41. B. Humpert and W.L. Van Neerven, Phys. Lett. 85B, 293 (1979); R.D. Field, private communication and lectures at this Summer School.

42. N. Sakai, Phys. Lett. 85B, 67 (1979).

43. G. Altarelli, R.K. Ellis, G. Martinelli and S.Y. Pi, MIT preprint, CTP-793.

44. R. Baier and K. Fey, Bielefeld preprint, BI-TP 79/11.

45. V.N. Gribov and L.N. Lipatov, Sov. J. Nucl. Phys. 15, 438, 678 (1972).

46. S.D. Drell, D. Levy and T.M. Yan, Phys. Rev. D187, 2159 (1969); L.N. Lipatov, Sov. J. Nucl. Phys. 20, 94 (1975); Yu. L. Dokshitser, Leningrad Preprint 330 (1977).

47. A. Zee, F. Wilczek and S.B. Treiman, Phys. Rev. D10, 2881 (1974).

48. S.J. Brodsky, T. Kinoshita and H. Terezawa, Phys. Rev. Lett. 27, 280 (1971).

49. E. Witten, Nucl. Phys. B120, 189 (1977). Witten's result has been rederived last year by Llewellyn Smith, Phys. Lett. 79B, 83 (1978) using ladder techniques. It has been also discussed in the Altarelli-Parisi approach[9] by De Witt et al., Phys. Rev. D19, 2046 (1979); Brodsky et al., Phys. Rev. D19, 1418 (1979), Frazer and Gunion, UCSD-78-5 and Kajantie, Helsinki University preprints HU-TFT-78-30, HU-TFT-79-5.

50. W.A. Bardeen and A.J. Buras, Phys. Rev. D20, 166 (1979). The parameter $\tilde{a}_n$ has also been discussed by J. Gunion and D. Jones, University Davis preprint UCD-79-2.

51. M. Dine and J. Sapirstein, SLAC-PUB-2332. Similar calculations are being now performed by D.A. Ross, A. Terrano and S. Wolfram, and W. Celemaster and R.J. Gonsalves.

52. R. Barbieri, E. d'Emilio, G. Curci and E. Remiddi, Nucl. Phys. B154, 535 (1979).

53. J. Kodaira et al., Phys. Rev. D20, 627 (1979) and Kyoto University preprint RIFP-360; S. Matsuda and T. Uematsu, Kyoto University preprint RIFP-376; V. Gupta, S.M. Paranjape and H.S. Mani, Tata Institute preprint TIFR/TH 79-37.

54. A.P. Contogouris and J. Kripfganz, Phys. Lett. 84B, 473 (1979) and McGill University preprint (1979); A.N. Schellekens and W.L. Van Neeruen, University of Nijmegen preprint, THEF-NYM-79.8.

55. T. Gottschalk and D. Sivers, ANL-HEP-PR-79-07; M. Furman, H. Haber and I. Hinchliffe, work in progress.

56. K. Bitar, P.W. Johnson and W.K. Tung, Phys. Lett. 83B, 114 (1979); P.W. Johnson and W.K. Tung, Illinois Institute of Technology preprint (1979).

57. M.K. Gaillard, Annecy preprint in preparation; J. Ellis, talk presented at the International Symposium on Leptons and Photons, Fermilab, 1979.

58. M.R. Pennington and G.G. Ross, Oxford University preprint 23/79.

59. A. Para and C.T. Sachrajda, CERN preprint TH-2702-CERN.

60. A. González-Arroyo, C. López and F.D. Yndurain, TH-2728-CERN.

THE MIT BAG MODEL

Carleton DeTar

University of Utah

Salt Lake City, Utah

Since its formulation in 1974, the MIT bag model in various versions has acquired a considerable following. It is neither possible nor desirable for me to attempt to review all of the bag-related developments of the past five years. A few works of a review nature exist. These are listed in Ref. 1-4. I shall give a brief introduction to the model in the static cavity approximation, since this approach has been the starting point for many applications. Following this is a discussion of deformations of the bag, applied to a calculation of the electric polarizability of the π meson, and applied to rotationally excited states. Some recent results regarding hadronic interactions are reviewed, in particular, for low energy nucleon-nucleon interactions and for meson-meson interactions. Finally, I shall discuss some recent efforts at treating low energy pion interactions in the bag model using PCAC.

To provide a balanced and reasonably comprehensive discussion it has been necessary to include much material that is not originally my own work. Any defects in this synopsis are, of course, my own responsibility.

1. INTRODUCTION AND THE STATIC CAVITY APPROXIMATION

The bag model for making hadrons from confined quarks and gluons was formulated at MIT in 1974 by Chodos, Jaffe, Johnson, Thorn, and Weisskopf.[5] It is based on quantum chromodynamics (QCD), but includes an explicit mechanism for producing extended objects with confined quarks and gluons. The bag action is

$$W = \int_{t_1}^{t_2} dt \int_V d^3x [\frac{i}{2} (\bar{q} \overleftrightarrow{\not{\partial}} q) - \bar{q} mq - \frac{1}{4} F^{a\mu\nu} F^a_{\mu\nu} + g \bar{q} A^a_\mu \gamma^\mu \lambda^a q - B]. \tag{1.1}$$

The spatial integration is over a finite volume V. The last term B is a constant. If we set $B = 0$ and let V be all of space, the action is the usual QCD action. The Dirac spinor q is the quark field; the tensor is defined as usual in terms of the colored vector gluon field by

$$F^a_{\mu\nu} = \partial_\mu A^a_\nu - \partial_\nu A^a_\mu + g f_{abc} A^b_\mu A^c_\nu \ , \tag{1.2}$$

and the color coupling constant is g.

If we regard the boundary of the volume V as a dynamical variable, then the above action is gauge invariant and Lorentz invariant. There are some rather complicated constraints which relate the surface orientation to the internal fields. Thus the full quantum theory of the three-dimensional bag has not yet been formulated. However, versions of the theory have been quantized in one space and one time dimension.[2,5] Most work has been based upon the static cavity approximation[6] in which the surface position is treated as a sharply defined classical variable (usually motionless). The static cavity approximation provides the simplest pedagogical approach to the bag model, so we will focus our attention upon it here. However it should be kept in mind that it is just an approximation to a more consistent underlying theory.

Let us consider a bag with only quarks. The action is

$$W = \int_{t_1}^{t_2} dt \int_{V(t)} d^3x [\frac{i}{2} \bar{q} \overleftrightarrow{\partial}_\mu \gamma^\mu q - \bar{q} mq - B]. \tag{1.3}$$

Stationarity under variations in q and V requires that

$$(i\partial_\mu \gamma^\mu - m) q = 0 \qquad \text{in V} \tag{1.4}$$

$$\left.\begin{aligned} n^\mu \gamma_\mu q &= 0 \\ \frac{i}{2}(\bar{q} \overleftrightarrow{\not{\partial}} q) - \bar{q} mq - B &= 0 \end{aligned}\right\} \text{on S.} \tag{1.5}$$

Here n^μ is the covariant inward normal to the surface. Chodos et al. (Ref. 5) argue that these surface boundary conditions would lead to an unphysical result. To get a physically sensible result, we must consider letting the quarks have a large mass M outside, solve the Dirac equation, match the solutions at the boundary, and take $M\to\infty$. The boundary conditions then become

$$\left.\begin{aligned} i n^\mu \gamma_\mu q &= q \\ \frac{1}{2} n^\mu \partial_\mu (\bar{q}\, q) &= B \end{aligned}\right\} \text{ on } S. \tag{1.6}$$

Physically the first (linear) boundary condition implies that $\bar{q}\, q = 0$ and therefore that

$$n_\mu J^\mu_b = n_\mu \bar{q} \gamma^\mu q = 0 \quad \text{on } S, \tag{1.7}$$

i.e., that the baryonic current does not flow across the surface. Hence the quarks remain inside the cavity. The second (quadratic) boundary condition implies that the pressure of the quark field at the surface is balanced by B, a sort of external confining pressure. An elegant way to get this result directly from a variational principle is described by Chodos and Thorn.[7] They add a total divergence to the action, giving

$$W = \int_{t_1}^{t_2} dt \int_{V(t)} d^3x \left[\frac{i}{2} \bar{q} \overleftrightarrow{\partial\!\!\!/}\, q - \bar{q} m q - B + \partial_\mu (\lambda^\mu \bar{q} q)\right]. \tag{1.8}$$

The field still satisfies the Dirac equation inside the bag, and on the surface

$$\frac{i}{2} n^\mu \gamma_\mu q + n^\mu \lambda_\mu q = 0 \tag{1.9a}$$

$$\frac{i}{2} (\bar{q} \overleftrightarrow{\partial\!\!\!/}\, q) - \bar{q} m q - B + \partial_\mu (\lambda^\mu \bar{q} q) = 0. \tag{1.9b}$$

Now $(i n^\mu \gamma_\mu)^2 = 1$, so from (1.9a),

$$\begin{aligned} 4 (n^\mu \lambda_\mu)^2 &= 1 \\ \bar{q} q &= 0. \end{aligned} \tag{1.10}$$

With the negative root $n^\mu \lambda_\mu = -1/2$, we get

$$i n^\mu \gamma_\mu q = q. \tag{1.11}$$

The quadratic boundary condition (1.9a) can be simplified using the equations of motion and (1.10) giving

$$\frac{1}{2} n_\mu \partial^\mu (\bar{q}q) = B. \tag{1.12}$$

For a cavity with a time-independent surface, the quadratic boundary condition can also be derived by minimizing the energy with respect to variations in the orientation of the surface. This observation is extremely useful in applications.

If the bag is a sphere of fixed radius, we can find solutions to the Dirac equation with the linear boundary condition as follows. For $j = \ell + 1/2$, the positive frequency solutions are

$$q_{j\ell mn} = N \begin{pmatrix} j_\ell(k_n r)\psi_{j\ell m} \\ \dfrac{k_n}{\omega_n + m_q} i\vec{\sigma}\cdot\hat{r} j_{\ell+1}(k_n r)\psi_{j\ell m} \end{pmatrix} e^{-i\omega_n t}, \tag{1.13}$$

and for $j = \ell - 1/2$

$$q_{j\ell mn} = N \begin{pmatrix} j_\ell(k_n r)\psi_{j\ell m} \\ -\dfrac{k_n}{\omega_n + m_q} i\vec{\sigma}\cdot\hat{r} j_{\ell-1}(k_n r)\psi_{j\ell m} \end{pmatrix} e^{-i\omega_n t}, \tag{1.14}$$

where the spinor harmonics are

$$\psi_{j\ell m} = \begin{pmatrix} Y_{\ell,m-1/2}(\theta,\phi) \langle \ell,m-1/2;1/2\ 1/2|jm\rangle \\ Y_{\ell,m+1/2}(\theta,\phi) \langle \ell,m+1/2;1/2\ -1/2|jm\rangle \end{pmatrix}, \tag{1.15}$$

and the symbol $\langle|\rangle$ is a Clebsch-Gordon coefficient. The conventional normalization $\int d^3x\, q^+ q = 1$ defines N. The linear boundary condition requires that

$$\begin{aligned} j_\ell(k_n R) &= \frac{k_n}{\omega_n + m_q} j_{\ell+1}(k_n R) \qquad & j = \ell + 1/2 \\ j_\ell(k_n R) &= -\frac{k_n}{\omega_n + m_q} j_{\ell-1}(k_n R) \qquad & j = \ell - 1/2\,. \end{aligned} \tag{1.16}$$

The anti-particle solutions can be found by multiplying by γ_5, putting $m_q \to -m_q$, and writing $\omega_n = -\sqrt{k_n^2 + m_q^2}$, i.e.,

$$q^c_{j\ell mn}(m_q, \omega_n) = \gamma_5 q_{j\ell mn}(-m_q, -\omega_n). \tag{1.17}$$

For massless quarks, the lowest three eigenmodes may be classified as the $S_{1/2}$, $P_{3/2}$, and $P_{1/2}$ levels (for ℓ = 0, 1, and 1 resp.) with frequencies

$$\omega = k = 2.043/R,\ 3.204/R,\ 3.812/R. \tag{1.18}$$

As the mass increases, the momentum k increases so that at $m_q = \infty$ the corresponding eigenvalues are

$$k = \pi/R,\ 4.493/R,\ 4.493/R. \tag{1.19}$$

Only the j = 1/2 states satisfy the quadratic boundary condition.

Let us consider constructing the low-lying baryon SU(3) octet and decuplet in the bag model. As in the traditional quark models, these states have all three quarks in the lowest orbital--in this case the $S_{1/2}$ orbital. The color configuration is singlet and the flavor configuration is dictated by the usual SU(3) considerations.[8] The energy of the state is found from the static hamiltonian

$$H = \int d^3x [q^\dagger(-i\vec{\alpha}\cdot\vec{\nabla})q + q^\dagger \beta m_q q + B]. \tag{1.20}$$

In the static cavity approximation, the field q is written in terms of creation and annihilation operators for the cavity eigenmodes. Thus

$$q = \sum_{j\ell mn} (q_{j\ell mn} b_{j\ell mn} + q^c_{j\ell mn} d^\dagger_{j\ell mn}). \tag{1.21}$$

(The flavor and color indices have been suppressed.) If we evaluate the expectation value of H on a three quark state we obtain

$$E = \langle H \rangle = \sum_{i=1}^{3} \Omega(m_i R)/R + \frac{4\pi}{3} BR^3 - \sum_{nj\ell m} \omega_{nj\ell m}, \tag{1.22}$$

where Ω is the coefficient of 1/R in (1.18) and (1.19). The first term is the kinetic energy of the quarks, the second the volume energy, and the third is the oscillator zero-point energy. It is divergent, and has an infinite term proportional to the volume, which represents a renormalization of B. It also has a finite term of the form $-Z_0/R$. The sum has so far resisted efforts at calculating it completely (Ref. 6), and so in practice it has simply been replaced by writing

$$E = \sum_{i=1}^{3} \Omega(m_i R)/R + \frac{4\pi}{3} BR^3 - Z_0/R, \tag{1.23}$$

where B is now the finite, renormalized constant and Z_0 is an adjustable constant to be found by fitting values of the masses.

The term $-Z_0/R$ plays another important function. Since the center of mass of the hadron is moving (the three quarks are in uncorrelated cavity eigenmodes), the mass of the state is less than its cavity energy. Johnson[10] estimates that for massless quarks this correction adds about 0.75 to Z_0.

The value of R is obtained by minimizing E with respect to variations in R. This operation is equivalent to imposing the quadratic boundary condition. Thus for n massless quark (Ω is independent of R),

$$\begin{aligned} R^4 &= (n\Omega - Z_0)/(4\pi B) \\ E &= \tfrac{4}{3}(n\Omega - Z_0)^{3/4}(4\pi B)^{1/4}. \end{aligned} \tag{1.24}$$

Suppose we let the up and down quarks have zero mass and the strange quark a finite mass. Since the masses of these states depend only upon the quark masses, states having the same number of strange quarks will be degenerate. Our nucleon and Δ will have the same mass, our Y*(1380), Σ, and Λ will have the same mass, etc. Similar considerations for the low-lying meson 0^- and 1^- octets apply: the π and ρ are degenerate, etc. What lifts the degeneracy is the spin-dependent gluon interaction.

When the gluon fields are included in the action (1.8) we have

$$\begin{aligned} W = \int_{t_1}^{t_2} dt \int_{V(t)} d^3x [\bar{q}(\tfrac{i}{2}\overleftrightarrow{\partial}_\mu + gA^a_\mu \lambda^a)\gamma^\mu q - \bar{q}mq - \tfrac{1}{4}F^{a\mu\nu}F^a_{\mu\nu} \\ - B + \partial_\mu(\lambda^\mu \bar{q}q)]. \end{aligned} \tag{1.25}$$

The equations of motion are now

$$\left.\begin{aligned} (i\partial_\mu + gA_\mu^a\lambda_a)\gamma^\mu q &= mq \\ \partial_\nu F^{a\mu\nu} &= g(\bar{q}\gamma^\mu\lambda^a q + f_{abc}F^{b\mu\nu}A_\nu^c) \equiv j^{a\mu} \end{aligned}\right\} \text{ in } V \qquad (1.26)$$

$$in^\mu\gamma_\mu q = q; \quad n_\mu F^{a\mu\nu} = 0 \quad \text{on } S \qquad (1.27)$$

$$\frac{1}{2}n_\mu\partial^\mu(\bar{q}q) - \frac{1}{4}F^{a\mu\nu}F^a_{\mu\nu} = B \quad \text{on } S. \qquad (1.28)$$

If we define the color electric and magnetic fields in the usual way in terms of the tensor $F^a_{\mu\nu}$ then the equations of motion become

$$\begin{aligned} &\vec{\nabla}\cdot\vec{E}^a = j^{ao};\ \vec{\nabla}\times\vec{B}^a = \dot{\vec{E}}^a + \vec{j}^a \quad \text{in } V \\ &\hat{n}\cdot\vec{E}^a = 0;\ \hat{n}\times\vec{B}^a = 0 \quad \text{on } S. \end{aligned} \qquad (1.29)$$

The conditions upon $\vec{E}^a$ obviously assure that the total color charge generators vanish

$$Q^a = \int d^3x\ j^{ao} = 0. \qquad (1.30)$$

(This charge is the sum of quark and gluon color charges.) Therefore, only color singlet states exist, a natural consequence of the requirement that W (1.25) be stationary.

Let us now return to the computation of the masses of the low-lying baryon octet and decuplet, now including the gluon field. Rather than solving the coupled equations (1.26), we follow the perturbative approach of DeGrand et al.[6] (DJJK). We assume that we can treat the gluon interaction energy inside the cavity in second order perturbation theory. The second order contribution involves simply the exchange of a single gluon with no gluon-gluon interactions. Therefore we drop the terms in $F^a_{\mu\nu}$ which are quadratic in A^a_μ, making the $F^a_{\mu\nu}$ simply eight Maxwell tensors. The approximate Hamiltonian for the static cavity is written in Coulomb gauge as

$$\begin{aligned} H &= H_o + H_I \\ H_o &= \int d^3x[q^\dagger(-i\vec{\alpha}\cdot\vec{\nabla}+\beta m)q + \tfrac{1}{2}(\vec{E}^a_\perp)^2 + \tfrac{1}{2}(\vec{B}^a_\perp)^2 + B] \qquad (1.31) \\ H_I &= -\int d^3x\ \vec{j}^a\cdot\vec{A}^a + \int d^3x d^3x'\ j^{oa}(x)D(x,x')j^{oa}(x'), \end{aligned}$$

where the last term gives the instantaneous Coulomb interaction and $\vec{E}^a_\perp$ is the transverse electric field. D(x,x') is the cavity Coulomb Green function.

The shift in energy to second order in g is found in the usual way:

$$\Delta E_2 = \sum_n \frac{\langle 0|H_I|n\rangle\langle n|H_I|0\rangle}{E_0 - E_n} + \langle 0|H_I|0\rangle. \qquad (1.32)$$

The sum is over all contributing gluon and quark modes.

The contributions to the energy of the three quark state are represented in Fig. 1.

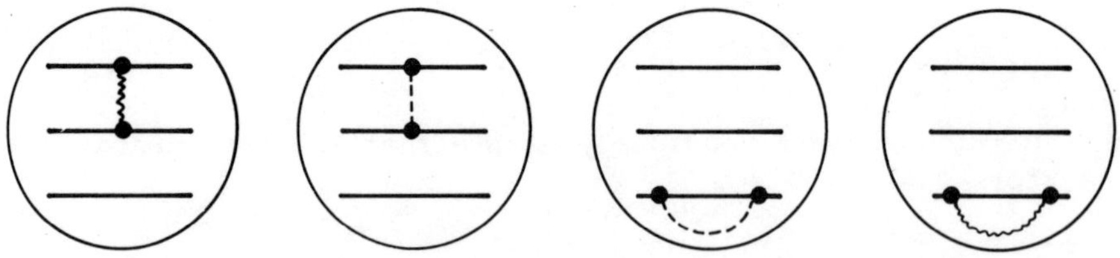

Fig. 1. Second order contributions to the energy of a three quark bag.

The wavy line represents the transverse field, the dashed line the Coulomb contribution. There are two types of terms--the exchange and self-energy terms (first two and last two figures, resp.). The exchange terms are easily calculated, since the associated currents are static. The transverse fields are magnetic only. The self-energy contributions on the other hand are difficult to compute. They produce both a mass renormalization and an energy renormalization (cavity Lamb shift). They have not yet been computed. However, it is essential to include at least one part of the self-energy graphs--namely the Coulomb term in which the quark remains in the same orbital. This term contains an electric monopole contribution, which because of the boundary conditions cannot be defined separately from the monopole contribution of the Coulomb exchange term. DJJK chose to include only this essential part of the self-energy.

For the exchange terms, finding the Green function D and summing over the gluon modes is equivalent to solving Maxwell's equations directly for the currents of the quarks, and then inserting the solutions into the normal ordered expression

$$\Delta E_{2\ \mathrm{exch}} = \langle :\frac{1}{2}\int (\vec{E}^a \cdot \vec{E}^a - \vec{B}^a \cdot \vec{B}^a) d^3x : \rangle, \tag{1.33}$$

where $\vec{E}^a$ and $\vec{B}^a$ satisfy Maxwell's equations (1.29) and are linear in the quark creation and annihilation operators. The only part of these operators which contributes in the low-lying baryons has the form

$$\begin{aligned} \int \vec{E}^a \cdot \vec{E}^a d^3x &= \sum_{f,f'} \varepsilon_{ff'} (b_f^\dagger \lambda^a b_f)(b_{f'}^\dagger \lambda^a b_{f'}) \\ \int \vec{B}^a \cdot \vec{B}^a d^3x &= \sum_{f,f'} \mu_{ff'} (b_f^\dagger \lambda^a \vec{\sigma} b_f) \cdot (b_{f'}^\dagger \lambda^a \vec{\sigma} b_{f'}), \end{aligned} \tag{1.34}$$

where f and f' refer to the flavor (the quark masses and the fields depend on these) and the color and spin labels have been suppressed. The constants $\varepsilon_{ff'}$ and $\mu_{ff'}$ are color and spin-independent electric and magnetic interaction strengths between the $S_{1/2}$ quarks. It is helpful to write this expression following DJJK in terms of the Pauli spinors $\vec{\sigma}_i$ and color spinors $\lambda_i{}^a$ for the individual quarks. Then we have

$$\Delta E_{2\ \mathrm{exch}} = \langle \sum_{i>j} (\varepsilon_{ij} \lambda_i^a \cdot \lambda_j^a - \mu_{ij} \lambda_i^a \vec{\sigma}_i \cdot \lambda_j^a \vec{\sigma}_j) \rangle . \tag{1.35}$$

The expectation value here is with respect to the color spin wave functions only. For massless quarks $\mu_{ij} = 0.47\alpha_c/R$ where $\alpha_c = g^2/4\pi$. The monopole part of ε_{ij} cannot be isolated from the self-energy contribution. Thus we must include this essential term to obtain a well-defined expression

$$\Delta E_{2\ \mathrm{exch}} + \Delta E^0_{2\ \mathrm{self}} = \langle \sum_i \frac{1}{2}\varepsilon_{ii} \lambda_i^a \cdot \lambda_i^a + \sum_{i>j} (\varepsilon_{ij} \lambda_i^a \cdot \lambda_j^a - \mu_{ij} \lambda_i^a \vec{\sigma}_i \cdot \lambda_j^a \vec{\sigma}_j) \rangle . \tag{1.36}$$

If all quarks have the same mass and orbital, $\varepsilon_{ii} = \varepsilon_{ij}$ and the electric terms drop out completely, since they are proportional to $(\sum_i \lambda_i{}^a)^2$ which vanishes on a color singlet state. If there are quarks of different masses, there is a net positive electric contribution. DJJK show that the electric contribution is in any case negligibly small for the low lying mesons and baryons. Thus the main contribution from gluon exchange is the color magnetic dipole interaction between the quarks. It lifts the degeneracy of states with different spin configurations. It is easy to see that it gives the correct result that $m_\Delta > m_N$ and $m_\rho > m_\pi$. Since for baryons $(\lambda_1^a + \lambda_2^a + \lambda_3^a)^2 = 0$, it follows that $\lambda_i^a \cdot \lambda_j^a = -\frac{1}{2}(\lambda_i^a)^2 = -8/3$.

For mesons $(\lambda_1^a + \lambda_2^a)^2 = 0$, so $\lambda_1^a \cdot \lambda_2^a = -16/3$. It is as though the interaction was between opposite charges in both cases! A similar computation for the spin content yields $\vec{\sigma}_i \cdot \vec{\sigma}_j = 1$ for the Δ, -1 for the N, 1 for the ρ, and -3 for the π. Therefore the mass splitting has the correct sign in each case. This result is good circumstantial evidence for colored gluon exchange. An abelian charge would give the wrong order of masses.

The final expression for the energy of the low lying baryons is, then

$$E = \sum_i \Omega(m_i R)/R - Z_o/R - \langle \sum_{i>j} \mu_{ij} \lambda_i^a \vec{\sigma}_i \cdot \lambda_j^a \vec{\sigma}_j \rangle + \frac{4\pi}{3} BR^3 , \qquad (1.37)$$

which must be minimized with respect to R. (The negligible electric contribution has been omitted.) DJJK show that this formula gives a quite respectable accounting for the masses of the low lying baryons and mesons. There are four adjustable parameters whose values are found to be these: $\alpha_c = 0.55$, the color coupling constant, $m_s = 279$ MeV, the mass of the strange quark $Z_o = 1.84$, $B^{1/4} = 146$ MeV. Other static parameters, such as the axial vector coupling constants for the baryons, the magnetic moments, and the mean square charge radius of the proton are also calculated and give reasonable values.

Two important approximations were made in the work of DJJK: that higher order gluon exchanges were small and that the dynamics of the surface motion can be neglected. The latter question has received the most attention. Rebbi has discussed the dynamics of small surface deformation of large bags[11] and Johnson has considered approximations for small bags.[2] Surface motions are always coupled with internal oscillations of the fields. One very important example is that of a uniform translation of the spherical bag. This motion is coupled to the symmetric excitation of one or more quarks in the P orbitals. Thus the energies of these P wave excited states are significantly altered when translated motion is considered.[12] These states are still not well understood in the bag model.

Other formulations of the bag model have been proposed. Hasenfratz and Kuti[3] introduce a surface tension in addition to the volume term. Johnson[13] showed that a very similar theory would result from a Lagrangian in which the fields were not restricted explicitly to a given volume, but were instead restricted by the fields themselves:

$$L = (L_{QCD} - B)\theta(\bar{q}q) \qquad (1.38)$$

Since $\bar{q}q$ changes sign on the boundary of the bag, this theory has features similar to the original theory. However, it would probably not allow "radially" excited states which have nodes in $\bar{q}q$ in the original theory and it would exclude bags with only gluons. Other, similar formulations permit pure gluon states.[13] Chodos and Thorn[7] and Callen, Dashen, and Gross,[14] concerned that the bag theory does not appear to preserve chiral symmetry, introduced an explicit pion field so as to restore PCAC. This field becomes involved in the bag dynamics and contributes to the pressure balance.[15] Lee[16] suggests introducing an additional scalar field which interacts with the quarks and gluons so as to reproduce in effect the bag boundary conditions. His Lagrangian is

$$\begin{aligned} L = & -\frac{1}{4}\left(1 - \frac{\sigma}{\sigma_{vac}}\right) F^{a}_{\mu\nu} F^{a\mu\nu} + i\bar{q}\gamma_{\mu}\partial^{\mu}q + gA^{a}_{\mu}\,\bar{q}\gamma^{\mu}\lambda^{a}q \\ & - \bar{q}(f\sigma + m)q - \frac{1}{2}(\partial_{\mu}\sigma)^{2} - U(\sigma). \end{aligned} \tag{1.39}$$

The self-interaction of the σ field is designed so that $\sigma = \sigma_{vac}$ is the vacuum field in the absence of quarks and gluons. In the presence of quarks and gluons $\sigma \to 0$ so that except for surface effects, the usual bag Lagrangian is obtained with $U(0) = B$ in the region where quarks and gluons exist. The coefficient of the $(F^{a}_{\mu\nu})^2$ term is interpreted as a dielectric constant or a reciprocal magnetic permeability. The region external to the bag has, in effect, a vanishing dielectric constant and infinite magnetic permeability. Therefore, one expects that near the surface of the bag $\hat{n} \cdot \vec{E}^{a} = 0$ and $\hat{n} \times \vec{B}_{a} = 0$. If $f\sigma_{vac}$ is sufficiently large, the quarks will have a large effective mass outside the bag. This property leads to the fermion linear boundary conditions.

It is a subject of intense interest whether a bag-like theory can be derived from unadorned QCD. Mandelstam[17] and the Princeton group[14] both propose that the traditional vacuum ($q = 0$, $A^{a}_{\mu} = 0$) is unstable and that the stable vacuum is impermeable to quarks and gluons. The Princeton scenario resembles in some ways the mechanism involving the dielectric constant outlined above. In both approaches it is believed that something like the bag model will emerge. The constant B is the energy difference per unit volume between the two vacua; and it is expected that the bag boundary conditions will also be obtained. However, the importance of the transition at the surface between vacua and other details, such as the role of instantons, may indeed require changes. But at present, the bag model remains as a plausible, simple and reasonably successful explanation for the static features of hadrons.

2. DEFORMATION FROM THE SPHERICAL ORBITALS: ELECTRIC POLARIZABILITY OF MESONS

Let us consider one of the most elementary deformations of spherical bag states, namely the distortion of orbitals in the spherical bag produced by an applied static electric field. The field causes the quarks of different charges to separate, thereby producing an induced electric dipole moment. One place to look for the induced dipole moment of the π^- might be in π-mesic atoms, where the π^- experiences the strong field of the nucleus.[18] As the quarks separate, one expects a deformation of the shape of the bag. For present purposes, we restrict our attention to spherical bag shapes, balancing the bag pressure only on average over the whole surface. Shape deformations are considered in Ref. 19 and do not modify the results significantly for small separations. Thus the distortion we consider affects the orbitals of the quarks.

A non-relativistic particle in a one-dimensional box of length L has the lowest eigenstates

$$\phi_S = \sqrt{\frac{2}{L}} \cos \pi z/L \ ; \quad \phi_A = \sqrt{\frac{2}{L}} \sin 2\pi z/L \tag{2.1}$$

for $L/2 \leq z \leq L/2$. The first is symmetric under $z \to -z$ and the second antisymmetric. From these one can form the hybrid "left" and "right" orbitals

$$\phi_L = \phi_S - \sqrt{\mu}\phi_A \ ; \quad \phi_R = \phi_S + \sqrt{\mu}\phi_A \tag{2.2}$$

in which the particle is pushed to the left (ϕ_L) or right (ϕ_R). As μ ranges from 0 to 1, the orbitals range from complete overlap to complete orthogonality. If an electric field is applied along the z-direction, the ground state wave function is distorted as in (2.2), depending upon its charge. (We neglect contributions from higher orbitals.) In the spherical bag, the two states of interest are the lowest symmetric and antisymmetric orbitals, $S_{1/2}$ and $P_{3/2}$

$$q_S = \frac{N_S}{\sqrt{4\pi}} \begin{pmatrix} j_0(\Omega_S r/R)\mathcal{U}_m \\ i\vec{\sigma}\cdot\hat{r} j_1(\Omega_S r/R)\mathcal{U}_m \end{pmatrix} ; \quad q_P = \frac{N_P}{\sqrt{4\pi}} \begin{pmatrix} j_1(\Omega_P r/R)\mathcal{U}_{P_m} \\ i\vec{\sigma}\cdot\hat{r} j_2(\Omega_P r/R)\mathcal{U}_{P_m} \end{pmatrix} \tag{2.3}$$

where

$$\mathcal{U}_{P_m} = \frac{3}{\sqrt{2}} \left(z - \frac{1}{3} \vec{\sigma}\cdot\vec{r}\sigma^3\right)\mathcal{U}_m/r. \tag{2.4}$$

With real normalization constants N_S and N_P, the left and right orbitals are formed with real $\sqrt{\mu}$. Correspondingly, we may define a state in terms of the creation operators for quarks

$$b_L^\dagger = b_S^\dagger - \sqrt{\mu}\, b_P^\dagger \;\; ; \;\; b_R^\dagger = b_S^\dagger + \sqrt{\mu}\, b_P^\dagger \, , \tag{2.5}$$

and similarly for antiquarks. The state formed from one quark on the right and one antiquark on the left is then

$$|R\bar{L}> = b_R^\dagger d_L^\dagger |0> = (b_S^\dagger d_S^\dagger - \sqrt{\mu} b_S^\dagger d_P^\dagger + \sqrt{\mu} b_P^\dagger d_S^\dagger - \mu b_P^\dagger d_P^\dagger)|0> \, . \tag{2.6}$$

This is the state which we shall use to describe induced polarization. Note that we are deliberately sending the quark as much to the right as the antiquark to the left. In this way we avoid exciting a spurious mode which corresponds to both fermions shifting in the same direction in the cavity.

The electric dipole moment of the quarks can be computed from the charge density operator

$$\rho = :q^\dagger e q: \tag{2.7}$$

where e is the charge matrix. Then

$$p = <\int \rho z d^3x>/<> \tag{2.8}$$

$$p = Rc_d<(b_S^\dagger b_P + b_P^\dagger b_S)e_q + (d_S^\dagger d_P + d_P^\dagger d_S)e_{\bar{q}}>/<> \tag{2.9}$$

where

$$Rc_d = \int d^3r \, q_S^\dagger(\vec{r}) q_P(\vec{r}) z = 0.34R \tag{2.10}$$

and e_q and $e_{\bar{q}}$ are the charges of the quark and antiquark, resp. Using (2.6) we find

$$p = \frac{2\sqrt{\mu}(e_q - e_{\bar{q}})}{1 + \mu} Rc_d \, . \tag{2.11}$$

The dipole moment is simply related to the average separation of the quark. If we use the analogy with classical point-like objects, we may define the separation δ by

$$p = \frac{e_q - e_{\bar{q}}}{2} \delta \, . \tag{2.12}$$

The energy of the distorted state is computed as before, following DJJK.[6] There are more gluon exchange terms because two orbitals appear. Gluons may produce transitions between the orbitals, as well as contributing interorbital interaction. The possible interactions are shown in Fig. 2 where each graph represents a contribution to the bag energy in second order perturbation theory. (It is necessary to regard the orbitals $S_{1/2}$ and $P_{3/2}$ as quasi-degenerate in computing the gluon energies.)

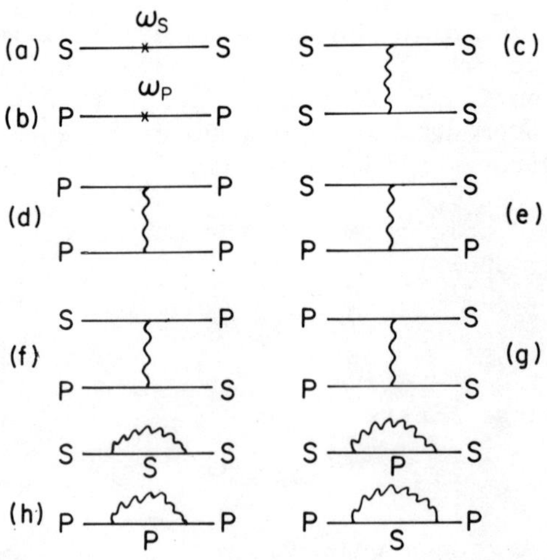

Fig. 2. Contributions to bag energy with two orbitals occupied.

The actual calculation of the terms contributing to the energy is too detailed for these notes. The procedure is described in Ref. 19. Only results for the pion are given here. The quark kinetic energy is given by

$$E_Q(\mu) = 2\frac{\Omega_S + \mu\Omega_P}{(1+\mu)R} , \tag{2.13}$$

where $\Omega_S = 2.043$ and $\Omega_P = 3.204$ are the cavity eigenfrequencies for the $S_{1/2}$ and $P_{3/2}$ modes. The gluon interaction energy for the pion is, in the notation of Ref. 20, Eq. (2.14),

$$E_M(\mu) = \frac{32}{3(1+\mu)^2}\{(W_{MSz} + 2W_{MS\perp}) + \mu^2(W_{MPz} + 2W_{MP\perp})$$

$$+ (2\mu W_{MSPz} + \mu W_{ED}) - [4\mu(W_{MX\perp} + W_{EX\perp}) + 2\mu(W_{MXz} + W_{EXz})$$

$$- 2\mu(W_{EX} + W_{MX})] - [4\mu(W_{MX\perp} - W_{EX\perp}) + 2\mu(W_{MXz} - W_{EXz}) \qquad (2.14)$$

$$- 2\mu(W_{EX} - W_{MX})]\} .$$

(The contributions from graphs (c)-(h) of Fig. 2 are grouped in order in the parentheses above.) The principle contributions come from the color electric dipole fields in graphs (e) and (f) as well as the color magnetic dipole terms. The color electric fields cause the energy to increase sharply with the separation of the $q\bar{q}$ pair. The full contribution to the bag energy is, then

$$E(R,\mu) = E_Q(\mu) + E_M(\mu) - Z_0/R + \frac{4\pi}{3}BR^3 \qquad (2.15)$$

and is summarized by

$$E(R,\mu) = \frac{c_0 + c_1\mu + c_2\mu^2}{(1+\mu)^2 R} + \frac{4\pi}{3}BR^3 \qquad (2.16)$$

where, for the pion (using bag parameters of DJJK and of Ref. 20) $c_0 = 0.73$, $c_1 = 16.17$, and $c_2 = 5.14$. The lowest energy state is obtained by minimizing the above expression with respect to both μ and R. The minimum occurs for $\mu = 0$ (the parameter μ cannot be negative in these expressions) and we are back to a spherical bag state.

We next consider the effect of applying a uniform external electric field ε. We shall do so in the same manner as in the two nucleon interaction (see Lecture 5 below). The new expression for the energy reads

$$E(R,\mu,\varepsilon) = E(R,\mu) - \varepsilon p$$

$$= E(R,\mu) - \frac{2\sqrt{\mu}}{1+\mu} c_d \varepsilon (e_q - e_{\bar{q}})R \qquad (2.17)$$

where we have used (2.11). The additional term causes the minimum in μ to shift away from $\mu = 0$, resulting in a separation of the quark-antiquark pair. For small μ, the minimum is easily found to occur (for fixed R) at

$$\sqrt{\mu} = \frac{c_d R^2 \varepsilon (e_q - e_{\bar{q}})}{c_1 - 2c_0} [1 + \mathcal{O}(\mu)] . \tag{2.18}$$

Thus to lowest order in ε, the shift in energy and in R is of order ε^2 as is to be expected. The minimum perturbed energy is therefore of the form

$$E_{min}(\varepsilon) = E_0 - \frac{1}{2}\alpha_0 \varepsilon^2 + \mathcal{O}(\varepsilon^4) , \tag{2.19}$$

where E_0 is the unperturbed spherical bag energy. The dipole moment is

$$p = -\frac{dE_{min}}{d\varepsilon} = \alpha_0 \varepsilon \tag{2.20}$$

where α_0 is the inverse electric polarizability. But from (2.11) the dipole moment is actually

$$p \approx 2\sqrt{\mu} c_d (e_q - e_{\bar{q}}) R . \tag{2.21}$$

Using (2.18) for $\sqrt{\mu}$, we have

$$\alpha_0 = p/\varepsilon \approx \frac{2c_d^2 R^3 (e_q - e_{\bar{q}})^2}{(c_1 - 2c_0)} . \tag{2.22}$$

Since we are neglecting terms in the energy of fourth order in ε, we are allowed to use the unperturbed bag radius for R. For the π^-

$$\alpha_0 \approx 2.1 \times 10^{-3} e^2 R_\pi^3 . \tag{2.23}$$

Estimates for R_π range from 3.4 to 4.1 GeV^{-1}. The "electric polarizability" is the inverse of α_0.

We can deduce from the same reasoning the static inter-quark potential. The term

$$\varepsilon p = \lambda \delta \tag{2.24}$$

can be regarded as a Lagrange variational constraint on the separation parameter δ. Here λ plays the role of the applied field. Thus we may consider

$$I(R,\mu,\lambda,\delta_0) = E(R,\mu) - \lambda(\delta - \delta_0) \tag{2.25}$$

as a variational quantity which, when minimized with respect to R, μ, and δ_0 gives the static quark-antiquark potential

$$\min_{R,\mu,\lambda} I(R,\mu,\lambda,\delta_0) = V(\delta_0) \, . \tag{2.26}$$

Since we have already evaluated, in effect, the minimum with respect to R and μ for fixed small λ, we may write

$$I(R,\mu,\lambda,\delta_0) = E_0 - \frac{1}{2}\beta\lambda^2 + \lambda\delta_0 \, , \tag{2.27}$$

where from (2.12) and (2.24)

$$\lambda = \varepsilon(e_q - e_{\bar{q}})/2 \, , \tag{2.28}$$

and from (2.19)

$$\beta = \frac{4\alpha_0}{(e_q - e_{\bar{q}})^2} \, . \tag{2.29}$$

The minimum of (2.27) with respect to λ is easily found, with the result

$$\beta\lambda = \delta_0$$

$$V(\delta_0) \approx E_0 + \frac{1}{2}E_0''\delta_0^2 \tag{2.30}$$

$$E_0'' = \frac{1}{\beta} = \frac{c_1 - 2c_0}{8c_d^2 R^3} \, .$$

For the pion $E_0'' \approx 13/R_\pi^3$. Using a low estimate $R_\pi \approx 3.4$ GeV^{-1}, we get $E_0'' \approx 8.5$ GeV fm^{-2}. For the ρ meson $E_0'' \approx 10/R_\rho^3$ and $R_\rho = 4.7$ GeV^{-1}. Therefore $E_0'' \approx 2.5$ GeV fm^{-2}. The color magnetic interaction between the $q\bar{q}$ pair is attractive in the ρ meson and repulsive in the π meson; this is reflected in the difference in size. We have not taken account of any effects of asymptotic freedom, which would, of course, make the $q\bar{q}$ pair in the pion slightly less bound than the above figure suggests. It is interesting to see the source of the attraction in the case of the pion. The coefficient 13 gets a contribution of about 2.5 from the $\Omega_P - \Omega_S$ level difference, about 5 from the color electrostatic attraction between the $q\bar{q}$ pair and the rest from the fact that the color magnetic interaction in the spin singlet state is much more attractive in the $S\bar{S}$ orbital than in the $S\bar{P}$, $\bar{S}P$, or $P\bar{P}$ orbitals.

Let us estimate the order of magnitude of the energy level shifts in a pionic atom which would be produced by the polarization induced in the pion by the nucleus. It is given by (2.19), where for ε we use the average field in the nth Bohr orbital:

$$\delta E \approx -\frac{1}{2}\alpha_0 \varepsilon^2 = -\frac{e^2}{72E_0''}\left(\frac{Ze}{R_n^2}\right)^2$$
$$= -\frac{Z\alpha}{2R_n}\,\frac{1}{36}\,\frac{(Z\alpha)^4 m_\pi^3}{E_0'' n^6} \approx -\frac{Z\alpha}{2R_n} \times 10^{-8} \tag{2.31}$$

where we have used Z = 30 and n = 2. This number is very small indeed.

3. ROTATIONALLY EXCITED STATES: LONG BAGS

When a quark and an antiquark are separated to distances much greater than those relevant to the previous lecture, the bag stretches.[3,19] Ultimately a long bag is formed (unless one allows quark-antiquark pairs to appear and initiate fission into two bags) with confined color electric flux linking the quark charges at the ends (Fig. 3).[21]

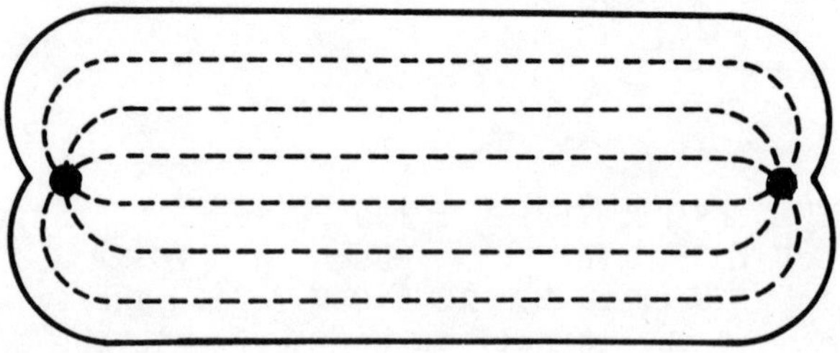

Fig. 3. Bag with a separated quark and antiquark.

If we assume that the electric flux is uniform on a cross-section transverse to the axis of deformation, then it is a simple matter to estimate the energy of the state. The field is parallel to the axis and has strength

$$E^a = \frac{g\lambda^a}{A} \tag{3.1}$$

where A is the cross sectional area and g is the color charge. Since the field pressure balances B on the surface, the area is determined by

$$B = \frac{1}{2}(E^a)^2 = \frac{g^2(\lambda^a)^2}{A^2} . \tag{3.2}$$

The total energy of the bag is then approximately

$$M = E_q + BAL + \frac{1}{2}(E^a)^2 LA \tag{3.3}$$

where L is the length and E_q is the energy contributed by the quarks. The volume term and the electric field terms are equal and with (3.2)

$$M = E_q + \sqrt{2Bg^2(\lambda^a)^2} L . \tag{3.4}$$

Therefore, for very long bags we expect E_q to become a constant and the energy to grow linearly with the length, in analogy with string models[22] and phenomenological potentials for charmonium.[23] At short distances geometrical considerations become more important. For the charmonium system a Coulomb-like short range attraction is expected[3] at least at distances larger than the Compton wavelength of the quarks. For light quarks, however, we found a quadratic dependence on effective separation at very short distances. If we compare the π and ρ mesons, we would expect that they would have the same asymptotic behavior (3.4) at large L, since the color magnetic interaction is short range. At short range we found that the deformation energy of the π-meson increased much more steeply with increasing separation. Thus we might expect the deformation energy to have the behavior shown in Fig. 4. (The result for the ρ was calculated in Ref. 19.)

Let us consider a long rotating bag. Johnson, Chodos, Thorn and Nohl[21,24] analyzed the bag state which has the minimum mass for a given angular momentum (leading Regge trajectory). They use a semi-classical approach. Let us consider the case in which the rotating quarks have equal masses, and review the calculation of the leading trajectory.

The state in question is a straight, long tube of color electric flux rotating at angular frequency ω, with quarks at the ends moving with velocity v. The energy of the state depends upon the

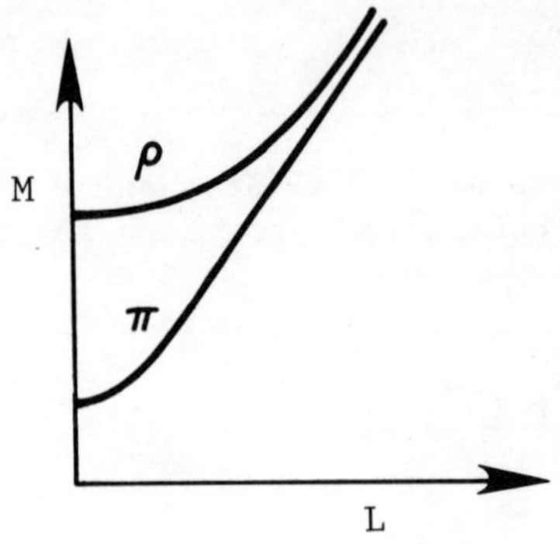

Fig. 4. Schematic comparison of π and ρ deformation energies.

energy density of the gluon fields, the quark kinetic energy and the bag volume energy. We need to take into account the effects of the rotation on the fields and the bag geometry in calculating the energy of the state. The angular momentum receives a contribution from the quarks and the gluon fields, but not from the BV term. In order to see this, it is necessary to consider the energy-momentum tensor for the bag.[5] It is

$$T^{\mu\nu} = (T^{\mu\nu}_{QCD} - Bg^{\mu\nu})\theta_V(x) \tag{3.5}$$

where θ_V is 1 inside and 0 outside the bag. The bag constant plays the role of a pressure term and so does not contribute to the momentum per se. It is also Lorentz invariant. The angular momentum tensor is

$$M^{\mu\nu\lambda} = [M^{\mu\nu\lambda}_{QCD} - B(x^{\mu}g^{\nu\lambda} - x^{\nu}g^{\mu\lambda})]\theta_V(x) \ , \tag{3.6}$$

so that the constant B does not contribute to the angular momentum J_k where

$$\varepsilon_{ijk}J_k = \int M^{ij0}d^3x \ . \tag{3.7}$$

As the stretched bag rotates, the electric flux lines induce a magnetic field. A point at distance x from the center moves with speed v = ωx. A part of the bag with velocity v develops a field

$$\vec{B}_a = \vec{v} \times \vec{E}_a \ , \tag{3.8}$$

which is found by carrying out a Lorentz boost upon the uniform field (3.1). Therefore, the quadratic boundary condition becomes

$$B = \frac{1}{2}[(E_a)^2 - (B_a)^2] = \frac{g^2(\lambda^a)^2}{A^2}(1 - v^2) \tag{3.9}$$

or

$$A - \sqrt{\frac{g^2(\lambda^a)^2(1 - v^2)}{2B}} \ . \tag{3.10}$$

Johnson and Thorn observe that this result corresponds to a Lorentz contraction of the area.

The total energy of the system is then

$$E = E_q + E_f + BV \ . \tag{3.11}$$

Johnson and Thorn approximate E_q by the "convective" energy so that for two quarks of velocity v_o and mass m,

$$E_q = \frac{2m}{\sqrt{1 - v_o^2}} \ . \tag{3.12}$$

The field energy is just

$$E_f = 2\int_0^{L/2} dx \frac{1}{2}[(\vec{E}^a)^2 + (\vec{B}^a)^2]A \tag{3.13}$$

$$= \frac{2}{\omega}\sqrt{\frac{1}{2}g^2(\lambda^a)^2 B} \int_0^{v_0} dv \frac{1 + v^2}{\sqrt{1 - v^2}} \tag{3.14}$$

and the volume energy is

$$BV = \frac{2}{\omega}\sqrt{\frac{1}{2}g^2(\lambda^a)^2 B} \int_0^{v_0} dv \sqrt{1 - v^2} \ . \tag{3.15}$$

The total energy of the state is then

$$M = \frac{2m}{\sqrt{1 - v_0^2}} + \frac{4}{\omega}\sqrt{\frac{1}{2}g^2(\lambda^a)^2 B} \int_0^{v_0} dv \frac{1}{\sqrt{1 - v^2}} \ . \tag{3.16}$$

This is the equation for the energy of a uniformly rotating massless string of proper tension

$$T_0 \equiv \frac{1}{2\pi\alpha'} = \sqrt{2g^2(\lambda^a)^2 B} \tag{3.17}$$

with point masses attached to the ends.[22] This identification of the tension T agrees with the equation (3.4) for the stationary bag. The speed v_0 is determined dynamically by the requirement that the masses accelerate according to the applied tension, i.e.,

$$T_0 = \frac{m\omega v_0}{1 - v_0^2} . \tag{3.18}$$

This equation fixes v_0 in terms of ω so that M is strictly a function of ω. Simplifying (3.16), we have

$$M = \frac{2m}{\sqrt{1 - v_0^2}} + \frac{1}{\pi\alpha'\omega} \sin^{-1} v_0 . \tag{3.19}$$

The angular momentum is

$$J = J_q + J_f \tag{3.20}$$

where the angular momentum of the quarks is

$$J_q = \frac{2(mv_0 L/2)}{\sqrt{1 - v_0^2}} \tag{3.21}$$

and of the gluon field,

$$\begin{aligned} J_f &= \int d^3x \sum_a \vec{r} \times (\vec{E}_a \times \vec{B}_a) \\ &= \frac{2}{2\pi\alpha'} \int_0^{L/2} dx \frac{x^2\omega}{\sqrt{1 - v^2}} = \frac{1}{\omega^2\pi\alpha'} \int_0^{v_0} dv \frac{v^2}{\sqrt{1 - v^2}} , \end{aligned} \tag{3.22}$$

so that

$$J = \frac{2mv_0^2}{\omega\sqrt{1 - v_0^2}} + \frac{1}{2\pi\alpha'\omega^2}(\sin^{-1} v_0 - v_0\sqrt{1 - v_0^2}) \tag{3.23}$$

where we have used

$$v_0 = \omega L/2 \ . \tag{3.24}$$

With the constraint (3.18) J is also a function of ω alone. Therefore J is implicitly a function of M, and we have an equation for the leading trajectory.

It is interesting to consider the case $m \to 0$. From (3.18) $v_0 = 1 + \mathcal{O}(m)$ and

$$M \approx \frac{1}{2\alpha'\omega} \ ; \quad J \approx \frac{1}{4\alpha'\omega^2} \tag{3.25}$$

so that

$$J \approx \alpha' M^2 \tag{3.26}$$

and we obtain a linearly rising trajectory. With the values of $B^{1/4} = 146$ MeV, $\alpha_c = 0.55$ from DJJK[6] and $(\lambda^a)^2 = 16/3$ substituted into (3.17) the slope α' is just

$$\alpha' = 0.88(\mathrm{GeV})^{-2} \tag{3.27}$$

which agrees nicely with the experimental value 0.9 $(\mathrm{GeV})^{-2}$ for the ρ trajectory and many others for $J \lesssim 5$ even though the equation (3.22) is expected to apply only asymptotically. Chodos and Thorn[24] quote the next correction to (3.26):

$$J = \alpha' M^2 [1 - \frac{8}{3\pi}(\frac{m\pi}{M})^{3/2} + \mathcal{O}((\frac{m}{M})^{5/2})]. \tag{3.28}$$

For very large masses and low velocities one obtains the classical formula for the angular momentum of two massive objects moving non-relativistically in a linear potential

$$J \approx J_0 + \frac{4\pi}{3^{3/2}}\alpha'\sqrt{m}\sqrt{(M-2m)^3} \ , \tag{3.29}$$

where a "quantum defect" J_0 has been added to allow for quantum corrections. Johnson and Nohl[21] show that this semiclassical formula agrees very well with the quantum levels. They also derive the result for the case of unequal masses. When one mass is large and the other zero, they have

$$J = 2\alpha'(M-m)^2[1 - (\frac{2}{\pi})^2 \frac{M-m}{\pi} + \ldots] \tag{3.30}$$

for $(M-m)/m \ll 1$. Of course ultimately in this case the trajectory also rises according to (3.21). They find excellent agreement between the more general relativistic expression and the known trajectory for the K^* system, which has one strange quark and one light quark.

One may also consider rotating bags with different combinations of quarks and gluons at either end. As long as the contents of each end do not form a color singlet, an electric flux tube is formed and one obtains a long bag at high angular momenta. The tension and the ultimate trajectory slope depend upon the color charge strength at the ends. The equation (3.17) is modified to read

$$T_0 = \frac{1}{2\pi\alpha'} = \sqrt{2g^2 C_3 B} \tag{3.31}$$

where C_3 is the quadratic Casimir operator $(\lambda^a)^2$ for the representation of SU(3) to which the end cluster belongs. Thus a bag with two quarks at one end and two antiquarks at the other $(q^2)(\bar{q}^2)$ occurs in two forms. Either the two quarks form a color $\bar{3}$ in which case $C_3 = 16/3$ as before, or they form a 6 in which case $C_3 = 40/3$. If three quarks are placed into a color octet at one end, then $C_3 = 12$. Thus, these higher representations of SU(3) yield a stronger color field, greater tension, and a smaller trajectory slope. The $(q^2)(\bar{q}^2)$ states can occur in nucleon-antinucleon collisions (they are called baryonium states) and have attracted much attention recently.[25] There are also various amusing combinations involving six quarks, divided as $(q)(q^5)$, $(q^2)(q^4)$ and $(q^3)(q^3)$ which could resonate in the two-nucleon channel.[27]

4. EXOTICS AND CRYPTOEXOTICS IN THE BAG MODEL

In addition to the conventional quark model states formed from $q\bar{q}$ and qqq the quark model also predicts states in the combinations $qq\bar{q}\bar{q}$, $qqqq\bar{q}$, etc. With additional quarks, it becomes possible to create mesons in SU(3) flavor multiplets other than 1 and 8 and baryons in flavor multiplets other than 1, 8, and 10. Hadrons not in these multiplets are called "exotic" hadrons. Hadrons with additional $q\bar{q}$ pairs which nevertheless still lie in the traditional flavor multiplets masquerading as $q\bar{q}$ and qqq systems are called "cryptoexotics."[25] Jaffe has analyzed extensively multiquark states formed in the bag model with all quarks in the cavity $S_{1/2}$ state.[25] Higher rotational states formed from $qq\bar{q}\bar{q}$ are of interest in baryon-antibaryon scattering (baryonium).[26]

Let us focus our attention for the moment upon the lowest $qq\bar{q}\bar{q}$ meson multiplet with all quarks in the $S_{1/2}$ state, following Jaffe.[25]

The energy levels of the $qq\bar{q}\bar{q}$ states are calculated in much the same way as the $q\bar{q}$ states. The chief difficulty lies in the group theoretical problem of classifying the states which diagonalize the hamiltonian. The general procedure is first to construct a state with the desired total angular momentum, isospin, strangeness, and color (singlet) then diagonalize quark kinetic energy and gluon interaction terms in the hamiltonian (1.37), and finally to adjust the radius of the bag so as to minimize the total energy. Jaffe first chose to simplify the gluon interaction term so that it was diagonal in the number of strange quarks. This is accomplished by making the replacement

$$R\mu_{ij} \equiv M(m_i R, m_j R) \rightarrow M\left(\frac{n_s m_s}{N}, \frac{n_s m_s}{N}\right) \tag{4.1}$$

where n_s/N is the number of strange quarks divided by the total number of quarks and m_s is the strange quark mass. This simplification makes the hamiltonian diagonal in the number of strange quarks. The eigenstates will then obey the "magic mixing" rule. [This is the rule which states, for example, that the $q\bar{q}$ vector flavor octet mixes with the flavor singlet so as to form a ϕ meson which is entirely $s\bar{s}$ and an ω meson which is entirely $1/\sqrt{2}(uu + dd)$.] The simplification, of course, introduces small errors in the mass calculation, but at an enormous savings in calculational effort.

It remains to find states which diagonalize the color magnetic operator

$$\begin{aligned} -\sum_a \sum_{i>j} \lambda_i^a \lambda_j^a \vec{\sigma}_j \cdot \vec{\sigma}_i &= -\sum_a \sum_{i>j\varepsilon q} \lambda_i^a \vec{\sigma}_i \cdot \lambda_j^a \vec{\sigma}_j \\ &\quad - \sum_a \sum_{i>j\varepsilon \bar{q}} \lambda_i^{a*} \vec{\sigma}_i^{\,*} \cdot \lambda_j^{a*} \vec{\sigma}_j^{\,*} - \sum_a \sum_{\substack{i\varepsilon q \\ j\varepsilon \bar{q}}} \lambda_i^a \sigma_i \cdot \lambda_j^{a*} \vec{\sigma}_j^{\,*} \end{aligned} \tag{4.2}$$

where $i\,\varepsilon\, q(\bar{q})$ means the index i refers to a quark (antiquark) and we have introduced explicitly the matrices $-\lambda^{a*}$ and $-\sigma^{k*}$ for the antiquark representation of $SU(3)_c$ and SU(2). To facilitate the evaluation of this operator Jaffe introduces an SU(6) "color-spin" group formed from $SU(3)_c \times SU(2)$. The 35 generators in the quark representation are

$$\{\alpha\} = \begin{cases} (\frac{2}{3})^{1/2}\sigma^k & k = 1,2,3 \\ \lambda^a & a = 1,2,\ldots,8 \ . \\ \sigma^k\lambda^a & \end{cases} \tag{4.3}$$

The generators for the antiquark representation are $-\alpha^*$. The α's are normalized so that $\mathrm{Tr}\alpha^2 = 4$. The $qq\bar{q}\bar{q}$ state is then constructed by first combining the quarks so as to form a particular color-spin and flavor multiplet, and likewise the antiquarks. The multiplets thus formed are then combined to make the full meson multiplet. The advantage of proceeding in this way is that the color magnetic operator can be expressed in terms of the various SU(n) Casimir operators for the quark cluster and antiquark cluster. In fact, it is easily shown that

$$\begin{aligned} -\sum_a \sum_{i>j} \vec{\sigma}_i \cdot \vec{\sigma}_j \lambda_i^a \lambda_j^a &= 8N + \tfrac{1}{2}C_6(\mathrm{tot}) - \tfrac{4}{3}S_{tot}(S_{tot}+1) \\ &+ C_3(Q) + \tfrac{8}{3}S_Q(S_Q+1) - C_6(Q) \\ &+ C_3(\bar{Q}) + \tfrac{8}{3}S_{\bar{Q}}(S_{\bar{Q}}+1) - C_6(\bar{Q}) \ , \end{aligned} \tag{4.4}$$

where

$$\begin{aligned} C_6 &= \sum_{\mu=1}^{35} (\sum_{i=1}^{N} \alpha_i^\mu)^2 \\ C_3 &= \sum_{a=1}^{8} (\sum_{i=1}^{N} \lambda_i^a)^2 \\ 4S(S+1) &= \sum_{k=1}^{3} (\sum_{i=1}^{N} \sigma_i^k)^2 \ , \end{aligned} \tag{4.5}$$

and the labels, Q, $\bar{Q}$, and tot refer to the quark cluster, the antiquark cluster, and the entire system, respectively. Note that the one gluon exchange hamiltonian is diagonal in the Casimir operators C_6, C_3 and $S(S+1)$ for the quark cluster, antiquark cluster, and the entire system.

Now a qq pair can be in a $\bar{3}$ or 6 of $SU(3)_c$ and 1 or 3 of SU(2) spin. In color-spin the quarks belong to a 6 and the possible $SU(3)_{cs}$ representations for qq are given by

$$6 \times 6 = 15 + 21 \; . \tag{4.6}$$

The 15 and 21 can be decomposed on the $SU(3)_c \times SU(2)$ basis with the result that

$$\begin{aligned} (\bar{3},3) + (6,1) &\subset 15 \\ (\bar{3},1) + (6,3) &\subset 21 \; . \end{aligned} \tag{4.7}$$

The complex conjugate results apply for the $\bar{q}\bar{q}$ pair. The representation 21 is symmetric and 15 is antisymmetric under interchange of the color and spin labels. To form a fully antisymmetric wave function of the qq pair, it is necessary to combine the symmetric color-spin wave function with the antisymmetric SU(3) flavor wave function and vice versa. For the quarks there are two possibilities denoted by

$$[21]\bar{3} \quad \text{and} \quad [15]6 \tag{4.8}$$

where the second label refers to SU(3) flavor. The complex conjugate choices apply to the $\bar{q}\bar{q}$ system. Therefore the possible color-spin, flavor choices for the entire system are given by

$$\begin{aligned} &[21]\bar{3} \times [\bar{21}]3 \\ &[15]6 \times [\bar{15}]\bar{6} \\ &[15]6 \times [\bar{21}]3 \\ &[21]\bar{3} \times [\bar{15}]\bar{6} \; . \end{aligned} \tag{4.9}$$

We now restrict our attentions to the lowest 0^+ mesons. Inspection of (4.7) shows that to get spin 0, we must consider either $15 \times \bar{15}$ or $21 \times \bar{21}$. This rules out the last two alternatives above. To get the lowest energy states, we examine Eq. (4.4). The lowest states will have the most negative (most attractive) value of the color magnetic operator. Jaffe has summarized the circumstances favoring attraction in two "Hund's rules":

1. "The quarks and antiquarks are separately in the largest possible representation of $SU(6)_{cs}$" [maximizing $C_6(Q)$ and $C_6(\bar{Q})$.]

2. $C_6(\text{tot})$ is as small as possible.

Therefore, we are led to considering only the first alternative above (4.9).

The decomposition of [21] × [$\bar{21}$] is

$$[21] \times [\bar{21}] = [1] + [35] + [405] \tag{4.10}$$

and both the [1] and the [405] contain a 0^+ color singlet meson. The meson is composed of qq and $\bar{q}\bar{q}$ pairs, which have spin 0 part of the time and spin 1 part of the time because of (4.7). Since the hamiltonian depends upon these spins through (4.4), it is necessary to know the projection of the [1] and [405] 0^+ states on the spin basis of the qq and $\bar{q}\bar{q}$ clusters. Jaffe and Mandula calculated

$$\begin{aligned} |0^+\underset{\sim}{9}[1]\rangle &= \sqrt{\tfrac{6}{7}}\,|(6,3)\underset{\sim}{\bar{3}};\ (\bar{6},3)\underset{\sim}{3}\rangle + \sqrt{\tfrac{1}{7}}\,|(\bar{3},1)\underset{\sim}{\bar{3}};\ (3,1)\underset{\sim}{3}\rangle \\ |0^+\underset{\sim}{9}[405]\rangle &= \sqrt{\tfrac{1}{7}}\,|(6,3)\underset{\sim}{\bar{3}};\ (\bar{6},3)\underset{\sim}{3}\rangle - \sqrt{\tfrac{6}{7}}\,|(\bar{3},1)\underset{\sim}{\bar{3}};\ (3,1)\underset{\sim}{3}\rangle\ , \end{aligned} \tag{4.11}$$

where the $\underset{\sim}{9}$ refers to the flavor representation $\underset{\sim}{\bar{3}} \times \underset{\sim}{3}$ and the (qq; $\bar{q}\bar{q}$) decomposition is displayed on the right in the notation of (4.7) and (4.9). The color magnetic operator (4.4) mixes these states with the result that

$$\begin{aligned} |0^+,\underset{\sim}{9}\rangle &= 0.972\,|0^+\underset{\sim}{9}[1]\rangle + 0.233\,|0^+\underset{\sim}{9}[405]\rangle \\ |0^+,\underset{\sim}{9}^*\rangle &= 0.233\,|0^+\underset{\sim}{9}[1]\rangle - 0.972\,|0^+\underset{\sim}{9}[405]\rangle\ . \end{aligned} \tag{4.12}$$

The lowest energy state is almost entirely in the [1] of color-spin, as expected from the Jaffe-Hund rule. Jaffe quotes the eigenvalue

$$\langle 0^+,\underset{\sim}{9}|\ -\sum_a \sum_{i>j} \vec{\sigma}_i \cdot \vec{\sigma}_j \lambda_i^a \lambda_j^a\, |0^+,\underset{\sim}{9}\rangle = -43.66\ . \tag{4.13}$$

When this result is inserted into the expression for the bag energy, the masses of the members of the nonet are readily calculated. Let us consider in more detail what quark content the foregoing analysis has produced. The combination qq forms a flavor $\bar{3}$ which is displayed in the weight diagram of Fig. 5(a). The $\bar{q}\bar{q}$ forms a flavor 3. With magic mixing, the resulting nonet is as shown. The masses are summarized in Table 1. Jaffe concluded that the familiar 0^+ mesons also listed in Table 1 are not the L = 1 $q\bar{q}$ states as had hitherto been claimed, but are in fact cryptoexotic $qq\bar{q}\bar{q}$ states. Although controversial at the time, his idea has gained wider acceptance and now represents a striking confirmation of the idea of colored gluon exchange.

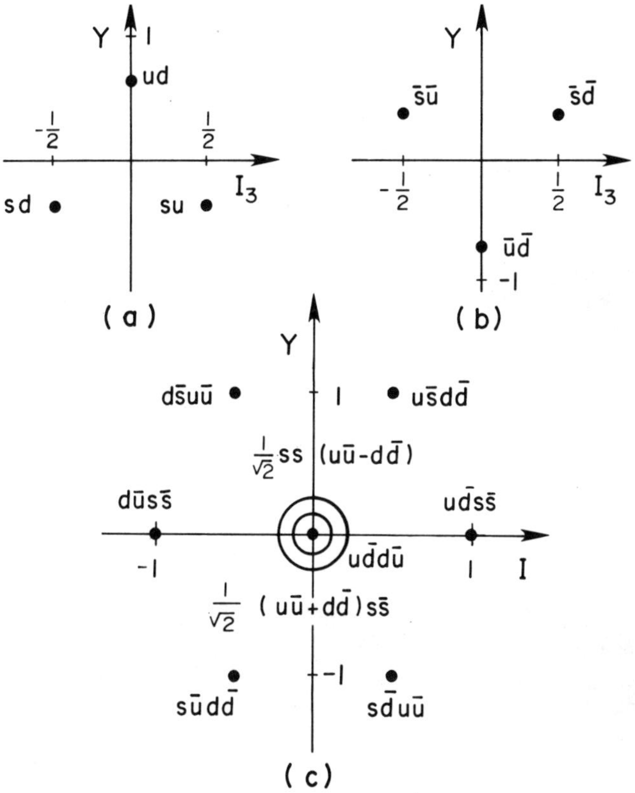

Fig. 5. (From Jaffe, Ref. 25) SU(3) weight diagram for (a) the $\underset{\sim}{3}$ formed from two quarks, (b) the $\underset{\sim}{\bar{3}}$ formed from two antiquarks, and (c) the cryptoexotic nonet.

Table 1. (From Jaffe, Ref. 25) Masses of the lowest-lying 0^+ $qq\bar{q}\bar{q}$ cryptoexotic nonet

	State I	State Y	Theoretical Mass	Experimental Mass	Width (Γ)
ε	0	0	650	660 ± 100	640 ± 140
S^*	0	0	1100	993.2 ± 4.4	40.5 ± 7.4
δ	1	0	1100	976.4 ± 5.4	46.9 ± 11.2
κ	1	1	900	1400 - 1450	200 - 300

5. THE NUCLEON-NUCLEON INTERACTION

It is not yet known how to carry out rigorous practical dynamical calculations of hadron-hadron interactions in the bag model. What we do know is based on the static cavity approximation extended by means of additional interpretation to hadronic interactions. There are two complementary approaches. In the case of the two-nucleon interaction, one might consider what happens to the six-quark bag as it is slowly pulled apart into two three-quark nucleon bags.[19] The assumption of this approach is that for the purposes of calculations at low c.m. energies, it is possible to speak of a collective motion of the system, e.g., a variation in the internuclear separation, which is slow compared to the rate at which the light degrees of freedom (quarks and gluons) accommodate to the changes in the collective coordinate(s). Thus one might consider using the static cavity approximation to adjust the quark and gluon configuration at each constrained value of the internuclear separation. This is an approach similar to the Born-Oppenheimer computation of the potential between two hydrogen atoms. There, the electrons are supposed to adjust instantaneously under slow variations of the interproton distance. We are led, therefore, to computing an energy of adiabatic deformation as a function of internucleon separation, and interpreting it as a potential energy of separation. Similar techniques are well-known in nuclear physics.[28]

The second approach was developed by Jaffe and Low and is particularly well suited for interpreting the highly unstable $qq\bar{q}\bar{q}$ spherical bag state.[29] They drew an analogy between the spherical bag calculation and the Wigner-Eisenbud[30] theory of scattering into compound nuclear states. They interpreted the restriction to spherical symmetry as corresponding to the artificial imposition of a boundary condition on the relative wave function of the communicating two-meson channel, namely that the wave function vanishes at a separation comparable to the bag radius. Their calculation provides an elegant link between properties of experimental phase shifts and parameters of the spherical bag calculation. We shall examine their calculation in more detail in a subsequent lecture. Here we shall focus on the two-nucleon interaction.

The most trivial interaction between two three-quark bags is a purely geometrical one in which the bags join and separate without gluon exchange.[31] It is not very dramatic. With gluon exchange, however, interesting effects are seen. The lowest order gluon exchange must be accompanied by a quark interchange so as to conserve the color singlet form of the nucleon, as in Fig. 6.

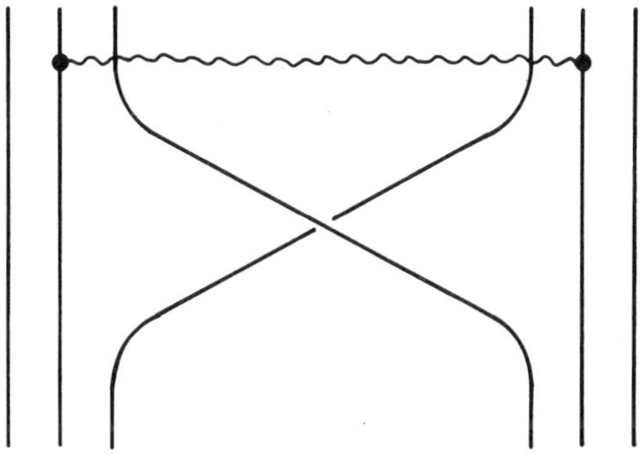

Fig. 6. Lowest order gluon interaction between two nucleons.

From the standpoint of quantum numbers, quark interchange is analogous to $q\bar{q}$ meson exchange as well as $qq\bar{q}\bar{q}$ meson exchange. (Because of the indistinguishability of nucleons, the final nucleons in Fig. 6 can be switched around giving a diagram with pairs of quarks exchanged.) Thus one hopes for a link between the lowest order quark interchange diagram and the conventional meson exchange picture of nuclear interactions. The quark interchange diagram with a single gluon exchanged produces a soft repulsive core at short distances and a region of strong attraction at intermediate distances. A brief description of the calculation is given below.

The next higher order interaction allows quarks to remain in the respective nucleons (as well as exchanging as before). We could have the process shown in Fig. 7, which takes place inside a single bag. There are other fourth order graphs which must be considered as well. The exchange of two gluons has been offered as a likely candidate for the "bare Pomeron" in high energy scattering.[32] At low energies it produces a sort of Van der Waals force between the nucleons.[33] The range of the force is quite limited. Willey[33] has estimated that it is less important than one pion exchange at 2 fm. It is not clear whether the two-gluon

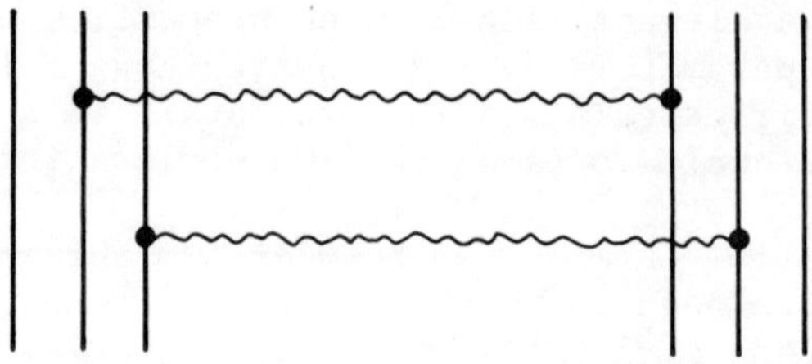

Fig. 7. Two gluon exchange between two nucleons.

exchange mechanism has a counterpart in the conventional meson exchange picture of the nuclear interaction.[34] At long range it would be associated with exchanges of objects with vacuum quantum numbers, perhaps even exchanges of pure gluon mesons, should they exist. The longest range component would presumably be coupled to the $\pi\pi$ S-wave enhancement, although the association with the $\pi\pi$ channel as an intermediate is not required per se, and the exchange certainly does not look like anything derivable from old fashioned meson-nucleon field theories. There have not been any detailed calculations of two gluon exchange between nucleons. The short range behavior is unknown, but of considerable interest.

Let us return to the exchange depicted in Fig. 6 and consider the deformation energy associated with single gluon exchange for the static six-quark bag. We would naturally want to study cavities of a variety of shapes ranging from a compact object to a shape which fissions into two bags. In the static cavity approximation the interaction ceases when the bags undergo fission. Thus we cannot hope to calculate the long range interaction. In the present lecture we shall limit our attention instead to a spherical bag shape. It turns out that we don't make any appreciable error in the short range calculation in so doing.[20]

The separation of the two nucleons in the six-quark configuration is accomplished in the same way that the quark and antiquark were separated in calculating the electric polarizability of mesons: three quarks for one nucleon are placed in the left orbital and three quarks for the other, in the right orbital. The state is fully antisymmetrized, of course. The internal symmetry wave function for the proton with spin-up is given by the following linear combination of creation operators

$$
\begin{aligned}
(18)^{1/2}p^{\dagger}(\uparrow) = {} & 2u_r^{\dagger}(\uparrow)u_y^{\dagger}(\uparrow)d_b^{\dagger}(\downarrow) + 2u_r^{\dagger}(\uparrow)d_y^{\dagger}(\downarrow)u_b^{\dagger}(\uparrow) \\
& + 2d_r^{\dagger}(\downarrow)u_y^{\dagger}(\uparrow)u_b^{\dagger}(\uparrow) - u_r^{\dagger}(\uparrow)u_y^{\dagger}(\downarrow)d_b(\uparrow) - u_r^{\dagger}(\uparrow)d_y^{\dagger}(\uparrow)u_b^{\dagger}(\downarrow) \\
& - d_r^{\dagger}(\uparrow)u_y^{\dagger}(\downarrow)u_b^{\dagger}(\uparrow) - u_r^{\dagger}(\downarrow)u_y^{\dagger}(\uparrow)d_b^{\dagger}(\uparrow) - u_r^{\dagger}(\downarrow)d_y^{\dagger}(\uparrow)u_b^{\dagger}(\uparrow) \\
& - d_r^{\dagger}(\uparrow)u_y^{\dagger}(\uparrow)u_b^{\dagger}(\downarrow) \quad , \qquad (5.1)
\end{aligned}
$$

where the quark creation operators create up or down quarks of color r,y, or b with spin up or down. If in addition these quarks are all assigned to the right-hand spatial orbital, [see Eq. (2.5)], i.e.,

$$
u_{rR}^{\dagger}(\uparrow) = u_{rS}^{\dagger}(\uparrow) + \sqrt{\mu}\,u_{rP}^{\dagger}(\uparrow) \qquad (5.2)
$$

etc., then we obtain a "proton" in the right-hand orbital. The deuteron channel is formed from the combination of proton and neutron creation operators given below:

$$
D^{\dagger}(m_s = +1) = p_R^{\dagger}(\uparrow)n_L^{\dagger}(\uparrow) + p_L^{\dagger}(\uparrow)n_R^{\dagger}(\uparrow) \ . \qquad (5.3)
$$

Other channels may be formed in a similar fashion. In this manner we obtain a six-quark configuration which, for $\mu = 0$ has all six quarks in the $S_{1/2}$ orbital and for $\mu = 1$ has the nucleons in orthogonal left and right orbitals. A more complete description of the state would necessarily also include configurations which separate into two Δ's and the various color octet baryon channels.

An internucleon separation distance δ is then defined. It is to be constrained so as to produce the appropriate deformation of the orbitals. The energy of the system $V(\delta)$ will be expressed as a function of δ. The definition of separation is rather arbitrary when the nucleons are overlapping. The resolution of the arbitrariness comes only when the kinetic energy associated with time variations in δ is obtained. The complete adiabatic hamiltonian would then be

$$
H = \frac{1}{2}m(\delta)\dot{\delta}^2 + V(\delta) \ . \qquad (5.4)
$$

A functional change in the definition of δ would produce a compensating redefinition of the inertial parameter $m(\delta)$. Since we will be considering only the deformation energy $V(\delta)$, this arbitrariness must be kept in mind. The results we shall present are based on the definition

$$\delta = \frac{2\sqrt{\mu}(1+\mu)}{1+\mu^2} \int q_S^\dagger(\vec{r}) q_P(\vec{r}) z d^3r = d(\mu)R \tag{5.5}$$

which is similar to (2.12) for the polarizing $q\bar{q}$ separation.

The evaluation of the deformation energy follows the same approach as in the calculation of the energy of separation of the $q\bar{q}$ system. It is just technically more difficult to evaluate the contributions from the various graphs of Fig. 2 (see Ref. 20). In the end one obtains a variational expression of the form

$$\begin{aligned} I(R,\mu,\lambda) &= E(R,\mu) - \lambda d(\mu)R \\ E(R,\mu) &= c(\mu)/R + \frac{4\pi}{3} R^3 B \end{aligned} \tag{5.6}$$

to be compared with (2.16), where the term $c(\mu)/R$ summarizes the contribution from the quark and gluon field energy, corrected by the "zero point energy," and the term with the Lagrange multiplier λ serves to fix a particular separation δ. Rather than writing the constraint term as $\lambda(\delta-\delta_0)$ and finding the extremum in R, μ, and λ, we can equally well simply look for the extrema of $I(R,\mu,\lambda)$ at fixed λ and see what we get for the total energy $E(R,\mu)$ and separation $d(\mu)R$. This pair of values gives us one point on the deformation energy curve $V(\delta)$ [where the slope is $\lambda = V'(\delta)$]. If there is more than one point with slope λ, there is more than one extremum in I at that fixed value of λ. By varying λ, the whole deformation energy curve can be mapped out. Clearly $\lambda = 0$ corresponds to the extrema of the deformation energy itself.

The bag energy for $\mu = 0$ (all six quarks in the S-orbital) was calculated by DJJK[6] and found to be about 270 MeV higher than twice the nucleon mass. Therefore the configuration is repulsive. The repulsion is entirely color-magnetic, the same effect which causes the $\Delta(1236)$ to be more massive than the nucleon. If we consider minimizing the energy in the absence of any constraint ($\lambda = 0$) we find that a minimum occurs for $\mu \approx 1/2$ with an energy about 290 MeV below the two-nucleon mass. This remarkably strong attraction comes chiefly from a strong color electric field. It can be interpreted as resulting from the binding of two clusters of three quarks to form color singlets. Since the color electrostatic interaction is attractive and Coulombic within a color singlet configuration, the gluon field energy is reduced by bringing together the attracting members of a color-singlet. The separation of color singlets is not complete, however, since localizing the quarks also increases their kinetic energy. In Fig. 8 are shown results for calculations in all two-nucleon channels.

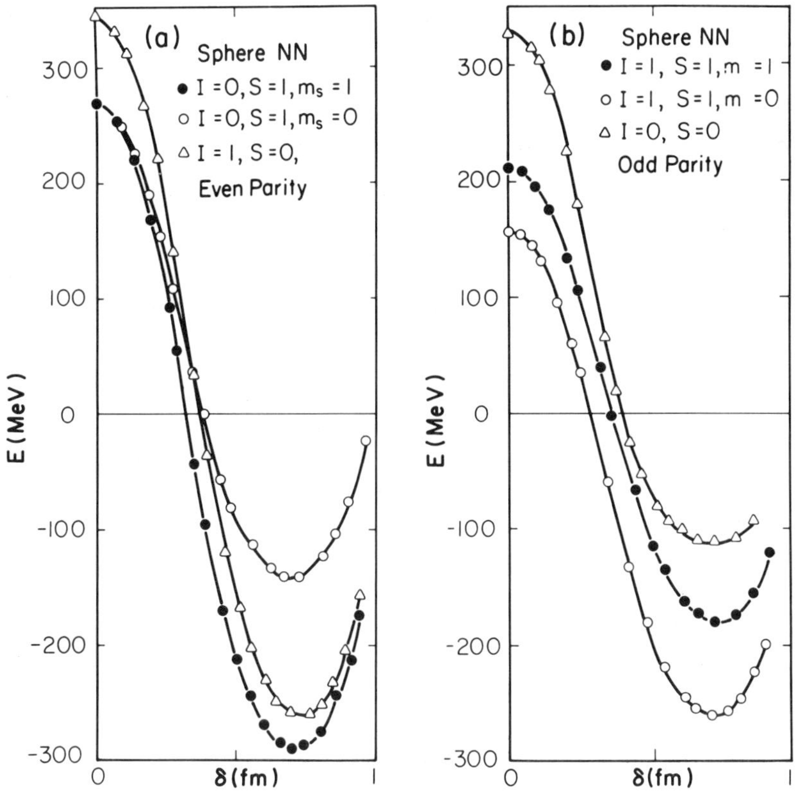

Fig. 8. Interaction energy for the two-nucleon configuration in a spherical bag as a function of the constrained separation parameter (a) for even parity and (b) for odd parity configurations.

Qualitatively, the results correspond to what is expected from phenomenological studies of the two-nucleon interaction. There remains the difficult problem of developing a full dynamics incorporating the kinetic energy of deformation. There is also the problem of making a connection with the long-range meson exchange theory. Because of its particularly long range, it may well be useful to introduce the pion as a field at least outside the static cavity, following the procedure of Chodos and Thorn[7] and Brown, Rho and Vento[15] (see Lecture 7). It would then contribute to the energy even after fission takes place. It remains to be seen how it would modify the short-range features of the calculation.

It is amusing to speculate about three-body and four-body nuclear interactions in the bag model. The repulsive core associated with placing three and four nucleons in the same orbital can be derived from a formula of DJJK. The color magnetic energy is expressed as a function of the number of quarks, n, and the total isospin and spin:

$$E_M = [n(n-6) + J(J+1) + 3I(I+1)]\frac{1}{2}M_{00} \tag{5.7}$$

where $M_{00} = 0.26/R$. For He^3 or H^3, $n = 9$, $I = 1/2$ and $J = 1/2$, so that

$$E_{M9} = 30(\tfrac{1}{2}M_{00}) \ . \tag{5.8}$$

Let us compare the repulsion among these nine quarks with what would be expected from the two-baryon interactions alone. There are three pairs of two-body interactions in He^3 or H^3. A given pair is equally likely to occur in an $I = 0$, $S = 1$ state as in an $I = 1$, $S = 0$ state. If we use (5.6) to calculate the interaction energy of the color singlet six-quark (two baryon) system, we get

$$\begin{aligned} E_{M9}^{(2)} &= [(\tfrac{1}{2})3 \times 2 + (\tfrac{1}{2})3 \times 6]\tfrac{1}{2}M_{00} \\ &= 12(\tfrac{1}{2}M_{00}) \ . \end{aligned} \tag{5.9}$$

Thus there is a substantial additional repulsion at short range attributable to a three-body interaction. Its actual magnitude depends on the size of the system. Taking the radius of the nine-quark spherical bag ($R = 1.5$ fm) we estimate $M_{00} \approx 30$ MeV so that the additional three-body repulsion is about 300 MeV. A similar calculation for He^4 shows an additional repulsion of about 700 MeV over what is expected from two-body interactions above! The strong repulsion is entirely due to the color-magnetic interaction. One could say that some departure from a two-body description is to be expected, since cramming all the quarks into the same orbital requires that the six-quark cluster, which might have been regarded as two interacting nucleons (therefore a color singlet) must be arranged part of the time in color non-singlet configurations. These configurations interact strongly with other clusters in non-color-singlet combinations. Whether the effect is of net attraction or repulsion depends entirely on the nature of the interaction. In QCD it is strongly repulsive. The range of repulsion is presumably not much different from what is observed in the two-body interaction, since the nucleons need merely to separate slightly to relax the color magnetic repulsion. Perhaps these strong multi-body forces can help in accounting for the saturation of nuclear forces in the higher nuclei.

6. LOW ENERGY SCATTERING AND QUARK-BAG EIGENSTATES

One of the intriguing puzzles in the quark-bag model has been the question of interpreting the S^4 $qq\bar{q}\bar{q}$ states which are formed by putting all quarks in the lowest $S_{1/2}$ orbital.[25] Many of these "states" presumably decay readily into a pair of $q\bar{q}$ mesons. Therefore, one might legitimately ask whether these bag model states, calculated in an approximation that ignores hadronic decays, have anything at all to do with reality. Indeed the bag model seems to predict exotic states which do not even occur in the data. One might consider applying the same analysis to the $qq\bar{q}\bar{q}$ states as was done for the two nucleon interaction. It is expected that such an analysis would show that the act of separating the color singlet $q\bar{q}$ clusters into left and right orbitals would lower the energy of the state. A deformation energy curve could be mapped out with the S^4 bag state corresponding to the value of the deformation energy at zero separation. The S^4 state would not be the ground state for spherical geometry--rather it would be a state of mixed configuration involving at least the P-orbital. From this viewpoint one would not expect the S^4 state to have any particular observable consequences.

Jaffe and Low[30] have suggested an intriguing interpretation of the S^4 states. They draw an analogy between the quark model calculation and the Wigner-Eisenbud theory of scattering into compound nuclear states.[30] The Wigner-Eisenbud theory analyzed low energy, two-body scattering by first considering the problem of constructing the energy levels of the two-body system constrained by imposing an artificial boundary condition on the wavefunction, namely that it's derivative vanish at a specified radius, $r = b$. A set of discrete eigenstates could be found. In the Jaffe-Low approach, the wave function instead is supposed to vanish at $r = b$. The boundary condition in effect places a wall around the two-body system. A knowledge of these eigenstates of the hamiltonian and their overlap with the external channels gives a great deal of information about the potential for $0 \leq r \leq b$. If the potential vanishes for $r > b$, the S-matrix can be constructed. We work out an example below.

Consider two-body S-wave scattering with $V(r > b) = 0$. We can write a solution for the scattering problem with asymptotic momentum k as follows

$$\psi_k(r) = \begin{cases} u_k(r) & 0 \leq r \leq b \\ \cos k(r - b) + P(k)/k \sin k(r - b) & r > b, \\ \text{or } A_k(e^{-ikr} - e^{2i\delta(k)}e^{ikr}) & r > b, \end{cases} \tag{6.1}$$

where the normalization is fixed by requiring the coefficient of cos k(r - b) to be one. It is clear that

$$e^{2i\delta} = -e^{-2ikb}\left(\frac{1 - iP/k}{1 + iP/k}\right)$$
$$\text{or } P = k\cot(kb + \delta)\ . \tag{6.2}$$

The matching condition at r = b requires u(b) = 1 and u'(b) = P(k). However, if P(k) is very large, then if $u_k(r)$ were normalized on the inveral [0,b], the renormalization of u would make u(b) very small. Thus the poles of P(k) are associated with the eigenstates of the internal hamiltonian with u(b) = 0. Jaffe and Low call these internal eigenstates "primitives." Knowing the poles and residues of P(k) gives considerable information about the S-matrix, $e^{2i\delta}$. The partial wave S-matrix contains a left-hand cut[36] in the c.m. energy squared, $s = (E_1(k) + E_2(k))^2$. It also contains threshold branch points for other channels. However, P(s) does not contain a branch point for the two-body channel of interest, since for reasonable potentials the boundary condition $u_k(b) = 0$ is associated with only discrete eigenvalues--hence only poles in P. The poles of P occur for real values of s and are associated with poles of the S-matrix at complex values of s. This can be seen by considering the behavior of P in the vicinity of a pole:

$$P \approx \tilde{a} + \frac{\tilde{b}}{s_p - s}\ . \tag{6.3}$$

This formula is considered to be good as long as $|s_p - s|$ is much smaller than the distance to the next pole in P. The pole in the S-matrix occurs where the denominator vanishes, i.e.,

$$\tilde{a} + \frac{\tilde{b}}{s_p - s} \approx ik\ , \tag{6.4}$$

which can be solved for complex s. However, since the formula (6.3) is valid only in the vicinity of $s = s_p$, one cannot guarantee the presence of a pole in the S-matrix if the complex solution of (6.4) is far from s_p. If the interaction is sufficiently weak that there are no poles in $e^{2i\delta}$, there are still poles in P, and the above analysis is invalid. On the other hand, a narrow resonance in the S-matrix is always associated with a pole in P, typically with a small residue.

Even if a pole is lacking in the S-matrix, a knowledge of $\delta(k)$ from a scattering experiment permits a reconstruction of P(k). Jaffe and Low give the example of a square well potential of radius b and strength V = -U/2m. In this case, in the notation of (6.1),

$$u_k(r) = \frac{\sin qr}{\sin qb} \; ; \qquad q = \sqrt{k^2 + U}$$
$$P(k) = q \cot qb = k \cot[kb + \delta(k)] \; . \tag{6.5}$$

[Solving the second equation above gives the phase shift $\delta(k)$.] The poles of P occur at the zeros of the (correctly normalized) internal wave function, i.e., at

$$b\sqrt{k_n^2 + U} = n\pi \qquad n = 1,2,\ldots \tag{6.6}$$

or at the zeros of $\sin[kb + \delta(k)]$, which depends on knowledge of b and $\delta(k)$. In this example, it is clear that even in the absence of interaction ($U = 0$, $\delta = 0$) there are still poles in P(k). These poles occur at $k_n^2 = n^2\pi^2/b^2$ with residue $2n^2\pi^2/b^3$. If the interaction is moderately attractive ($U > 0$) in the square-well example, the pole occurs at a slightly lower value of k and of total energy; if it is repulsive, a slightly higher value. This relationship between the positions of poles in P with large residues and the values k_n^2 may be used to predict whether the phase shift should be positive (attractive) or negative (repulsive).

Jaffe and Low proposed that the S^4 bag states coupled strongly to two-body channels be interpreted as eigenstates of the internal two-body hamiltonian, i e., whose wave functions vanish at a particular radius $r = b$, to be associated approximately with the bag radius. More precisely, b is defined by determining the mean separation between clusters of quarks

$$\bar{r}^2 = \int d^3r_1 d^3r_2 d^3r_3 d^3r_4 \left(\frac{\vec{r}_1 + \vec{r}_2}{2} - \frac{\vec{r}_3 + \vec{r}_4}{2}\right)^2 \rho(\vec{r}_1,\vec{r}_2,\vec{r}_3,\vec{r}_4) \; , \tag{6.7}$$

where ρ is defined in terms of the baryon number density, and then by equating $\bar{r}^2$ to be the value which is obtained from a free meson-meson wave function

$$\psi(r) = \frac{1}{\sqrt{2\pi b}} \, \frac{\sin \pi r/b}{r} \; . \tag{6.8}$$

The result is that

$$b = 1.4R \tag{6.9}$$

where R is the bag radius. Contact with experiment is then made by reconstructing P(k) from the measured phase shift $\delta(k)$, using

(6.2) and the predetermined value of b, and checking that poles correspond in energy to the theoretical S^4 bag state energies.

Before we look at the comparison with data, let us develop the multichannel formalism. Following Jaffe and Low we define (for $\ell = 0$, still) a complete set of stationary eigenstates of the hamiltonian. If there are n channels, there are n states which can be written, in analogy to (6.1) for state i and channel j as

$$\psi_{i,j}(r) = \begin{cases} u_{ij}(r) & r < b \\ \left.\begin{aligned} &\delta_{ij} \cos k_j(r-b) + \frac{P_{ji}}{k_j} \sin k_j(r-b) \\ &= \sum_\ell \frac{A_{\ell i}}{\sqrt{k_j}} \left(\delta_{j\ell} e^{-ik_j r} - S_{j\ell} e^{ik_j r}\right) \end{aligned}\right\} & r > b \,. \end{cases} \tag{6.10}$$

Solving for S,

$$S = -e^{-ikb} \left(\frac{I - i/\sqrt{k}\,P\,1/\sqrt{k}}{I + i/\sqrt{k}\,P\,1/\sqrt{k}}\right) e^{-ikb} \,. \tag{6.11}$$

Since S is unitary and symmetric for real values of k^2, P must be real and symmetric. It can also be shown that P has no threshold branch points at $k_i = 0$. Its only branch point singularities are at the "left-hand" branch points of S for the appropriate angular momentum and at the normal threshold branch points for those channels not included in the analysis.[36] It is customary to use the channel invariant s in place of the variables k_i. Jaffe and Low report a useful theorem which gives significance to the residues of poles in P. They observe that by direct differentiation of (6.11), using the fact that the S is independent of b, it can be shown that

$$-\frac{dP}{db} = P^2 + k^2 \,. \tag{6.12}$$

Thus if P has a pole in s at $s_0(b)$, so that

$$P \approx c + \frac{r_0(b)}{s - s_0(b)} \tag{6.13}$$

(where c is a constant background matrix), it follows from (6.12) that

$$-\frac{r_0(b)}{[s-s_0(b)]^2}\frac{ds_0}{db} = \frac{r_0^2(b)}{[s-s_0(b)]^2} \tag{6.14}$$

and therefore

$$r_0(b)_{ij} = -\frac{ds_0}{db}\xi_i\xi_j \tag{6.15}$$

where $\sum_i \xi_i^2 = 1$; i.e., the residue is factorizable and its normalization is given by the dependence of the location of the pole on the boundary position b.

We have seen that the residue of the pole in P is determined by the way the position of the primitive s_0 depends upon the boundary position b. To determine this dependence in the bag model requires further interpretation. We expect that if b is increased, s_0 will decrease, since the wave function is less constrained. What does increasing b mean in the bag model? Jaffe and Low argue that it means enlarging the cavity so that the quarks can "expand" into their open channels. This is accomplished in effect by removing the bag pressure in the open channels by writing

$$H = H_B(R) - \Lambda\frac{4\pi}{3}BR^3 \tag{6.16}$$

where Λ projects onto open channels. Then since $H_B'(R) = 0$ (the condition fixing the radius of the spherical state), we have

$$\frac{dE}{dR} = -4\pi BR^2\langle\Lambda\rangle \ . \tag{6.17}$$

From this result and (6.9) Jaffe and Low derive

$$\frac{1}{s_0}\frac{ds_0}{db} = \frac{2}{E}\frac{dE}{db} = \frac{2}{E}\frac{dE}{dR}\frac{R}{b} = -\frac{3}{2b}\langle\Lambda\rangle \ . \tag{6.18}$$

The projection onto open channels is to be determined by the fractional parentage coefficients of the state in question.

As an example of the comparison with experiment, Jaffe and Low consider the low lying cryptoexotic nonet formed from a spherical $q^2\bar{q}^2$ state. In particular they consider the I = 0 $\pi\pi$ S-wave state in this nonet at energy 650 MeV and radius $R_0 = 4.4$ GeV^{-1}. A pole is predicted in the P matrix at

$$s_0 = (0.65)^2 \mathrm{GeV}^2 \tag{6.19}$$

where

$$b \cong 1.4 R_0 = 6.2 \ \mathrm{GeV}^{-1} \ . \tag{6.20}$$

Since only the $\pi\pi$ channel is involved, the residue is just $-ds_0/db$. The projection onto the open channel is found from the expansion

$$|C^0(9)0^+\rangle = 0.64|\pi\pi\rangle - 0.37|\vec{\pi}\cdot\vec{\pi}\rangle + 0.19|\eta\eta\rangle - 0.11|\vec{\eta}\cdot\vec{\eta}\rangle + 0.29|\rho\rho\rangle - 0.34|\vec{\rho}\cdot\vec{\rho}\rangle + \ldots \tag{6.21}$$

given by Jaffe and Low. The vector over the meson symbol refers to a color octet combination of the quarks. Since the $\pi\pi$ channel is involved here,

$$\langle\dot{\Lambda}\rangle = (0.64)^2 \ . \tag{6.22}$$

The predicted residue is therefore (from 6.18)

$$-\frac{ds_0}{db} = 0.04 \ \mathrm{GeV}^3 \ . \tag{6.23}$$

An analysis of the $\pi\pi$ S-wave phase shift shows that the P-matrix for the same value of b [Eq. (6.20)] has a pole at $(.69)^2$ GeV and residue 0.064 GeV^3. Such agreement in the pole positions is very gratifying. In view of the approximations involved in deriving (6.17) an agreement in order of magnitude in the residues is all that might have been hoped for. In fact it is much better than expected.

Jaffe and Low give several other examples. A particularly interesting case is that of the predicted I = 2 $q^2\bar{q}^2$ $\pi\pi$ S-wave state in the spherical bag model at 1150 MeV. No resonance appears in the measured phase shift at these energies. However, if the P matrix is constructed from the measured phase shifts, a pole appears at 1040 MeV! The predicted residue is half the experimental value. Jaffe and Low argue that the negative phase shift in this channel, which is necessary in order to produce a pole at such a high mass, is direct evidence for the repulsive one-gluon exchange interaction in the exotic channel, and is, as such, further evidence in support of QCD.

7. PCAC AND THE BAG MODEL: THE PION MASS

A. Chiral Symmetry and Zero Mass Pions

It is widely believed that confinement occurs in conventional QCD without the need for introducing a bag pressure in the Lagrangian at the outset. Now if the masses of the up and down quarks vanish in the original QCD Lagrangian, the theory possesses chiral symmetry.[37] The kinetic energy

$$\frac{1}{2}\bar{q}i\gamma\cdot\overleftrightarrow{\partial}q \qquad (7.1)$$

is invariant under the transformation (chiral rotation)

$$q \to \exp(i\vec{\theta}\cdot\vec{\tau}/2\ \gamma_5)q \qquad (7.2)$$

where $\vec{\theta}$ is a constant and $\vec{\tau}$ acts on the SU(2) basis labeled by up and down. The mass term

$$\bar{q}Mq = \bar{q}_u q_u m_u + \bar{q}_d q_d m_d + \bar{q}_s q_s m_s + \ldots$$

is not invariant under chiral rotations unless the masses of the up and down quarks vanish. In the case that they do vanish, Noether's theorem[38] gives us a conserved axial vector current,

$$\vec{A}_\mu = \bar{q}\,\frac{\vec{\tau}}{2}\,\gamma_\mu\gamma_5 q\ , \qquad (7.3)$$

which is supposed to be the same current appearing in weak interactions involving hadrons. Now the confinement of the quarks apparently destroys chiral symmetry. This is most evident if we consider that on the surface of the bag the fermion field satisfies

$$\bar{q}q = 0\ . \qquad (7.4)$$

This condition is not invariant under chiral rotations. The Goldstone theorem[39] tells us that in similar cases in which a solution to a theory breaks an inherent symmetry, we should expect to find a massless boson.[40] Modern experience has shown us exceptions to this expectation--in particular in the Weinberg-Salam model of the weak interactions in which the massless boson is eliminated when a gauge vector field acquires a mass, and in the instanton mechanism proposed as a resolution of the celebrated U(1) problem.[38] However, let us suppose that we have a Nambu-Goldstone boson. The pion is the traditionally accepted candidate. In reality it is not massless, of course, but it is the lightest of the hadrons. It is presumably massive because the up and down quarks do not both have exactly zero mass, so that the chiral symmetry is not perfect.

B. PCAC and the Phenomenological Pion

The pion decays through the weak axial vector current. Thus

$$<0|A_\mu^{1+i2}(x)|\pi^-> = i\sqrt{2}p_\mu f_\pi e^{-ip\cdot x} \tag{7.5}$$

where p_μ is the on-shell four momentum of the pion and f_π is the pion decay constant. From the observed lifetime of the pion

$$f_\pi \cos\,\theta_c \approx 96\ \text{MeV} \tag{7.6}$$

where the Cabbibo angle is $\theta_c \approx 15°$. The isospin component of the axial vector current appearing above is

$$A_\mu^{1+i2}(x) = A_\mu^1(x) + iA_\mu^2(x)\ . \tag{7.7}$$

In the bag model it is natural to write the axial vector current in lowest order in the strong interactions

$$\vec{A}_{\mu Q}(x) = :\bar{q}(x)\frac{\vec{\tau}}{2}\gamma_\mu\gamma_5 q(x) :\ \theta_R(x) \tag{7.8}$$

where the quark fields are the free cavity fields. We have included the subscript Q to emphasize the fact that this current is not renormalized by strong interactions, i.e., it is based only on the free cavity quark fields.

Several authors[6,7] have calculated the axial vector current for bag states making the approximation

$$\vec{A}_\mu = \vec{A}_{\mu Q}\ . \tag{7.9}$$

If the free cavity fields are massless, $\partial^\mu A_{\mu Q}$ vanishes inside the cavity. There is only a surface contribution

$$\begin{aligned}\partial^\mu\vec{A}_{\mu Q}(x) &= :\bar{q}(x)\frac{\vec{\tau}}{2}\gamma_\mu\gamma_5 q(x) :\ n^\mu\delta_S \\ &= :\bar{q}(x)\frac{\vec{\tau}}{2}i\gamma_5 q(x) :\ \delta_S\ ,\end{aligned} \tag{7.10}$$

where we have used the bag boundary condition (1.27). Therefore, the axial vector current is not conserved. In conventional field theoretic models of spontaneously broken symmetry[40,41] there is still a conserved current after the symmetry is broken, because of the Nambu-Goldstone mechanism. We shall consider this possibility in the next lecture.

The current (7.8) was used by Johnson to calculate f_π. He begins by expanding the static pion bag state on a momentum space basis

$$|\pi,\text{bag}\rangle = \int \frac{d^3p}{\omega_p}\,\psi(p)\,|\pi,\vec{p}\rangle \; . \tag{7.11}$$

Then if $\vec{A}_{\mu Q}$ can be used in place of $\vec{A}_\mu$,

$$\langle 0|A_{\mu Q}^{1+i2}(x)|\pi^-,\text{bag}\rangle = i\sqrt{2}f_\pi \int \frac{d^3p}{\omega_p}\,p_\mu e^{-ip\cdot x}\psi(p) \; . \tag{7.12}$$

It is further necessary to assume something about the overlap between the true vacuum and the empty bag state, which results from the annihilation of the quark and antiquark. Johnson proposes

$$\langle 0|\text{empty bag}\rangle \approx 1 \; . \tag{7.13}$$

Then it is straightforward to compute

$$\begin{aligned} &\int \langle \text{empty bag}|:\bar{u}(\vec{x},0)\gamma_0\gamma_5 d(\vec{x},0) : |\pi^-,\text{bag}\rangle e^{-i\vec{p}\cdot\vec{x}}d^3x \\ &\quad = i\sqrt{2}f_\pi\psi(p)(2\pi)^3 \; . \end{aligned} \tag{7.14}$$

This procedure permits a direct calculation of the wave function $\psi(p)$. In fact, from (1.13)

$$\psi(r) \propto j_0^2(\omega_S r) - k^2/(\omega + m_q)^2 j_1^2(\omega_S r) \; . \tag{7.15}$$

From the usual normalization condition for $\psi(p)$, f_π can be determined to be

$$f_\pi = 0.501/R_\pi \; . \tag{7.16}$$

We shall see what answer we get after considering Johnson and Donoghue's recent modification to the static bag model of the pion.

C. Zero Mass Pion in the Bag Model

Johnson and Donoghue have investigated the possibility that a pion constructed from massless quarks would in fact have zero mass in the bag model.[10,42] The DJJK[6] pion was too massive. Johnson and Donoghue introduce two modifications to the original static bag calculation: a more careful treatment of the center of mass motion of the bag and an allowance for the effects of asymptotic freedom upon the value of the color coupling constant. They observe first that the static bag model of the pion describes a localized hadron with indefinite total momentum. The quark and antiquark move independently (at least to lowest order in QCD) in the cavity with the same momentum distribution. Thus the energy calculated is actually the expectation value,

$$E_{static} = < \sqrt{m^2 + p^2} > \tag{7.17}$$

where p is the total three momentum of the quarks.

Following Donoghue and Johnson, the expectation value (7.17) can be evaluated using $\psi(p)$ as determined from pion decay (7.15). Then in the limit of small quark masses

$$E_{static} \approx <p + \frac{m_\pi^2}{2p}> = A/R + m_\pi^2 R/C \tag{7.18}$$

where A and C are constants. Using (7.15)

$$A = 2.3; \quad C = 2.9 . \tag{7.19}$$

On the left side of (7.18) we use the expression from DJJK. However, we must be careful about the value of Z_0 used in E_{static}. We noted in Lecture 1 that the value $Z_0 = 1.84$, obtained by DJJK from the observed masses of the N, Δ, and ρ, includes a correction for non-relativistic c.m. motion. Johnson and Donoghue estimate that this correction added about 0.75 to Z. In the present case we want E_{static} to be the energy before any c.m. correction. Therefore, we should use $Z_0 \approx 1.1$ here.

In summary, the mass of the pion bag state is given by

$$m_\pi^2 = \frac{C}{R}\left(\frac{2\Omega}{C} - \frac{\alpha_S(R)}{R} G_0 + \frac{4\pi}{3} BR^3 - Z_0/R - A/R\right) \tag{7.20}$$

where $\Omega = 2.04$ and $G_0 = 0.7$. The radius and mass are determined by minimizing m_π^2 with respect to R. It is clear that requiring $\partial m_\pi^2/\partial R = 0$ with α_S constant gives zero mass only when the coefficient of the $1/R^2$ term vanishes--and then R also vanishes. This result is incompatible with chiral symmetry and Eq. (7.16) for f_π, since the chirally symmetric limit, f_π is supposed to be non-zero as m_π approaches zero. Donoghue and Johnson argue that it is necessary to consider the strong rescaling of α_S at small R. If α_S approaches zero for small R, then it is possible to obtain a zero mass at a non-zero radius. They choose

$$\alpha_S(R) \approx \frac{2\pi}{9\ell og(A + R_0/R)} \tag{7.21}$$

with a range of values of A from 0.3 to 1.0. They then undertake a revision of the calculation of DJJK so as to produce the correct masses for the ρ, nucleon, and Δ and a vanishing m_π when the quark masses are zero. The value for α_S for the nucleon is only slightly changed, with a best value at A = 1, R_0 = 2.13 GeV^{-1}, $B^{1/4}$ = 135 MeV, and Z_0 = 1.01. In the range of parameter values they considered, they get R_π = 3.3-3.5 GeV^{-1}. When the quarks are given a mass

$$m_u = m_d = 33 \text{ MeV}, \tag{7.22}$$

then the pion mass has the correct value. They also discuss the relationship between this mass and current algebra determinations of quark masses.[43]

The value of f_π from (7.16) is, then, about 150 MeV, to be compared with the experimental value of 96 MeV. Considering the delicate way in which the pion bag is constructed, particularly its sensitivity to the dependence of α_S on R and on the c.m. correction, this result is quite reasonable.

It is possible, at least in principle, to calculate the strong rescaling of α_S in the bag model. Thus we should regard the choice (7.21) as an indication of what is necessary in order to reconcile the adjustment of the static parameters of the light baryons and vector mesons with the requirements of PCAC and the current phenomenology regarding asymptotic freedom.

8. PCAC AND THE BAG MODEL: THE AXIAL VECTOR CURRENT

A. The Axial Vector Current of the Nucleon

We saw in the previous lecture that the identification of the current $\vec{A}_{\mu Q}(x)$ with the full axial vector current, gave a reasonable approximation to the pion decay constant. If we use it to compute form factors for the nucleon, we shall see that it is deficient.

Let us begin by reviewing some well known facts about the axial vector form factors for β-decay of the nucleon.[37] They are defined so that

$$<p|A_\mu^{1+i2}(0)|n> = \sqrt{\frac{m}{E_p}}\sqrt{\frac{m}{E_n}}\frac{1}{(2\pi)^3}\bar{u}_p\{g_A(q^2)\gamma_\mu\gamma_5 + g_p(q^2)q_\mu\gamma_5\}u_n \tag{8.1}$$

where $q = p_p - p_n$, the difference between the proton's and neutron's four momentum. We consider also the matrix element of the four-divergence of the current.

$$<p|\partial^\mu A_\mu^{1+i2}(0)|n> = \sqrt{\frac{m^2}{E_pE_n}}\frac{1}{(2\pi)^3}\bar{u}_p i\gamma_5 u_n[-2mg_A(q^2) - g_p(q^2)q^2] . \tag{8.2}$$

In the limit $q^2 \to m_\pi^2$, both expressions are dominated by the pion pole as diagrammed below

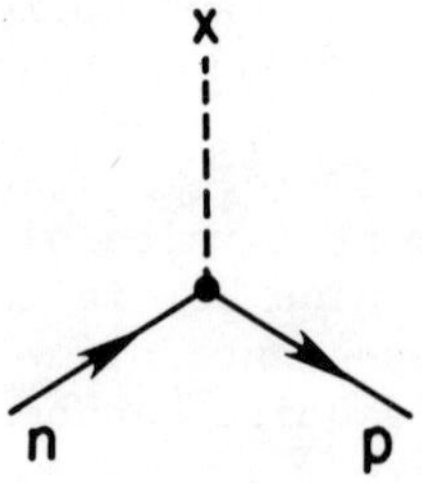

Thus

$$<p|A_\mu^{1+i2}(0)|n> \underset{q^2 \to m_\pi^2}{\sim} \frac{<p\pi|n><0|A_\mu^{1+i2}(0)|\pi>}{m_\pi^2 - q^2}$$

$$\sim \frac{(\bar{u}_p i\gamma_5 u_n\sqrt{2}g_r)(q_\mu f_\pi\sqrt{2})}{m_\pi^2 - q^2} , \tag{8.3}$$

where g_r is the pion-nucleon coupling constant and the $\sqrt{2}$ is an isospin coupling coefficient. Thus we conclude that

$$g_p(q^2) = \frac{2g_r f_\pi}{m_\pi^2 - q^2} + \tilde{b}(q^2) , \tag{8.4}$$

and similarly

$$2mg_A(q^2) + g_p(q^2)q^2 = \frac{2g_r f_\pi m_\pi^2}{m_\pi^2 - q^2} + b(q^2) \tag{8.5}$$

where b and $\tilde{b}$ are "background" functions which do not contain the pion pole. Therefore

$$g_A(q^2) = \frac{g_r f_\pi}{m} + \bar{b}(q^2) \ . \tag{8.6}$$

If we put $q^2 = 0$ and assume $\bar{b}(0) = 0$, we get the famous Goldberger-Treiman relation $mg_A(0) \approx g_r f_\pi$. The relation is approximate only, since we don't know how small $\bar{b}(0)$ really is; however, the Goldberger-Treiman relation is accurate to about 15%. We also observe that the pion pole occurs only in $g_P(q^2)$, basically because it must be associated with the factor $\gamma_5 q_\mu$ in the form factor. Three pion states and higher states contribute to both g_A and g_P, of course.

Armed with these observations, we now proceed to evaluate the nucleon axial vector form factors in the bag model. Suppose we use $\vec{A}_{\mu Q}$ for the axial vector current. We can treat the nucleon bag as a superposition of momentum eigenstates if we wish. We write

$$\langle p,bag|A_{\mu Q}^{1+i2}(0)|n,bag\rangle = \int d^3p_p\, d^3p_n \psi^*(p_p)\psi(p_n)\langle p|A_{\mu Q}^{1+i2}(0)|n\rangle. \tag{8.7}$$

where $\int|\psi(p)|^2 d^3p = 1$. On the left side we use the free cavity quark current; on the right we fill in the phenomenological form factors from (8.1) which we wish to determine. After a Fourier transform, we get

$$\int d^3x\, e^{i\vec{q}\cdot\vec{x}} \langle p,bag|A_{\mu Q}^{1+i2}(\vec{x},0)|n,bag\rangle$$

$$= \int d^3p_n \psi^*(p_n+q)\psi(p_n)\sqrt{\frac{m^2}{E_n E_p}}\, \bar{u}_p \{g_{AQ}(q^2)\gamma^\mu\gamma_5 + g_{PQ}(q^2)q^\mu\gamma_5\} u_n . \tag{8.8}$$

We use the Bjorken-Drell normalization for the states. An added subscript Q reminds us that these form factors are based on the current $\vec{A}_{\mu Q}$. If we had displayed the time dependence of the left and right side of (8.8) we would have found disagreement. The non-vanishing contributions to the matrix element on the left have no time dependence whereas the usual factor $\exp i(E_n - E_p)t$ for on-shell values of E_n and E_p appears on the right. This is a defect of the approximate treatment of the c.m. motion embodied in the momentum space expansion and requires us to avoid calculations far from $q = 0$. For small q we do not need to know $\psi(p)$ in detail. If we insert the bag fields (1.13) into (8.8) and evaluate the left side, we get for the three-vector matrix element

$$\int \langle p,\text{bag}|A_{jQ}^{1+i2}(x)|n,\text{bag}\rangle e^{i\vec{q}\cdot\vec{x}}d^3x$$

$$= \frac{5}{3}N^2 4\pi u_p^\dagger \sigma_j u_n \int_0^R \{[j_0^2(\omega r) - j_1^2(\omega r)]j_0(qr) + \frac{2j_1(qr)}{qr} j_1^2(\omega r)\} r^2 dr$$

$$+ \frac{5}{3}N^2 4\pi u_p^\dagger \vec{\sigma}\cdot\vec{q} q_j u_n \; 2\int_0^R \frac{1}{q}\frac{d}{dq}\left[\frac{j_1(qr)}{qr}\right] j_1^2(\omega r) r^2 dr \; . \tag{8.9}$$

We might like to see some indication of a pole at $\vec{q}^2 = -m_\pi^2$ in this expression. Clearly the integrals over Bessel functions are all well-behaved, and no pole will appear. Perhaps the integrals give a large contribution. Let us evaluate the expression in the limit $q \to 0$, keeping the coefficients of only the lowest powers of q. We get

$$u_p^\dagger \sigma_j u_n \bar{g}_A + u_p^\dagger \vec{\sigma}\cdot\vec{q} q_j u_n \bar{g}_p R^2 \; ,$$

where

$$\bar{g}_A = \frac{5}{3}N^2 4\pi R^3 \int_0^1 [j_0^2(\Omega r) - \frac{1}{3}j_1^2(\Omega r)] r^2 dr$$

$$\bar{g}_p = -\frac{2}{15}\left(\frac{5}{3}N^2 4\pi R^3\right) \int_0^1 j_1^2(\Omega r) r^4 dr \; . \tag{8.10}$$

The same small q approximation to the right side yields

$$g_{AQ}(0) \int d^3p |\psi(p)|^2 \left(\frac{E+2m}{3E}\right) u_p^\dagger \vec{\sigma} u_n$$

$$+ g_{PQ}(0) \int d^3p |\psi(p)|^2 u_p^\dagger \vec{\sigma}\cdot\vec{q}\vec{q} u_n \left(\frac{2E+m}{6E^2}\right) . \tag{8.11}$$

The first integral can be done if we use a non-relativistic approximation for E and take Johnson's estimate of the correction for center of mass motion

$$\int d^3p |\psi(p)|^2 \frac{p^2}{2m} \approx 0.75/R \; , \tag{8.12}$$

where R is the bag radius. Thus we have

$$g_{AQ}(0) = \frac{\bar{g}_A}{1 - \frac{2}{3}(\frac{0.75}{mR})} \; ; \quad g_{PQ}(0) = 2\bar{g}_p mR^2 \; . \tag{8.13}$$

The quantity $\bar{g}_A = 1.09$ was calculated by DJJK for massless quarks; we have $g_{AQ}(0) = 1.21$ compared with the experimental value of 1.24.[44] It is nice that correcting for the c.m. motion gives a better value, but the relatively close agreement is probably fortuitous in light of the approximations of the static cavity. The value of $g_{PQ}(0)$ is found to be about 0.2 GeV^{-1}. If we use the pole term (8.3) to estimate $g_P(0)$, we get about 130 GeV^{-1}, about three orders of magnitude bigger than $g_{PQ}(0)$. The discrepancy is obviously due to the fact that g_{PQ} does not contain the pion pole.

B. PCAC and a Phenomenological Pion Coupling

How can we compensate for the absence of the pion pole in $\langle\vec{A}_{\mu Q}\rangle$? Chodos and Thorn and others[7,14,15] suggested adding an explicit pion field. Without going into technicalities about the introduction of a σ meson and restricting the pion field to a particular volume, we shall just write

$$\vec{A}_\mu(x) = \vec{A}_{\mu Q}(x) + f_\pi \partial_\mu \vec{\pi} \; . \tag{8.14}$$

Now from PCAC[37] we have

$$\partial^\mu \vec{A}_\mu(x) = -f_\pi m_\pi^2 \vec{\pi}(x) \tag{8.15}$$

so that the equation of motion for the pion is

$$\begin{aligned}(\partial^2 + m_\pi^2)\vec{\pi} &= -\frac{1}{f_\pi}\partial^\mu \vec{A}_{\mu Q}(x) = -\frac{1}{f_\pi}\delta(r-R)\bar{q}\frac{\vec{\tau}}{2}\gamma_5\gamma^\mu n_\mu q \\ &= -\frac{1}{2f_\pi}\delta(r-R)\bar{q}\,\vec{\tau}\, i\gamma_5 q \; .\end{aligned} \tag{8.16}$$

What is appealing about introducing the pion in this way is that it not only puts the pion pole into the form factor g_P, it also restores the partial conservation of the axial vector current which was lost by writing $\vec{A}_\mu = \vec{A}_{\mu Q}$. Indeed, if $m_\pi^2 = 0$, the current is exactly conserved when the pion field satisfies the equation of motion. With the addition of the extra field, we are assured a pion pole in $g_P(q^2)$, since

$$\langle p|f_\pi \partial_\mu \pi^{1+i2}(0)|n\rangle \sim \frac{(\bar{u}_p i\gamma_5 u_n \sqrt{2} g_r) q_\mu f_\pi \sqrt{2}}{m_\pi^2 - q^2} . \tag{8.17}$$

Furthermore, we can now compute g_r for the bag directly since the equation of motion links the pion field to the free cavity quark fields. We should get a result for g_r that is at least as good as that which PCAC and our previously calculated value of g_A permits. Indeed we do recover the Goldberger-Treiman relation with the bag value of g_A.

As another exercise, we can calculate the $N\pi$ width of the $\Delta(1236)$ resonance. This can be done following the same approach as for the NN coupling. It is not surprising that we get the SU(6) symmetric result, except for a kinematical correction coming from the quark wave functions. If the momentum spread of the bag states is ignored (no recoil correction) the result obtained by Chodos and Thorn[7] is

$$\Gamma_{\Delta^{++} \to p\pi^+} = \frac{48}{25} \frac{\bar{f}^2}{4\pi} \frac{q^3}{m_\pi^2} \frac{E_p}{m_\Delta} \eta^2 , \tag{8.18}$$

where q = 227 MeV is the pion momentum and

$$2\bar{f}/m_\pi = \bar{g}_r/m = \bar{g}_A/f_\pi , \tag{8.19}$$

and g_A is given by (8.10). The SU(6) symmetric result has $\eta = 1$. Here, however,

$$\eta = \frac{3j_1(qR)}{qR} \approx 1 - \frac{1}{10}(qR)^2 \approx 0.87 . \tag{8.20}$$

Therefore

$$\Gamma_\Delta \approx 34 \text{ MeV}$$

compared with an experimental value of about 120 MeV. The result can be improved if we take into account the recoil correction. The matrix element which was calculated was based on particular localized states for N and Δ with a spread in c.m. momenta. Therefore it is smaller than the plane wave result in which the nucleon travels with the correct final momentum. Suppose the momentum wave functions are gaussian:

$$\psi(p) \propto \exp(-3p^2/4\sigma_p^2) \tag{8.21}$$

where, using Johnson's estimate

$$\langle \frac{p^2}{2m} \rangle = \frac{\sigma_p^2}{2m} = 0.75/R \ . \tag{8.22}$$

Then the overlap factor is

$$\exp(-3q^2/8\sigma_p^2) \approx 1 - \frac{3}{8}\frac{(qR)^2}{1.5\ mR} \approx 1 - \frac{1}{20}(qR)^2 \ . \tag{8.23}$$

This factor is close numerically to the geometrical factor η and so nearly cancels it. Without the factor η and with the correct value for g_r, we get the non-relativistic SU(6) value of $\Gamma \approx 74$ MeV--still not very good, however.

C. What is the Pion Field Doing in the Axial Vector Current?

For all the successes of the device of introducing a pion field, there are hazards as well. The most obvious is that it leads to a serious ambiguity in the description of the pion. This is seen in the calculation of pion decay. One must decide first whether the pion is really a quark-antiquark state or an elementary field. If the former, then is it like the bagged pion as described by Donoghue and Johnson.[10,41] If we accept their approach the pion field term in the axial vector current is not only superfluous, it is unwanted in the calculation of the pion decay. Because of the equation of motion for the axial vector current, we can write

$$\begin{aligned} \langle 0|\vec{A}_\mu|\pi\rangle &= \langle 0|\vec{A}_{\mu Q} + f_\pi \partial_\mu \vec{\pi}|\pi\rangle \\ &= \langle 0|\vec{A}_{\mu Q} - \frac{f_\pi \partial_\mu \partial^\nu \vec{A}_{\nu Q}}{\partial^2 + m_\pi^2}|\pi\rangle . \end{aligned} \tag{8.24}$$

The equation of motion for the pion field tells us that the part of $\vec{A}_{\mu Q}$ which annihilates the quarks in the pion bag acts also as a source of the pion field. Thus both terms in (8.24) contribute to π decay, and the f_π which multiplies $\partial_\mu \pi$ is not the value of the decay constant which is calculated via the method of Ref. 41.

It appears that we are faced with the alternative or rejecting the bag description of the pion or of rejecting the fundamental and universal validity of introducing an explicit pion field into the axial vector current.* Recall that the arguments in favor of doing so were to restore PCAC and to introduce a pion pole into the baryon axial vector form factors. Although $\vec{A}_{\mu Q}$ was unable to produce a pion pole in those form factors, it does a fair job annihilating the bag pion state itself.

If we wish to accept the bag description of the pion, therefore, we must regard the introduction of the pion field as a heuristic device for generating results of PCAC in particular cases, but not as an act of fundamental significance.

D. The Pion and Bag Dynamics

Various authors have suggested incorporating the pion field explicitly into the bag Lagrangian, despite the possible ambiguities mentioned above.[7,14,15] One example is the following

$$W = \int d^4x \,\{\theta_R(x)(\mathcal{L}_{QCD}-B) - \tfrac{1}{2}(\partial_\mu\sigma\partial^\mu\sigma + \partial_\mu\vec{\pi}\cdot\partial^\mu\vec{\pi})\theta_{\bar{R}}(x)\}$$
$$+ \int_S d^3x\, n^\mu \lambda_\mu \bar{q}(\sigma + i\vec{\tau}\cdot\vec{\pi}\gamma_5)q \ , \tag{8.25}$$

where $\theta_R(x)$ vanishes outside the bag volume and $\theta_{\bar{R}}$ vanishes inside. The surface boundary condition is now chirally invariant since (1.11) is replaced by

$$in_\mu\gamma^\mu q = -\frac{\sigma + i\vec{\tau}\cdot\vec{\pi}\gamma_5}{\sqrt{\sigma^2 + \pi^2}}q \ , \tag{8.26}$$

and the σ and $\vec{\pi}$ fields transform in the usual way[37] under chiral rotations. The surface term in (8.25) provides a coupling between the external pion and sigma fields and the internal quark fields. In the approximation $\sigma \approx f_\pi$ we recover (8.16) for the pion field outside the bag. In addition, of course, the external pion field acts as a source for quarks inside the bag.

---

*There is a way out of this dilemma, however. The pion field can be defined outside the bag in such a way that it represents the escaping component of an initially compact pion wave packet. This interpretation requires some modification of the approach described above.

Let us consider the effect of a classical pion field upon the static nucleon bag. It is easy to estimate its effect upon the bag energy. Since the field couples with a strength $1/f_\pi$ [see Eq. (8.16)] and its energy is quadratic in the field, if we ignore the mass of the field,

$$E_\pi \propto -1/f_\pi^2 R^3 \tag{8.27}$$

by simple dimensional reasoning. It contributes negatively, since if we consider doing perturbation theory in the πNN coupling constant, this term is a second order energy shift.

If we now minimize the energy of the bag plus field with respect to R, we find that the nucleon bag collapses because of the strong negative contribution of E_π. The classical theory must break down before this happens, however. In particular, there are $\pi\pi$ interactions that need to be considered.[15] These may be incorporated into the Lagrangian (8.25) according to one's fancy. In addition, the strong pion field must excite quark-antiquark pairs inside the cavity, which would tend to increase the internal pressure and prevent a complete collapse. Obviously the situation is complicated and needs further theoretical work. It is premature to conclude that the original static bag picture needs drastic revision in the presence of pion interactions.[15] However, it is clear that we are beginning to develop a framework for soft-pion physics[37] in the baryon sector.

REFERENCES AND FOOTNOTES

1. K. Johnson, Acta Phys. Polonica B6, 865 (1975).

2. K. Johnson, Proc. of the Seventeenth Scottish Universities Summer School in Physics, St. Andrews (1976), ed. by I.M. Barbour and A.T. Davis.

3. P. Hasenfratz and J. Kuti, Phys. Reports 40C, 75 (1978).

4. A.T.M. Aerts. The MIT Bagmodel and Some Spectroscopic Applications (Krips Repro, Meppel, The Netherlands, 1979) (Ph.D. thesis, Nijmegen).

5. A. Chodos, R.L. Jaffe, K. Johnson, C.B. Thorn, and V.F. Weisskopf, Phys. Rev. D9, 3471 (1974).

6. A. Chodos, R.L. Jaffe, K. Johnson, and C.B. Thorn, Phys. Rev. D10, 2599 (1974). T.A. DeGrand, R.L. Jaffe, K. Johnson, and J. Kiskis, Phys. Rev. D12, 2060 (1975) referred to as DJJK.

7. A. Chodos and Charles B. Thorn, Phys. Rev. D12, 2733 (1975).

8. J.J.J. Kokkedee. The Quark Model (Benjamin, New York, 1969).

9. C. Bender and P. Hays, Phys. Rev. D14, 2622 (1976).

10. K. Johnson, Coral Gables talk, MIT-CTP 766 (1979).

11. C. Rebbi, Phys. Rev. D12, 2407 (1975); Phys. Rev. D14, 2362 (1976). T.A. DeGrand and C. Rebbi, Phys. Rev. D17, 2358 (1978).

12. T.A. DeGrand and R.L. Jaffe, Ann. Phys. (NY) 100, 425 (1976). T.A. DeGrand, Ann. Phys. (N.Y.) 101, 496 (1976).

13. K. Johnson, Phys. Letters 78B, 259 (1978).

14. C. Callen, R. Dashen, and D. Gross, Phys. Rev. D19, 1826 (1979).

15. G.E. Brown and Mannque Rho, Phys. Letters 82B, 177 (1979); G.E. Brown, Mannque Rho, and Vincent Vento, SUNY Stony Brook report (1979).

16. T.D. Lee, Phys. Rev. D19, 1802 (1979); R. Friedberg and T.D. Lee, Phys. Rev. D18, 2623 (1978). See also J. Kogut and L. Susskind, Phys. Rev. D9, 3501 (1974), M. Creutz and K.S. Soh, Phys. Rev. D12, 443 (1975), and K. Huang and Daniel R. Stump, Phys. Rev. D14, 223 (1976).

17. S. Mandelstam, Physics Rep. 23C, 245 (1976).

18. Daniel Lu, private communication (1979).

19. C. DeTar, Phys. Rev. D17, 302, 323 (1978); 19, 1028(E) (1979).

20. C. DeTar, Phys. Rev. D19, 1451 (1979).

21. K. Johnson and C.B. Thorn, Phys. Rev. D13, 1934 (1976); K. Johnson and C. Nohl, Phys. Rev. D19, 291 (1979).

22. C. Rebbi, Phys. Rep. 12C, 1 (1974).

23. E. Eichten, K. Gottfried, K. Lane, T. Kinoshita, and T.-M Yan, Phys. Rev. D17, 3090 (1978).

24. A. Chodos and C.B. Thorn, Nucl. Phys. B72, 509 (1974). See also K. Kikkawa, Phys. Rev. D18, 2606 (1978); K. Kikkawa, Tsuneyuki Kotani, Masa-aki Sato, and Masakatsu Kenmoku, Phys. Rev. D19, 1011 (1979).

25. R.L. Jaffe, Phys. Rev. D15, 267, 281 (1977).

26. R.L. Jaffe, Phys. Rev. D17, 1444 (1978). Chan-Hong Mo et al., Phys. Lett. 76B, 634 (1978).

27. P.J.G. Mulders, A. Th. M. Aerts, and J.J. DeSwart, Phys. Rev. Letters 40, 1543 (1978).

28. M. Baranger and M. Veneroni, Ann. Physics (N.Y.) 114, 123 (1978).

29. R.L. Jaffe and F.E. Low, MIT-CTP 747 (1979).

30. E.P. Wigner, and L. Eisenbud, Phys. Rev. 72, 29 (1947).

31. G.T. Fairley and E.J. Squires, Nucl. Phys. B93, 56 (1975).

32. F.E. Low, Phys. Rev. D12, 163 (1975).

33. R.S. Willey, Phys. Rev. D18, 270 (1978). See also P.M. Fishbane and M.T. Grisaru, Phys. Rev. Letters 40, 931 (1978).

34. M.M. Nagels, T.A. Rijken and J.J. DeSwart, Phys. Rev. D17, 768 (1978).

35. For another approach, see C.E. Carlson, F. Myhrer, and G.E. Brown, SUNY Stony Brook report (1979).

36. See, for example, G.F. Chew, The Analytic S-Matrix (Benjamin, New York, 1966).

37. S.L. Adler and R.F. Dashen. Current Algebras (Benjamin, New York, 1968). H. Pagels, Phys. Reports 16C, 219 (1975).

38. S.B. Treiman, R. Jackiw, and D.J. Gross. Lectures on Current Algebra and Its Applications (Princeton University Press, Princeton, 1972), p. 101 ff.

39. J. Goldstone, A. Salam, and S. Weinberg, Phys. Rev. 127, 965 (1962).

40. Y. Nambu and G. Jona-Lasinio, Phys. Rev. 122, 345 (1961).

41. M. Gell-Mann and M. Levy, Nuovo Cim. 16, 705 (1960).

42. John F. Donoghue and K. Johnson, MIT-CTP 802 (1979).

43. S. Weinberg, Harvard preprint HUTP-77/A057; J. Gunion et al., Nucl. Phys. B23, 445 (1977).

44. Donoghue and Johnson (Ref. 42) give a slightly different estimate of the c.m. correction, namely $\langle p^2 \rangle / m^2 \approx 10/(mR)^2$, and find $g_A = 1.27$. They also calculate an improved magnetic moment for the proton: $2m_p\mu_p = 2.5$ and charge radius $\langle r^2 \rangle = (0.82)^2 fm^2$. Experimental values for the last two are 2.79 and $(0.88)^2 fm^2$.

SEMI-CLASSICAL METHODS IN QCD AND HADRONIC STRUCTURE

David J. Gross

Princeton University

Princeton, New Jersey

These lectures dealt with semi-classical methods in quantum field theory and their application to Quantum Chromodynamics. Most of the material presented has appeared already in published form, including two reviews (see References 1.1g and 1.3b). In lieu of lecture notes I present below a list of references to the relevant literature as well as a collection of problems that may serve as a pedagogical guide to semi-classical treatment of QCD.

1. LIST OF REFERENCES

The following is a list of references to the literature on semi-classical methods and their application to Quantum Chromodynamics. This list is not meant to be complete, but rather to serve as a guide to the relevant literature. Further references can be found in the works cited below. (CDG = C. Callan, R. Dashen and D. Gross.)

1.1 Instantons and Vacuum Tunneling

a. A. Polyakov, Phys. Lett. 59B, 82 (1975) ((2+1) dimensional model of confinement due to instantons).

b. A. Belavin, A. Polyakov, A. Schwartz and Y. Tyupkin, Phys. Lett. 59B (1975) (Discovery of instanton solution in Yang Mills theory).

c. G. 't Hooft, Phys. Rev. D14, 3432 (Calculation of instanton determinent, U(1) problem).

d. CDG, Phys. Lett. 63B, 334 (1976) (θ vacua, mechanism for chiral symmetry breaking).

e. R. Jackiw and C. Rebbi, Phys. Rev. Lett. 37, 172 (1976) (θ vacua).

f. Y. Basilov and S. Podrovsky, Nucl. Phys. B143, 431 (1978); C. Barnard, Phys. Rev. D19, 3013 (1979). (Calculation of instanton determinent of SU(N).)

g. S. Coleman, "The Uses of Instantons," Erice lectures 1977 (Review of instanton physics up to 1977).

h. L. Yaffe and H. Levine, "Higher Order Instanton Effects," Phys. Rev. D19, 1225 (1978).

i. H. Levine, "Regularization and Renormalization of Semiclassical QCD," Nucl. Phys. B157, 237 (1979).

j. L.S. Brown, R. Carlitz, D. Creamer and C. Lee, Phys. Rev. D17, 1583 (Calculation of scalar and fermion propagators in background instanton field).

k. CDG, Phys. Rev. D16, 2526 (1977) "Pseudo Particles and Massless Fermions in Two Dimensions".

l. R. Jackiw, K. Nohl and C. Rebbi, Phys. Rev. D15, 1642 (1977) (Discussion of 't Hooft ansatz multi-instanton solution.)

m. M. Atiyah, V. Drinfeld, N. Hitchin and Y. Manin, Phys. Lett. 65A, 185 (1978); E. Corrigan, D. Fairlee, P. Goddard and S. Templeton, Nucl. Phys. B140, 31 (1978), B151, 93 (1979); N. Christ, E. Weinberg and N.K. Stanton, Phys. Rev. D18, 2013 (1978) (Construction and properties of the general multi-instanton solution).

1.2 The Effects of Instantons

a. CDG, "Towards a Theory of the Strong Interactions," Phys. Rev. D17, 2717 (1978) (Systematic study of vacuum tunneling, instanton effects, chiral symmetry breaking, instanton interactions, coupling constant renormalization and merons).

b. F. Wilczek and A. Zee, Phys. Rev. Lett. 40, 83 (1978) "Instanton and Spin Forces Between Massive Quarks"; C, D, G, Wilczek and Zee, Phys. Rev. D18, 4684 (1978) "The Effect of Instantons on the Heavy Quark Potential".

c. N. Andrei and D. Gross, Phys. Rev. D18, 468 (1978) "The effect of instantons on the short distance structure of hadronic currents".

d. D. Caldi, Phys. Rev. Lett. 39, 121 (1977); R. Carlitz, Phys. Rev. D17, 3225 (1978) (Chiral symmetry breaking due to instantons).

1.3 The QCD Bag

a. CDG, Phys. Rev. D19, 1826 (1979) "A Theory of Hadronic Structure".

b. CDG, "Semiclassical Methods in QCD: Toward a Theory of Hadron Structure," La Jolla Summer School 1978.

c. A. Jevicki, "Collective Behavior of Instantons in QCD," I.A.S. preprint, July 1979.

d. P. Hasenfratz and J. Kuti, "The Quark Bag Model," Phys. Reports 40, 75 (1978) (A review of the MIT bag model).

1.4 Lattice Theories

a. K. Wilson, Phys. Rev. D10, 2445 (1974) (Lagrangian lattice gauge theory).

b. J. Kogut and L. Susskind, Phys. Rev. D11, 395 (1975) (Hamiltonian lattice gauge theory).

c. L. Kadanoff, Rev. Mod. Phys. 49, 267 (1977); J. Kogut, Rev. Mod. Phys. 51, 659 (1979) (Reviews of lattice gauge theory).

d. CDG, "Instantons as a bridge between weak and strong coupling in QCD," Princeton preprint, July 1979.

e. J. Kogut, R. Pearson and J. Shigemitsu, "QCD β Function at Intermediate and Strong Coupling," Phys. Rev. Lett. 43, 4841 (1979).

2. PROBLEMS

These problems are intended to serve as a guide to the semi-classical treatment of QCD. Most are treated somewhere in the literature. Starred problems are difficult. Doubly starred problems are unsolved. If you can solve them — publish!

* a. The functional integral

$$\int_{x(-\frac{T}{2})=0}^{x(\frac{T}{2})=0} [x(t)] \exp\left\{-\frac{1}{2}\int_{-T/2}^{+T/2} dt\,[\dot{x}^2(t) + U(t)x^2(t)]\right\}$$

is proportional to $\det\{-\partial_t^2 + U(t)\} \equiv D(U)$. Prove that

$$\frac{D(U_1)}{D(U_2)} = \frac{\Psi^{(1)}(T/2)}{\Psi^{(2)}(T/2)}, \text{ where } \Psi(t) \text{ satisfies}$$

$$\Psi(-T/2) = 0,\ \partial_t\Psi(-T/2) = 1,\ (-\partial_t^2 + U(t))\Psi(t) = 0.$$

b. Using the above calculate the energy levels of the simple harmonic oscillator.

* c. Calculate the splitting of the parity even and odd ground states of the double-well potential: $L = \frac{1}{2}\dot{x}^2 - V(x)$ $V(x) = V(-x) \geq 0$, $V(a) = V(-a) = 0$, by (i) Euclidean path integral methods, (ii) ordinary Schrödinger WKB methods.

d. Prove that there are no stable minima of the Euclidean action for a scalar field theory, $L(\phi) = \frac{1}{2}(\partial_\mu\phi)^2 - U(\phi)$, except $\phi = \text{const}$, for $d > 1$. (d = number of space-time dimensions.)

e. Assume the 't Hooft ansatz for the gauge field: $A_\mu^a = \underset{\sim}{\eta}_{a\mu\nu}\partial_\nu \ln\rho$. For SU-2 show that the self dual equation, $F = \tilde{F}$, reduces to $\Box\rho/\rho = 0$. Show that $S = c\Box\Box\ln\rho$ and determine the value of c.

f. Show that an instanton, for the gauge group SU(N), has 4N degrees of freedom.

* g. Calculate the instanton determinant for the case of SU(2).

h. Calculate the free energy, F, of the one dimensional Ising model:

$$H = J\sum_{i=1}^{L}\sigma_i\sigma_{i+1},\quad Z = \exp(-LF) = \frac{1}{2^L}\sum_{(\sigma_i=\pm 1)}\exp[-\beta H],$$

(i) By the dilute gas approximation. (ii) exactly.
(iii) Explain the reasons for the difference.

i. Prove that $G_\mu \equiv L\mathrm{Tr}\,\{\varepsilon_{\mu\nu\lambda\sigma}A_\nu(\partial_\lambda A_\sigma + 2/3A_\lambda A_\sigma)\}$ where $A_\mu = iA_\mu^a(\tau^a/2)$ (for SU_2), satisfies: $\partial^\mu G_\mu = \mathrm{Tr}\, F_{\mu\nu}\tilde{F}^{\mu\nu}$.

**j. Develop a completely systematic expansion for the ground state energy of the double well potential to all orders in g and e^{-1/g^2}.

* k. Define the fermionic measure $\prod_x d\Psi d\bar{\Psi}(x)$ as $\prod_{n,m} d\,\bar{b}_m\, da_n = d\mu$ where a_n, $\bar{b}_n$ are elements of a Grassman algebra, and $\Psi = \Sigma\, a_n\phi_n(x)$, $\bar{\Psi} = \Sigma\, \phi_n^+(x)\bar{b}_n$; where ϕ_n are a complete set of eigenfunctions of $\not{D} = \gamma_\mu(\partial^\mu - A^\mu)$: $\not{D}\,\phi_n = \lambda_n\phi_n$ $(\phi_n,\phi_m) = \delta_{nm}$. Prove that under an axial baryon number transformation: $\Psi(x) \to e^{i\alpha\gamma_5}\Psi(x)$, that $d\mu \to d\mu\, \exp\left(\frac{2i}{8\pi^2}\int \mathrm{Tr}\, F\tilde{F}\, d^4x\right)$.

* l. Show that the number of right handed (left handed) solutions of the Euclidean Dirac equation $\not{D}(A)\psi = 0$ in a background field A_μ satisfies

$$n_R - n_L = \frac{1}{8\pi^2}\int \mathrm{Tr}\, F\tilde{F}(A)d^4x .$$

* m. Construct the zero-mode solution, ψ_0, i.e. $D(A)\psi_0 = 0$, in the presence of a single instanton.

n. Show that the "interaction action" of an instanton-anti instanton separated by distance R in the presence of N_f massless fermions is $S_{int} \underset{R\to\infty}{=} 6N_f \ln R$.

o. Consider a theory of two massless quarks, u and d, with the following 4-fermion interaction

$$L_{int} = K\int \prod \frac{d^4\rho_i}{(2\pi)^4}\, \bar{\Psi}_u(p_1)\Psi_u(p_2)\bar{\Psi}_d(p_3)\Psi_d(p_4)\prod_i e^{-|p_i|\rho}\,\delta^{(4)}(\Sigma p_i) .$$

Show that the theory is invariant under γ_5 transformations. Set up a self consistent equation (for small K) for the quark propagator. Determine the minimal value of K for which a massive solution exists.

p. Derive the value of $\varepsilon(R)$, the interaction energy of a quark-antiquark pair separated by distance R, in the "dilute gas approximation":

$$\varepsilon(R) = 2\int \frac{d\rho}{\rho^2}D(\rho)W(R/\rho),\quad W(R/\rho) = -\frac{1}{3\rho^3}\int d^3x_I\, \mathrm{tr}[U(\bar{R}-\bar{x}_I)U^+(\bar{x}_I)-I]$$

where $U(\bar{x}) = \exp\,[2\pi i\, x_a\, R_{a\alpha}\,\lambda^\alpha/\sqrt{x^2+\rho^2}]$. Evaluate $\varepsilon(R)$

for small R. Evaluate $\varepsilon(R)$ for large R and show that it yields a quark mass renormalization of $\delta m = 18.5 \int (d\rho/\rho^2)D(\rho)$ and a coupling constant renormalization ($g^2 \to g^2\mu$) of

$$\mu = 1 + \frac{\pi^2}{2} \int \frac{d\rho}{\rho} D(\rho) \left(\frac{8\pi^2}{g^2}\right) .$$

q. Calculate the gluon propagator in the diluate gas approximation. Evaluate it for small momenta and derive the same expression for μ as in problem p.

r. Show, by using the result of problem p, that the density of instantons at $\vec{x}$ in the presence of heavy quarks at $\vec{x}_1$ and $\vec{x}_2$ is reduced by an amount approximately given by $\exp\left[\frac{2\pi^4\rho^2}{g^2} \vec{E}_x^2\right]$, where $\vec{E}_x = \vec{E}_1 + \vec{E}_2$ and $\vec{E}_i$ is the electric field produced by the ith quark at $\vec{x}$, i.e. $\vec{E}_i = g\sqrt{4/3}\, \frac{\vec{x}-\vec{x}_i}{|x-x_i|^3}$.

**s. The fact that matter fields screen color can be understood as being due to the fact that matter loops behave as four dimensional dipoles with an imaginary dipole moment. Give an analogous explanation for the fact that gluon Gaussian fluctuations anti-screen color (and thus produce asymptotic freedom).

t. Instantons in the Weinberg-Salam Model: (i) Consider the following (SU-2) Lagrangian: $L = L_{\text{Yang-Mills}} + |D_\mu\phi|^2 + V(\phi)$, where ϕ is an isotriplet and V is minimal at $|\phi| = F$. Show that instantons of scale size ρ (fixed by a constraint) exist, and for small F their action is $S = (8\pi^2/g^2)+4\pi^2F^2\rho^2$. (ii) Determine the value of ρ for which the instanton density peaks. (iii) Consider now the Weinberg-Salam model and estimate the maximum density of instantons.

**u. Derive the bag from the dense ($f = 1$) instanton vacuum by evaluating the (spatially dependent) instanton density in the presence of a Wilson loop.

v. Surface Effects: Consider a (large) bag (or cylinder in 4 dimensions) of diluate phase in equilibrium with the dense (take $\mu = \infty$ outside) phase. An instanton placed inside the bag doesn't satisfy the boundary condition ($n \cdot F = 0$) on the surface. Show that this can be corrected by superimposing the instanton with an "image anti-instanton" on the other side of the surface. Calculate the reduction in action of the instanton as a function of its distance from the surface and use this to estimate the surface thickness.

* w. Derive a phenomenological chirally symmetric bag model, by adding to the MIT bag model Lagrangian σ & $\vec{\pi}$ fields, coupled to the quarks in a chirally symmetric fashion, and adjusting the parameters of a potential, $V(\sigma,\vec{\pi})$, so that in the bag they decouple (are infinitely massive) and outside $m_\sigma = \infty$, $m_{quark} = \infty$ and $f_\pi < \infty$.

SUPERGRAVITY

Julius Wess

University of Karlsruhe

Karlsruhe, German Federal Republic

1. INTRODUCTION

In the same way that a relativistic field theory is a field theory based on the Poincaré algebra, a supersymmetric theory is a field theory based on the supersymmetry algebra

$$\{Q_\alpha, \bar{Q}_\beta\}_+ = 2i\gamma^m_{\alpha\beta} P_m$$
$$[P_m, Q_\alpha]_- = 0 \ , \quad [P_m, P_n]_- = 0 \ . \tag{1}$$

P_m is the energy momentum four vector, Q_α is a Majorana spinor.

The surprising thing is that such theories exist and that they meet all requirements of a renormalizable quantum field theory.[1] In such theories, the spinor charges Q_α can be obtained from local currents,

$$Q_\alpha = \int d^3x \, J_{\alpha 0} \ . \tag{2}$$

The currents satisfy conservation laws,

$$\partial^m J_{\alpha m} = 0 \ , \tag{3}$$

and are local functions of field operators. The algebra (1) is obtained due to the canonical commutator relations of the field operators.

Field theories which possess such conservation laws are less divergent than one would expect from power counting without taking into account the relations which follow from the conservation law of $J_{\alpha m}$. Due to these relations divergences of bosonic and fermionic loops cancel, a cancellation which cannot take place in the usual gauge theories.

There is a second reason why supersymmetric gauge theories are of great interest. Due to 'no-go' theorems,[2] the symmetry group of a reasonable field theory should be the direct product of the Poincaré group and a compact internal symmetry group. The gauge theories which are candidates for unification schemes are theories in which the internal symmetry group is gauged. These theories belong to the class of renormalizable quantum field theories. Gauging the Poincaré group leads to Einstein's theory of gravitation, a theory which is invariant under general coordinate transformation but which cannot be interpreted as a quantum field theory because it is too divergent. The gravitational interaction alone, without any matter fields, is one-loop renormalizable on mass-shell, but all couplings of the gravitational field to matter fields of spin 0, 1/2 or 1 are not even one-loop renormalizable.

Supersymmetry has the remarkable property that the Poincaré algebra is tied into a larger algebra — the algebra (1) — in a non-trivial way. The 'no-go' theories are circumvented by the use of anti-commutators in the defining relations. Gauging of supersymmetry will lead to a theory invariant under general coordinate transformations. The gravitational field will couple to matter fields in a well-defined way and the spin 3/2 currents $J^A_{\alpha m}$ will serve as sources of spin 3/2 fields in the same way as the energy-momentum tensor serves as a source of the graviton.[3]

The gauge theory of supersymmetry is called supergravity. We have good reasons to expect that supergravity will be less divergent than gravity. We therefore have some hope that it will be possible to combine gravity with some internal gauge group structure. At present, this might seem to be an academic problem but let me remind you that the energy which is characteristic of the grant unification scheme is $10^{16} - 10^{17}$ GeV, compared to Planck's energy of 10^{19} GeV.

It has been shown that supergravity,[4] where the graviton couples to a spin 3/2 field in a consistent way, is at least two-loop renormalizable on mass shell. It is just the presence of the spin 3/2 field that improves the situation in comparison to pure gravity. However, this is true only for the self-coupling of the minimal supersymmetry multiplet. Any coupling to another multiplet ruins the renormalization properties, even if done in an invariant way.

It has been shown also that supersymmetry is the only way to combine the Poincaré group with an internal symmetry group without violating some basic postulates of quantum field theory.[5] Supersymmetry and extended supersymmetry ($N > 1$) are therefore the only possible models in which a unification of gravitation with other interactions can be studied in a serious way, i.e. in a way which is not from the very beginning in conflict with the axioms of quantum field theory.

Let us study the on-mass shell representation of the algebra (1) for $P^2 = 0$. We use helicity states and find that the helicity ranges from J to $J + N/2$, the binomial coefficients giving the multiplicity of the states. For a field-theoretical description, we also have to add the states with opposite helicity. If we do not want to have a state with spin larger than two we are restricted to $N = 4$ if we start from $J = 0$. It is however possible to start from $J = -2$ and to arrange the theory in such a way that this representation already contains the states with opposite helicity. In this case we can go up to $N = 8$. This is certainly the most interesting model in which to study a unification scheme. In terms of physical fields it contains a field with spin 2 (the graviton), 8 spin 3/2 fields, 28 vector fields, 56 spin 1/2 Majorana fields and 70 fields with spin zero.

Nevertheless the simpler $N = 2$ model is very interesting. It contains one spin 2, 2 spin 3/2 and one vector field. This is a theory in which the graviton and the Maxwell field interact in a fully covariant way. Its field-theoretical properties are remarkable — divergences cancel in the one-loop and two-loop approximation. The spin 3/2 fields seem to be capable of compensating the short-range singularities of a gravitational theory.

2. METHODS OF DIFFERENTIAL GEOMETRY IN GAUGE THEORIES AND GRAVITATIONAL THEORY

2.1 Geometrical Concepts

Let me briefly summarize the essential geometrical concepts[6] which allow a formulation of gauge theories, of Einstein's gravitational theory and of supergauge theories and supergravity from a common point of view. I will not be very precise with the definitions, since you can find them and many more in mathematical textbooks.

a) Manifolds: they allow us to introduce the concept of differentiability.

b) Tensor fields: a linear space which admits a local basis $\sigma^i(x)$ and which serves as a representation space of a Lie group,

$$\begin{aligned} \sigma'^i(x) &= \sigma^\ell(x) G_\ell^i(x) \quad , \\ G(x) &= e^{i\Lambda(x)} \quad , \qquad \Lambda(x) = \sum_r \lambda_r(x) T^r \quad . \end{aligned} \tag{4}$$

T^r are the generators of the group. The tensor product

$$\sigma^{i_1}(x) \times \sigma^{i_2}(x) \ldots \times \sigma^{i_r}(x)$$

should be defined.

A special tensor field is the tangent space of a manifold and its dual, the cotangent space. We denote the basis of the tangent space by $e_a(x)$ and the basis of the cotangent space by $e^a(x)$. We can use the tensor product,

$$e_{a_1}(x) \times \ldots \times e_{a_r}(x) \times e^{b_1}(x) \ldots \times e^{b_s}(x) \quad ,$$

as a basis for the r-fold contravariant and s-fold covariant tensor space.

c) Forms: a p-form is an antisymmetric, p-fold covariant tensor field. If we express the basic relations of a theory in terms of differential forms we automatically guarantee that these relations are independent of the particular coordinates we choose in our manifold. The antisymmetrization of the tensor product is easily taken into account if we introduce the exterior product:

$$e^{a_1} \wedge e^{a_2} = -e^{a_2} \wedge e^{a_1} \quad . \tag{5}$$

We write a p-form as follows,

$$\omega = e^{a_1} \wedge e^{a_2} \ldots \wedge e^{a_p} \omega_{a_1 \ldots a_p}(x) \quad .$$

Differentiation is introduced through the exterior derivative. This is a mapping of a p-form into a p+1 form with the defining properties:

$$d\omega = \phi, \quad \omega \text{ is a p form, } \phi \text{ is a p+1 form;}$$

$$d(\omega_1 + \omega_2) = d\omega_1 + d\omega_2 \quad ;$$

$$d(\omega_1 \wedge \omega_2) = \omega_1 \wedge d\omega_2 + (-)^{p_2} d\omega_1 \wedge \omega_2 \quad ; \tag{6}$$

$$d\,d = 0 \quad .$$

Exterior derivatives of local coordinate lines, dx^m, form a natural basis in the cotangent space. This basis can be related to the one above via the vier-(viel-) bein field $e_m^a(x)$,

$$e^a = dx^m e_m^a \quad . \tag{7}$$

The vielbein field is assumed to have an inverse,

$$e_m^a \tilde{e}_a^n = \delta_m^n \quad . \tag{8}$$

d) Connection: the exterior derivative maps a p-form into a p+1 form but it does not map a tensor field into a tensor field. If $\sigma'(x)$ transforms like a tensor field $\sigma' = \sigma G$, we find for its exterior derivative an inhomogeneous transformation law,

$$d\sigma' = \sigma dG + d\sigma G \quad . \tag{9}$$

We want to define a derivative which maps a p-form into a p+1 form and a tensor field into a tensor field. This can be done with the help of the connection form which transforms as follows,

$$\phi' = G^{-1}\phi G + G^{-1} dG \quad . \tag{10}$$

The connection form can be chosen to be Lie-algebra valued,

$$\phi = dx^m \phi_{mr} T^r \quad . \tag{11}$$

The transformation law will not change this property. The covariant derivative is defined as

$$\mathcal{D}\sigma \equiv d\sigma - \sigma\phi = dx^m \mathcal{D}_m \sigma \quad , \tag{12}$$

and it transforms like a tensor field,

$$(\mathcal{D}\sigma)' = (\mathcal{D}\sigma)G \quad .$$

e) Curvature: it is possible to construct a tensor field in terms of the connection and its derivatives,

$$F = d\phi - \phi \cdot \phi = \frac{1}{2}\, dx^m \wedge dx^n\, F_{nm} \quad . \tag{13}$$

F is a 2-form and it is Lie-algebra valued if ϕ is. F is a tensor field

$$F' = G^{-1} F G \quad . \tag{14}$$

The covariant derivatives of tensor fields and the curvature tensor F are in general all the independent tensorial quantities which can be used to formulate covariant differential equations (field equations). If we try to construct new tensorial quantities through the repeated use of the exterior derivative we obtain identities instead. This is due to the fact that $d \cdot d = 0$. These identities are called Bianchi identities.

f) Bianchi identities: from (12) follows

$$d\mathcal{D}\sigma = (d\sigma)\phi - \sigma(d\phi) = \mathcal{D}\sigma\phi - \sigma F \quad , \tag{15}$$

and from (13) follows

$$dF = \phi F + F\phi \quad . \tag{16}$$

Within the scheme developed so far we have constructed a set of tensor quantities in terms of which it is possible to describe the dynamics of the corresponding physical system in a covariant way. The usual procedure is to postulate a Lagrangian and to derive the equations of motion from it. This Lagrangian should satisfy a certain set of postulates. It should be invariant under the structure group. We have developed the tensor calculus essentially with the aim of listing all the possible invariants.

Furthermore, the differential equations derived from the Lagrangian should be not higher than second order; they should be local and should give rise to a causal propagation. Ghosts — negative contributions to the energy — should not appear as physical degrees of freedom; they should rather be gauge degrees of freedom and they should not contribute to any physical process. These requirements seem to make the Lagrangian unique. Let me now explain these concepts within the respective theories.

2.2 Gauge Theories

The manifold is R^4, the tensor fields are 0-forms, i.e. fields that transform under a compact Lie group like

$$\sigma'^{i}(x) = \sigma^{j}(x) G^{i}_{j}(x) \quad . \tag{17}$$

The connection is Lie-algebra valued,

$$\phi = dx^m A_m(x) = dx^m A_{mr}(x) T^r \quad . \tag{18}$$

The component fields $A_{mr}(x)$ are called Yang-Mills potentials. The curvature F_{mn} is called the Lie-Yang-Mills field,

$$F_{nm} = \partial_n A_m - \partial_m A_n + [A_n, A_m] \quad . \tag{19}$$

The covariant derivative is the covariant Yang Mills derivative,

$$\mathcal{D}_m \sigma^i = \partial_m \sigma^i - \sigma^j T_j^{ri} A_{mr}(x) \quad , \tag{20}$$

and the Bianchi identity (15) expresses the fact that covariant derivatives do not commute — this commutator yields the Yang Mills field,

$$(\mathcal{D}_m \mathcal{D}_n - \mathcal{D}_n \mathcal{D}_m)\sigma = -\sigma F_{nm} \quad . \tag{21}$$

The Bianchi identity (16) becomes the well-known cyclic condition,

$$\mathcal{D}_\ell F_{mn} + \mathcal{D}_m F_{n\ell} + \mathcal{D}_n F_{\ell m} = 0 \quad .$$

The Lagrangian which meets all the requirements is

$$L = -\frac{1}{4} \mathrm{Tr}\,(F_{mn} F^{mn}) \quad . \tag{22}$$

2.3 Gravitational Theory

The manifold is locally isomorphic to R^4. The basic tensor fields are the cotangent vector fields. They are 1-forms,

$$e^a = dx^m e_m^a(x) \quad . \tag{23}$$

The structure group is the Lorentz group,

$$e'^a = e^b L_b^a \quad . \tag{24}$$

The transformation acts on tangent space indices only. We will reserve the first half of the alphabet for tangent space indices. They can be raised and lowered by the Lorentz metric η^{ab}.

Space indices, which transform under general coordinate transformations, will be denoted by letters from the second half of the alphabet. Space indices and tangent space indices are related via the vielbein.

The connection is Lie-algebra valued; this means

$$\phi_{m\,ab} = -\phi_{m\,ba} \quad . \tag{25}$$

The covariant derivative of the cotangent vector field is a 2-form; it is called the torsion form,

$$de^a - e^b \phi_b^a = T^a = \frac{1}{2} dx^m \wedge dx^n T^a_{nm} \quad . \tag{26}$$

The curvature tensor is the Riemannian tensor,

$$F_a^b = \frac{1}{2} dx^m dx^n R^b_{nma} \quad . \tag{27}$$

F is a 2-form and is Lie-algebra valued. This implies for R

$$R_{mn\,ab} = -R_{nm\,ab} = -R_{mn\,ba} \quad . \tag{28}$$

The Bianchi identities yield the following relations between the curvature tensor and the torsion,

$$\sum_{\text{cycl}} \{\mathcal{D}_e T^a_{dc} - R^a_{edc} + T^f_{ed} T^a_{fc}\} = 0 \quad , \tag{29}$$

$$\sum_{\text{cycl}} \{\mathcal{D}_e R^b_{dca} + T^f_{ed} R^b_{fca}\} = 0 \quad . \tag{30}$$

Einstein's theory of gravitation can now be formulated as follows.

The metric tensor is expressed in terms of the vielbein,

$$g_{mn} = e_m^a e_{na} \quad . \tag{31}$$

The torsion is set equal to zero. This is a constraint equation and allows us to express the connection in terms of the vielbein field and its derivatives,

$$T = 0 \quad ,$$

$$\begin{aligned}\phi_{mn,\ell} = +\tfrac{1}{2}\{ & e_\ell{}^a(\partial_m e_{na} - \partial_n e_{ma}) \\ & + e_n^a(\partial_\ell e_{ma} - \partial_m e_{\ell a}) \\ & + e_m^a(\partial_\ell e_{na} - \partial_n e_{\ell a})\} \quad .\end{aligned} \tag{32}$$

The vielbein is the only dynamically independent field left. Einstein's Lagrangian is

$$L = \frac{1}{2K} R_{mn\,ab}\, e^{ma} e^{nb} \det e \quad . \tag{33}$$

The Bianchi identities reduce to the well-known set of equations

$$\sum_{cycl} R^a_{edc} = 0 \quad , \tag{34}$$

and

$$\sum_{cycl} \mathcal{D}_e R^b_{dca} = 0 \quad . \tag{35}$$

2.4 Supersymmetric Yang-Mills Theories

The underlying manifold is the superspace. We denote its elements by

$$z^M \approx (x^m, \theta^\mu_M\, \bar{\theta}^L_{\dot\mu}) \quad . \tag{36}$$

The algebraic properties of z^M are

$$z^M z^N = (-)^{mn} z^N z^M \quad . \tag{37}$$

n is a function of the index N. It is zero if $N = n$ (vectorial) and it is one if $N = \mu$ or $\dot\mu$ (spinorial).

The tensor fields are superfields; these are functions of z^M and should be understood as power series in θ, $\bar\theta$.

The notion of forms can be generalized by defining the exterior product with an additional sign factor,

$$E^{A_1} \wedge E^{A_2} = -(-)^{a_1 a_2} E^{A_2} \wedge E^{A_1} \quad . \tag{38}$$

The definition of the exterior derivative is unchanged. Covariant derivatives and the curvature tensor are introduced in the same way. The Bianchi identities hold as well.

It is convenient to choose a basis in the cotangent space which is adjusted to supersymmetry transformations. In this basis the coordinate lines become curves which are generated by supersymmetry transformations,

$$e^A = dZ^M e_M^A \quad . \tag{39}$$

With the help of the inverse vielbein it is possible to define a suitable basis in the tangent space,

$$D_A = e_A^M \frac{\partial}{\partial Z^M} \quad . \tag{40}$$

These derivatives have the property that they commute or anticommute with the generators of supersymmetry.

The constraint equation and the Lagrangian are known for N = 1 and N = 2.

For N = 1 the constraint equations are

$$F_{\alpha\beta} = F_{\dot\alpha\dot\beta} = F_{\alpha\dot\beta} = 0 \quad . \tag{41}$$

The Bianchi identities can be solved subject to these constraint equations and they tell us that

$$F_a^\alpha = -\frac{1}{8} \bar\sigma_a^{\dot\beta\alpha} \bar W_{\dot\beta} \tag{42}$$

and that all the other curvature components can be expressed in terms of this W superfield which is chiral,

$$\mathcal{D}_{\dot\alpha} W_\beta = 0 \quad . \tag{43}$$

The Lagrangian is found to be

$$L \sim T_r(W^\alpha W_\alpha + \bar W_{\dot\alpha} \bar W^{\dot\alpha}) \quad . \tag{44}$$

In the case of supersymmetric theories, the Lagrangian has to be integrated over a superspace volume.

For N = 2 the constraints are different;[7]

$$F^{B}_{\dot{\alpha}A,\beta} = 0\ , \qquad F^{AB}_{\alpha\beta} + F^{AB}_{\beta\alpha} = 0 \ ,$$

$$F_{\dot{\alpha}A\ \dot{\beta}B} + F_{\dot{\beta}A\ \dot{\alpha}B} = 0 \ . \tag{45}$$

The Bianchi identities yield

$$F^{ij}_{\alpha\beta} = \varepsilon_{\alpha\beta} g^{ij}\, W \ ,$$

$$\mathcal{D}^{i}_{\alpha}\, W = 0 \ . \tag{46}$$

Again, W is the only remaining independent superfield. The Lagrangian is

$$L \sim \mathrm{Tr}(WW + \bar{W}\bar{W}) \ . \tag{47}$$

2.5 Supergravity

The manifold is locally superspace. The basic tensor fields are the cotangent vector fields,

$$E^{\mathcal{A}} = dZ^{\mathcal{M}} E^{\mathcal{A}}_{\mathcal{M}}(Z) \ . \tag{48}$$

The structure group is the direct product of the Lorentz group and an internal symmetry group. The transformation acts on the tangent space indices in a reducible form,

$$E'^{a} = E^{\mathcal{B}} L^{A}_{\mathcal{B}} \ . \tag{49}$$

The connection is Lie-algebra valued, $\phi_{\mathcal{M}A}{}^{\mathcal{B}}$,

$$\phi_{M\alpha\dot{\alpha}\beta\dot{\beta}} = -2\varepsilon_{\alpha\beta}\phi_{M\dot{\alpha}\dot{\beta}} + 2\varepsilon_{\dot{\alpha}\dot{\beta}}\phi_{M\alpha\underline{\beta}} \ ,$$

$$\phi^{A\beta}_{M\alpha B} = \delta^{A}_{B}\phi^{\beta}_{M\alpha} + \delta^{\beta}_{\alpha}\phi^{A}_{MB} \ ,$$

$$\phi^{\dot{\alpha}B}_{MA\dot{\beta}} = \delta^{B}_{A}\phi^{\dot{\alpha}}_{M\dot{\beta}} + \delta^{\dot{\alpha}}_{\dot{\beta}}\phi^{B}_{MA} \ , \tag{50}$$

$$\phi^{A}_{MB} = -\phi^{B}_{MA} \ : \quad \text{unitary group,}$$

$$\phi^{A}_{MA} = 0 \ : \quad \text{unimodular group.}$$

The curvature form

$$\frac{1}{2}\, dZ^{M} \wedge dZ^{N} R_{NMA}{}^{B}$$

is Lie-algebra valued as well.

For N = 1 the constraints are[8]

$$T^{\gamma}_{\underline{\alpha}\underline{\beta}} = 0 \quad ,$$

$$T^{c}_{\dot{\alpha}\beta} = T^{c}_{\beta\dot{\alpha}} + 2\, i\sigma^{c}_{\dot{\alpha}\beta} \quad ,$$

$$T^{c}_{\alpha\beta} = T^{c}_{\dot{\alpha}\dot{\beta}} = 0 \quad , \tag{51}$$

$$T^{c}_{\underline{\alpha}b} = T^{c}_{a\underline{\beta}} = 0 \quad ,$$

$$T^{c}_{ab} = 0 \quad .$$

From the Lagrangian

$$L = \mathrm{Det}\, E \quad , \tag{52}$$

the variation of which is subject to the constraint equations, we derive the field equations

$$T^{\beta}_{d\underline{\alpha}} = 0 \quad . \tag{53}$$

They satisfy the specified requirements.

For N = 2 we can put different restrictions on the structure group.[9]

For $\phi_{MA}{}^{B} = 0$, no internal group, the constraints are

$$T^{\dot{\delta}\dot{\beta}\alpha}_{DBA} = T^{DBA}_{\delta\beta\dot{\alpha}} = 0 \quad ,$$

$$T^{DBa}_{\delta\beta} = T^{\dot{\delta}\dot{\beta}a}_{DB} = 0 \quad ,$$

$$T^{D\dot{\beta}a}_{\delta B} = 2i\delta^{a\dot{\beta}}_{\delta}\ \delta^{D}_{B} \quad ,$$

$$T^{Da}_{\delta b} = T^{\dot{\delta}a}_{Db} = 0 \quad , \qquad T^{a}_{db} = 0 \quad , \tag{54}$$

$$T^{D\dot{\beta}B}_{\delta B\dot{\alpha}} = T^{\dot{\delta}B\alpha}_{D\beta B} = 0 \quad ,$$

$$\oint_{ABC} g^{CB} T^{D\dot{\beta}A}_{\delta B\dot{\alpha}} = \oint_{ABC} g_{CB} T^{\dot{\delta}B\alpha}_{D\beta A} = 0 \quad ,$$

$$\mathcal{D}^{\delta}_{D} T^{D\dot{\beta}A}_{\delta B\dot{\alpha}} = \mathcal{D}^{D}_{\dot{\delta}} T^{\dot{\delta}B\alpha}_{D\beta A} = 0 \quad .$$

The Lagrangian is not known; the field equations are

$$T^{D\dot{\beta}A}_{\delta D\dot{\alpha}} = T^{\dot{\delta}D\alpha}_{D\beta A} = 0 \quad . \tag{55}$$

For $\phi^{B}_{MA} = -\phi^{A}_{MB}$, $\phi^{A}_{MA} = 0$, SU(2) structure group, the constraints are the same as before, with in addition:

$$R^{DCA\alpha}_{\delta\gamma\alpha B} = -R^{DC\dot{\alpha}B}_{\delta\gamma A\dot{\alpha}} = 0 \quad ,$$

$$R^{\dot{\delta}\dot{\gamma}\dot{\alpha}B}_{DCA\dot{\alpha}} = -R^{\dot{\delta}\dot{\gamma}A\dot{\alpha}}_{DC\dot{\alpha}B} = 0 \quad , \tag{56}$$

$$T^{D\alpha}_{b\alpha A} = T^{\dot{\alpha}A}_{bD\dot{\alpha}} = 0 \quad .$$

The field equations are as before,

$$T^{D\dot{\beta}A}_{\delta D\dot{\alpha}} = T^{\dot{\delta}D\alpha}_{D\beta A} = 0 \quad . \tag{57}$$

3. FORMULAE FOR THE SUPERSPACE FORMULATION OF SUPERGRAVITY

In this lecture I would like to derive some of the formulae which are frequently used in the superspace formulation of supergravity.

Let me start from the definition of an element in superspace,

$$z^{M} \sim (x^{m}, \theta^{\mu}_{M}, \theta^{L}_{\dot{\mu}}) \quad . \tag{58}$$

Under a "general coordinate transformation" in superspace, this element will transform as

$$\delta z^{N} = G^{N}(z) \quad , \tag{59}$$

and due to the definition of the exterior derivative we find

$$\delta\, d\, Z^N = d\, Z^M\, \partial_M G^N \quad . \tag{60}$$

The vielbein transforms accordingly,

$$\delta E_N^{\ A} = -(\partial_N G^M) E_M^{\ A} - G^M \partial_M E_N^{\ A} \quad . \tag{61}$$

The vielbein transforms under the structure group as well,

$$\delta E_N^{\ A} = E_N^{\ B} L_B^{\ A} \quad . \tag{62}$$

The infinitesimal transformation $L_B^{\ A}$ is restricted by the requirement that the structure group should be the direct product of the Lorentz group and a compact internal symmetry group. This requirement was stated previously in Eq. (50).

It is useful to express the transformation law (61) also in terms of

$$G^A = G^M E_M^{\ A} \tag{63}$$

because it is possible to formulate supergravity in such a way that the gauged supersymmetry transformations are parametrised with field-independent G^A parameters. In this case $G^M = G^A \tilde{E}_A$ will be field-dependent.

We compute

$$\partial_N G^A = (\partial_N G^M) E_M^{\ A} + (-)^{nm} G^M \partial_N E_M^{\ A} \quad . \tag{64}$$

Therefore, we can write (61) as

$$\partial E_N^{\ A} = -\partial_N G^A - G^M (\partial_M E_N^{\ A} - (-)^{nm} \partial_N E_M^{\ A}) \quad . \tag{65}$$

From the definition of the torsion (26) it follows that we can rewrite this formula as

$$\delta E_N^{\ A} = -\mathcal{D}_N G^A + G^M T_{MN}^{\ \ A} - G^M \phi_{MN}^{\ \ A} \quad . \tag{66}$$

The last term in this equation is a structure group transformation. It can be compensated by such a transformation if we choose $L_B^{\ A}$ in (62) to be

$$L_B^{\ A} = G^M \phi_{MB}^{\ \ A} \quad . \tag{67}$$

From the definition of the connection, this L is Lie-algebra valued. Under the combined transformation the vielbein transforms,

$$\delta E_N^{\ A} = -\mathcal{D}_N G^A + G^M T_{MN}^{\ \ \ A} \quad . \tag{68}$$

Any tensorial quantity which transforms under general coordinate transformations and structure group as

$$\delta V^A = -G^M \partial_M V^A + V^B L_B^{\ A}$$

will under the combined transformations transform as

$$\delta V^A = -G^M \partial_M V^A + V^B G^M \phi_{MB}^{\ \ \ A} = -G^B \mathcal{D}_B V^A \quad . \tag{70}$$

The gauged supersymmetry transformations will be of the type (68) and (70). Therefore, it is of interest to compute the commutator (anticommutator) of two covariant derivatives. We proceed as follows.

$$\mathcal{D}\mathcal{D}V^F = \mathcal{D}\{E^A \mathcal{D}_A V^F\} = E^A E^B \mathcal{D}_B \mathcal{D}_A V^F + T^A \mathcal{D}_A V^F \quad . \tag{71}$$

On the other hand we compute $\mathcal{D}\mathcal{D}V^F$ to be

$$\mathcal{D}\mathcal{D}V^F = d\{dV^F - V^C \phi_C^{\ F}\} - \{dV^C - V^{\mathcal{D}} \phi_{\mathcal{D}}^{\ C}\} \phi_C^{\ F} = -V^C R_C^{\ F} \quad . \tag{72}$$

A comparison of (71) and (72) gives the result

$$\{\mathcal{D}_B \mathcal{D}_A - (-)^{ab} \mathcal{D}_A \mathcal{D}_B\} V^F = -(-)^{c(a+b)} V^C R_{BAC}^{\ \ \ \ F} - T_{BA}^{\ \ \ C} \mathcal{D}_C V^F \quad . \tag{73}$$

In an abstract way, this result can be written

$$\{\mathcal{D}_B \mathcal{D}_A\}_{\pm} = -R_{BA\cdot}^{\ \ \ \ \cdot} - T_{BA}^{\ \ \ F} \mathcal{D}_F \quad , \tag{74}$$

and it has the form of a structure equation of an algebra. It can be used as a starting point for a supergravity theory — the task being to construct a representation of this algebra. We have started from a representation in terms of $E_B^{\ A}$ and $\phi_{MA}^{\ \ \ B}$. This representation is highly reducible and our task is to reduce the representation. This will be done by imposing certain covariant constraints. These constraints can be formulated in terms of the torsion,

$$T^A = dE^A - E^B \phi_B^{\ A} = \frac{1}{2} E^B E^C T_{CB}^{\ \ A} = \frac{1}{2} dZ^M dZ^N T_{NM}^{\ \ \ A} \quad . \tag{75}$$

By comparing the coefficients of the 2-forms we find

$$T_{CB}^{\ \ A} = \left[(-)^{c(b+m)} \tilde{E}_B^{\ M} \tilde{E}_C^{\ N} - (-)^{b\,m} \tilde{E}_C^{\ M} \tilde{E}_B^{\ N} \right] \partial_N E_M^{\ \ A} - \phi_{CB}^{\ \ A} + (-)^{bc} \phi_{BC}^{\ \ A} \ . \tag{76}$$

4. LINEARIZED THEORY

The dynamics of the respective physical system should be described in terms of differential equations which have to satisfy a certain number of requirements. The equations should be Lorentz covariant and supersymmetric. They should be not higher than of second order, they should be local and give rise to causal propagation. Ghosts — negative contributions to the energy — should not appear as physical degrees of freedom; they should rather be gauge degrees of freedom and they should not contribute to any physical process.

To show that these requirements are actually satisfied for the exact solutions is impossible. We have to resort to perturbation theory and start with the lowest order, the free field content of the theory. This is done by making a linear approximation to the theory and we have to sort out those theories which satisfy our requirements on this level. It is interesting to note that given the base space and the structure group, the theory with the least number of physical degrees of freedom seems to be quite unique.

We are going to expand the torsion around the flat space vielbein $E_M^{(o)A}$. This guarantees Lorentz covariance and supersymmetry. Because the curvature tensor can be expressed in terms of the torsion and its covariant derivatives[10] we do not have to look at the curvature tensor separately.

For the torsion we have

$$T_{DB}^{\ \ A} = \{(-)^{(b+m)n} \tilde{E}_D^{\ N} \tilde{E}_B^{\ M} - (-)^{db+(d+m)n} \tilde{E}_B^{\ N} \tilde{E}_D^{\ M}\} \partial_N E_M^{\ \ A}$$
$$+ \phi_{DB}^{\ \ A} - (-)^{db} \phi_{BD}^{\ \ A} \quad , \tag{77}$$

and for the linearized vielbein $H_B^{\ A}$,

$$E_{\mathcal{M}}^{\mathcal{A}} = E^{(o)\mathcal{A}}_{\mathcal{M}} + E^{(o)\mathcal{B}}_{\mathcal{M}} H_{\mathcal{B}}^{\mathcal{A}} ,$$

$$\tilde{E}_{\mathcal{A}}^{\mathcal{M}} = \tilde{E}^{(o)\mathcal{M}}_{\mathcal{A}} - H_{\mathcal{A}}^{\mathcal{B}} \tilde{E}^{(o)\mathcal{M}}_{\mathcal{B}} . \tag{78}$$

The flat space vielbein is given by

$$E^{(o)\mathcal{A}}_{\mathcal{M}} = \begin{pmatrix} \delta_m^a & 0 & 0 \\ i\theta^M{}_{\mu}\bar{\sigma}^{a\dot{\mu}}{}_{\mu} & \delta_\mu^\alpha \delta_A^M & 0 \\ i\theta_M^\mu \sigma^a{}_\mu{}^{\dot{\mu}} & 0 & \delta^{\dot{\mu}}_{\dot{\alpha}} \delta^A_M \end{pmatrix} , \tag{79}$$

and the flat space torsion by

$$T^{(o)A\dot{\beta}c}{}_{\alpha B} = +2i\sigma^c{}_\alpha{}^{\dot{\beta}} \delta^A_B \; ; \tag{80}$$

with all other components zero.

The linearized torsion is determined by

$$\begin{aligned} T^{\mathcal{A}}_{\mathcal{DB}} = T^{(o)\mathcal{A}}_{\mathcal{DB}} &- H_{\mathcal{D}}^{\mathcal{C}} T^{(o)\mathcal{A}}_{\mathcal{CB}} + (-)^{bd} H_{\mathcal{B}}^{\mathcal{C}} T^{(o)\mathcal{A}}_{\mathcal{CD}} \\ &+ T^{(o)\mathcal{C}}_{\mathcal{DB}} H_{\mathcal{C}}^{\mathcal{A}} + D_{\mathcal{D}} H_{\mathcal{B}}^{\mathcal{A}} - (-)^{bd} D_{\mathcal{B}} H_{\mathcal{D}}^{\mathcal{A}} \\ &+ \phi^{\mathcal{A}}_{\mathcal{DB}} - (-)^{bd} \phi^{\mathcal{A}}_{\mathcal{BD}} . \end{aligned} \tag{81}$$

The structure group is assumed to be the direct product of the Lorentz group and an internal group. The torsion, therefore, splits into various reducible components. We list the components which are reduced with respect to the Lorentz group:

1 $\quad T^{\dot{\delta}\dot{\beta}\alpha}_{DBA} = D^{\dot{\delta}}_D H^{\dot{\beta}\alpha}_{BA} + D^{\dot{\beta}}_B H^{\dot{\delta}\alpha}_{DA} ,$

2 $\quad T^{DBA}_{\delta\beta\dot{\alpha}} = D^D_\delta H^{BA}_{\beta\dot{\alpha}} + D^B_\beta H^{DA}_{\delta\dot{\alpha}} ,$

$$7 \qquad T_{\delta\beta}^{DBa} = D_\delta^D H_\beta^{Ba} + D_\beta^B H_\delta^{Da} - 2i\sigma_\beta^{a\dot\gamma} H_{\delta\dot\gamma}^{DB} - 2i\sigma_\delta^{a\dot\gamma} H_{\beta\dot\gamma}^{BD} ,$$

$$8 \qquad T_{DB}^{\dot\delta\dot\beta a} = D_D^{\dot\delta} H_B^{\dot\beta a} + D_B^{\dot\beta} H_D^{\dot\delta a} - 2i\sigma_\gamma^{a\dot\beta} H_{DB}^{\dot\delta\gamma} - 2i\sigma_\gamma^{a\dot\delta} H_{BD}^{\dot\beta\gamma} ,$$

$$10 \qquad T_{\delta b}^{D\ \ a} = 2iH_{b\dot\gamma}^{D}\sigma_\delta^{a\dot\gamma} + D_\delta^D H_b^{\ a} - \partial_b H_\delta^{Da} + \phi_{\delta b}^{D\ \ a} ,$$

$$11 \qquad T_{DB}^{\dot\delta\ \ a} = 2i\sigma_\gamma^{a\dot\delta} H_{bD}^{\ \ \gamma} + D_D^{\dot\delta} H_b^{\ a} - \partial_b H_D^{\dot\delta a} + \phi_{Db}^{\dot\delta\ \ a} ,$$

$$4 \qquad T_{D\beta\dot\alpha}^{\dot\delta BA} = +2i\sigma_\beta^{c\dot\delta} H_{c\dot\alpha}^{\ \ A}\delta_D^B + D_D^{\dot\delta} H_{\beta\dot\alpha}^{BA} + D_\beta^B H_{D\dot\alpha}^{\dot\delta A} + \phi_{\beta D\dot\alpha}^{B\dot\delta A} , \tag{82}$$

$$3 \qquad T_{D\beta A}^{\dot\delta B\alpha} = 2i\sigma_\beta^{c\dot\delta} H_{cA}^{\ \ \alpha}\delta_D^B + D_D^{\dot\delta} H_{\beta A}^{B\alpha} + D_\beta^B H_{DA}^{\dot\delta\alpha} + \phi_{D\beta A}^{\dot\delta B\alpha} ,$$

$$9 \qquad T_{\delta B}^{D\dot\beta a} = +2i\sigma_\delta^{c\dot\beta}\delta_B^D(\delta_c^a + H_c^{\ a}) + D_\delta^D H_B^{\dot\beta a} + D_B^{\dot\beta} H_\delta^{Da} -2iH_{\delta B}^{D\gamma}\sigma_\gamma^{a\dot\beta} - 2iH_{B\dot\gamma}^{\dot\beta D}\sigma_\delta^{a\dot\gamma} ,$$

$$16 \qquad T_{db}^{\ \ a} = \partial_d H_b^{\ a} - \partial_b H_d^{\ a} + \phi_{db}^{\ \ a} - \phi_{bd}^{\ \ a} .$$

All these components can be used individually for covariant restrictions of the theory. If we assume all of them to be equal to the flat space torsion, then we learn from the Bianchi identities that all components of the curvature are zero as well. The only solution of these equations is the flat space. Therefore we have to relax the condition in order to allow for a certain number of physical degrees of freedom — actually we are interested in the least nontrivial number of physical fields possible. These, we hope, will be an irreducible representation of the supersymmetry algebra with one spin 2 (graviton) field only.

The set of equations we are looking for are of two different types. The first set of equations will serve to reduce the number of independent fields without restricting the x dependence of the fields. Such equation we shall call constraints. The second type of equations will be equivalent to x-space field equations — these we shall call mass shell conditions.

To find the proper set of constraints and mass shell conditions is still an art;[11] we are only led by experience.

Let us try

$$T^{\dot{\delta}\dot{\beta}\alpha}_{DBA} = D^{\dot{\delta}}_{D} H^{\dot{\beta}\alpha}_{BA} + D^{\dot{\beta}}_{B} H^{\dot{\delta}\alpha}_{DA} = 0 \quad , \tag{83}$$

and the complex conjugate

$$T^{DBA}_{\delta\beta\dot{\alpha}} = D^{D}_{\delta} H^{BA}_{\beta\dot{\alpha}} + D^{B}_{\beta} H^{DA}_{\delta\dot{\alpha}} = 0 \quad . \tag{84}$$

Both of these equations can be solved by introducing potentials which are not restricted in x-space,

$$H^{\dot{\beta}\alpha}_{BA} = D^{\dot{\beta}}_{B}\, \Lambda^{\alpha}_{A} \, , \quad H^{BA}_{\beta\dot{\alpha}} = D^{B}_{\beta} \Sigma^{A}_{\dot{\alpha}} \quad . \tag{85}$$

Next, we postulate

$$\begin{aligned} T^{DBa}_{\delta\beta} &= D^{D}_{\delta}(H^{Ba}_{\beta} - 2i\sigma^{a\dot{\gamma}}_{\beta}\Sigma^{B}_{\dot{\gamma}}) + D^{B}_{\beta}(H^{Da}_{\delta} - 2i\sigma^{a\dot{\gamma}}_{\delta}\Sigma^{D}_{\dot{\gamma}}) = 0 \quad , \\ T^{\dot{\delta}\dot{\beta}a}_{DB} &= D^{\dot{\delta}}_{D}(H^{\dot{\beta}a}_{B} - 2i\sigma^{a\dot{\beta}}_{\gamma}\Lambda^{\gamma}_{B}) + D^{\dot{\beta}}_{B}(H^{\dot{\delta}a}_{D} - 2i\sigma^{a\dot{\delta}}_{\gamma}\Lambda^{\gamma}_{D}) = 0 \quad . \end{aligned} \tag{86}$$

From these equations we can conclude

$$\begin{aligned} H^{Ba}_{\beta} &= D^{B}_{\beta} Y^{a} + 2i\sigma^{a\dot{\gamma}}_{\beta}\Sigma^{B}_{\dot{\gamma}} \quad , \\ H^{\dot{\beta}a}_{B} &= D^{\dot{\beta}}_{B} X^{a} + 2i\sigma^{a\dot{\beta}}_{\gamma}\Lambda^{\gamma}_{B} \quad . \end{aligned} \tag{87}$$

We have solved the constraint equations,

$$T^{\dot{\delta}\dot{\beta}\alpha}_{DBA} = 0, \quad T^{DBA}_{\delta\beta\dot{\alpha}} = 0, \quad T^{\dot{\delta}\dot{\beta}a}_{DB} = 0, \quad T^{DBa}_{\delta\beta} = 0 \quad , \tag{88}$$

by introducing the potentials Λ^{α}_{A}, $\Sigma^{A}_{\dot{\alpha}}$, X^{a} and Y. These potentials may be changed without introducing a change in the respective vielbein component:

$$\begin{aligned} \Lambda^{\alpha}_{A} &\to \Lambda^{\alpha}_{A} + \ell^{\alpha}_{A} \, , & D^{\dot{\beta}}_{B}\ell^{\alpha}_{A} &= 0 \quad , \\ \Sigma^{A}_{\dot{\alpha}} &\to \Sigma^{A}_{\dot{\alpha}} + S^{A}_{\dot{\alpha}} \, , & D^{B}_{\beta} S^{A}_{\dot{\alpha}} &= 0 \quad , \\ Y^{a} &\to Y^{a} + y^{a} \, , & D^{B}_{\beta} y^{a} + 2i\sigma^{a\dot{\gamma}}_{\beta} S^{B}_{\dot{\gamma}} &= 0 \quad , \end{aligned} \tag{89}$$

$$X^a \to X^a + x^a \ , \qquad D_B^{\dot\beta} x^a + 2i\sigma^{a\dot\beta}_{\gamma} \ell^{\gamma}_B = 0 \ .$$

This freedom of "gauge" will be considerable use when we write explicitly the independent field components of the theory.

In addition to the gauge transformations above, the theory is also invariant under general coordinate transformations,

$$\delta E_M^{\ A} = \mathcal{D}_M G^A + G^{\mathcal{B}} T_{\mathcal{B}M}^{\ \ \ A} \ , \tag{90}$$

which, for the linearized vielbein, become

$$\delta H_M^{\ A} = D_M G^A + G^{\mathcal{B}} T_{\mathcal{B}M}^{(o)A} \ . \tag{91}$$

The potentials, introduced above, have to transform as follows:

$$\delta\Lambda^{\alpha}_{A} = G^{\alpha}_{A} \ , \quad \delta\Sigma^{A}_{\dot\alpha} + G^{A}_{\dot\alpha} \ , \quad \delta X^a = \delta Y^a = G^a \ . \tag{92}$$

With the help of the potentials, it is possible to combine the remaining components of the vielbein into quantities which are invariant under general coordinate transformations:

$$\begin{aligned}
\mathcal{H}_b^{\ a} &= H_b^{\ a} - \tfrac{1}{2}\partial_b (X^a + Y^a - Z^a) \ , \\
\mathcal{H}^{D\gamma}_{\delta B} &= H^{D\gamma}_{\delta b} - D^D_{\delta}\Lambda^{\gamma}_{B} \ , \\
\mathcal{H}^{\dot\beta D}_{B\dot\gamma} &= H^{\dot\beta D}_{B\dot\gamma} - D^{\dot\beta}_{B}\Sigma^{D}_{\dot\gamma} \ , \\
\mathcal{H}_{b\dot\kappa}^{\ \ D} &= H_{b\dot\kappa}^{\ \ D} - \partial_b \Sigma^{D}_{\hat\kappa} \ , \\
\mathcal{H}_{bD}^{\ \ \gamma} &= H_{bD}^{\ \ \gamma} - \partial_b \Lambda^{\gamma}_{D} \ , \\
Z^a &= X^a - Y^a \ .
\end{aligned} \tag{93}$$

All the remaining equations can be written entirely in terms of these invariant expressions. We continue with the torsion component $T^{D\dot\beta a}_{\delta B}$, which still does not contain the connection explicitly,

$$T^{D\dot{\beta}a}_{\delta B} = 2i\sigma^{c\dot{\beta}}_{\delta}\delta^{D}_{B}(\delta^{a}_{c} + H^{a}_{c}) + D^{D}_{\delta}D^{\dot{\beta}}_{B}Z^{a}$$

$$- 2i(\sigma^{a\dot{\gamma}}_{\delta}H^{\dot{\beta}D}_{B\dot{\gamma}} + \sigma^{a\dot{\beta}}_{\gamma}H^{D\gamma}_{\delta B}) \tag{94}$$

$$= 2i\sigma^{a\dot{\beta}}_{\delta}\delta^{D}_{B} \quad .$$

In spinor notation, this equation reads

$$-2i\delta^{D}_{B}H_{\beta\dot{\beta}\alpha\dot{\alpha}} = D^{D}_{\beta}D_{B\dot{\beta}}Z_{\alpha\dot{\alpha}} - 4i(\varepsilon_{\alpha\beta}H_{B\dot{\beta}\dot{\alpha}}^{\;D} + \varepsilon_{\dot{\beta}\dot{\alpha}}H^{D}_{\beta\alpha B}) \quad . \tag{95}$$

Taking the trace in DB we obtain

$$H_{\beta\dot{\beta}\alpha\dot{\alpha}} = \frac{i}{2N}D^{D}_{\beta}D_{\dot{\beta}D}Z_{\alpha\dot{\alpha}} + \frac{2}{N}(\varepsilon_{\dot{\beta}\dot{\alpha}}H^{D}_{\beta\alpha D} + \varepsilon_{\alpha\beta}H_{D\dot{\beta}\dot{\alpha}}^{\;D}) \quad . \tag{96}$$

The traceless part becomes

$$(D^{D}_{\beta}D_{B\dot{\beta}} - \frac{1}{N}\delta^{D}_{B}D^{C}_{\beta}D_{C\dot{\beta}})Z_{\alpha\dot{\alpha}} = 4i(\varepsilon_{\alpha\beta}\tilde{H}_{B\dot{\beta}\dot{\alpha}}^{\;D} + \varepsilon_{\dot{\beta}\dot{\alpha}}\tilde{H}^{D}_{\beta\alpha B}) \quad . \tag{97}$$

Here we have introduced $\tilde{H}$ for the traceless part of H. This equation has three separate pieces. Symmetrisation over $\beta\alpha$, $\dot{\beta}\dot{\alpha}$ tells us that

$$\sum_{P(\alpha\beta,\dot{\alpha}\dot{\beta})}(D^{D}_{\beta}D_{B\dot{\beta}} - \frac{1}{N}\delta^{D}_{B}D^{C}_{\beta}D_{C\dot{\beta}})Z_{\alpha\dot{\alpha}} = 0 \quad . \tag{98}$$

Symmetrisation over $\beta\alpha$ and antisymmetrisation in $\dot{\beta}\dot{\alpha}$ tells us

$$\tilde{H}^{D}_{\beta\alpha B} + \tilde{H}^{D}_{\alpha\beta B} = -\frac{i}{8}\sum_{\alpha\beta}(D^{D}_{\beta}D^{\dot{\alpha}}_{B} - \frac{1}{N}\delta^{D}_{B}D^{C}_{\beta}D^{\dot{\alpha}}_{C})Z_{\alpha\dot{\alpha}} \quad , \tag{99}$$

and vice versa,

$$\tilde{H}_{B\dot{\beta}\dot{\alpha}}^{\;D} + \tilde{H}_{B\dot{\alpha}\dot{\beta}}^{\;D} = -\frac{i}{8}\sum_{\dot{\alpha}\dot{\beta}}(D^{D}_{\beta}D_{B\dot{\beta}} - \frac{1}{N}D^{C}_{\beta}D_{C\dot{\beta}})Z_{\alpha\dot{\alpha}} . \tag{100}$$

Antisymmetrisation in $\alpha\beta$ and $\dot{\beta}\dot{\alpha}$ yields

$$\tilde{H}^{\dot{\alpha}D}_{B\dot{\alpha}} + \tilde{H}^{D\alpha}_{\alpha B} = -\frac{i}{8}(D^{D}_{\beta}D^{\dot{\beta}}_{B} - \frac{1}{N}\delta^{D}_{B}D^{C}_{\beta}D^{\dot{\beta}}_{C})Z^{\beta}_{\dot{\beta}} \quad . \tag{101}$$

This exhausts all of the Eqs. (98).

Next we consider those components of the torsion which allow us to compute the connection $\phi_{\mathcal{D}a}{}^{b}$ in terms of the vielbein. The connection component $\phi_{da}{}^{b}$ is obtained from the torsion $T_{da}{}^{b}$,

$$T_{db}{}^{a} = \partial_d H_b{}^{a} - \partial_b H_d{}^{a} + \phi_{db}{}^{a} - \phi_{bd}{}^{a} = 0 \quad . \tag{102}$$

Because ϕ_{dba} is Lie-algebra valued,

$$\phi_{dba} = -\phi_{dab} \quad , \tag{103}$$

we can solve this equation for ϕ_{dba},

$$\phi_{bda} = \frac{1}{2}\{\partial_d(H_{ba} + H_{ab}) - \partial_a(H_{bd} + H_{db}) + \partial_b(H_{ad} - H_{da})\}. \tag{104}$$

Next we consider the torsion components,

$$T^{D}_{\delta b}{}^{a} = D^{D}_{\delta}H_b{}^{a} - 2i\sigma^{a}_{\delta\dot{\kappa}}H^{\dot{\kappa}D}_{b} + \phi^{D}_{\delta b}{}^{a} = 0 \quad , \tag{105}$$

$$T^{\dot{\delta}}_{Db}{}^{a} = D^{\dot{\delta}}_{D}(H_b{}^{a} - \partial_b Z^{a}) + 2iH_{bD}{}^{\gamma}\sigma^{a\dot{\delta}}_{\gamma} + \phi^{\dot{\delta}}_{Db}{}^{a} = 0 \quad . \tag{106}$$

These equations can be used to compute the torsion components $\phi^{D}_{\delta b}{}^{a}$ and $\phi^{\dot{\delta}}_{Db}{}^{a}$. They have to be Lie-algebra valued,

$$\begin{aligned} \phi^{D}_{\delta\ \beta\dot{\beta}\ \alpha\dot{\alpha}} &= -2\varepsilon_{\beta\alpha}\phi^{D}_{\delta\underline{\dot{\beta}\dot{\alpha}}} + 2\varepsilon_{\dot{\beta}\dot{\alpha}}\phi^{D}_{\delta\underline{\beta\alpha}} \\ \phi^{\dot{\delta}}_{D\ \beta\dot{\beta}\ \alpha\dot{\alpha}} &= -2\varepsilon_{\beta\alpha}\phi^{\dot{\delta}}_{D\underline{\dot{\beta}\dot{\alpha}}} + 2\varepsilon_{\dot{\beta}\dot{\alpha}}\phi^{\dot{\delta}}_{D\underline{\beta\alpha}} \quad . \end{aligned} \tag{107}$$

This, in turn, leads to new equations for the vielbein. From Eq. (107) follows

$$\begin{aligned} \phi^{D}_{\delta\ \beta\dot{\beta}\ \alpha\dot{\alpha}} &= -D^{D}_{\delta}H_{\beta\dot{\beta}\ \alpha\dot{\alpha}} - 4i\varepsilon_{\alpha\beta}H_{\beta\dot{\beta}\dot{\alpha}}{}^{D} \\ \phi_{\dot{\delta}D,\beta\dot{\beta}\ \alpha\dot{\alpha}} &= -D_{\dot{\delta}D}(H_{\beta\dot{\beta}\ \alpha\dot{\alpha}} - \partial_{\beta\dot{\beta}}Z_{\alpha\dot{\alpha}}) + 4i\varepsilon_{\dot{\alpha}\dot{\delta}}H_{\beta\dot{\beta}\ \alpha D} \quad . \end{aligned} \tag{108}$$

Symmetrising and antisymmetrising in $\alpha\beta$ and $\dot{\alpha}\dot{\beta}$ alternatively yields

$$\phi^{D}_{\delta\underline{\alpha\beta}} = \frac{1}{4}(D^{D}_{\delta}H_{\beta\dot{\beta}\alpha}{}^{\dot{\beta}} + \varepsilon_{\alpha\delta}D^{D}_{\beta}H^{\sigma\dot{\sigma}}_{\dot{\sigma}\sigma}) \tag{109}$$

$$\phi^{D}_{\delta\dot\alpha\underline{\dot\beta}} = \frac{1}{8}D^{D}_{\delta}\sum_{P\{\dot\alpha\dot\beta\}} H^{\dot\sigma}_{\beta\sigma\dot\alpha} - \frac{1}{24}D^{\alpha D}\sum_{P\{\alpha\delta,\dot\alpha\dot\beta\}} H_{\alpha\dot\alpha\delta\dot\beta}$$

$$\phi^{\dot\delta}_{D\underline{\beta\alpha}} = -\frac{1}{8}D^{\dot\delta}_{D}\sum_{P\{\alpha\beta\}}(H^{\dot\sigma}_{\beta\alpha\dot\sigma} - \partial^{\dot\sigma}_{\beta}Z_{\alpha\dot\sigma}) - \frac{1}{24}D_{D\dot\alpha}\sum_{P(\alpha\beta,\dot\alpha\dot\delta)}(H^{\dot\delta\dot\alpha}_{\beta\alpha} - \partial^{\dot\delta}_{\beta}Z^{\dot\alpha}_{\alpha})$$

$$\phi^{\dot\delta}_{D\underline{\beta\dot\alpha}} = -\frac{1}{4}\{D_{D}(H_{\sigma\beta\dot\alpha}{}^{\sigma} - \partial_{\sigma\dot\beta}Z^{\sigma}{}_{\dot\alpha}) + \delta^{\dot\delta}_{\dot\alpha}D_{\dot\beta D}(H^{\sigma\dot\sigma}_{\dot\sigma\sigma} - \partial^{\dot\sigma}_{\sigma}Z^{\sigma}_{\dot\sigma})\} \quad . \qquad (109)$$

Symmetrisation or antisymmetrisation in $\alpha\beta$ and $\dot\alpha\dot\beta$ simultaneously yields

$$H_{\beta\dot\beta\dot\alpha}{}^{D} = -\frac{i}{24}D^{\alpha D}\sum_{P\{\alpha\beta,\dot\alpha\dot\beta\}} H_{\alpha\dot\alpha\;\beta\dot\beta} + \frac{i}{8}\varepsilon_{\dot\beta\dot\alpha}D^{D}_{\beta}H^{\sigma\dot\sigma}_{\dot\sigma\sigma}$$

$$(110)$$

$$H_{\alpha\dot\beta\;D\beta} = \frac{i}{24}D^{\dot\alpha}_{D}\sum_{P\{\alpha\beta,\dot\alpha\dot\beta\}}(H_{\beta\dot\beta\;\alpha\dot\alpha} - \partial_{\beta\dot\beta}Z_{\alpha\dot\alpha}) - \frac{i}{8}\varepsilon_{\alpha\beta}D_{\dot\beta D}(H^{\dot\sigma\sigma}_{\sigma\dot\sigma} - \partial^{\dot\sigma}_{\sigma}Z^{\sigma}_{\dot\sigma}) \; .$$

Let me summarize the results up to now. We have expressed the covariant vielbein ($H_{\beta\dot\beta\;\alpha\dot\alpha}$, $\tilde H^{D}_{\delta\beta B}$, $\tilde H_{B\dot\beta\dot\delta}{}^{D}$, $H_{b\dot\alpha}{}^{D}$ and $H_{bD}{}^{\alpha}$) and the connection components (ϕ_{bda}, $\phi^{Da}_{\delta b}$ and $\phi^{\dot\delta}_{Db}{}^{a}$) in terms of $Z_{\alpha\dot\alpha}$, $H^{D\delta}_{\delta D}$, $H^{\dot\delta D}_{D\dot\delta}$, $H^{D}_{\underline{\delta\alpha D}}$, $H_{D\dot\delta\dot\alpha}{}^{D}$ and $\frac{1}{2}R_{B}{}^{D} = \tilde H^{D\gamma}_{\gamma B} - H^{\dot\gamma D}_{B\dot\gamma}$. $Z_{\alpha\dot\alpha}$ is subject to the condition (98) which for N=2 can be evaluated explicitly, and we see that it does not lead to mass shell conditions. For $N > 2$ this condition implies mass shell equations.

The components $H^{D}_{\underline{\delta\alpha D}}$ and $H_{D\dot\delta\dot\alpha}{}^{D}$ can be considered as gauge degrees of freedom under local Lorentz transformations.

To proceed one has to analyse the equations component-wise. This has been done for N=1 and N=2 and the result has been stated in the second lecture.

Knowing the constraints allows us to solve the Bianchi identities in terms of a restricted number of superfields.[12] The components of these superfields can be identified with the fields of the component approach.[13] This is done by using the transformation laws which have been derived in the third lecture.

ACKNOWLEDGMENT

The formulas for extended supergravity have been derived in collaboration with N. Dragon and R. Grimm. They will be published in a forthcoming paper.

I would like to thank R. Ecclestone for reading the manuscript.

REFERENCES

1. J. Wess and B. Zumino, Nucl. Phys. B70, 39 (1974); J. Iliopoulos and B. Zumino, Nucl. Phys. B76, 310 (1974).

2. S. Coleman and J. Mandula, Phys. Rev. 159, 1251 (1967).

3. S. Deser and B. Zumino, Phys. Lett. 62B, 335 (1976); D.Z. Freedman, P. van Nieuwenhuizen and S. Ferrara, Phys. Rev. D13, 3214 (1976).

4. M.T. Grisaru, P. van Nieuwenhuizen and J.A.M. Vermaseren, Phys. Rev. Lett. 37, 1662 (1976); P. van Nieuwenhuizen and J.A. Vermeseren, Phys. Lett 65B, 263 (1976) and Phys. Rev. D16, 768 (1977); S. Deser, J. Kay and K. Stelle, Phys. Rev. Lett. 38, 527 (1977); S. Deser and J. Kay, Phys. Lett. 76B, 400 (1978); J.G. Taylor, Proc. R. Soc. London A362, 493 (1978), and Talk given at the EPS Conference, Geneva, 1979.

5. R. Haag, J.T. Lopuszański and M. Sohnius, Nucl. Phys. B88, 257 (1975).

6. H. Flanders, Differential Forms (Academic Press, 1963); J. Wess in Topics in Quantum Field Theory and Gauge Theories, ed. J.A. de Azcárraga, Lecture Notes in Physics 77 (Springer-Verlag, 1978).

7. R. Grimm, M. Sohnius and J. Wess, Nucl. Phys. B133, 275 (1978).

8. J. Wess and B. Zumino, Phys. Lett. 66B, 361 (1977); R. Arnowitt, P. Nath and B. Zumino, Phys. Lett. 56B, 81 (1975); V.P. Akulov, D.V. Volkov and V.A. Soraka, JETP Letters 22, 396 (1975) (English 187); L. Brink, M. Gell-Mann, P. Ramond and J.H. Schwarz, Phys. Lett. 74B, 336 (1978).

9. N. Dragon, R. Grimm and J. Wess, to be published.

10. N. Dragon, Z. Phys. C2, 29 (1979).

11. S. James Gates Jr. and W. Siegel, preprint HUTP -79/A034.

12. R. Grimm, J. Wess and B. Zumino, Nucl. Phys. B152, 255 (1979).

13. J. Wess and B. Zumino, Phys. Lett. 79B, 394 (1978).

List of Participants

AERTS, ADRIAN T., *Institute for Theoretical Physics, Toernooiveld 1, 6525 ED Nijmegen, The Netherlands*

AFEK, YACHIN, *Department of Physics, McGill University, Montreal H3A 2T8, Canada*

ALTARELLI, G., *Istituto di Fisica, Universita di Roma, Piazzale della Scienze 5, I-00185 Rome, Italy*

AMBJORN, JAN, *Niels Bohr Institute, Blegdamsvej 17, 2100 Kbh Ø, Denmark*

BANKS, JAY, *Loomis Lab. of Physics, University of Illinois, Urbana, Illinois 61801 USA*

BARTLETT, DAVID, *Department of Physics, University of Colorado, Boulder, Colorado 80309 USA*

BHANOT, GYAN, *Newman Hall, Cornell University, Ithaca, NY 14853 USA*

BRICKNER, RALPH G., *Physics Dept., University of Connecticut, Storrs, CT 06268 USA*

BURAS, A., *Theory Group, Fermilab, P.O. Box 500, Batavia, IL 60510 USA*

BURROWS, C.J., *D.A.M.T.P., Cambridge University, Cambridge CB3 9EW U.K.*

CALOGERACOS, ALEXANDER, *Maps P/G P/H, University of Sussex, Brighton BN1 9QH, U.K.*

CHANG, VAN, *Institute of Theoretical Science, University of Oregon, Eugene, OR 97403 USA*

CHIU, TING-WAI, *Department of Physics, University of Utah, Salt Lake City, UT 84112 USA*

CLARK, THOMAS E., *T-8 Group, Los Alamos Scientific Laboratory, Los Alamos, NM 87544 USA*

COOK, GALE, *Department of Physics, University of Colorado, Boulder, CO 80309 USA*

DANIELE, DOMINICI, *Istituto Nazionale di Fisica Nucleare, Largo E. Fermi 2 50125 Firenze, Italy*

DAVIDSON, AHARON, *Department of Physics, Syracuse University, Syracuse, NY 13210 USA*

DEMIRLIOGLU, DUYGU, *Physics Department, Bogazici University, Bebek, Istanbul, Turkey*

DE TAR, C., *Department of Physics, University of Utah, Salt Lake City, UT 84112 USA*

DUERKSEN, GARY L., *Department of Physics, University of Chicago, Chicago, IL 60637 USA*

DREITLEIN, JOSEPH, *Department of Physics, University of Colorado, Boulder, CO 80309 USA*

ECCLESTONE, RALPH E., *Department of Physics, The Univeristy, Southampton, U.K.*

ELLIS, S., *Department of Physics, University of Washington, Seattle, WA 98105 USA*

FIELD, R., *Department of Physics, California Institute of Technology, Pasadena, CA 91125 USA*

FLEISHON, NEIL L., *Lawrence Berkeley Laboratory, University of California, Berkeley, CA 94720 USA*

FORD, WILLIAM T., *Department of Physics, University of Colorado, Boulder, CO 80309 USA*

GOFFIN, VINCENT, *Physics Department, University of Pennsylvania, Philadelphia, PA 19104 USA*

GOTTLIEB, STEVEN A., *362-HEP, Argonne National Laboratory, Argonne, IL 60439 USA*

GRADY, MICHAEL P., *Department of Physics, Rockefeller University, New York, NY 10021 USA*

GREENSITE, JEFFREY P., *Division of Natural Sciences II, University of California, Santa Cruz, CA 95064 USA*

GROSS, D., *Department of Physics, Princeton University, Princeton, NJ 08540 USA*

GUERIN, FRANCOISE N., *Physics Department, Brown University, Providence, RI 02912 USA*

GUHA, ARUNABHA, *Department of Physics, City College, New York, NY 10031 USA*

GUPTA, SUBHASH, *Pupin Physics Labs, Columbia University, New York, NY 10027 USA*

HABER, HOWARD, *Lawrence Berkeley Laboratory, Berkeley, CA 94720 USA*

HENDRICK, EDWARD R., *Physics Department, St. Bonaventure University, St. Bonaventure, NY 14778 USA*

HILLER, JOHN, *Department of Physics and Astronomy, University of Maryland, College Park, MD 20742 USA*

HUGHES, RICHARD J., *Department of Theoretical Physics, Oxford University, Oxford OX1 3NP, U.K.*

IZATT, DALE, *Physics Department, University of Utah, Salt Lake City, UT 84112 USA*

JANAH, ARJUN, *Department of Physics, University of Maryland, College Park, MD 20742 USA*

KANG, ILWON, *Department of Theoretical Physics, Oxford University, Oxford OX1 3NP, U.K.*

KENWAY, RICHARD D., *Physics Department, Brown University, Providence, RI 02912 USA*

KING, STEPHEN F., *Department of Theoretical Physics, University of Manchester, Manchester, U.K.*

LAURSEN, MORTEN L., *Oklahoma State University, Stillwater, OK 74074 USA*

LEVINE, ROBERT Y., *Department of Physics, University of Michigan, Ann Arbor, MI 48109 USA*

LLEWELLYN SMITH, C., *Department of Theoretical Physics, 1 Keble Road, University of Oxford, Oxford OX1 3NP, U.K.*

MAHANTHAPPA, K.T., *Department of Physics, University of Colorado, Boulder, CO 80309 USA*

MALTMAN, KIM R., *Department of Physics, University of Toronto, Toronto, Ontario, M5S 1A7 Canada*

MERCU, MICHAEL, *Physikalisches Institut der Universität Bonn, Nussallee 12, 5300 Bonn, West Germany*

MOTTOLA, EMIL, *Physics Department, Columbia University, New York, NY 10027 USA*

MULDERS, PETER J., *Institute for Theoretical Physics, Toernooiveld 1, 6525 ED Nijmegen, The Netherlands*

NAUENBERG, URIEL, *Department of Physics, University of Colorado, Boulder, CO 80309 USA*

NEWLAND, DAVID, *Chadwick Laboratory, University of Liverpool, P.O. Box 147, Liverpool, L69 3BX U.K.*

NICOLE, DENNIS A., *Department of Physics and Astronomy, University of Pittsburgh, Pittsburgh, PA 15260 USA*

ONOFRI, ENRICO, *Universita Degli Studi Di Parma, Istituto Di Fisica, Via M D'Azeglio, 85, 43100 Parma, Italy*

OZER, MURAT, *Department of Physics, University of Maryland, College Park, MD 20742 USA*

OVRUT, BURT A., *Department of Physics, Brandeis University, Waltham, MA 02154 USA*

PARAGA, NESTOR, *Department of Physics, University of California, Santa Barbara, CA 93106 USA*

PAVER, NELLO, *ICTP, P.O. Box 586, 34100 Trieste, Italy*

POMPONIU, CONSTANTIN, *Department of Physics, Carnegie-Mellon University, Pittsburgh, PA 15213 USA*

PRAMUDITA, ANGGRAITA, *Department of Physics, University of Hawaii, Honolulu, HI 96822 USA*

PUHALA, MICHAEL, *Physics Department, University of Illinois, Urbana, IL 61801 USA*

RAGIADAKOS, CHRISTOS, *Physics Department, Simon Fraser University, Burnaby BC V5A 1S6 Canada*

RANDA, JAMES, *Department of Physics, University of Colorado, Boulder, CO 80309 USA*

REISS, DAVID B., *Physics Department, California Institute of Technology, Pasadena, CA 91125 USA*

RNO, JUNG S., *R. Walters College, University of Cincinnati, Cincinnati, OH 45236 USA*

ROBERTS, LYNN E., *Physics Department, Adelphi University, Garden City, NY 11530 USA*

RODRIGUEZ-VARGAS, A.M., *Universidad De Los Andes, Departamento De Fisica, Bogota, D.E. - Colombia*

SHER, MARC A., *Department of Physics, University of Colorado, Boulder, CO 80309 USA*

SMITH, JAMES, *Department of Physics, University of Colorado, Boulder, CO 80309 USA*

SOKOLOFF, MICHAEL D., *Department of Physics, University of California, Berkeley, CA 94720 USA*

STELTE, NORBERT, *Physics Department, Mainzer Gasse 33,3550 Marburg, West Germany*

STERN, ALLEN B., *Department of Physics, Syracuse University, Syracuse, NY 13210 USA*

TRAHERN, CHARLES G., *Department of Physics, Syracuse University, Syracuse, NY 13210 USA*

TSE, TSUN-YAN, *Physics Department, California Institute of Technology, Pasadena, CA 91125 USA*

UNGER, DAVID G., *Department of Physics, University of Colorado, Boulder, CO 80309 USA*

VALANJU, PRASHANT, *Department of Physics, University of Texas, Austin, TX 78712 USA*

VERGADOS, JOHN D., *Physics Department, University of Ioannina, Ioannina, Greece*

VIOLINI, GALILEO, *Universita Degli Studi -Roma, Istituto Di Fisica "Guglielmo Marconi" Piazzale Delle Scienze, 5, Roma, Italy*

WARNER, ROLAND C., *School of Physics, University of Melbourne, Parkville 3052, Victoria, Australia*

WATSON, PETER, *Department of Physics, Carleton University, Ottawa, Ontario, Canada*

WESS, J., *Institut für Theoretische Physik, der Universität Karlsruhe (TH), Kaiserstrasse 12. Physikhochhaus, 75 Karlsruhe 1, West Germany*

WOLIN, ELLIOT, *Physics Department, University of Washington, Seattle, WA 98195 USA*

YOUNG, BING-LIN, *Department of Physics, Iowa State University, Ames, IA 50011 USA*

ZAHIR, MUHAMMAD S., *Institute of Theoretical Science, University of Oregon, Eugene, OR 97403 USA*

INDEX